Geoheritage, Geoparks and Geotourism

Conservation and Management Series

Series editors

Wolfgang Eder, Munich, Germany
Peter T. Bobrowsky, Burnaby, BC, Canada
Jesús Martínez-Frías, Madrid, Spain

Spectacular geo-morphological landscapes and regions with special geological features or mining sites, are becoming increasingly recognized as critical areas to protect and conserve for the unique geoscientific aspects they represent and as places to enjoy and learn about the science and history of our planet. More and more national and international stakeholders are engaged in projects related to "Geoheritage", "Geo-conservation", "Geoparks" and "Geotourism" and are positively influencing the general perception of modern Earth sciences. Most notably, "Geoparks", have proven to be excellent tools to educate the public about "Earth Sciences". And shown to be areas for recreation and significant sustainable economic development through geotourism. In order to develop further the understanding of earth sciences in general and to elucidate the importance of earth sciences for Society the Geoheritage, Geoparks and Geotourism Conservation and Management Series has been launched together with its sister GeoGuides series. "Projects" developed in partnership with UNESCO, World Heritage and Global Geoparks Networks, IUGS and IGU, as well as with the "Earth Science Matters" Foundation, are welcome. The series aims to provide a place for in-depth presentations of developmental and management issues related to Geoheritage and Geotourism as well existing and potential Geoparks. Individually authored monographs as well as edited volumes and conference proceedings are welcome in this series. This book series is considered to be complementary to the Springer-Journal "Geoheritage".

More information about this series at http://www.springer.com/series/11639

Ismar Borges de Lima · Ronda J. Green
Editors

Wildlife Tourism, Environmental Learning and Ethical Encounters

Ecological and Conservation Aspects

Springer

Editors
Ismar Borges de Lima
School of Business and Tourism
Southern Cross University
Gold Coast, QLD
Australia

and

MultiAmazon & RECINATUR
State University of Roraima (UERR)
Boa Vista
Brazil

and

Federal University of Southern and Southeastern Pará (UNIFESSPA)
Marabá
Brazil

Ronda J. Green
Griffith University
Nathan, QLD
Australia

and

Wildlife Tourism Australia (WTA)
Rathdowney, QLD
Australia

ISSN 2363-765X ISSN 2363-7668 (electronic)
Geoheritage, Geoparks and Geotourism
Conservation and Management Series
ISBN 978-3-319-55573-7 ISBN 978-3-319-55574-4 (eBook)
DOI 10.1007/978-3-319-55574-4

Library of Congress Control Number: 2017934325

Cover photography credits: Baby Koala on tree (Cover). Credits to: Erik Veland (under Creative Commons License attribution CC BY-SA 2/3). Elephant, flying bat, lizard. Courtesy: Ronda J. Green

Printed on acid-free paper

This Springer imprint is published by Springer Nature
The registered company is Springer International Publishing AG
The registered company address is: Gewerbestrasse 11, 6330 Cham, Switzerland

Preface

Wildlife tourism is a broadly defined sector of tourism involving a wide range of traveler experiences. It includes watching and photographing animals in their natural habitats or in the zoo, swimming with, petting or feeding animals in a variety of settings, hunting anything from quails to elephants and any other activity involving non-domesticated animals when traveling away from home. Although the term 'wildlife' tends to conjure up images of mammals such as lions or kangaroos, or the larger birds (cassowaries, eagles) or reptiles (crocodiles, marine turtles), it is frequently taken to mean all animals and even plants and fungi. Wildlife tourism thus includes all levels of bird-watching, visits to glow worm colonies or night walks through forests to see luminous toadstools. In this volume however we focus mainly on tourism involving vertebrate animals (chiefly mammals and birds), and mainly on non-consumptive tourism involving free-ranging wildlife, although other taxa (e.g. turtles), hunting and captive settings are also discussed by some authors.

Proponents of wildlife tourism often focus on such factors as the educational value of introducing people from all walks of life to a variety of animals and their ecological needs, the preservation or restoration of wild habitats, monetary contributions to conservation projects, the breeding of vulnerable species and the alleviation of poverty in developing countries. Opponents point to many welfare problems such as harassment of animals in the wild for 'action photos', animals kept in small enclosures or mistreated for financial gain involving, for instance, photographs of tourists feeding or holding captive animals, and businesses developed around the taking of trophies of hunted creatures either in the wilderness setting or as 'canned hunting' in captive settings. There is criticism also of the amount of fossil fuels used in traveling to far-flung places, the possible conservation problems arising from interrupting breeding or feeding activities, habitat modification or favouring populations of some species (e.g. by feeding) at the expense of others (e.g. their prey species or competitors), of zoos which breed animals that will never be released into the wild and tourism operations that do not benefit local residents or interfere with their privacy, livelihoods and even their safety. There is much emotion involved and often far too few facts, and reality is generally not as simple as many with extreme views may imagine.

The pros and cons of wildlife tourism vary greatly from one situation to another, and benefits or otherwise tend to lie along a spectrum rather than falling clearly into discreet categories. Sometimes compromises must be found between what is ideal for different stake-holders (including the wildlife itself) to maximise or at least optimise overall positive impact. It is best when decisions can be based on the best available information, even if important decisions, due to time constraints, must still be made on a 'best guess' basis, while research simultaneously continues, to inform future decisions. As a discipline for study, wildlife tourism involves applied ecology (e.g. possible negative effects of wildlife-viewing in natural areas and mitigation of same, monitoring techniques for habitat restoration sites), animal behaviour and physiology (e.g. signs of stress), psychology (e.g. motivations of tourists, effective learning experiences), politics and economics. Researchers specialised in their own fields often fail to understand the complexities involved in other relevant disciplines, and there seems much to be gained by more communication of information and brainstorming of ideas relevant to maximising the positive impacts of wildlife tourism and minimising the negative.

There is thus an enormous field of actual and potential research and a range of philosophical perspectives from different stake-holders such as wildlife-loving tourists prepared to go to great lengths to see a variety of species, general tourists who like to include brief wildlife or wilderness experiences in their travels, tour operators and staff (including guides, accommodation owners, zoo keepers, etc.) who entered the industry out of a love of animals, tour operators and staff primarily focussed on income but seeing opportunities for value-adding to their products by adding a wildlife component, professional academics and students working in the fields of tourism or environmental sciences, environmental educators, conservation managers and people involved in all levels of government.

This volume cannot hope to cover all aspects of wildlife tourism across the world, but does offer an important contribution to understanding some of the problems and to insights and recent research leading to possible solutions.

Gold Coast, Australia Ronda J. Green

Contents

Contributors

Reyhan Arslan Ayazlar holds a Ph.D. from Adnan Menderes University. She is an Assistant Professor at the Department of Hospitality Management, Mugla Sıtkı Kocman University. Her research interests include tourism marketing and tourist behaviour. She has some studies focusing on tourist experience, local residents' roles in the tourism industry and alternative tourism types such as rural tourism, VFR tourism and wildlife issues. She also has published national and international conference papers in Turkish and in English.

Ismar Borges de Lima is a Ph.D. holder in Geography and Tourism from the University of Waikato, New Zealand. He did his postdoc studies at the School of Business and Tourism, Southern Cross University, SCU, Australia, where he is currently an Adjunct Faculty. Dr. Lima is an active member of the MultiAmazon/UERR Lab for Research on Amazon Rainforest, and he is academically linked to IEDAR, at Para South and Southeast University, UNIFESSPA. He holds a Master's degree in International Relations from the International University of Japan, IUJ. His past professional and academic experiences include a professional training course at Radio and TV Nederland – RNTC, in Hilversum, Holland, and an internship at the Environment Division, UNESCAP, in Bangkok, Thailand. Currently, he is the director in Brazil for the International Foundation for Research on Science, Nature & Tourism, RECINATUR. He was the editor-in-chief in the following volumes: '*Wildlife Tourism book*'; '*Tourism and Ethnodevelopment: Inclusion, Empowerment and Self-determination*', and '*Rare Earths Industry book*'. Author of several book chapters and Journal articles. Member of the Editorial Board of eight prestigious Journals. His research interests are: regional development, ethnodevelopment, tourism resources, wildlife tourism, tourism planning, ecotourism, environmental governance, environmental interpretation and education, and human-nature related topics within the Human and Social Sciences realms, particularly related to Amazon rainforest biodiversity and its traditional peoples. Websites: CV Lattes: http://lattes.cnpq.br/1226725450970361 (Portuguese & English) www.multiamazon.weebly.com; SCU: www.scu.edu.au; www.recinaturbrasil.weebly.com. Emails: ismarborgeslima@gmail.com, ismarlima@yahoo.com.br.

Georgette Leah Burns holds a Ph.D. in Tourism from Murdoch University and is a foundation member of the Environmental Futures Research Institute at Griffith University. She is a senior lecturer in the Griffith School of Environment and has previously held positions as Head of the Tourism Department at Hólar University College and Head of Tourism research at the Icelandic Seal Centre. As an environmental anthropologist, Leah's research on wildlife tourism interactions spans topics such as ethics, coexistence, sustainability, resilience, interpretation and management.

Rita Gabriela Araujo Carvalho has a degree in Tourism, from the Federal University de Pelotas and holds a Master's degree in Tourism, from Caxias do Sul University, Brazil. Her tourism experiences and research interests include: Tourism and Ecology; Geotourism; Lake Tourism; Lacustrine Tourism System; and Valuation of Ecosystem Services.

S. Emre Dilek holds a Ph.D. from the Department of Gastronomy & Culinary Arts, Tourism and Hotel Management High School, Batman University, Batman, Turkey. His doctoral thesis was entitled, '*Evaluation of Moral Responsibility of Tourism Regarding Animals in the Context of Meta-Criticism*'. He also holds a Master's degree with a dissertation entitled, '*Green Marketing Applications in Tourism Businesses: A Field Research*'. Dilek's research interests are: Sociology of Tourism, Tourism and Animal Ethics, Philosophy, and Green Marketing.

Lee Foote is a Professor of Conservation Biology and the Director of the University of Alberta Botantic Garden, both in Edmonton, Alberta, Canada. He completed his Ph.D. in Wetland Ecology at Utah State University, and worked for 8 years as a research ecologist for the US Fish and Wildlife Service and the US Geological Survey before returning to academia. His applied research focus is on wetland ecology topics but includes threatened species, wildlife utilization, African wildlife habitat, and fisheries. He is the author or co-author of three books, and many journal articles, and writes a column for the magazine *Conservation*.

Carolin Funck holds a Ph.D. in Geography from Albert-Ludwig-University Freiburg, Germany. She is a professor of Tourism Geography at the Department of Integrated Arts and Sciences, Hiroshima University, Japan. Her research focuses are: the development of tourism in Japan and the rejuvenation of mature tourist destinations. She is the author of '*Tourismus und Peripherie in Japan*' and co-author of '*Japanese Tourism*'.

Ronda J. Green is an Adjunct Research Fellow at the Environmental Futures Research Institute (EFRI) Griffith University. She holds a Ph.D. in Zoology (Monash University), and has conducted much ecological and behavioural research over the years, especially seed dispersal by fruit-eating animals, the effects of habitat alteration on wildlife and various aspects of wildlife tourism, as well as play behaviour of chimpanzees. Her publications include three books (Bush Outings and Holidays in Victoria, Guidelines to the Preparation of Nature Trails, and Wildlife Tourism: a Handbook for Guides, Tour Operators, Job-seekers and Business Start-ups), various reports for the Sustainable Tourism CRC, ecology research articles and book chapters. She has had extensive experience in nature interpretation for all ages, having run a holiday farm, worked as ranger in charge of interpretive activities in a national park, lectured at various universities and adult education classes, run many field excursions for all ages, and for twenty years run her own company, Araucaria Ecotours, specialising in wildlife tourism. She is currently chair of Wildlife Tourism Australia, (WTA: website http://www.wildlifetourism.org.au) and of the Scenic Rim branch of the Wildlife Preservation Society of Queensland, Australia, (website: http://scenicrim.wildlife.org.au/). Email: platypuscorner@bigpond.com.

Serhat Harman holds a Ph.D., Master's and Bachelor's degrees from the Institute of Social Sciences, Canakkale Onsekiz Mart University, Turkey. He is an Associated Professor at Batman University. His main research interests are: tourist behavior; wildlife tourism; and marine tourism.

Howard W. Harshaw holds a Ph.D. in Forestry from the University of British Columbia - UBC, Canada. He is a professor of outdoor recreation in the Faculty of Physical Education & Recreation at the University of Alberta. Howard (Howie) is an associate editor of the Journal of Outdoor Recreation & Tourism and of Leisure/Loisir; he also a member of the Human Dimensions Working Group of the North American Waterfowl Management Plan. Howie examines the human dimensions of natural resources in an effort to understand the relationships that people have with nature, and to investigate the interactions of resource development and quality of life. The focus of Howie's research is outdoor recreation management and planning, and public participation in natural resources decision-making.

Azizul Hassan holds a Ph.D. degree from the Cardiff Metropolitan University. He is a member of the Tourism Consultants Network of the UK Tourism Society. His main areas of research are: technology-supported marketing in tourism; innovative marketing dynamics;

destination branding in tourism; cultural heritage tourism; heritage interpretation; and sustainable management/marketing alternatives for cultural heritage industries. He is a regular reviewer of Tourism Management Perspective, Tourism Analysis, the International Journal of Human Resource Management, the International Journal of Ecotourism, the eReview of Tourism Research (eRTR) and The International Interdisciplinary Business-Economics Advancement Journal.

Pekka Iivari holds a Ph.D. in Geography from the University of Oulu, Finland. He works as a Senior Specialist at the Multidimensional Tourism Institute in Rovaniemi, Finland. He has published books and papers on tourism safety, incident communication and corporate security management. His current research interests include occupational health and safety in the context of tourism industry.

Rosane Maria Lanzer holds a Ph.D. in Biogeography from the University of Saarland, UDS, Germany, and a post-doctor in Toxicology from the German Research Center for Environmental Health, Helmholtz Zentrum München. She is a full-time Professor of Ecology and Tourism, and develops research on the following themes: freshwater ecology, aquatic toxicology and ecotourism.

Margie Maccoll is a Ph.D. student at the University of Queensland, Australia, conducting research in the areas of wildlife tourism and conservation. She is a freelance journalist and a former newspaper editor.

Gayle Mayes holds a Ph.D. in Arts from the University of the Sunshine Coast, Australia, and she is currently a member of the Sustainability Research Centre at the University. She is a lecturer in the Faculty of Arts, Business and Law Tourism, and teaches courses in the Marketing and Tourism, Leisure and Event Management disciplines. She is on the editorial board of Tourism in Marine Environments (TIME) and develops research on the following themes: island and marine tourism/recreation/sport.

Aline Correa Mazzoni holds a bachelor's degree in Biology and a Master's degree in Biotechnology, Caxias do Sul University. She has carried out studies and research on Zoology and Ecology, with an emphasis on Freshwater Ecology, working mainly on the following subjects: benthic macroinvertebrates, ecological integrity of lotic and lentic ecosystems, bioindicators and biomonitoring of water resources.

Farhad Moghimehfar earned his doctorate degree from the University of Alberta, Canada. He is a professor of Outdoor Recreation and Tourism Management at the University of Northern British Columbia, Canada. Farhad is an associate editor of the Journal of Ecotourism. He examines social and psychological aspects of visitors' behavior, with an emphasis on outdoor recreation and nature-based tourism in an effort to understand factors that influence people's pro-environmental behavior.

Jasmine Cardozo Moreira has a degree in Tourism and she holds a Ph.D. in Geography. She is a Professor at the Department of Tourism, at Ponta Grossa State University, Brazil. She is the Head of LABTAN, a University Lab with a focus on tourism, outdoor recreation and public use of protected areas. Her expertise is on human dimensions of tourism planning, National Parks and Geoparks worldwide. She is also an adjunct faculty member at West Virginia University's Recreation, Parks, and Tourism Resources Program, and a member of the International Academy for the Development of Tourism Research in Brazil, ABRATUR and of TAPAS Group, The Tourism and Protected Areas Specialist Group, IUCN. Website: https://uepg.academia.edu/JasmineMoreira.

Naomi Moswete is a Senior Lecturer in the Department of Environmental Science, University of Botswana. Her research interests include tourism as a strategy for rural development, community based natural resources management, community-based ecotourism; transfrontier

Park-based tourism, climate and tourism, community conservation, natural and heritage resource management and gender-based empowerment through tourism, etc.

Kenosi Nkape holds a Bachelor's Degree in Ecology from the University of Botswana. He is a Senior lecturer at the Botswana Wildlife Training Institute (BWTI), at the Department of Wildlife and National Parks (DWNP). His research interests are wildlife management on rangelands, biodiversity conservation and nature tourism in National parks and Reserves.

Dirk Reiser is a professor of sustainable tourism management at the Rhine-Waal University of Applied Sciences in Kleve, Germany. He is also a visiting lecturer at the European Overseas Campus in Thailand and Bali. His research interests are sustainable tourism, in particular wildlife tourism, CSR, marketing and environmental management. He is a member of the German Society for Tourism Science (DGT-Deutschen Gesellschaft für Tourismuswissenschaft e.V.) and the International Association of Scientific Experts in Tourism (AIEST).

Rafael Azevedo Robles holds a degree in Geography from PUC-Sao Paulo, Brazil, and a Park ranger professional course from SENAC-SP; he also has a degree in Environmental Law, from FAAP-SP. He has worked for many years with tourism and ecotourism as a tool for community development. He was the Coordinator of the Fernando de Noronha Base of the TAMAR Project, in Pernambuco, Brazil, 2010–2016, website: www.tamar.org.br.

Anukrati Sharma holds a Ph.D. in Tourism Marketing from the University of Rajasthan, India. She has two postgraduate degree specialties, a Master's degree in International Business and a Master's degree in Commerce and Administration. She is currently an Associate Professor in the Faculty of Commerce and Management, University of Kota, Kota Rajasthan, India. Dr. Sharma has worked as an internal trainer and teacher in the management arena. In 2015 she got a Research Award by UGC, New Delhi, on the topic of Analysis of the Status of Tourism in Hadoti and Shekhawati Region/Circuit (Rajasthan): Opportunities, Challenges and Future Prospectus. She has written 10 books on different subjects: Tourism Marketing; Organizational Behaviour; Principles and Practices of Management; Business Communication; Business Law; Advertising; Management; Company Law; Human Resource Management; Strategic Management; International Business and Sales and Distribution. She is an active member of the editorial board of 26 reputed national and international Journals.

Eric J. Shelton holds a Ph.D. from University of Otago, Dunedin, New Zealand, and he is a Trustee of the Yellow-eyed Penguin Trust, a whole-of-ecosystem environmental NGO situated in Dunedin. Eric's research interests include theories of nature and the relationship between conservation and tourism.

Andrew Tribe was a senior lecturer in Wildlife at the University of Queensland, for the past 24 years. He has worked with companion animals and wildlife for more than 30 years and is recognised for his research and teaching in animal welfare and behaviour. From 2016, Andrew has been managing the $5m Wildlife Centre at Spicers Hidden Vale, near Grandchester in South East Queensland, Australia. This privately funded facility runs in as a cooperative venture with the University of Queensland to support conservation initiatives through captive breeding, teaching and research.

Mpho Tseme holds a Certificate and Diploma in Wildlife Management and Conservation from the Botswana Wildlife Training Institute (BWTI). He is a conservation officer, working for the Anti-Poaching Unit (APU) in the Department of Wildlife and National Parks in Kasane, Northern Botswana. His research interests are wildlife management, biodiversity conservation and visitor management in national parks and reserves.

Hazel Tucker is an Associate Professor in the Department of Tourism at the University of Otago, New Zealand. Her research currently focuses on tourism and post-colonialism, tourism encounters, emotion, heritage and gender.

Rie Usui is pursuing her Ph.D. at Graduate School of Integrated Arts and Sciences, Hiroshima University in Japan, focusing on human and non-human animal relationships at tourism destinations. Her current research sites include Miyajima Island, Ohkunoshima Island and Yakushima Island which free-ranging wild/semi-wild animals inhabit. Prior to Hiroshima University, she obtained a M.Sc. degree in Primate Behavior and Ecology from Central Washington University in U.S.A. She carried out her master's thesis research at a wild monkey park in Anhui Province, China.

Karel Werdler holds a MA in History from the University of Amsterdam. He is a senior lecturer and researcher at Inholland University of Applied Sciences for the faculty of Tourism Management. He was a coordinator for Inholland of the Nuffic Rwanda Capacity Strengthening Program, at Rwanda Tourism University College, and nowadays he is a coordinator at Inholland for the Hospitality Program for the Amsterdam Metropolitan Area. He has done research and published on Dark Tourism and Tourism in Africa and is a Fellow of the Institute for Dark Tourism Research at University of Central Lancashire and board member of Atlas Africa. Before becoming part of the academic world he was active as a tourism professional in Europe, Asia and Africa.

Jundan (Jasmine) Zhang is a Ph.D. holder from the University of Otago, New Zealand, and is working as a post-doctoral researcher at the Department of Geography and Economic History, Umeå University, Sweden. Jundan's main research interests lie in the relationships between humans and nature in the context of tourism, from a political ecology perspective. Most recently Jundan contributed a chapter to the book *Political Ecology of Tourism (2016).*

Introduction: Wildlife Tourism Management and Phenomena: A Web of Complex Conceptual, Theoretical and Practical Issues

1

Ismar Borges de Lima and Ronda J. Green

Abstract

This introductory chapter highlights the major conceptual and practical issues regarding wildlife tourism worldwide. A series of events have brought concerns that the status and conditions of wild animals in the tourism needs further critical discussion, with current study cases being in the spotlight for analysis. There is a web of complexities permeating the field of wildlife tourism in terms of planning and management, not to mention the ethical issues. The current state of wildlife tourism draws attention to the need of in-depth reflections and insights on the use of animals as attractions as well as the needs and attitudes of tourism personnel and visitors. A change in perception of the natural world on the whole is needed, from a fully utilitarian view to a more compassionate one. The Earth is not home only for humans, so we need to break away from a predominantly anthropocentric view in our society. Indeed, within these epistemological and philosophical frames, 'ecological' and 'conservation' aspects have been regarded as fundamental for bringing a certain consensus to the equation on a morally acceptable human-nature relation for the 21st Century. This introductory chapter begins by presenting conceptual and disciplinary approaches to environmental social sciences, as well as human and political ecology, pertinent to this volume. It then presents some of the polemic cases involving wildlife and visitors, such as Cecil the lion, the tigers in the Thai Buddhist Temple, and, the killing of gorilla Harambe. The chapter concludes by presenting a summary of each chapter providing unique and original content to making this volume an exciting reading experience to update the readers' knowledge and understanding of the current state of wildlife tourism and issues facing it, as part of the bigger picture of our practical and ethical viewpoints of humans and the rest of nature on our planet.

Keywords

Wildlife tourism • Tourists • Environmental ethics • Wildlife ecology • Human/nature relations • Animal welfare

This book gathers a great selection of case studies that fill gaps in the literature on wildlife tourism, by critically and insightfully informing the readers on theoretical and practical issues with regards to human and wild animal encounters, and the ways to approach, understand and manage this complex and intricate relationship.

I. Borges de Lima (✉)
School of Business and Tourism,
Southern Cross University, Gold Coast, Australia
e-mail: ismarborgeslima@gmail.com; ismarlima@yahoo.com.br; ismar.borges.de.lima@scu.edu.au

I. Borges de Lima
MultiAmazon lab, Recinatur Intl Foundation, and UNIFESSPA, Roraima State University, UERR, Boa Vista, Brazil

R.J. Green
Environmental Futures Research Institute, Griffith University, Nathan, Australia
e-mail: chair@wildlifetourism.org.au

R.J. Green
Wildlife Tourism Australia Inc., Rathdowney, Australia

I. Borges de Lima and R.J. Green (eds.), *Wildlife Tourism, Environmental Learning and Ethical Encounters*, Geoheritage, Geoparks and Geotourism, DOI 10.1007/978-3-319-55574-4_1

1.1 Introduction

Past and current episodes at a global scale involving wild animals and visitors have been the motivation for proposing a book with updated critical reviews on wildlife tourism from an ecological, conservationist and educative perspective at a conceptual and operational level.

This chapter brings a collection of tragic and good events related to visitors and wild animal encounters in the wild and in captivity that took place in the last years. Since the killing of Cecil the lion in Zimbabwe, a series of bad events have been strikingly appearing on news shocking people with regards to wildlife resources whether they are in the wild or in captive settings. Bad news is occasionally permeated by good ones; great news that surrenders apprehension to hope, expectations that some changes may benefit the beasts. Wildlife tourism should generally be a case of contemplation, rather than touching. However, things in this field are not so deterministic and the complexities involving wildlife tourism require thorough investigations and critical insights. For example, we cannot say *all* touching is bad, and there are other forms of wildlife not involving touching that can be much worse, such as scaring birds from their nests while trying to photograph or identify them, or interfering with a predator's chance of catching prey. In this Volume, the conceptual and practical approaches to better understand and elucidate the human and wild animal relations in the wildlife tourism field, which include (un)ethical and conservation, and educative and educational aspects (components), come from several contributors drawing from different disciplinary perspectives and academic backgrounds. The words 'educational' and 'educative' are very close in meaning, and according to Merriam Webster, 'educational' refers to 'action or process of being educated'; the knowledge and development resulting from an educational process; and, as for the term 'educative', it refers to 'tending to educate'; an educative experience. The former is more about a 'process' while the latter is more about means, the outcomes. For this Volume, both terms will be used interchangeably. As for 'environmental learning', it refers to ways of integrating environmental science and social sciences with a multi-disciplinary approach, which, by the way, is one of the core missions of this book.

The terms 'visitor' and 'tourist' are used interchangeably throughout this Volume. Moscardo and Saltzer (2005) define 'visitors' as "actual tourists, excursionists, and local residents on day trips" (p. v). As 'wildlife tourism' has a plethora of definitions and approaches in the literature, and it is not the mission of this introductory chapter reviewing it all over again, thus wildlife tourism is defined as 'a nature-based tourism niche on interactions and viewing of wild animals in either their natural habitats, in semi-captivity or captivity'.

This book is a valuable contribution to the literature with a collection of case studies that approach the core themes of the book from all sides. Chapters draw attention to educational and learning opportunities in wildlife tourism, to conservation of turtles in Brazil, waterfowl hunting tourism in Canada, Parks and Wildlife in Australia, not being obviously limited to these issues by greatly approaching ethical and unethical aspects of interest for readers of all kinds and backgrounds. Ecological, biological, behavioural and habitat aspects of wild animals are also on the spotlight enriching this multidisciplinary volume within the realms of environmental sciences.

Before revisiting the main concepts relevant to this work and outlining each chapter, the next paragraphs will focus on reminding the readers of events involving wild animals and visitors that remarkably spotted 2015 and 2016; events that have been in essence a replication and recurrence of what has been witnessed throughout the last decades in (dis)advantages especially to iconic animals of tourism interest. Before outlining these events, some basic concepts related to wildlife tourism need to be elucidated as a starting point. Subsequently, in the second part of this Chapter, other conceptual approaches will be addressed serving as a basic theoretical foundation that anticipates more in-depth analyses and understandings of theories provided by contributors that accepted the invitation to be part of this work. In the third part, the main approaches and themes on wildlife tourism are outlined for each chapter.

The term 'wildlife' is defined from the very beginning in here, and it can be equated to 'nature resource(s) to collectively "relating to non-domesticated species of plants, animals or microbes", but in some academic writings 'wildlife' has been used to refer to mammals and birds, as well as small and large reptiles (e.g. pythons, in Australia), but not including invertebrates (Usher 1986, p. 4). It is also necessary to define 'wildlife conservation' and 'wildlife management' as these disciplines have been used in this Volume. 'Wildlife conservation' deals with "the preservation and protection of species and their habitat in the face of threats from human development"; as for 'wildlife management', it is concerned with the management that "seeks sustainable strategies to exploit wild species while ensuring their persistence and availability for future use … is often oriented toward specific objectives for one or a few species of economic interest", and it includes fisheries (Festa-Bianchet and Apollonio 2003, p. 3–4).

Conservation means different things to different people depending on their perspective, it can mean "prevention of waste, maximum development, efficient use, sustained yield, preservation, and non-use" (Reynold et al. 1974, p. 111). Usher (1986) explains that 'conservation' is not easy to define, and its definition ranges from non-interventionist; to wise use of the country's resources, of land and water and wildlife for every purpose (p. 4). 'Environmental conservation' is also an all-encompassing concept and it is concerned with "an integral part of any development process" so resources can be used, but through efforts for keeping them to future generations likewise", and under this understanding, "its simplest and most readily understandable form is conservation of wildlife, e.g. all fauna and flora species and their habitat, the forested areas" (Melkania 1998, p. 89). How bad is an ecological imbalance for an ecosystem and local communities (both human and non-human ones)? The case of whitetail deer in the USA illustrates it, in the 1940s most Americans fell in love with 'Bambi', but nowadays such a love has given space for complaints as the grace of America's whitetail deer has caused human health, social, and ecological problems; the deer have destroyed crops, and become a real threat to drivers, besides spreading Lyme disease across the country (Cambronne 2013).

One issue not extensively debated in the wildlife tourism concerns about the controversies that reside in the role of hunting tourism as a control management tool to help control 'ecological imbalances'. Hunting wildlife tourism may lie in this spectrum, though it has been given room to 'stormy' debates. For example, Novelli and Humavindu (2005) examine the 'trophy hunting tourism' in Namibia by evaluating whether it may be regarded as a sustainable form of niche tourism and whether it can serve as an environmental management 'tool'; they take into account social, economic and environmental implications this type of tourism may generate, particularly on poor undeveloped regions, as a way of instigating insightful debates on this theme. For Franklin (1999, p. 110) hunters make their contribution to conservation by culling and, consequently helping to restore the ecological balance staying aligned to environmental programmes' targets. In this Volume, Moghimehfar, Harshaw and Foote present a very interesting case study on the Prairie waterfowl hunting tourism in Canada by analysing it from the perspectives of the hunters. Their studies show the hunters' perception while developing and applying new theoretical approaches to the theme.

How much is it ethically acceptable to have hunting tourism or massive cull of certain types of species as a way of keeping an ecological balance in a geographic area or region? The super-population of a certain species can create severe ecological problems to other species and their habitats. Ecological balance is the term used to refer to "a state of dynamic equilibrium within a community of organisms in a habitat or ecosystem. It can also be explained as a stable balance in the numbers of each species in an ecosystem … [it] is often disturbed by human interferences [disturbances]" (Khullar 2016, p. 221), e.g. with the introduction of non-native species to an ecosystem, and this species can overwhelmingly dominate a habitat to the disadvantage of other species. There are examples worldwide where hunting tourism, as a modality of wildlife tourism, has been officially allowed to restore an ecological balance, but in some cases this type of permission is controversial from a social and ecological perspective. Holden (2016) brings an example of a culling tour package that may raise long-lasting debates,

> In Norway, the involvement of tourists in seal culling was set to begin in January 2005, with one company advertising culling tour packages on the internet. The company's website had photos of hunters posing with their kill, and the trips included accommodation, food and guidance on how to cut and preserve the seal carcasses. The Norwegian Fisheries Minister said that 'this move' would restore the ecological balance between fish and seals along Norway's coast; conversely, environmental groups say that over-fishing is indeed the cause of devastated fish stocks not the seals (Holden 2016, p. 63).

Notwithstanding, the notions of 'ecological balance' have been refuted by most research ecologists in the last decades as they feel that to talk about ecological balance is rather naïve and simplistic. Kricher (2009) has reinforced this perception, and has harshly criticised it; he says that the theory that nature is permanently in balance has been largely discredited and is a dominant rhetoric of Western philosophy that endured in the public imagination, and even today it persists among some ecologists. According to Kricher (2009), a balance of nature does not exist indeed nor has it ever been at any stage in the Planet's history. The idea is said to live on in the minds of general public rather than scientists. We can't entirely take for granted the definition without acknowledging that it has somewhat fallen out of favour. However, it is not an intention in this volume to provide a literature review on this specific ecological topic, or to argue for either side of the debate; but, what concerns wildlife hunting tourism is whether it is possible to affirm that there are examples of hunting tourism, as a modality of wildlife tourism, restoring something akin to the original balance between species within an ecosystem. It is certainly not always the case, and there are often doubts as to its merits from both a social and ecological perspective (Bauer and Giles 2002).

While culling of introduced species is often necessary, and sometimes even of native animals to spare them a lingering death by starvation, to counteract the effects of one species increasing its numbers at the expense of prey or competitor species, or to preserve habitat features (trees, grasses, riverbanks) important to other, sympatric species, decisions on what or how much to cull have not always been based on adequate scientific research. As Reardon (2012) remarks, "Some misconceptions are repeated so often that they become self-perpetuating." For instance, many members of the public 'knew' that the Kruger's ecological carrying capacity for elephants is 7000, although that figure was never declared policy or based on much scientific evidence. There is also much philosophical conflict over whether such culling should form part of recreation and tourism, or be undertaken as a serious and unpleasant necessity by carefully selected professionals. The debate becomes more complicated when part of the revenue from hunting tourism contributes to certain aspects of wildlife conservation or local economies in poverty areas, with too little research into how this might vary between situations, which may range from a response to genuine problems to excuses for scoring points in competitions involving numbers of species shot or trophies on a hunter's wall. The examples presented below represent just a very few of the more sensational and well-publicised incidents involving deaths of animals in the wild or captivity involving some form of wildlife tourism. There are very many other examples of conflict between pro-hunting and anti-hunting groups, or between anti-zoo lobbies and those who point to conservation, research and education services performed by zoos. There is often much emotion and too little research on actual outcomes regarding biodiversity conservation, animal welfare or contribution to local economies. The next section will provide an overview of the main current occurrences involving visitors and wild animal encounters in captivity or in the wild.

1.2 Poor Press: Visitor and Wildlife Interactions in the News

1.2.1 Cecil the Lion, an Emblematic Killing in Zimbabwe!

For many decades, the magnificent wildlife across Africa has been a great draw and choice for tourists wishing to have a glimpse of roaring big cats, huge elephants and crocodiles. For several African nations, wildlife is more than a tourism masterpiece; the sector is also a major source of revenue. An UNWTO's report with figures and data compiled from government tourism agencies and tour operators reveals that 80% of international travel sales to Africa are linked to wildlife tourism (CNN online). International tourist arrivals in Africa are estimated to have decreased by 3% in 2015 as the region continued to struggle with health and security challenges, as well as slower economic growth due to lower oil and commodity prices. Africa welcomed 53 million international tourists and earned US$33 billion in international tourism receipts. Zimbabwe been a country with one of the most robust growths in tourism reaching 9% in 2015 (UNWTO 2016). A recent report by the International Fund for Animal Welfare revealed that some tourists will pay from US$24,000 to US$71,000 to chase and kill lions in Africa, and roughly 8200 African lion trophies were imported to other nations in ten years, from 2004 to 2014 (Actman 2016).

However, wildlife tourism is not only 'wildlife watching and contemplation'; it also includes 'hunting tourism'. And in 2015, the violent killing of Cecil—a famous black maned lion beloved by nearby dwellers in Zimbabwe—sparkled a series of outrage worldwide (Fig. 1.1). The lion, an animal of 13 years, was one more victim of the arrows shot by an American dentist, Walter Palmer, 55 years-old, who has as part of his hobby posed pretentiously for pictures beside the defeated beasts before butchering them as witnessed in his picture collection. Cecil was lured off the protection area of Hwange National Park and injured by the hunter who then chased him for an exhaustive 40 h to finally get the lion killed and beheaded. Zimbabwean officials say Cecil was attracted outside the Park and brutally slaughtered. Palmer had paid US $55,000 for some local guides to help to carry out his unmeritorious African bloody adventure. Currently, still eight African countries permit exports of lion body parts, among them Tanzania, Namibia, Mozambique, Zimbabwe. They are countries that retain nearly half of all wild lions in the continent. The amount of wild lions has been drastically declining over the years (Actman, National Geographic, June 30, 2016). The killing of Cecil rapidly sparked thousands of compassionate messages and public manifestations on the social networks in what has been regarded as the largest global response to a wildlife event ever. Most people agreed that "the killing, apparently for fun, of the majestic lion by a Minnesota dentist visiting Zimbabwe was a perplexing act of pointless cruelty and cowardice (Fig. 1.1). It spawned millions of posts on Facebook and Twitter—a kind of outrage tsunami" as pointed by Frida Ghitis, in her comments at Point, on CNN (2015, online). Cecil the Lion died one year ago—But, what has happened since the slaying of Cecil after one has passed? What has changed during one year period time for the wildlife? According to Wildlife Watch (December 2016, online), an aftermath over one year of his killing of his resulted in the following:

- **Laws have been changed by several nations to avoid lion-trophy-hunting**: Australia, the USA and France have modified their laws to avoid 'wild animal trophies' entering the countries. This serves as import barriers to discourage hunting overseas, particularly in Africa. The USA included new clauses to its Endangered Species Act for a more encompassing protection for lions. Trophy-hunting bans have been placed long ago before Cecil's killing by Botswana, Kenya and Zambia. But, Zimbabwe only suspended the hunting for 10 days.
- **Airlines have banned wild animal trophies. No transport allowed!** Big Airlines such as Delta, Air Canada, JetBlue and British Airways totalling 40 airlines worldwide have reinstated their bans on transporting trophies from leopards, elephants, Cape buffaloes, rhinos, and surely lions.
- **Globally, people have been more aware of the trophy hunting and its impact on wildlife**. Wild life hunting if not banned, should be effectively managed. For example, scientists recommend strict enforcement of low quotas and only allowing hunts of older lions.

Fig. 1.1 Cecil the lion (pictured in 2012). *Source* Nature.com/NATURE News. Headline—Use License Number: 4117870597019. License date: May 28, 2017. Licensed content Publisher: Nature Publishing Group

- The best and greatest news over Cecil's tragedy is that his **seven young cubs are doing just fine**;
- **Walter Palmer was not arrested, and never charged allegedly** because he was a holder of permissions to hunt in Zimbabwe explained some country's authorities; Palmer has said he didn't know the lion hunted was the beloved Cecil. Where is he? He returned to his routines at the dental office in Minnesota.

Although many hunters are undoubtedly highly responsible and skilled at identifying species, not all fall under this description, and one of the concerns often expressed about hunting is the effect on species other than allowed targets. According to the Australian Broadcasting Commission:

> While shooters deny rarely—if ever—getting it wrong, the issue of misidentification is incontrovertible. Rare and protected species are killed each season, and last year the list of protected species found shot and abandoned included swans, grebes, coots, magpies, spoonbills, stilts, cormorants, parrots, owls, birds of prey, and the threatened Blue-billed Duck and Freckled Duck. There are only two plausible explanations to account for the killing of these non-game species: either hunters cannot be trusted to accurately tell the difference between a duck and the very distinctive (not to mention nocturnal) Barn Owl or they have deliberately shot non-target species in a fit of excitement…

Of course, journalists are not always accurate, but the concerns should be taken seriously enough for further studies on the scarcity or otherwise of 'mistakes'.

1.2.2 Tigers of a Buddhist Temple in Thailand: A Shadowy Uncovered Case

In 2016, people around the world got breaking news on TV and the internet about the government intervention on the

Fig. 1.2 A Monk walks close to tigers at Thailand's Temple. *Source* Wat Phra Luang Ta Bua, Kanchanaburi Province, Thailand. Date 13 June 2004. Creative Commons license (CC BY-SA 3.0). Credits to: Michael Janich

Thailand's Tiger Temple with the presence of more than 500 wildlife officials, veterinarians and police to rescue 137 tigers (by Ramsey, June 2016, Aljazeera Online) (Figs. 1.2 and 1.3), scientifically named *Panthera tigris tigris.* For decades, the Temple has kept the tigers under the status of a wild animal conservation and protection, but using them to steadily cash out as major tourism attractions alluring thousands of visitors every year, massively formed by foreigners, "with an entrance fee of anything from 600 baht ($17) to 5000 baht ($140) per person, millions of dollars have flowed into the temple over the years" (by Ramsey, June 2016, Aljazeera Online). After years of allegations of animal abuses by multiple non-governmental organizations against the Buddhist Temple, the Thailand's Department of National Parks, Wildlife and Plant Conservation—DNP, decided to act on behalf of the "big cats" (Fig. 1.2). With regards to the tigers' situation in Asia, animal groups have been alleging for years on cruelty, illegal wildlife trafficking and breeding for supplying a market with animal body parts. According to a news features on Aljazeera online,

> Tiger Temple has long been a staple attraction for tourists and millions of dollars have flowed into the temple over the years., backpackers looking for the perfect photo-op. A romantic picture was painted of ochre-clad monks and endangered tigers living together in a relationship of numinous unity. The message: "You too can partake in the harmony" - for a price (Ramsey, Aljazeera online, June 2016).

The raid on the Buddhist temple in Kanchanaburi, a province located west of Bangkok, in Thailand, unveiled that "apart from 137 live tigers, they found a laboratory, suggesting that the monks were using tiger parts to make wines and medicines—as well as the carcasses of 40 cubs stored in a freezer", such scenes that obviously had been kept out of sight of the visitors and shocked wildlife investigators (by Vidal 2016, The Guardian online). From the Temple's side, some explanation was given by Tanya Erzinclioglu, a tiger

Thailand Tiger Temple: 40 tiger cub bodies found in freezer at controversial tourist attraction. (source: Independent)

Thailand's controversial Tiger Temple shut down for keeping animals without a permit (DailyMail)

The temple is a tourist attraction but had been investigated for suspected links to wildlife trafficking and abuse (source: Independent)

Thai Officials to Rescue 147 Tigers from Monastery After Investigations (source: Earthisland.org)

Conflict unfolds during last relocation from Thailand's Tiger Temple (source: Borneobulletin.com)

Getting to the Truth Behind Thailand's Infamous Tiger Temple (source: Time.com)

Zoo License Awarded To Thailand's Controversial Tiger Temple (source: worldanimalnews.com)

WWF Applauds Removal of Tigers from Tiger Temple and Encourages Thai Government to Permanently Bar the Temple from Keeping Tigers (source: panda.org)

Fig. 1.3 A Collection of News Headlines in 2016 from different media sources show how polemic was worldwide the rescue of tigers in a Thai Temple very popular for captive Wildlife Tourism

caretaker, working for many years there, and interviewed by Aljazeera for a feature news on wildlife tourism. According to her, "these finds were actually the easiest to account for … These were the policy of Dr Somchai [*Visasmongkolchai*, a former vet at the temple] since 2010 […] The DNP has been in our freezer loads of times in the past. Why are they acting surprised? Tanya emphasises that the even months before the raid, The Temple page on Facebook had commented on a post on this Policy, "which claimed they were keeping carcasses as proof they were not being sold on", and she also explained that the revenues got with entrance fees were used to build tiger island,

> The temple's pride and joy: its almost five-hectare "Tiger Island" enclosure. Completed in 2011 at an estimated cost of 90 million baht, (slightly more than $2.5 m) the 28 enclosures in Tiger Island meant that, for the first time, the tigers were able to experience outside spaces, albeit on a rotating schedule. The open areas are strikingly different from other tiger zoos in the country. At the frenetic Sriracha Zoo, activities include tiger shows that seem to, in part, include having the cats jump through flaming hoops. When ex-DNP director general Damrong Pidech visited the Tiger Temple in 2012, he actually praised the tiger's living conditions, telling the Bangkok Post: "Frankly speaking, their living conditions are better than those in state-owned zoos (by Ramsey 2016, Aljazeera online).

In Kanchanaburi, Thailand, the former caretakers of the 137 tigers removed from the Tiger Temple are deeply concerned that the tigers would face a worse fate in government hands, but DNP's staff declared that the Department has been discussing the possibility of creating a new sanctuary for the tigers aware that the animals will suffer neglect because the DNP's structure is not ideal for keeping several big wild felines, and the plans include assistance and support from Four Paws, an animal welfare charity organisation, to manage the new tiger sanctuary (by Ramsey 2016, Aljazeera online). As a result of the NDP's intervention, the Tiger Temple, also known as *Wat Pha Luang Ta Bu Yannasampanno*, was closed for visitation and police charged 22 people with illegal activities of wildlife trafficking of body parts; three Buddhist monks were among them (Fig. 1.3). This ends a long history of controversies with the Monks being repeatedly accused of mistreating the tigers and of illegal breeding (BBC online, June 1, 2016).

Globally, trade of tiger parts feed an ever-increasing, quenchless and rapacious trade in China and has threatened the few remaining tigers in the wild. Current population of wild tigers have decreased from 100,000 in 1900 to 3200 as an estimated number. According to the 2014 report of the Commission for the Convention on International Trade in Endangered Species of Wild Fauna and Flora—CITES, "an increasing number of live animals and frozen bodies are being detected, with more than 50% of seizures over the past 14 years occurring since 2010. It is suspected that many of these are of captive origin … seizures of suspected captive-origin tigers have risen in Laos, Thailand and Vietnam, and evidence suggests that such trade is also taking place in Indonesia." (Vidal 2016, The Guardian online). The Environmental Investigation Agency (EIA) and other wild animal protection organisations suggest that more than 5000 tigers are being farmed in China, 1450 in Thailand, 180 in Vietnam and possibly 400 in Laos. In addition, there are private collections and zoos in most other Asian countries, yet according to the same news source published on The Guardian (by Vidal 2016):

- 3200 tigers live in the wild. It is estimated that 2200 remain in India, 500 in Russia, 50 in Vietnam and only 30 in Vietnam;
- In Southeast Asia, it believed that 7000 have been bred in captivity spread in 240 farms. China leads the number of this type of tiger farm with 5000 animals, Thailand 1450, and Laos 400;
- Tigers' skin, bones, meat and claws are well-known to serve an industry of "luxury goods" and Chinese traditional medicine production;
- Roughly 30% of illegal trade of tigers' body parts are believed to come from captive breeding in farms;
- Between 2000 and 2014 is estimated that about 1600 tigers have been seized and killed in the world.

1.2.3 But, Not Everything is About Bad News Regarding the Asian Tigers: Indian Tiger Protection Reserves

Chapter 9 of this current Volume discusses Royal Bengal tigers in two nations, India and Bangladesh, and presents key aspects of the context of the tiger within the wildlife tourism perspectives and the opportunities for enhancing visitors' wildlife experiences by developing educative and educational tours. In Bangladesh, a Zoo is part of the investigation. In India, the Sundarbans region is the geographic object of the case study. Sundarbans forest, in West Bengal, was established in 1973–1974 and is the second largest tiger reserve of India with a total area of 2585.10 km^2 (IISC 2016), and it is estimated that about 270 tigers live there. Apart from the tigers, fishing cat, spotted deer, wild boar, Ganges river dolphin, water monitor, estuariane crocodile, river terrapin, olive ridley turtle, ground turtle, hawks bill turtle and king crab are other wild animals found in Sundarbans. The estuarine of Sundarbans has high salinity, no erosion and daily overflow by high tides. Wildlife conservation is managed by government and organisations protection "from poaching and external influences such as prawn fishing is a primary concern. Boats patrol the area, soil conservation is practised and man-animal conflicts are being addressed" (WPSI, Online 2016). India has 103 National Parks, and in 1936 the Hailey National Park (currently called, Corbett National Park) was the first National Park to be created in the country (IISC 2016, online). India has 28 areas specifically established as Tiger Reserves created by Project Tiger launched by the government in 1973 to protect endangered species of tiger; the reserves correspond to 1.09% of the total area of the country (IISC 2016, online).

Many wild animals who live their lives in sanctuaries, or in elephant camps, due to illegal trade, poaching, or because they retire from a life working for humans, like former working elephants. A sanctuary is expected to be a safe haven where animals should be allowed to live out their lives with little or no captivity, but this is not a reality indeed for most elephant sanctuaries in Thailand as they are not totally free; though many elephants seem to have a much better life if compared to their nearly slaved life in the logging sector, still the elephant camps have raised ethical concerns to the way some of them manage the animals for the visitors' interest (Fig. 1.4). The work of Kontogeorgopoulos corroborates this notion, "although elephants working in camps tend to fare better than those toiling away in circus venues, or in illegal logging operations, most still live far from ideal, natural lives. Camp elephants face several problems" such as injuries, artificial environments e.g. a barrier from natural formation of herds and animal family bonds, low caloric diet —some of them don't have enough food, and harsh training and abusive disciplines methods (2009, p. 430).

Some sanctuaries depend on outside funding to provide care to animals and to maintain their structure; others need to generate revenues themselves through programs and activities, which, by the way, usually involve the elephants being major tourism attractions with visitors direct or indirectly being in contact with them, and interactions involve rides, feeding and photographs very close 'beside' or 'between' the beasts, as observed by Ismar Lima in his field work on Elephant Tourism in Thailand, in 2015. In Chiang Mai, visitors usually pay a certain amount of dollars, which is not really a bargain as a voluntary day experience may set them back a hefty US$100 fee per person, to help the Mahouts (elephant carers and tamers). But, Brando (2016) underlines that even though is worthy knowing the elephants have been removed from abusive treatments and taken to sanctuaries, their participation in exhaustive, intense and highly exploratory shows, displays and series of performances, such as "lifting the trunk, leg or let people sit or ride on their back," [painting] and all this should be object of scrutiny to ensure the elephants' welfare.

In Thailand, since the Chakri Dynasty, A.D. 1800, elephants have been under protection by Thai law; in 1960 the Wild Elephant Protection Act was endorsed, and in 1980, the Wild Animals Reservation and Protection Act was also approved to help safeguarding the elephants likewise (Humphrey and Bain 1990, p. 370). Currently, Thai government authorities responsible for elephants have been the Department of Livestock, Department of Transport and the Forest Industry Organisation, rather than the Department of

Fig. 1.4 Chinese visitors enjoying a ride on the Elephant's back guided by a Thai Mahout at Maesa Elephant Camp (*right picture*), in Chiang Mai province, Northern Thailand, and Elephants pulling logs at Thai Conservation Centre, in Lampang, near Chiang Mai. In both places, elephants are attractions for visitors with interactive situations, shows, displays, feeding, and rides, but as the reader can note the elephants look healthy, and they are not all the time available for the visitors. There is a schedule for shows, feeding and bath. *Source* Ismar Lima, field work on Elephant Tourism in Chiang Mai (*right*) and in Lampang (*left*), Thailand

National Parks or Ministry of Environment as Elephants are considered 'working animals' rather than wildlife; and "Elephants are probably the only animals employed by man that have never been bred selectively, and being for all intents and purposes wild animals they should receive greater consideration than more domesticated animals. Actually the very reverse is usually the case" (Lair 1999, online). According to Cohen, Thailand has about 3000 elephants and 2000 of them are privately owned either by a mahout—an elephant driver, tamer, and lifelong carer and keeper. (Duffy and Moore 2011, p. 596) point that,

> Captive working elephants are important in long-term elephant conservation in Thailand. Without them, the long-term survival of Thailand's elephants would be at risk. Any attempt to close down the elephant trekking industry would have to grapple with the possibility that it would spell the end for Thailand's elephants. Without careful consideration of how captive elephants would be re-wilded (including where they would live and who would pay for re-wilding and management).

Captivity, or even semi-captivity of elephants (Kontogeorgopoulos 2009), is thus perceived as an alternative means to ensure their conservation, rather than having free in the wild as it occurs in some African nations. This is because elephants, due to their former captivity situation as working beasts [e.g. logging industry, etc.], may not easily get adapted to the wild again, "captive working elephants are not always suitable for 're-wilding'" (Duffy and Moore 2011, p. 594); moreover, if they are released they can return to highly abusive working imposed by private owners without the means to properly look after them. Pickover (2005) estimates that about 70% of the elephants in European zoos are wild-caught and "those born in captivity can never be released into the wild" (p. 63). Captivity and semi-captivity of elephants as wildlife tourism attractions are certainly a long-lasting debatable issue across the disciplines, especially related to wild animal and human encounters and interactions, wild animal welfare, and animal ethics.

1.2.4 The Case of the Cincinnati Zoo: The Killing of Gorilla Harambe

On 28 May 2016, on a Saturday holiday at the Zoo and Botanic Garden of Cincinnati city, State of Ohio, USA, people around the world were outraged after watching the news that a 3-year-old boy fell into a gorilla enclosure, and as a result the zoo staff needed to kill the gorilla. Harambe was a rare adult gorilla, 17-year-old male, and was shot dead allegedly to avoid a likely attack from the gorilla against the child. Two female gorillas were also in the enclosure. In a press interview after the incident, the Cincinnati Zoo President Thane Maynard explained that the little boy spent only 10 min in the enclosure together with gorilla when Dangerous Animal Response Team decided for its killing, considering the situation extremely risky to the boy's life and demanding immediate action.

A press release from the Cincinnati Fire Department, stated that the boy was between the gorilla's legs at the moment it was shot, and two fire-fighters entered the enclosure to rapidly rescue the child who, by reason of the fall, had some injuries and was treated in the Cincinnati Hospital Medical Centre. In the press release, the Fire Department reported that its staff "witnessed a gorilla who was violently dragging and throwing the child" on the scene. It was said that in 28 years since its creation, it was the first time the Cincinnati Zoo had an emergency situation. According to Maynard, "the decision to shoot Harambe instead of tranquillising was made in the interest of the boy's safety … In an agitated situation, it may take quite a while for the tranquilliser to take effect …". Harambe hadn't attacked the little boy, but the gorilla's size would pose danger. The spokesman of the Police Department, Steve Saunders, explained that no charges were made against the boy's parents (Knight and Sullivan, Cincinnati News online, June 18, 2016).

Brittany Nicely of Dayton was visiting the zoo with her two children and four other children on that fatalistic Saturday, at Gorilla World, where the incident happened. She gave an interview to local News stating that,

> "I saw the little boy in the bushes past the little fence area. I tried to grab for him. I started yelling at him to come back…Everybody started screaming and going crazy," she said. "It happened so fast…The gorilla rushed toward the boy and led him by the arm through the water in the enclosure. She said initially the gorilla seemed protective and only alarmed by all the screaming. The area was then evacuated by zoo staff. Nicely stood with her group outside the exhibit…About four or five minutes later we heard the gunshot…We were pretty distraught. All the kids were crying…It's a very traumatising experience for anybody involved. The kids, the zoo-keepers, the other gorillas that now don't have him [Harambe] there any more" (by Knight and Sullivan, Cincinnati News online, June 18, 2016).

Figure 1.5 shows two snapshots of Harambe. The image on the left is a Mirro's website snapshot that shows the gorilla Harambe protecting the little boy by holding his hand minutes before being killed, and the second image from Zoo Cincinnati shows Harambe in its 17th birthday.

But Harambe gorilla was not an isolated case of visitors getting into Zoo enclosures and having its wild animals killed to rescue them. On 23 May 2016, just about a week before the incident involving Harambe, in a Zoo in Santiago, capital of Chile, a naked suicidal man tried to kill himself by entering the lions' lair. In order to save the man, two lions—a male and a female—had to be killed. According to Zoo's director, the protocols on visitors' safety need to be followed strictly because human lives are the priority, and the tranquillisers would not help saving the man's life immediately as they takes a bit of time to make a full effect. The animals needed to be sacrificed (by Roterman, May 23, 2016, Latin Times online). On November 2012, at the Pittsburgh Zoo, a two-year-old boy fell into African painted dogs' enclosure as he slipped from his mother's hand. Unfortunately the little biy was killed by a pack of wild dogs. In 2007, an incident involving a four-year-old Siberian tiger, known as Tatiana, at the San Francisco Zoo, managed to escape from her setting and attacked two men who were teasing the animal and throwing rocks. One of the men died on scene, and the Siberian tiger was shot to death by the local police (Eco Watch, June 2016, online).

Some environmentalists, animal lovers, animal protection organisations and people around the world concerned with animal welfare, deeply touched by these tragic events and amid the debates over who was at fault in the death of the wild animals, have come to a common ground that these tragedies should never taken place. The CRC Research Report on Captive Wildlife Tourism in Australia, produced by Tribe (2001), found at that time that zoos in Australia were changing their structure and function looking for ways of developing three important justifications for keeping wild animals in captivity: conservation, education and research (2001, p. i). But Tribe (2001) highlighted that despite setting an agenda for re-modelling themselves, the zoos have been the object of "philosophical accusations that they are irrelevant and wrong, and some in the community even advocate their abolition" (p. 27). In 2013, Costa Rica government officials announced their intention of shutting down public zoos throughout the country, "the Simon Bolivar Zoo in the capital of San Jose—which currently houses hundreds of animals—will be transformed into a botanical garden, and the Santa Ana Conservation Center west of the city will be turned into a park", and the plans are to have wild animals released into the wild or sent to animal sanctuaries (Romo and Sholchet 2013), but this is not actually happening. The Australian and New Zealand Federation of Animal Societies is opposed to keeping wild animals in captivity due to stressful living conditions. On the other hand, zoos can also contribute to conservation through educational programs and services and the captive breeding, management and display of the wildlife (Tribe 2001).

The circumstances through which Harambe was killed made him an instant worldwide celebrity; an event that in some aspects resembles—in real life—the movie King Kong, a giant beast kept in captivity to entertain humans. At the time of the kill of Harambe the gorilla, Marc Bekoff, wrote a brief article on the blog of Scientific American arguing that the discussion should move far beyond the point of regardless who was guilty or innocent. "Opinions vary as to whether the boy was really in danger and who was to blame, e.g. the zoo (why was the boy able to get into the enclosure and why wasn't Harambe tranquillised?) Playing the blame game will not bring Harambe back" (Marc Bekoff, Scientific American blog, June 1, 2016). According to

Fig. 1.5 Some News Headlines on the Harambe killing at Cincinnati Zoo, in USA, on May 2016. The gorilla images below are only illustrative. It is not Harambe. The pictures show how warm and lovely the gorillas can be with their babies. *Source* Photo credits: Sabine bresser, Creative Commons (CC BY 2.0), 31 May 2013. Illustration/Figures—assembled by Ismar Lima

Bekoff's views the discussion should be more insightfully reinstated by questioning, 'why was Harambe in a zoo in the first place?' And, he gives a lead that the debates should also take place within the anthropozoology realms.

1.3 A Theoretical and Disciplinary Approach in Human-Animal Encounters: From Anthropozoology (Anthrozoology), Environmental Social Sciences to Human and Political Ecology

With all these tragic and controversial events related to encounters and interactions between humans and captive (and semi-captive) wild animals, it seems that there has been a global sensibility moved by a heightened awareness about the nature of human-animal relationships, particularly with regards to the wildlife tourism sector (Semeniuk et al. 2010, p. 2699; Hughes and Carlsen 2008) and animal used on display or in shows for entertainment. Human and animal encounters lie within a scientific and academic domain that has shown evidence of growing rapidly, the Anthropozoology. With a combination of 'anthro', human(s), and zoology, study of animals, the term was first coined in 1987 as a Journal title *Anthrozoos*; literally 'anthrozoology' would be etymologically more correct to be use (Mills and Marchant-Forde 2010). Anthrozoology is much wider encompassing as it deals with studies and concepts regarding applied animal ecology, conservation science and animal welfare, but it also embraces "studies of associations, and especially relationships, between individual humans and individual animals, rather than to interactions that take place at the population level; in this sphere, anthropozoology is about the understanding of the human-animal bonds." (Mills Mills and Marchant-Forde 2010, p. 28). But, according to Tyler (2009), the names for this study field focussed on human-animal encounters (or relationships) are various and it lacks consensus in the literature as the field itself is debatable, should it be 'Animal Studies' … 'Human-Animal Studies'? (Shapiro 2002; Shapiro and Copland, 2005), 'Anthrozoology' (Rowan et al. 1997), or should it be anything else such as Anthropozoology? It surely lacks of consensus even with regards to nature or bounds of the object (or subject) of this study zoo area, 'the animal' (Ingold 1988).

With so wide a scope, this fast growing and varying scientific field "denote genuine and entirely healthy multidisciplinary engagement" (Tyler 2009, p. 2), and these incongruous approaches rather than evidencing a weakness it denotes a strength of the field, if it is taken into account that varied—multiple—disciplines gather specific knowledge that contributes to the issues that permeate the human [visitor] and animal encounters, including ecological, biological, sociological, anthropological, political ecology, restoration ecology, animal welfare and ethical ones (not limited to them) with researchers coming from distinct fields to investigate the animal-human interactions, moved by dissimilar scientific and academic interests and concerns willing to develop and apply different methodologies and methods (Taylor and Goldsmith 2003; Tyler 2009; Newman 2013); it is a promising investigation field, given its amplitude of applications within the anthropozoology, paving ways to make human and non-human interactions less controversial and problematic; it is in fact "an open contested field, with no clear canon, Animal Studies is a meeting point where different species of researcher gather" (Tyler 2009, p. 2).

Attention is usually given to taming animals and its sociological implications throughout the history, domesticated species as companions—and also used as part of therapeutic treatments; animal-assisted therapy, for example, with horses in equine therapy that can contribute to enhancing skills of impaired people with physical, emotional, behavioural and emotional disabilities (Chandler 2012; Altschiller 2011; Engel et al. 1994; Cusack 1988). Moreover, animals have been kept as transport means, e.g. camels have been domesticated in Southern Arabia since 2500 BC; and animals have served as pets and companions and "have furnished bone and pearl ornaments, and have been subject to scientific experimentation" (Tyler 2009, p. 7). As for the wildlife, visitors, and zoos and sanctuaries (captive animals) two more recent releases have a contribution to the literature, Zoos and Tourism, edited by Frost (2011), a volume that deals with the commodification of wildlife, captive wildlife management and ethical issues in the zoos with regard to visitors and wild animal interactions; another contribution in this field with a selection of case study is Animals and Tourism, edited by Markwell (2015). For Ed Stewart, president of the Performing Animal Welfare Society, there is no ethical way to keep elephants in captivity because zoos have very dissimilar environments (and settings) and philosophies as compared to sanctuaries, particularly with regards to 'captivity itself', that is, "there presently exists no state-of-the-art keeping [wild animals] in captivity, e.g. elephants [in zoos] (Stewart 2013).

As for the incidents involving visitors and wild animals in the zoos, and a predatory and exploratory relation to wildlife as attractions, for example, in the case of the Tiger Temple in Thailand, much of anthropozoology can be used to explain those events and phenomena, but further contributions can be made by combining anthropozoology with behavioural biology and ethology as disciplines that study human-animal behaviour and social organisation from a biological perspective (Archer 1988). Some approaches of concern include the use of sociology for the emancipation of animals (Taylor 2011), and the studies on animals as performers, that is, the representation of animal actors in the animal kingdom (Szarycz 2011; Armstrong 2011) and mapping and theorising human animal relations (Beatson 2011; Kemmerer 2011).

With regards to wildlife tourism, attention is drawn to wildlife-human encounters and interactions, and in this field 'human ecology' (Semeniuk et al. 2010, p. 2699), 'political ecology' and 'environmental social sciences' can be used to advance the understanding and insights on the current wildlife tourism phenomena, and consequently to help point to feasible solutions to the complexities that this field holds, for example, with regards to hunting wildlife tourism, and captive wild animal as tourism attractions, wildlife management, conservation and ethics. According to Moran (2010), the attempts to integrate social and environmental sciences have not been an easy task, "nor has the cacophony of competing theories and paradigms helped to promote collaboration between the social sciences and the natural sciences" (p. 6) due to a latent, but fundamental ideological divide which is centred on two ideological streams: realism and constructionism; two ideological positions that permeate the social environmental sciences (Stoddart 2012). Notwithstanding, the mission of this chapter is to contemplate both conceptual and practical solutions (Vaccaro et al. 2010) for dealing with the wildlife tourism dilemmas. In this sphere, it is desirable that decision making and planning processes can lead to more ethically reasonable ways of managing the controversial issues that permeate wildlife tourism practices, particularly related to cruel, and overly exploratory—if not predatory—practices.

From both conservation and welfare points of view, it can be claimed that wild animals should not be kept in captivity solely for fulfilling human entertainment, neither should they be object of cruelties, e.g. in the practices of hunting tourism and animal-made products; this requires a shift from an anthropocentric view to ecocentric and biocentric ethics that can enlighten human actions towards wild animal with actual preoccupation for their welfare and "can contribute to valuation for conservation" (Lindenmayer and Burgman 2005, p. 21). A new understanding in this field should harness social and environmental sciences to the benefit of wild animals, but "we should be careful not to reproduce the dualism that exists also within the social environmental sciences between micro analyses of environmental (un) friendly behaviours on the one hand and the macro analyses of institutional developments on the other" (Spaargaren 2000, p. 58–59). Most scientists have concluded that we cannot begin to understand global [and local] environmental change and to deal with environmental challenges without orchestrating and harnessing efforts that can contemplate both biophysical and social sciences, that is, "Human agency (i.e. actions of individuals) is implicated in most of our current dilemmas, and must play a part in solving them" (Moran 2010, p. 1). This is an indisputable arena, and "the rhetoric of sustainable utilisation to bring political standpoints and moral questions to centre stage in the conservation debate" (Smith and Duffy 2003, p. 196), but there are oftentimes anthropocentric discourses that situate nature as a utilitarian 'entity' to 'solely' serve humans (Lima 2009).

Thus, the sustainable utilisation and conservation in wildlife tourism demands pragmatic solutions to manage wild life resources of a certain geographic area. Ecology plays a crucial role for understanding and managing a series of activities such as crop rotation, weed control, management of grasslands, forestry, biological surveys, fishery biology, conservation of soil, wild life, forest, water supplies; and ecotourism and wildlife resources in tourism studies (Holden 2016; Raina 2005). Applied ecology is a scientific field that studies "concepts, theories, models and methods to solving of environmental problems, including the management of natural resources, such as land, energy, food or biodiversity" (Bertelsmeier et al. 2012, p. 52). Other disciplines and approaches are relevant to managing wildlife and habitats and they can also serve as foundations in environmental interpretation and education, such as: social dimensions of resource use; perception of environmental change; environmental risks monitoring; environmental justice; environmental decision making and planning; politics of natural resources; environmental policies (Cunningham and Cunningham 2005; Hastings and Gross 2012). But, Moran (2010) alerts that many other incentives than just ensuring good environmental management such as "political pressures, misvaluation of the resources, self-interest, and corruption" (p. 20) can negatively affect successful outcomes in wildlife conservation and management.

Human ecology also plays a pivotal role in helping mitigate negative impacts that tourism, particularly wildlife tourism (Hughes and Carlsen 2008; Higginbottom 2004; Green and Higginbottom 2001), may cause to ecosystems. A human ecology approach seeks to praise a holistic understanding of significant social issues for critically assessing the continuing evolution of the human-environment interface (Miller et al. 2002, p. 30), and wildlife tourism is highly characterised by this interface,

> Wildlife tourism attractions are characterised as having intricately coupled human-wildlife interactions. Accordingly, the ability to mitigate negative impacts of tourism on wildlife necessitates research into the ecology of the system and of the human dimensions, since plans aimed at optimising wildlife fitness must also be acceptable to tourists" (Semeniuk et al. 2010, p. 2699)

As for the political ecology, Chap. 2 written by Shelton, Tucker and Zhang in this Volume brings a thorough conceptual discussion on 'political ecology' issues and applies it to the wildlife tourism field to shed light on the yellow-eyed penguins in Southern New Zealand.

1.4 Wildlife Tourism's Potential for Positive Outcomes

While much of the above has emphasised problems with wildlife tourism, it must not be forgotten that the industry has great potential to contribute to wildlife conservation (e.g. Higginbottom et al. 2001). Habitats which might otherwise have been cleared may be left intact if governments and local residents can see a monetary reward in attracting tourists to see wild animals. The animals themselves may also be viewed more favourably instead of being regarded as pests or simply ignored. Tour operators and tourists might actively help to restore habitat or contribute donations to conservation projects. Frequent visits to wildlife areas can have a deterrent effect on would-be poachers. Contact of many thousands of tourists each year with knowledgeable guides can raise the awareness and appreciation of wildlife by the public and an understanding of current conservation problems. Tour operators and volunteer tourists can contribute to wildlife research (Green and Wood 2015) leading to more effective conservation management plans.

In 2015 Wildlife Tourism Australia Inc. (http://www.wildlifetourism.org.au) held a conference entitled "Wildlife Tourism: a Force for Biodiversity Conservation and Local Economies?" involving tour operators, zoo and ecolodge staff, conservation groups, government representatives and academic researchers. Many interesting papers were presented exploring this theme, and round table discussions covered many aspects. The lack of research into many issues was noted, and it will probably be some years before some of the major knowledge gaps are filled, but delegates heard many examples of conservation, research and educational projects and evidence of assistance to local economies in both developed and developing nations, and few if any doubted that wildlife tourism has already conveyed many positive advantages and has the potential to contribute far more in the future.

The current volume is important in bringing together some of the research and philosophical ideas pertaining to wildlife tourism, but there is still much to be investigated, and much discussion yet needed between people from different backgrounds and perspectives, to formulate effective plans to facilitate the synergies between wildlife tourism, wildlife conservation, wildlife research, habitat restoration, animal welfare and poverty alleviation.

1.5 Overview of the Chapters in This Volume

'Wildlife tourism' is a term that covers a wide range of activities throughout the world, from long treks through wilderness to see rare species or simply experience relatively untouched ecosystems, through comfortable rides in safari vehicles or boats or easy strolls from ecolodges, to interacting with semi-wild animals at feeding stations or viewing them in zoos or wildlife parks. This wide variety of tourist experience is reflected in the spread of topics in this book. The geographic spread of wildlife tourism is also reflected in chapters from Australia, New Zealand, Brazil, Canada, Botswana, Rwanda, Turkey, Finland, Germany, India, Bangladesh, Thailand and Japan. Wildlife especially considered in the book includes tigers, kangaroos, deer, whales, dolphins, penguins, other birds and turtles.

It is often suggested (or hoped) that wildlife tourism can enhance conservation of biodiversity and local economies. Tribe (Chap. 3) describes a partnership between business and a research institute with involvement of local residents and conservation groups to reintroduce rare and threatened species into a large luxury tourism property in a mountainous region of Australia. Moreira and Robles speak of income for local residents generated by turtle-based tourism in Brazil. Shelton et al.'s chapter provides an example of an integrated approach to reintroduction of a species. Some districts most in need of assistance for local biodiversity protection and poverty alleviation are in areas away from the usual tourist trails and lacking in the kinds of facilities that international travellers tend to expect, and may need some innovative ways of attracting the more adventurous or experienced traveller, and finding key points of difference (rare species, access to different kinds of wilderness or novel experiences etc.) their district can provide. The problems of poor road quality and lack of transport and accommodation for tourists in some regions are also noted by both Moswete et al. and Hassan and Sharma, in Africa and Asia respectively.

The role of ecology in wildlife tourism is to determine what tourist activities might have an effect on population numbers, both of the animals the tourists are seeking and of other species that share their habitats, and on disruptions of ecological processes. Controversial topics such as hunting, fishing and feeding of wildlife need to be looked at not just in abstract general terms but also in particular local situations if final management decisions are to be made. There is already a large knowledge base that does not always become known to planners, and also much that remains to be determined. There is still a lack of ecological research into effects of wildlife tourism on population numbers of wildlife. Green refers to some of the literature that does exist, and suggests some avenues for future research. Moswete et al. (Chap. 6) review literature to identify a number of factors to moderate impacts of tourist activities on wildlife. Moreira and Robles (Chap. 10) speak of research into turtle ecology by TAMAR at tourist destinations in Brazil. Usui and Funck (Chap. 16) make the point that not much ecological research has been conducted in highly-modified habitats in relation to tourism, and discuss this in relation to management of tourists and semi-tame deer in Japan.

If wildlife tourism is to expand and flourish, the needs and interests of tourist must be understood, and possibilities of diversification explored. The chapter by Werdler (Chap. 4) explores the potential for bird-watching in Rwanda, a tropical African country best known in the industry as a primate-trekking destination. Moswete et al. (Chap. 6) consider what is needed to reduce congestion in one region and encourage use of alternative routes in another African country, Botswana. Lanzer et al. (Chap. 5) explore the potential for diversifying lake-based wildlife tourism in Brazil, including science tourism, and Hassan and Sharma (Chap. 9) for diversifying the tiger tourism experience in India and Bangladesh. Ayazlar (Chap. 12) speaks of the need for wildlife tourism in Turkey, which already has a variety of wildlife experiences, to be more clearly defined and catered for. Mayes, (Chap. 7) Werdler (Chap. 4), Lanzer et al. (Chap. 5), Hassan and Sharma (Chap. 9), Moreira and Robles (Chap. 10), and de Lima (Chap. 8) point to the importance of wildlife interpretation, including missed opportunities for interpretation in some tour operations, the training of guides and the role of interpretation in promoting appropriate behaviour. Mayes (Chap. 7) warns against too much interpretation while high-intensity wildlife encounters are actually in progress but offers research-based advice on quality interpretation before and after such experiences. Hassan and Sharma (Chap. 9) speak of the importance of appropriate marketing and the potential for learning experiences in tourism. Moswete et al. (Chap. 6) explore tourism personnel and wildlife officer perspectives on tourist usage of a Botswanan park with a view to understanding tourist (and tour operator) preferences and suggesting procedures to spread visitor usage and avoid over-congestion of some areas. Harman and Dilek (Chap. 11) explore visitor responses to whale-watching experiences. Usui and Funck (Chap. 16) remind us that cultural differences must be considered when managing human-wildlife interactions.

There can be conflicts of interest associated with wildlife tourism, both within the industry and with other stakeholders. Burns points to growing public concern about animal welfare, and Reiser (Chap. 17) explores whether zoos can still justify a place in modern-day wildlife tourism. Ivari (Chap. 15) describes a conflict between wildlife tourism and reindeer herding, and reviews a conflict resolution framework in relation to this. Hunting tourism or feeding stations may not be compatible with conservation management, animal welfare considerations, needs of land-owners, or other forms of wildlife tourism such as treks to observe wild animals behaving naturally. Moghimehfar et al. (Chap. 18) make a case that responsible hunting tourism supports conservation, and lament that the sport appears to be declining in Canada. Ayazlar (Chap. 12) points to the revenue brought into Turkey by hunters. These advantages of hunting tourism need to be tempered within an ethical framework and ecological studies of effects on both target and non-target animal populations (Burns, Chap. 13, Green, Chap. 14). Political will does not always mesh easily with ecological needs of wildlife. Shelton et al. (Chap. 2) provide an interesting case study involving reintroduction of the yellow-eyed penguin in southern New Zealand and the associated development of wildlife tourism, explaining the concept of 'political ecology' and its necessarily interdisciplinary approach, and pointing out that accounts of "political-ecology-of-tourism studies" of developed countries are to date very sparse. They discuss a range of stake-holders (governments, Indigenous, NGOs, tour operators, tourists and others) and problems with definitions of terms such as 'nature' and 'environment' and the concept of 'equilibrium.'

Ethics in wildlife tourism involves environmental ethics (e.g. does the operation minimise its impact on the environment, including biodiversity loss, or better still make a positive contribution to conservation?), animal welfare (e.g. do animals suffer pain or undue stress as a result of tourist activities, whether injured or seriously disturbed in the wild, or mistreated or kept in inadequate enclosures in captive settings?), tourists (e.g. safety, enjoyable experiences, valid interpretation), other tour operators (e.g. not impacting negatively on other businesses, and forming mutually-beneficial partnerships) and the local communities (e.g. assisting local economies and conservation efforts, not exploiting or intruding on residents' lifestyles). Burns (Chap. 13) reviews the ethical frameworks relevant to wildlife tourism, with a view to assisting decisions for effectively managing wildlife tourism for the benefit of both humans and wildlife, laments that "scholarship in wildlife tourism has yet to incorporate environmental ethics in any substantial manner," and comments that "increases in the numbers of tourists visiting remote and rural areas can have profound social and ecological consequences that require sound ethical guidance to ensure effective management". Green (Chap. 14) also explores ethical implications of disturbance to wildlife to the animals themselves, population numbers, and human stakeholders. Maccoll and Tribe (Chap. 3) provide an encouraging example of how tourism, conservation and research can be combined: also de Lima (Chap. 8). Once again, the more controversial activities such as consumptive (hunting, fishing, collecting) tourism and alteration of natural behaviours (e.g. hand-feeding, other close approaches) need to be examined in view of all stakeholders (including the animals themselves) in local situations to arrive at optimal solutions.

Communication between stakeholders is important to determine what research is most urgently needed for future planning. Such communication also assists in disseminating the knowledge we do already possess both from the results of academic research and the long practical experience of operators to decision-makers in government and industry, and other tourism operations. Such information benefits both major tourism attractions such as zoos and well-established tour companies to enhance their educational, conservation and community roles, and also small groups off the usual tourist trails, including those in developing countries, who may be struggling to find the best way to help their communities or local ecosystems while making enough income to keep their projects going. It is hoped that this volume will assist such dissemination of ideas and knowledge already held and point the way to much-needed future research.

References

Actman J (2016) Cecil the lion died one year ago—here's what's happened since. Wildlife Watch News. National Geographic (May, 2016). Retrieved from, http://news.nationalgeographic.com/2016/06/cecil-african-lion-anniversary-death-trophy-hunting-zimbabwe/. Accessed on 8 Dec 2016

Altschiller D (2011) Animal-assisted therapy. Health and medical issues today. Greenwood, Oxford, UK

Armstrong P (2011) The gaze of animals. In: Taylor N, Signal T (eds) Theorizing animals: re-thinking humanimal relations. Brill, Leiden & Boston, pp 175–201

Archer J (1988) The behavioural biology of aggression. Cambridge University Press, Melbourne

Bauer J, Giles J (2002) Recreational hunting: an international perspective. Wildlife Tourism Research Report Series n. 13. Cooperative Research Centre for Sustainable Tourism—CRC, Queensland, Australia

BBC (2016) Thai Tiger Temple's long history of controversy. Asia section (June 1, 2016). Retrieved from http://www.bbc.com/news/world-asia-36423321. Accessed on 08 Dec 2016

Beatson P (2011) Mapping human animal relations. In: Taylor N, Signal T (eds) Theorizing animals: re-thinking humanimal relations. Brill, Leiden & Boston, pp 21–58

Bertelsmeier C, Bonnaud E, Gregory S, Courchamp F (2012) Applied ecology. In: Hastings A & Gross L (eds), Encyclopedia of theoretical ecology. University of California Press, Berkeley and Los Angeles
Brando S (2016) Wild animals in entertainment. In: Bovenkerk B, Keulartz FWJ (eds) Animal ethics in the age of humans: blurring boundaries in human-animal. Springer, Cham, pp 295–318
Cambronne A (2013) Deerland: American hunt for ecological balance and the essence of wilderness. Nature. Lyon Press, New York
Chandler CK (2012) Animal assisted therapy in counseling. Routledge, New York
Cunningham W, Cunningham M (2005) Environmental Science. McGraw-Hill Education, New York
Cusack O (1988) Pets and mental health. Routledge, London
de Lima IB (2009) Utilitarian and abstract rhetorics in ecotourism and social constructionism: the power of language. Int J Manag Decis Mak 10(231):2009
Duffy R, Moore L (2011) Global regulations and local practices: the politics and governance of animal welfare in elephant tourism, Journal of Sustainable Tourism, 19(4-5):589–604
EcoWatch (2016) Five times animals have been killed in zoos due to human encounters. Animals (June 3, 2016). Retrieved from http://www.ecowatch.com/5-times-animals-have-been-killed-in-zoos-due-to-human-encounters-1891162359.html. Accessed on 08 Dec 2016
Engel TB, Galloway ML, Bull MP (1994) The horse, the handicapped and the riding team in a therapeutic program. A training manual for volunteers. University of Virginia, USA
Festa-Bianchet M, Apollonio M (eds) (2003) Animal behaviour and wildlife conservation. General introduction. Island Press, London, pp 3–12
Franklin A (1999) Animals and Modern Cultures: A sociology of human-animal relations in modernity. Sage Publications, London
Frost W (ed) (2011) Zoos and tourism: conservation, education, entertainment? Channel View Publications, Ontario
Green RJ, Higginbottom K (2001) The Negative Effects of Wildlife Tourism on Wildlife. Wildlife Tourism Research Report Series No. 5, status assessment of wildlife tourism in australia series, CRC for sustainable tourism, Queensland, Australia
Green RJ, Wood P (2015) Australian wildlife and research network. In: Slocum SL, Kline C, Holden A (eds) Scientific tourism: researchers as travellers. Taylor and Francis, New York, pp 164–180
Hastings A, Gross L (eds) (2012) Encyclopedia of theoretical ecology. University of California Press, Berkeley and Los Angeles
Higginbottom K (2004) Wildlife tourism: impacts, management and planning. CRC sustainable tourism & common ground, UniversityPress.com, Altona, Victoria, Australia
Higginbottom K, Northrope C, Green RJ (2001) The positive effects of wildlife tourism on wildlife. wildlife tourism research report series No. 6, status assessment of wildlife tourism in australia series, CRC for sustainable tourism, Queensland, Australia
Holden A (2016) Environment and tourism. Routledge, New York
Hughes M, Carlsen (2008) Human-wildlife interaction guidelines in Western Australia. J Ecotourism 7(2–3):147–159
Humphrey SR, Bain JR (1990) Endangered animals of Thailand. Sandhill Crane Press Inc, Gainesville, Florida, USA
IISC (2016) Indian Institute of Science. Centre for Ecological Science. National Parks, Wildlife Sanctuaries and Biosphere Reserves in India. Retrieved from, http://ces.iisc.ernet.in/envis/sdev/parks.htm. Accessed on 08 Dec 2016
Ingold T (1988) Introduction. In: Ingold T (ed) What is an animal? Unwin Hyman, London, pp 1–16
Kemmerer L (2011) Theorizing others. In: Taylor N, Signal T (eds) Theorizing animals: re-thinking humanimal relations. Brill, Leiden & Boston, pp 59–86
Khullar DR (2016) Geography: fundamentals of physical geography and physical environment. New Saraswati House, New Delhi
Knight C, Sullivan M (2016) Gorilla killed after 3-year-old falls into zoo enclosure. Cincinnati.com. Retrieved from, http://www.cincinnati.com/story/news/2016/05/28/police-child-taken-hospital-after-falling-into-gorilla-pen/85095094/. Accessed on 08 Dec 2016
Krichner J (2009) The balance of nature: ecology's enduring myth. Princeton University Press, Princeton
Kontogeorgopoulos N (2009) Wildlife tourism in semi-captive settings: A case study of elephant camps in Northern Thailand. Current Issues in Tourism 12:429–44
Lair RC (1999) Gone astray: the care and management of the Asian elephant in domesticity. FAO. Forestry Department, Rome, Italy, and Regional Office for Asia and the Pacific (RAP). Dharmasam Co, Bangkok, Thailand. Available online. Retrieved from, http://www.fao.org/docrep/005/AC774E/AC774E00.HTM. Accessed on 11 Dec 2016
Lindenmayer D, Burgman MA (2005) Practical conservation biology. Csiro Publishing, Collingwood
Markwell K (2015) Animals and tourism: understanding diverse relationships. Channel View Publications, Ontario
Melkania NP (1998) Towards an integrated conservation strategy. In: Singh V, Sharma ML (eds) Mountain ecosystems: a scenario of unsustainability. Indus Publishing Company, New Delhi, pp 89–100
Miller JR, Lerner RM, Schiamberg LB, Anderson PM (eds) (2002) Encyclopedia of human ecology A-H. ABC Clio, Santa Barbara, CA, USA & Oxford, UK
Mills DS, Marchant-Forde JN (eds) (2010) The encyclopedia of applied animal behaviour and welfare. CABI, London
Moran EF (2010) Environmental social science: human—environmental interactions and sustainability. Wiley, Oxford
Moscardo G, Saltzer R (2005) Understanding tourism wildlife interactions: visitor market analysis. Technical Report. Cooperative Research Centre for Sustainable Tourism—CRC, Queensland, Australia
Newman JL (2013) Encountering gorillas: a chronicle of discovery, exploitation, understanding and survival. Rowman & Littlefield Publishers, Lanham
Novelli M, Humavindu MN (2005) Wildlife tourism: wildlife use vs local gain: trophy hunting in Namibia. In: Novelli M (ed) Niche tourism: contemporary issues, trends and cases. Elsevier Buttlerworth-Heinemann, Burlington, pp 171–182
Pickover M (2005) Animal rights in South Africa. Double Storey, Cape Town
Raina AK (2005) Ecology, wildlife and tourism development: principles, practices and strategies. Sarup & Sons, New Delhi
Ramsey A (2016) What really happened at Thailand's Tiger Temple? News Features online. Aljazeera. Retrieved from, http://www.aljazeera.com/indepth/features/2016/06/happened-thailand-tiger-temple-160605074332073.html. Accessed on 08 Dec 2016
Reardon M (2012) Shaping Kruger: animal behaviour, ecology and management in Africa's Premier Game Park. Struick Nature. Random House, Cape Town
Reynolds JE, Redfern JM, Shulstad RN (eds) (1974) Readings in natural resource economics. Committee on Soil and Water Conservation of the Agricultural Board. MSS Information Corporation, New York, pp 111–133
Romo R, Sholchet C (2013) Costa Rica to close zoos, release some animals. CNN Online. Regions. Retrieved from, http://edition.cnn.com/2013/08/07/world/americas/costa-rica-zoo-cages/. Accessed 12 Dec 2016
Roterman N (2016) Two lions killed to save suicidal young man in Chilean Zoo. Retrieved from, http://www.latintimes.com/two-lions-killed-save-suicidal-young-man-chilean-zoo-386300. Accessed on 4 Dec 2016
Rowan R, Knowlton N, Baker A, Jara J (1997) Landscape ecology of algal symbionts creates variation in episodes of coral bleaching. Nature 388:265–269

Semeniuk AD, Haider W, Cooper A, Rothley KD (2010) A linked model of animal ecology and human behavior for the management of wildlife tourism. Ecol Model 221(2010):2699–2713
Shapiro K (2002) The state of human-animal studies: solid, at the margin! Soc Anim 10:330–337
Shapiro K, Copeland MW (2005) Toward a critical theory of animal issues in fiction. Soc Anim 13:343–347
Smith M, Duffy R (2003) The ethics of tourism development. Routledge - Taylor & Francis: London and New York
Spaargaren G (2000) Ecological modernization theory and the changing discourse on environment and modernity. In: Spaargaren G, Mol APJ, Buttel FH (eds) Environment and global modernity. Studies in International Sociology 50. SAGE, New Delhi, India, pp 41–72
Stewart E (2013) No ethical way to keep elephants in captivity. A voice for elephants (May 3, 2013). National Geographic. Retrieved from, http://voices.nationalgeographic.com/tag/ed-stewart/. Accessed 10 Dec 2016
Stoddart MCJ (2012) Making meaning out of mountains: the political ecology of skiing. UBC Press, Vancouver
Szarycz G (2011) The representation of animal actors: theorizing performance and performativity in the animal kingdom. In: Taylor N, Signal T (eds) Theorizing animals: re-thinking humanimal relations. Brill, Leiden & Boston, pp 149–174
Taylor AB, Goldsmith ML (eds) (2003) Gorilla biology: a multidisciplinary perspective. Cambridge University Press, Cambridge
Tribe A (2001) Captive wildlife tourism in Australia. Wildlife Tourism Research Report Series n. 14. Cooperative Research Centre for Sustainable Tourism—CRC, Queensland, Australia
Tyler T (2009) Introduction; the case of the camel. In: Tyler T, Rossini MS (eds) Animal encounters. Brill, Leiden and Boston
Usher MB (ed) (1986) Wildlife conservation evaluation. Chapman and Hall, London
UNTWO (2016) World Tourism Organization - UNTWO Highlights 2016 Edition. Available online, http://www.e-unwto.org/doi/pdf/10.18111/9789284418145, accessed on November 2016
Vaccaro I, Smith EA, Aswani S (2010) Environmental social sciences: methods and research design. Cambridge University Press, New York
Vidal J (2016) Tiger temple scandal exposes the shadowy billion-dollar Asian trade. The Guardian online (4th June 2016). Online content. Retrieved from, https://www.theguardian.com/environment/2016/jun/04/tiger-temple-scandal-billion-dollar-trade-china-thailand. Accessed on 08 Dec 2016
WPSI (2016) Wildlife protection society of India. Tiger Reserves. Sundarbans. Online content. Retrieved from, http://www.wpsi-india.org/tiger/sundarban.php. Accessed on 08 Dec 2016

Part I

Wildlife Tourism and Conservation: Conceptual and Practical Issues

A Political Ecology of the Yellow-Eyed Penguin in Southern New Zealand: A Conceptual and Theoretical Approach

2

Eric J. Shelton, Hazel Tucker, and Jundan (Jasmine) Zhang

Abstract

Here, we engage with the political and ecological story of the yellow-eyed penguin (*Megadyptes antipodes*), a major tourist attraction, during four years of dramatically declining numbers of breeding pairs (New Zealand Department of Conservation in Unpublished census of yellow-eyed penguin breeding pairs 2015–16, 2016). One site, Long Point, is useful for presenting the possibilities of thematic integration since, using the principles of reintroduction biology (Seddon et al. in Conserv Biol 21(2):303–312, 2007; Armstrong and Seddon in Trends Ecol Evol 23:20–25, 2008), it is being used specifically to produce habitat for seabirds, rather than the more traditional restoration ecology approach. Also, the demands of tourism, for example to show respect through product offering (Zhang and Shelton in Tourism Anal 20(3):343–353, 2015) are, from the outset, being reinterpreted and integrated into the design and management of the site. Political ecology of tourism (Mostafanezhad et al. in Political ecology of tourism: communities, power and the environment. Routledge, London, pp 1–22, 2016) potentially is a fruitful analytic tool for formulating such thematic integration of 'wildlife tourism', 'applied ecology', and 'environmental education and interpretation'. Political ecology emerged as a critique of an allegedly *apolitical* cultural ecology and ecological anthropology, and illustrates the unavoidable entanglement of political economy with ecological concerns (Zimmerer in Prog Hum Geogr 32(1):63–78, 2006). Also, political ecology has been described as 'an urgent kind of argument or text … that examines winners or losers, is narrating using dialectics, begins and/or ends in a contradiction, and surveys both the status of nature and stories about the status of nature' (Robbins in Political ecology: a critical introduction. Wiley-Blackwell, New York, 2004, p. viii). Relevant examples of such narratives include Shelton and Tucker's (Tourism Rev Int 11(3):205–212, 2008, p. 198) text that constituted 'the restoration narrative … central to the long-term viability of tourism in New Zealand because environmental preservation, conservation and restoration facilitate the continuation, and possible expansion, of nature-based tourism' and Reis and Shelton's (Tourism Anal 16(3):375–384, 2011, p. i) demonstration that 'nature-based tourism activities are highly modulated by how Nature has been constructed in modern Western societies.' It is

E.J. Shelton (✉)
Yellow-eyed Penguin Trust, Dunedin, New Zealand
e-mail: ericjacobshelton@gmail.com

H. Tucker
Department of Tourism University of Otago, Otago, New Zealand
e-mail: hazel.tucker@otago.ac.nz

J. (Jasmine) Zhang
Institute of Geography and Economic History, Umeå University, Umeå, Sweden
e-mail: vein621@gmail.com

I. Borges de Lima and R.J. Green (eds.), *Wildlife Tourism, Environmental Learning and Ethical Encounters*, Geoheritage, Geoparks and Geotourism, DOI 10.1007/978-3-319-55574-4_2

this textual, discursive approach that differentiates political ecology from other approaches to issues surrounding 'natural area tourism', for example, the impacts approach of Newsome et al. (Natural Area Tourism: Ecology, impacts and management. Channel View Publications, Bristol, 2013).

2.1 Background

2.1.1 Long Point and the Yellow-eyed Penguin

The Long Point project has been described briefly elsewhere (Shelton 2013, pp. 192–194) as exemplifying neoliberal environmentality (Fletcher 2010, p. 172), a managerial and economic approach to conservation, but, in that description, there was no formulation of the site fitting more broadly within a political ecology framework, even though the phenomena described in that article gestured in that direction. Mostafanezhad et al. (2016, pp. 1–21) provide a broad introduction to the nature of political ecology. Long Point is in the Catlins region of the South Island of New Zealand, which is experiencing a rapid increase in guided and self-drive visitation. Before the coastal road was sealed, rental car companies would not offer insurance on their vehicles, since damage from flying stones was common. Now that the Southern Scenic Route road-sealing project is complete, it is estimated the area may attract 70,000 visitors annually. This growth, and the promise of the area simultaneously allowing visitors to 'get away from it all' means there will be a marked increase in the number of vehicles on secondary roads also, particularly where any of these roads leads to a beach.

Long Point, and its beach, lies at the end of one such road and, at first glance, looks similar to much other local grazed farmland. The promontory is well known for its surf break and local, national and international surfers have come to expect vehicle access over farmland, with the permission of the farmer. Historically, over the period 1790–1839, from Cook's voyage of exploration until the signing of tiriti o waitangi (the Treaty of Waitangi), European activity along New Zealand's southeast coast increased, largely unregulated (Church 2008). This increase took place alongside the establishing and consolidation (1650-) of power in a single, dominant Maori tribe, Kai Tahu (Anderson 1998).

The 'conservation and control' narrative, where conservation legislation is viewed as a tool to regulate or ban cultural harvesting of resources by indigenous people, is tempered by the fact that 'New Zealand is unique to the extent that there is one treaty, tiriti o waitangi, that permeates all interactions between the indigenous Maori people and the Crown (the government)' (Shelton and Tucker 2008, p. 202). This formal, bicultural, relationship between indigenous and settler society overtly recognizes the political nature of the use of land, including the beach, foreshore and seabed, which is yellow-eyed penguin habitat. Maori never have given up their claim to some land currently designated National Park, resulting in various, recent, co-management arrangements, for example with the *Tuhoe* people of *Te Urewera.*

Later in the European settlement process, during the 1860s and 1870s, the political economy of the Catlins included ship-building, (McPhee 2009), and *Manuka* was wrecked on an inshore reef at Long Point (Collins 2004), giving the site a European historical cultural attraction. For fewer than 100 years (1879–1971) the Catlins branch line of the national railway operated, (Tyrrell 1996) primarily to service the logging industry, as part of a larger story of that aspect of settler society commonly labeled pioneering (Tyrrell 1989). Long Point's 'existing forest was converted into poor-quality pasture … until 1984 often under direct or indirect government subsidy' (Shelton 2013, p. 193).

In 2009, the farm which included the promontory became available for purchase and, through generous support from government, various individuals and conservation-minded organisations for example, the New Zealand Forest and Bird Protection Society, the promontory and some adjacent land was divided-off and sold to the Yellow-eyed Penguin Trust, an environmental non-governmental organisation (ENGO) dedicated to protecting remnant coastal assemblages of flora and fauna, particularly those refugia, places where small ecosystems persist, involving yellow-eyed penguins (*Megadyptes antipodes*) (Fig. 2.1). Some of these refugia operated still at a whole-of-ecosystem scale of complexity. The Long Point project is part of a larger coastal seabird habitat production project. The latest governmental approach to conservation has been labeled a 'partnership' model and this project illustrates how this new approach is intended to work. This project may be used as a model for many such habitat restoration projects throughout the country.

Early in the ecological restoration project, the Trust invited various experts to suggest broadly how best to rehabilitate the site, and commissioned reports which formed the basis of the current management regimen (Yellow-eyed Penguin Trust 2008, 2012; Wildlands 2014). When planning to undertake ecosystem rehabilitation, typically there is a choice to be made between ecological restoration and reintroduction biology (Armstrong and Seddon 2008). The Trust

Fig. 2.1 Yellow-eyed penguin in New Zealand. *Credits* Yellow-eyed Penguin Trust

Fig. 2.2 Long Point site, New Zealand. *Credits* Otago Daily Times

chose to pursue the latter approach, accepting that reintroduction strategies deliver highly visible, tangible conservation outcomes, easily grasped in the short term by project participants and bystanders alike; a very productive way to mobilise public support (Ewan et al. 2008). This project is occurring during a time of significant changes in government environmental conservation policy, the roles of ENGOs, nationwide engagement with ecosystem services and increasing indigenous tribal aspirations both for the ownership and management of various protected areas.

The Long Point site is large enough (50 ha) that, for the foreseeable future, it will require grazing by sheep and the adjoining farmer pay a grazing lease (Fig. 2.2). Gradually, suitable habitats will be produced through earthworks, the provision of nesting boxes, and deliberately-dug tunnels, working through the nesting needs of a list of twelve seabirds, species-by-species. In such a degraded environment, a difficult early management task is deciding how to rank the different species in order optimally to allocate conservation effort.

Three ways of doing this is: by privileging the World Conservation Union (IUCN) threat status as a way of ordering or, by ease-of re-establishment or, by focusing on increasing the numbers of birds of a species already nesting in the area. The ease-of-re-establishment approach achieves relatively easy and rapid results and, within a few years, should be able to form the basis for wildlife tourism, appropriately monitored (Hadwen et al. 2007). Already, local entrepreneurs have approached the Trust, enquiring if and when concessions to operate wildlife tours are likely to be granted.

The initial 'raw material' for any such tours would be sites earmarked for transformation into seabird and, later, lizard and invertebrate habitat. Having tracks and hides built first, around which are the constructed nesting sites, increases the opportunity to use habituation as a deliberate management technique (Shelton et al. 2004; Higham and Shelton 2011). This 'production of protection' (Shelton 2012) is intended to be applied to the various endemic and native flora and fauna of Aotearoa/New Zealand which will in the future inhabit Long Point. It is the addition of this process to the political economy of Long Point and the Catlins that invites a political ecology formulation of its being inextricably bound up in a universal web of connectedness and power relations.

The yellow-eyed penguin is a species the IUCN, in the 2010 Red List, has labeled as 'Endangered' (Seddon et al. 2013). Ensuring a future for this iconic bird is enmeshed within the usual myriad ecological, economic and political positions, processes and contexts. Over the last four years, yellow-eyed penguin numbers at Long Point, and at most of its other breeding areas, have decreased significantly.

2.1.2 Long Point, and the Yellow-eyed Penguin, Within Political Ecology

A political ecology approach to engaging with the Long Point project is warranted since, in contrast with the outcomes of natural processes, 'land change' at Long Point is 'something people do', a condition of the political ecology approach, and that historical land management at this site has been what Robbins labels a 'chaotic seesaw' (Robbins 2004, p. xvi).

If we were to follow an 'impacts' line of inquiry, we would discuss, not completely ironically, 'the impact of the spread of tourism habitat' (Mostafanezhad et al. 2016, p. 2) as a metaphorical way of highlighting the connectedness of human and nonhuman species. Throughout their range, yellow-eyed penguin habitat and human habitat frequently overlap but what, though, compels us to write differently about this penguin at this time, and what is this political ecology within which we are operating?

Robbins (2004, pp. 5–7, 2012, pp. 15–16) provides a concise intellectual history of the concept political ecology and outlines three characteristics of a political ecology approach to the production of knowledge, in our case knowledge of a particular penguin species, at a particular site, at a particular historical juncture, and the role in wildlife tourism of these phenomena. First, there is the notion of action; 'political ecology as something people do' (Robbins 2012, p. 4). Announcing 'I do political ecology' may be viewed as being similar to a scientist announcing 'I do ecology', or 'I do physics' where *do* means to engage in the theory and practice of that field of study.

Second, in contrast to any encouragement that all environmental restoration projects should in some way be reported, Robbins argues that for such reporting to fit within a political ecology approach to the production of knowledge there must be more than simply a collection of 'separate and distinct cases' but also consideration of 'the common questions that underlie them' (Robbins 2012, p. 4). New Zealand currently is confronting such common questions about the lag between a significant increase in tourist numbers, now over 3 million annually and expected to reach 5 million (Tourism New Zealand 2016), and provision of the infrastructure required to manage them. This topic, until recently, was of limited interest to the lay public but now, with a nationwide debate occurring about the desirability of 'freedom campers', rental vehicles parked at the side of the road overnight and thus paying no fees, it is to the forefront, especially in areas like the Catlins.

Third, political ecology 'constitutes a community of practice and characterizes a certain kind of text' (Robbins 2012, p. 5). Our position is that the practice involved in this community of practice is the production of political/ecological knowledge through the overt demonstration of the elements and processes involved in the wider notion of political economy; in other words, praxis. Texts that present and represent the overtly politicized ecological knowledge produced, that is, discourse, then qualify as political ecological texts. We hope this chapter, itself text, fulfills these requirements.

This third characteristic appears to make available a rubric, a set of instructions, to apply to the question, if posed; does any particular knowledge, produced by a political economy approach to study, qualify as political ecology in a way analogous to how connectedness may characterize the knowledge produced by a community of scientists involved with biological ecology?

> 'In contrast (to political economy's focus on commodity chains and globalization), poststructuralism and neo-Marxism have come to the fore in an analysis of how people remake nature through their everyday interactions and broader societal understanding of the relationship between people and nature. Tourism … programs bring together people that have very different understandings of nature and society. Considering these nuanced understandings, a contextual analysis of political, economic, social, and ecological relations … has the potential to provide a broader understanding of the power structures concerning people and nature. As such, the conceptual framework of political ecology provides a contextual lens for analyzing the problems and potentials of sustainable tourism in the context of people, nature, and power' (Douglas 2014, p. 12).

Such a focus on context is entirely compatible with a view of political ecology as an interdisciplinary field of study that examines ecological matters from a broadly defined political economy perspective (Blaikie and Brookfield 1987), involving the entanglement of political economy with ecological concerns (Stonich 1998, p. 28). Political ecological perspectives illustrate how power and structural relations at different scales have implications for local people's natural resource and land use practices. In addition to various scales of analysis, political ecological analysis also is diachronic, it has developed over time, through its attention to historical factors that contribute to land use change and variability, as well as being involved in human–environment relations (Stonich 1998, p. 29). As an aside, O'Riordan (1976), Morton (2007) and Reis and Shelton (2011), offer reflexivity-based, bi-directional, critiques of such a cause *and* effect term as 'human-environment' and, it should be noted that amongst political ecology texts, there is a persistent lack of conceptual and philosophical clarity on this issue. Mostafanezhad et al.'s (2016) subtitle, 'Community, power and *the environment*', serves only to perpetuate the confusion.

With respect to 'community of practice', penguin tourism projects may be problematised by attending to layers of context, a characteristic way of engaging with neoliberal thought. We raise and respond to issues in ways able to be applied to other projects engaged in the production of ecological assemblages explicitly involving, if not overtly privileging, human visitation, and each ultimately enmeshed within the late capitalist economic system mentioned above. New Zealand offers a good opportunity for such an analysis through being an almost fully developed country. We say 'almost' because, within a public health and income context, the indigenous Maori population, and the social groups of Pacific Island ethnicity, experience enduring 'diseases of poverty', for instance rheumatic fever, not experienced to the same level by the rest of the society.

Academic publications dealing with political ecology of tourism typically involve developing countries and their aspirations for sustainable development through nature-based tourism, often involving wildlife viewing. Much less common are political-ecology-of-tourism studies situated within developed western economies. Through the observance of *te tiriti o Waitangi*, New Zealand is a bicultural (Maori, non-Maori) society, and through legislation is a multilingual (Maori, English and signing), Westminster-style small democracy. Tourism, comprising largely sightseeing, is the largest export industry by value. Whatever affects actual or potential successful land use is perceived to have the potential also to affect tourism and thus the political economy of the country. We use the term political economy since '(t)o invoke political economy or historical materialism is to recognise that economies can't be explained in economic terms alone' (Kunkel 2010, p. 18). This claim is true particularly when applied to the 'conservation economy' introduced below. Nonetheless, it is the case also that certain economic precepts can enter widespread lay thought, certainly about how to value wildlife. Kunkel (2010), reviewing Piketty's (2014) *Capitalism in the 21st Century*, comments that:

> 'he (Piketty) is one of very few contemporary economists eager to revive the old-fashioned spirit of political economy … economic life as a matter of individuals harmonising their preferences … has filtered into common sense … The biggest difference between the marginalists and the political economists concerned the question of economic value … for the marginalists, *value was a function of marginal utility*' (Kunkel 2010, p. 17, italics ours).

For the visitor to New Zealand, what is the marginal utility of yet another spectacular view, uncut forest or body of clean water? The then Minister of Conservation made it clear that:

> '(w)hen I talk of the conservation economy, the danger here is that some will incorrectly read into that phrase a lack of appreciation of the traditional and intrinsic conservation values —running the whole gamut from the preservationist view (and there must be a place in this wonderful country for the preservationist view to hold sway) to more mainstream public views … The government will work to protect the resources that tourism providers rely on-clean air, clean water, and unique landscape … The logic is simple enough. Healthy natural biodiversity means healthy ecosystems, and healthy ecosystems deliver well-functioning ecosystem services. Together these things form natural capital' (Groser 2009, p. 2).

The Minister desired a 'broadening of the long-term level of public support for conservation' (Groser 2009, p. 2), achieved through a mixture of ecosystem services and tourism. Every subsequent Minister of Conservation has made similar statements. In New Zealand, 'where conservation and tourism are inextricably linked' (Shelton and Tucker 2008, p. 198) the linking of ecosystem services and tourism then makes environmental protection inseparable from the functioning of the late capitalist economic system (Felluga 2016), of which tourism, with its typically poor wages and job insecurity, is an exemplar.

The scene for the presentation of the conservation economy had been set over a decade earlier when '(t)he Brundtland Commission (1987), invoking natural environments as a set of natural resources, drew ecotourism to a position within political economy' (Mostafanezhad et al. 2016, p. 2) but, within the 'sustainable development' project, the report presented gave scant attention to the idea of connectedness that permeates political ecology.

Another analytic layer further down, now as a part of late capitalism, environmental protection, as promised by the conservation economy, then falls within the business model of adding value and extracting revenue; in other words, the production and consumption of protection. At a national scale, contestably transforming the status of land and sea into protected areas constitutes the process of adding value. Revenue is then extracted by charging the tourists who are attracted by the protected status. Many NGOs have embraced this 'neoliberal economic model' (Palley 2005), of adding value through the production of protection, whether or not they list tourism as a primary activity of their organization, or simply as a by-product of the application of some more intrinsic set of values they hold. These intrinsic values often are representative of a public service model of the production and delivery of protection, usually through a government agency. Some NGOs want this model to be retained, and lobby against, in particular, the Department of Conservation, the government agency responsible for conservation on Crown land, divesting certain core conservation tasks to NGOs, who are all too eager to step up to the plate.

These actions, merging conservation and economics, made explicit a process that has been in train implicitly since the neoliberal economic reforms of the Fourth Labour Government of 1984 and the formation of the Department of Conservation (DOC) in 1987. DOC's function was, at its inception and has ever since been, a mixture of conserving and making available; 'fostering recreation and allowing tourism on conservation land, providing the use is consistent with the conservation of the resource' (Department of Conservation 2000). The recent overt championing of various iterations of *The Conservation Economy* (Groser 2009), above, whatever words are used to describe it, makes it now unavoidable 'to recognize that capitalist policies and values, and often neoliberal policies and values, pervade conservation practice' (Brockington et al. 2008, p. 3).

NGOs need what Rappaport (1977) called 'loot and clout'; how much money, time and expertise is required for an environmental group to be effective and, effective on whose terms? There is another, very recent loot-producing economic activity emerging; crowd-sourcing, using some form of the notion of directly 'giving-a-little', or, in some cases a lot, and it has had an almost instant impact on the funding of all sorts of projects. Every one of these projects originates from some sort of relatively unfiltered emotional response to experiencing some aspect of the human condition.

Does each dollar raised and spent in this way impact negatively on the allegedly more rational current methods of gathering and distributing of conservation dollars? Must NGOs change the way in which they obtain and distribute funds? As NGOs otherwise move to adopting more business-like corporate structures do they, in this age of connectivity, risk estranging themselves from 'the new givers', people who have no particular loyalty to the NGO and who will donate project-by-project? The political ecology question is whether performing this act of giving is simply another form of ecocriticism, that is, nature writing, fuelled by the Romantic aesthetic and the ideology of charity, and is not ecocritique, that is, ideological engagement and enactment, that first step in developing a truly ecological, fully connected, future (Morton 2007). At Long Point, this progression requires an important shift of focus, from merely performing the donation, that is, writing money as text, to engaging with the subject/object the money is spent on, and all its connections; that is, acting politically. Consequently, all NGOs perform within a particular politics, whether or not that politics is overtly acknowledged.

Ollman (1993, p. 11), argues that such processes, as in moving from ecocriticism to ecocritique, act to form a dialectic, which means:

> '... replacing the common sense notion of 'thing,' as something that has a history and has external connection with other things, with notions of a 'process,' which contains its history and possible futures, and 'relation,' which contains as part of what it is its ties with other relations' (Robbins 2012, p. 94).

The four-year decline in yellow-eyed penguin numbers at Long Point may usefully be viewed, not so much as the fates of a group of individuals, but as a process, the determinants of which remain unknown.

2.1.3 The Yellow-eyed Penguin Within Nature

Wildlife tourism, treated here as an element of applied ecology, needs a location in which to occur. The setting most frequently proposed to host these notions is nature. We subscribe to the view that, in order to be justifiable, the notion of nature needs to be rigorously interrogated. Nature and its conservation is a problematic concept economically and socially (Scandrett 2010), philosophically (Soper 1995; Jamieson 2008), linguistically (Morton 2007, 2010a, b) and as a basis for environmental analysis (Castree 1995; Mels 2009). We acknowledge how it would seem important to recognize:

> 'the multiple roles which 'nature' can be called upon to play in ecological discussion ... the 'metaphysical', the 'realist' and the 'lay' (or 'surface') ideas of nature. Employed as a metaphysical concept, which it mainly is in the argument of philosophy, 'nature' is the concept through which humanity thinks its difference and specificity ... One is invoking the metaphysical concept in the very posing of the question of humanity's relation to nature. Employed as a realist concept, 'nature' refers to the structures, processes and causal powers that are constantly operative within the physical world ... Employed as a 'lay' or 'surface' concept, as it is in much everyday, literary and theoretical discourse, 'nature is used in reference to ordinarily observable features of the world: the 'natural' ... This is the

> nature of immediate experience and aesthetic appreciation; the nature we have destroyed and polluted and are asked to conserve and preserve' (Soper 1995, p. 156).

Soper goes on to submit that:

> '(W)hen the Green Movement speaks of nature, it is most commonly in this third 'lay' or 'surface' sense: it is referring to nature as wildlife … (b)ut when it appeals to humanity to preserve nature … it is also of course employing the idea in a metaphysical sense to designate an object in relation to a subject (humanity), with the presumption being that subject and object are clearly differentiable and logically distinct. At the same time, by drawing attention to human transformation (destruction, wastage, pollution, manipulation, instrumental use of) nature, it is, at least implicitly, invoking the realist idea of nature' (1995, p. 156).

Wildlife tourism requires wildlife as its raw material; what the Brundtland Commission termed a natural resource. The yellow-eyed penguin, within its 'natural' environment, engages with all three of Soper's concepts of nature, as does applied ecology and environmental education and interpretation. Each of the concepts requires language to represent (or construct) it, and language is tricky. Morton (2007, p. 14) refers to the 'metonymic list' of figurative language that constitutes 'nature, a transcendental term in a material mask.' If everything in the universe is able to stand-in for nature, which is what metonymy implies, then nature becomes everything, and, simultaneously, nature becomes nothing.

> 'Nature' occupies at least three places in symbolic language. First, it is a mere empty placeholder for a host of other concepts. Second, it has the force of law, a norm against which deviation is measured. Third, 'nature' is a Pandora's box, a world that encapsulates a potentially infinite series of disparate fantasy objects' (Morton 2007, p. 14).

Also, Morton (2007, p. 1) proposes the concept of 'properly ecological forms of culture, philosophy, politics, and art' rather than ones based on some aspect of reified nature.

Picking up on the political aspect of Morton's vision of 'ecology without nature' leads to political ecology and its concerns with 'claims about the state of nature and claims about claims about the state of nature' (Robbins 2012, p. 87). Nature seems to be central to many claims; for example,'(a)ny sophisticated political ecology must contain a phenomenology of nature' (Watts and Peet 2004, p. 20). This claim does not address the 'everything and therefore nothing' objection to nature but does introduce the notion of multiple natures. For example, Fletcher (2014, p. 6) claims there is 'a long-standing tradition of research in political ecology exploring the complex and multidimensional relationship among political-economic institutions, cultural practices, and nonhuman natures'. This suggestion, that there are human and nonhuman natures, implies that humans exist outside of nonhuman nature; another version of O'Riordan's (1976) reflexivity problem, raised above.

From this very brief discussion, it seems clear that situating the yellow-eyed penguin within some notion of nature, although naively appealing, is deeply problematic since any attempt to use 'nature' as a descriptor, or analytic tool, rapidly produces no more than a circular argument, or tautology; what is nature, everything and nothing. Where then, conceptually, should the penguin be situated?

2.1.4 The Yellow-eyed Penguin Within Nature/Society/Environment

Fletcher's (2014, p. 6) claim revisits the notion of a human-environment binary, a claim extensively and deeply contested since O'Riordan's (1976) book *Environmentalism*.

Douglas explains:

> 'Broadly speaking, political ecology scholars seek to understand how the human–environment relationship is produced, reproduced, and altered through discursive and material articulations of nature and society' (Douglas 2014, p. 9).

This suggests discursive articulations of nature, and material articulations of nature, may profitably be combined with society to form 'the environment'. Douglas again:

> 'The production of nature thesis touts a negotiated understanding of environment and society as an unremitting space of interaction, portraying the relationship of people and nature through the myriad processes of production. However, this relationship goes beyond that of a material nature to one of people's conceptual understanding of the natural world' (Douglas 2014, p. 9).

Clearly, Douglas and Morton are at loggerheads over 'material nature'.

> 'Finally there is the very important question of the environment in political ecology … since so much of political ecology in the last decade has turned increasingly to nature itself. The questions are, of course, what passes for the environment? What form nature takes as an object of scrutiny? … political ecology rests on the dialectic of Nature and Society in which environment can be approached in a number of ways … what political ecology has done obviously is to open up the category of the environment itself and explore its multiform representations. Knowledge of the environment itself is examined—why particular forms of knowledge predominate, circulate and how' (Watts and Peet 2004, p. 19).

We accept this point, and work within the notion that any phenomenology of environment should include assemblages of human and nonhuman subjects and objects, acknowledging that the human, or any other, body is not a discrete entity but is, in itself, another assemblage; material or linguistic (Morton 2010a, b).

2.2 Thematic Integration at Long Point of 'Wildlife Tourism', 'Applied Ecology', and 'Environmental Education and Interpretation': A Political Ecology Approach

Fairhead and Leach (1996, p. 483) locate forest quality and biodiversity in the influence of past land use practices … 'vegetation patterns are the unique outcomes of particular histories not predictable divergences from characteristic climaxes'. This is true of Long Point; pre-contact Maori and European settlers' 'past use practices' altered Aotearoa/New Zealand's flora and fauna to the point where it never can be recreated. Callicott's concern is with how:

> 'fields of endeavor that have been informed by ecology will have to take account of the paradigm shift in ecology (from a 'balance of nature' [e.g. Suzuki 1999] to a 'flux of nature' paradigm) that is now virtually complete' (Callicott 2008, p. 571).

This concern underlines a fear that a combination of social constructivism, and a lack of any credible scientific model of stable and unchanging ecosystems, the notion of fixity, removes authoritative support for environmental protection. If everything around us is changing, why preserve or conserve what is here currently?

In New Zealand, a good example of a challenge to 'a scientific model of fixity' is a process at work in the 'beech (*Nothofagus* sp.) gap'.

> 'Beech forest is absent from south-central Westland today but is widespread to the north and to the south of this region. Previous pollen records from Westland have suggested that this 'beech gap' was narrower prior to the Last (sic) interglacial than today … (and) it has been suggested that suppression of beech in this region is due to the combination of severe conditions during glacial stages and the competitive dominance of podocarp forest during interglacial stages … showing successional development towards a podocarp forest climax' (Newnham et al. 2007, pp. 527–8).

This process of succession and (temporary) climax fits within a 'flux of nature' paradigm. At Long Point, the pre-1984 refugia ready to re-colonise land currently in pasture will not produce the assemblages characteristic of the pre-clearing state since an uncounted number of ecological niches have been destroyed.

> 'Notwithstanding Worster's (1977) warning that disequilibria can easily function as a cover for legitimating environmental destruction … the rethinking of ecological science can be effectively deployed in understanding the complexities of local management (for example … pest management)' (Watts and Peet 2004, p. 16).

2.2.1 Narratives

The persistence of the notion of a balance of nature, the equilibrium model, has consequences; for example how the lay members of the Long Point Management Committee envisage ecological 'best practice'. Equally, restoration ecology, if applied at Long Point, implies some sort of equilibrium will eventuate. Presenting this balance/flux tension as part of guided wildlife tourism may well be challenging both for the guides and for the clients since the 'balance' narrative, which underpins 'restoration ecology', is powerful (Shelton and Tucker 2008).

Robbins (2012, p. 21) identifies five dominant narratives in political ecology. The first is the 'degradation and marginalization narrative', where: '(t)he first assumption is that degradation of environmental systems, especially after passing an unidentified threshold, tends to require as much or more energy and investment to restore to its former state as was expended in its initial transformation' (Robbins 2012, p. 160). There is no dispute that (re)creation of seabird and other habitat eventually will cost far more than ever was spent clearing the land of forest.

2.2.2 Texts

Political ecology 'characterizes a certain kind of text' (Robbins 2012, p. 5) so it is important to note the form of such text. Morton (2007) reminds us that mimetic writing about nature, for example a standard commentary provided to wildlife tourism guides, known as ecocriticism, remains primarily and inescapably an act of writing, and is therefore a work of art, informed by an aesthetic, which in the case of nature remains the Romantic. This aesthetic, as with any aesthetic, is itself generated by an ideology, albeit often one that remains unacknowledged. The balance/flux narrative and the 'ecological restoration/reintroduction biology' narratives constitute such texts.

> 'To move from ecocriticism to ecocritique, a first step in developing a truly ecological future, the influences at play in these narratives must be acknowledged and made transparent. This first step requires an important shift of focus, from merely performing the work of art, as noted above with respect to donating money, that is, writing text, to engaging with the subject/object of the work, that is, acting politically. Political ecology texts are mimetic only to the extent that any political text that engages in critique must engage with ecocritique and never with ecocriticism, so, the texts must be political texts and not mimetic ecocritical texts. This division is important in any attempt fully to contextualise the yellow-eyed penguin. Peter Fritzell delineated a difference between naively mimetic and self

reflexive forms of nature writing. In the latter, ''what nature was really like' is often not what nature was really like (or, for that matter, what it is)' (Morton 2007, p. 14).

'Timothy Luke employs the term ecocritique to describe forms of left ecological criticism … Ecocritique is permeated with considerations common to other areas in the humanities such as race, class, and gender, which it knows to be deeply intertwined. Ecocritique fearlessly employs deconstruction in the service of ecology … In the name of all that we value in the idea of 'nature,' it thoroughly examines how nature is set up as a transcendental, unified, independent category' (Morton 2007, p. 13).

The ecological subject/object, here the yellow-eyed penguin, is engaged in this truly ecological restoration but, for it to be effective, '(s)ubject and object require a certain environment, in which they can join up together' (Morton 2007, p. 22). As stated above, such an environment may be simultaneously material and linguistic both.

2.2.3 The Kinds of New Zealand Texts Characterized by Political Ecology

It is the nature of these texts that separates political ecology from any insistence that all acts of conservation should be the subjects of allegedly straightforward reporting; political ecology texts must instead reflect 'the politicized state of the environment and the politicized nature of accounts about the state of the environment' (Robbins 2012, p. 6). These texts then, themselves polemical, may be understood as narratives; accounts of material and political processes that occur over time. Shelton and Tucker (2008), with their at-first-glance oxymoronic title *Managed to be Wild*, identified the politicized state of the New Zealand protected-areas spatial environment, claiming that; 'describing boundaries is an act that takes place *within the context of power relations*' (p. 202, emphasis ours). This focus on power relations and 'tensions between the restoration narrative and the multiple-use narrative' (Shelton and Tucker 2008, p. 203) invite a political ecology reading of these texts.

Attempting to answer these, and other closely related questions, posed in different language, has led to the most significant restructuring of DOC since it was established. Thirteen years after being established, DOC produced *The New Zealand Biodiversity Strategy: Our chance to turn the tide* (New Zealand Department of Conservation 2000), a strategy and vision document later supplemented by *Adapting to a Changing Climate: A proposed framework for the conservation of terrestrial native biodiversity in New Zealand* (Christie 2014). Over recent years, DOC's *Annual Report to Parliament* was characterized by a general worsening of the biodiversity situation nationally. This decline instigated a response from government that the state alone could not supply all the conservation effort required; there needed to be increased community and business involvement. Also, there was to be an increased focus on recreational and tourist use of conservation land (Groser 2009). The proposed model was presented by the Minister of Conservation and DOC senior staff at a national conference organized by the Yellow-eyed Penguin Trust in 2013 specifically for that purpose.

The government intended increased community involvement to be delivered, at least in part, by such ENGOs, using a contestable model for the allocation of funding. The Minister made it clear that, ideally, government hoped to be lobbied by only one, united, conservation voice. With tourism becoming ever more integrated with conservation, both at a local and national scale, the government believes it is reasonable to situate the visitor at the centre of the planning and delivery of conservation effort.

Under the new DOC structure, all things being equal, there is be more support for such conservation effort, directed at sites situated close-by human habitation, and visited by tourists, rather than being directed at sites that are distant and not often visited. This move from total state control of, and delivery of, conservation, to retained state legislative and regulatory control of conservation but mixed-agency delivery, involves a rearrangement and renegotiation of the power relationships involved. In particular, negative feelings previously directed by various groups, in an almost ritualized fashion, exclusively toward DOC because of its legislated authority under various Acts of Parliament, now may be targeted also at whatever ENGO is engaging in projects that, to some other individuals and groups, are unwelcome.

The documents the yellow-eyed penguin exists within fulfill the political function of, for example, 'providing good science' to inform 'best practice' or giving evidence of due process, for example: brainstorming solutions, developing management documents, creating a habitat advisory committee that keeps minutes of meetings, a willingness to change geographical boundaries through purchase and sale and, indigenous consultation.

2.2.4 The Act of Integration

One way to address the issue of whether or not it is legitimate to contextualize the Long Point project within political ecology is to use Robbins (2012) as a sort of checklist.

First: is there the notion of action? can the Long Point project be viewed as 'something people do' (Robbins 2012, p. 4)? Our answer is yes; the site is being prepared for wildlife tourism based on the yellow-eyed penguin, for whom this site is natal, and a range of seabird species which previously inhabited the site and will now be reintroduced to habitat produced to meet their individual needs. Tourism infrastructure will be in place before the birds are returned and habituation will be used as an active management tool.

Second: is the project more than simply one of a collection of 'separate and distinct cases'; does it also consider 'the common questions that underlie them' (Robbins 2012, p. 4). Again yes; we argue that currently New Zealand in general is confronting such common questions as; how best may we represent wildlife? Are the terms 'nature' or 'the environment' useful when being used to represent projects involving applied ecology? Also, should wildlife guides' commentaries attempt to explore the broader issues of representation or should they be restricted to explaining phenomenology through using some form of conservation narrative; for example, balance and/or flux.

Third: since political ecology constitutes a community of practice, is that the case for this project, within the national 'conservation community'? We say yes; the Long Point project offers insights of interest and importance to any other 'land-use-change' project nationally and internationally. Choosing to adopt restoration ecology or reintroduction biology is a complex process which, while offering some general components, must end up being tailored to the unique demands of every specific project.

Fourth: political ecology 'characterizes a certain kind of text' (Robbins 2012, p. 5). Do the textual elements of the Long Point project; its scoping document, its management plan, the minuted meetings of its advisory committee meetings, including visitor performance, qualify as just such texts? Again, we say yes; all of these texts are brought together to act within a matrix of power relations. What is the effect when textual material is granted post hoc status of being legitimately available for a political ecology reading? If, through reading Robbins (2012), the reader comes to believe their *oeuvre*, either written or performed, sits well within political ecology, what are the implications of this realization? The purposeful adoption of political ecology (Robbins 2004, p. 11) involves approaching an ecological issue expecting politics, inequality and the local effect of global economic forces.

The yellow-eyed penguin as subject/object is brought in to being by these texts.

> 'Concern for the subject in political ecology …means seeking to explain the way people's environmental actions and identities fit together, and the way these are together the products of power' (Robbins 2012, p. 76).

The 2009 introduction to New Zealand of the conservation economy, 'signals a move from intrinsic valuation of the (conservation) estate to extrinsic valuation: the question being, what are the ecosystem services delivered and how is tourism serviced?' (Shelton 2013, p. 184). Features of 'conservation for a new generation' (Knight and White 2009) include decentralization of resource governance to local authorities and non-state actors such as NGOs (Fletcher 2010, p. 172). This decentralisation more easily allows the community of practice of political ecology to offer critique of any given project, rather than being forced to confront a monolithic state agency.

2.3 Conclusion

Currently, most visitors to Long Point stumble upon it. There are, as yet, no interpretation material and no tourist operators; only sheep, some wildlife and geomorphology commonly labeled sublime or spectacular. This will change. It is intended that the development of wildlife tourism:

> 'facilitates a process of ceaseless capital accumulation via the body by selling an experience that withholds final fulfillment and thus leaves tourists constantly wanting more' (Fletcher 2014, p. 6).

Perhaps such capital accumulation will occur but, that is only one possible consequence out of many. We appreciate 'tourism … (is) not merely rooted in (such) developmentalism, but (is) fundamentally political, economic, social, and ecological' (Douglas 2014, p. 11). The vision remains:

> 'The science of seabird conservation has made very significant advances, placing Long Point on the research map nationally and internationally. Leading researchers in all conservation disciplines visit regularly, staying at the well-appointed Long Point research station' (Long Point Vision Document 2008).

Our political ecology reading of the Long Point project, then, is situated within soft constructivism, where the material world, as metaphor, is to be read as a collection of texts. Robbins (2012) makes the point that a problem with 'soft constructionism' is that it focuses on:

> 'social constructions or political influences that are responsible only for misunderstandings of the environment (but) it doesn't allow social influences to also account for correct understandings of the environment … Yet the political ecological world is filled with entanglements of knowledge, power, and landscape that are fully symmetrical' (Robbins 2012, p. 11).

This is a positive sentiment for the Long Point project; it may, through robust, ongoing, critique bring about such symmetry. Definitely, there are overlapping arguments about the nature of political ecology and there is the problem of dealing with these overlapping arguments while doing political ecology, especially where the doing involves labeling the enactment of wildlife tourism as performance, just as constrained as seabird behavior, but at another level of analysis; and to see both as suitable for a textual metaphor. In Long Point's:

> 'very human landscape, complicated ecological interactions create a world of unintended consequences and surprises defying even the most careful political assessments or predictions … In a curious way, political actors create the ecology of (Long Point) but not the (Long Point) of their choosing' (Robbins 2004, p. xvi).

This is an ongoing tension; Robbins (2004) claims political ecology to be; 'a field that seeks to unravel the political forces at work in the environment; access, management, and transformation' (p. xvi). With respect to the yellow-eyed penguins of Long Point, we concur. Definitely, there are political forces at work, at all levels of government, forming conservation policy and needing to be unraveled. These policies then influence the nature-based tourism industry, especially in terms of access to wildlife viewing and the construction of interpretation narratives surrounding iconic species, of which the yellow-eyed penguin is one. Unraveling the political dimensions of site and species management is an endless task and is situated within ongoing transformations of land use. It is this ability to engage politically in a multi-faceted way that makes political ecology such a useful analytic approach to the study of wildlife tourism, applied ecology and environmental education and interpretation.

References

Anderson A (1998) The welcome of strangers: an ethnohistory of southern Maori A.D. 1650–1850. University of Otago Press, Dunedin

Armstrong D, Seddon P (2008) Directions in reintroduction biology. Trends Ecol Evol 23:20–25

Blaikie P, Brookfield H (1987) Land degradation and society. Methuen, New York

Brockington D, Duffy R, Igoe J (2008) Nature unbound: conservation, capitalism and the future of protected areas. Earthscan, London

Brundtland Commission (1987) Report for the world commission on environment and development. Geneva, United Nations

Callicott J (2008) The implications of the 'shifting paradigm' in ecology for paradigm shifts in the philosophy of conservation. In: Nelson M, Callicott JB (eds) The wilderness debate rages on: continuing the great new wilderness debate. The University of Georgia Press, Athens, pp 571–600

Castree N (1995) The nature of produced nature: materiality and knowledge construction in marxism. Antipode 27(1):12–48

Christie J (2014) Adapting to a changing climate: a proposed framework for the conservation of terrestrial native biodiversity in New Zealand. New Zealand Department of Conservation, Wellington

Church I (2008) Gaining a Foothold: historical records of Otago's Eastern Coast 1770–1839. Heritage Books, Dunedin

Collins B (2004) The wreck of the manuka. Craig Printing Company, Invercargill

Douglas J (2014) What's political ecology got to do with tourism? Tourism Geogr 16:8–13

Ewan JG, Armstrong D, Parker K, Seddon P (2008) Avian reintroduction biology: current issues for science and management. Avian Reintroduction Biology. Proceedings of an international symposium held at Zoological Society of London, 8–9 May 2008

Fairhead J, Leach M (1996) Misreading the African landscape: society and ecology in a forest-savanna mosaic. Cambridge University Press, Cambridge

Felluga D (2016) Modules on Jameson: on late capitalism. Introductory guide to critical theory. Available at: http://www.purdue.edu/guidetotheory/marxism/modules/jamesonlatecapitalism.html. Accessed 28 Sept 2016

Fletcher R (2010) Neoliberal environmentality: towards a poststructuralist political ecology of the conservation debate. Conserv Soc 8 (3):171–181

Fletcher R (2014) Romancing the wild: cultural dimensions of ecotourism. Duke Universty Press, Durham, NC

Groser T (2009) The conservation economy. Research cluster for natural resources law newsletter two, pp 2–3

Hadwen W, Hill W, Pickering C (2007) Icons under threat: why monitoring visitors and their ecological impacts in protected areas matters. Ecol Manage Restor 8:177–181

Higham J, Shelton EJ (2011) Tourism and wildlife habituation: reduced population fitness or cessation of impact? Tour Manage 32 (6):1290–1298

Jamieson D (2008) Ethics and the environment. Cambridge University Press, Cambridge

Knight R, White C (2009) Conservation for a new generation: redefining natural resources management. Island Press, Washington

Kunkel B (2010) How much is too much? Lond Rev Books 33(3):9–14

McPhee M (2009) Catlins bound. Catlins Wildlife Ecotours, Owaka

Mels T (2009) Analysing environmental discourses and representations. In: Anderson K, Braun B (eds) Environment: critical essays in human geography. Ashgate, Aldershot, pp 385–399

Morton T (2007) Ecology—Without nature: rethinking environmental aesthetics. Harvard University Press, Massachusetts

Morton T (2010a) The ecological thought. Harvard University Press, Cambridge, Massachusetts

Morton T (2010b) Ecology as text, text as ecology. Oxford Literary Rev 32(1):1–17

Mostafanezhad M, Norum R, Shelton, EJ, Thompson-Carr A (2016) Introduction. In: Mostafanezhad M, Norum R, Shelton EJ, Thompson-Carr A (eds) Political ecology of tourism: communities, power and the environment. Routledge, London, pp 1–22

Newnham RM, Vandergoes MJ, Hendy CH, Lowe DJ, Preusser F (2007) A terrestrial palynological record for the last two glacial cycles from southwestern New Zealand. Quatern Sci Rev 26(3–4):517–535

Newsome D, Moore S, Dowling R (2013) Natural area tourism: ecology, impacts and management. Channel View Publications, Bristol

New Zealand Department of Conservation (2000) The New Zealand biodiversity strategy. Available from: https://www.biodiversity.govt.nz/pdfs/picture/nzbs-whole.pdf. Accessed 28 Sept 2016

New Zealand Department of Conservation (2016) Unpublished census of yellow-eyed penguin breeding pairs 2015–16

Ollman B (1993) Dialectical investigations. Routledge, New York

O'Riardon T (1976) Environmentalism. Pion, London

Palley TI (2005) From Keynesianism to neoliberalism: shifting paradigms in economics. Neoliberalism: a critical reader, pp 20–29. Available at: http://www.thomaspalley.com/docs/articles/macro_policy/keynsianism_to_neoliberalism.pdf. Accessed 28 sept 2016

Piketty T (2014) Capital in the 21st century. Harvard University Press, Cambridge, Massachusetts

Rappaport J (1977) Community psychology: values, research and action. Holt, Rinehart and Winston, New York

Reis A, Shelton EJ (2011) The nature of tourism studies. Tourism Anal 16(3):375–384

Robbins P (2004) Political ecology: a critical introduction. Wiley-Blackwell, New York

Robbins P (2012) Political ecology: a critical introduction, 2nd edn. Wiley-Blackwell, New York

Scandrett E (2010) Environmentalism of the poor and the political ecology of prophecy: a contribution to liberation ecotheology. Doctoral dissertation, University of Birmingham

Seddon P, Armstrong D, Maloney R (2007) Developing the science of reintroduction biology. Conserv Biol 21(2):303–312

Seddon PJ, Ellenberg U, an Heezik Y (2013) Yellow-eyed penguin (Megadyptes antipodes). Penguins: natural history and conservation, pp 97–100. Available at: http://nzbirdsonline.org.nz/species/yellow-eyed-penguin. Accessed 28 Sept 2016

Shelton EJ (2012) Unpublished PHD thesis at University of Otago, New Zealand

Shelton EJ (2013) Conservation, biodiversity, and tourism in New Zealand: engaging with the conservation economy. In: Bekoff M (ed) Ignoring nature no more: the case for compassionate conservation. University of Chicago Press, Chicago, pp 192–194

Shelton EJ, Higham JES, Seddon P (2004) Habituation, penguin research and ecotourism: some thoughts from left field. New Zealand J Zool 31(1):119 (abstract)

Shelton EJ, Tucker H (2008) Managed to be wild: species recovery, island restoration and nature-based tourism in New Zealand. Tourism Rev Int 11(3):205–212

Soper K (1995) What is nature?. Blackwell Publishers, Oxford

Stonich SC (1998) Political ecology of tourism. Ann Tourism Res 25:25–54

Suzuki D (1999) The sacred balance: rediscovering our place in nature. Bantam Books, Vancouver

Tourism New Zealand (2016) Tourism 2025. Available at: http://www.tourism2025.org.nz/. Accessed 11 Sept 2016

Tyrrell A (1989) Catlins pioneering. Heritage Books, Otago, Dunedin

Tyrrell A (1996) Catlins Rail: the story of the Catlins River branch railway 1879–1971. Catlins Historical Society Inc., Owaka

Watts M, Peets R (2004) Liberating political ecology. In: Peets R, Watts M (eds) Liberation ecologies: environment, development and social movements. Routledge, New York, p 20

Wildlands Consultancy (2014) Seabird re-establishment and habitat restoration at long point-Irahuka, Catlins. Report commissioned by Yellow-eyed Penguin Trust

Worster D (1977) Nature's economy: the roots of ecology. Sierra Club Books, San Francisco

Yellow-eyed Penguin Trust (2008) Long point/Irihuka (sic) restoration project discussion document. Yellow-eyed Penguin Trust, Dunedin

Yellow-eyed Penguin Trust (2012) Long point/Irahuka management plan. Yellow-eyed Penguin Trust, Dunedin

Zhang J, Shelton EJ (2015) Ordering the disordered subject: a critique of Chinese outbound tourists as New Zealand seeks to become China ready. Tourism Anal 20(3):343–353

Zimmerer KS (2006) Cultural ecology: at the interface with political ecology-the new geographies of environmental conservation and globalization. Prog Hum Geogr 32(1):63–78

Wildlife Tourism and Conservation: The Hidden Vale Project

3

Margie Maccoll and Andrew Tribe

Abstract

Wildlife tourism encompasses a wide range of tourism experiences that involve encounters for tourists with non-domesticated animals which are either free ranging or captive and includes zoos and nature parks. Wildlife tourism is a growing industry, which is reflected in the number and variety of activities on offer and the increase of tourism businesses offering them. Its growing popularity can contribute to local economies and have major impacts on wildlife and their habitats. It can thus benefit conservation by direct wildlife management, supporting research and educating visitors on conservation, while its revenue is considered to be a growing method of conserving wildlife, particularly for those species threatened or endangered, and particularly where government expenditure on conservation is limited or unavailable. However, while sustainability is a wildlife tourism goal which is widely accepted and adopted by governments and businesses, this requires management processes to identify negative effects, implement actions to correct them as well as conducting ongoing monitoring, ideally from the outset of the project. This chapter describes and discusses the Hidden Vale Project—a long term endeavour on private land which aims to combine wildlife conservation with a range of other activities including livestock production, wildlife tourism/ecotourism and adventure activities.

3.1 Introduction

This chapter describes and discusses the Hidden Vale Project—a long term endeavour to combine conservation with livestock production, ecotourism, wildlife tourism and adventure activities on a private property—Hidden Vale. Hidden Vale Nature Refuge comprises 3091 ha of a 4000-ha working cattle station. It also includes the Spicers Hidden Vale Retreat, a luxury resort, and is located on the Little Liverpool Ranges, in Southeast Queensland, one hour west of Brisbane. It provides suitable habitat for a range of native wildlife, including rare and threatened species such as the koala, glossy black-cockatoos, powerful owls, Albert's lyrebird, and the square-tailed kite.

In 2016, Hidden Vale entered into a long-term cooperative venture with the University of Queensland to enhance the wildlife on the property through a number of activities, including:

- Managing and rehabilitating the natural habitat
- Breeding and releasing local endangered species into suitable habitat
- Rehabilitating and releasing wildlife endemic to the area
- Developing wildlife activities and information for Spicers Retreat visitors

This project is intended to support conservation by providing a more natural balance of wildlife on the land, while

M. Maccoll (✉)
School of Agriculture and Food Sciences, University of Queensland Gatton, Gatton 4343, Australia
e-mail: margiemaccoll@gmail.com

A. Tribe
The Gainsdale Group, PO Box 108 Knapp Street, Fortitude Valley 4006, Australia
e-mail: a.tribe@uq.edu.au

I. Borges de Lima and R.J. Green (eds.), *Wildlife Tourism, Environmental Learning and Ethical Encounters*, Geoheritage, Geoparks and Geotourism, DOI 10.1007/978-3-319-55574-4_3

also enhancing the experiences of retreat visitors by allowing them to observe and to learn about Australian wildlife. This chapter will discuss ecotourism in general (and wildlife tourism in particular) and its relationship to conservation, illustrating this discussion with a more detailed review of the Hidden Vale Project.

3.2 A Short Review on Ecotourism

The World Conservation Union (IUCN) defines ecotourism as environmentally responsible travel and visitation to relatively undisturbed natural areas, in order to enjoy and appreciate nature (and any accompanying cultural features, both past and present) that promotes conservation, has low visitor impact and provides for beneficially active socio-economic involvement of local populations (IUCN 2016). This is not a universally held definition.

Similarly though, Ecotourism Australia defines ecotourism as ecologically sustainable tourism with a primary focus on experiencing natural areas that fosters environmental and cultural understanding, appreciation and conservation (Ecotourism Australia 2016). Weaver and Lawnton say ecotourism should constitute three criteria: attractions should be mostly nature-based, visitor interactions should be focused on education and experiences should follow principles and practices that are ecological, socio-cultural and economically sustainable (Weaver and Lawton 2007).

While there are broadly held understandings on the issues relevant to ecotourism that there is no universally agreed definition Ballantyne attributes to disagreements between the interests amongst academics and practitioners, which has, in turn, led to difficulties in creating legislation and conducting research (Ballantyne and Packer 2013).

The situation has enabled tourist operators to engage in greenwashing, a form of communication that misleads people into forming overly positive beliefs about an organisation's environmental practices or products (Lyon and Montgomery 2015). Consequently, operations that regard themselves as ecotourism range from ones that clearly generate net gains for conservation and communities as well as clients and shareholders to cases that fail to comply with any of the criteria for ecotourism but claim the label (Ballantyne and Packer 2013). Where legislation fails to rein in practises of greenwashing, activities have met in some cases with opposition through the mechanisms of social media, use of ecolabels and government restrictions on deceptive marketing practices (Lyon and Montgomery 2015).

That the term ecotourism seeks to do so much adds to both the difficulty in definition and in research. Ecotourism is an activity, a philosophy and a model of development (Stronza and Pegas 2008). Because of a lack of a precise definition, researchers have resorted to measuring aspects of it and discussing it through explorations of case studies. Some have looked at its impacts on local communities and natural resources. Others have measured how and why ecotourism leads to conservation and some have evaluated how economic changes associated with ecotourism advantage or disadvantage conservation outcomes (Tisdell 2003). Still others look at social changes associated with ecotourism, such as whether local involvement in ecotourism helps foster conservation (Stronza and Pegas 2008).

Despite its shortfall in definition ecotourism has become an increasingly used term since it became popularised in English academic literature in the late 1980s. Over 20 years its significance has grown as demonstrated in 2002 when the United Nations declared the International Year of Ecotourism and the specialized peer-reviewed Journal of Ecotourism was established (Weaver and Lawton 2007).

3.3 Key Features of Wildlife Tourism

Wildlife tourism and ecotourism are often regarded as overlapping forms of tourism with some differences. Wildlife tourism encompasses a wide range of tourism experiences that involve encounters for tourists with non-domesticated animals which are either free ranging or captive and includes zoos and nature parks (Higginbottom 2004). Wildlife tourism includes both consumptive (such as hunting and fishing) and non-consumptive (such as wildlife watching and photography) forms while ecotourism is generally associated with non-consumptive forms (Tisdell 2003). While the core element of wildlife tourism is the encounter between tourists and the wildlife, like ecotourism, the goal in managing wildlife tourism lies in its sustainability, the benefits it brings to communities and the consequence for the tourist of receiving messages of conservation along with the experience (Kutay 1993).

In 2003, the non-profit organisation Wildlife Tourism Australia (WTA) was formed to promote the sustainable development of a diverse wildlife tourism industry that supports conservation. The organisation has become the peak body representing the interests of sustainable wildlife tourism in Australia and has members that include wildlife tourism operators, government land management and conservation agencies (state and federal), non-government conservation groups, researchers and members of the general public with an interest in wildlife (STCRC 2009).

Wildlife tourism is important not least because of the numbers of people it involves. In 2001, in the US alone, more than 66 million adults engaged in the activity of wildlife watching (Higginbottom 2004). A 2012 survey conducted by the United States Fish and Wildlife Service puts the annual economic value generated by wildlife watchers at around US$32 billion in the United States alone

(UN 2012). Another study of non-captive wildlife watching in the US found total expenditures on wildlife watching activities which included purchasing of equipment including cameras generated $95.8 billion and more than a million jobs (Valentine and Birtles 2004). Wildlife tourism is a growing industry, which is reflected in the number and variety of activities on offer and the increase of tourism businesses offering them. Its growing popularity can contribute to local economies and have major impacts on wildlife and their habitats. A study of whale watching around Sydney, Australia between 2002 and 2004, for example, showed tourism numbers had doubled over that time and expenditure increased four-fold (Tapper 2006). The growing interest has been attributed at least in part to media coverage and social media interest. A subsequent increase in awareness of conservation issues and risks of species extinction has motivated tourists to view wildlife ranging freely in their natural habitat (UN briefing paper 1999).

3.4 Who Are Ecotourists/Wildlife Tourists?

While ecotourism has no clear definition, there is also no easy way to describe an ecotourist or a wildlife tourist. Ecotourism and wildlife tourism allow for wide-ranging attractions and activities from five-star luxury excursions to volunteer stays that may attract a correspondingly broad range of motivations and socioeconomic characteristics amongst tourists. Ballantyne says wealthy guests of upmarket lodges such as North Island in the Seychelles are effectively funding a conservation program for critically endangered birds, the Seychelles White eye but that is not why they visit and many may not even know it exists. Independent travelers to small-scale ecolodges run by NGOs and local communities in developing countries may be motivated by a desire to contribute to these initiatives but that does not guarantee these enterprises will yield the intended outcomes ((Ballantyne and Packer 2013).

Wearing and Neal (1999) categorise tourists into two streams, the traditional tourist who enjoys lounging beside swimming pools and being taken on organised sightseeing tours, and the ecotourist or wildlife tourist who is more interested in visiting protected areas, seeing wildlife, experiencing new lifestyles and contributing towards conservation. One way of identifying types of tourists is by finding out what motivates them.

Generally, ecotourists seek an understanding of the environment and want their impact on it to be minimal (Wearing and Neal 1999, p. 126). Hvenegaard also categorises tourists into two groups; the interactionalists who want to interact with their destination area, and the cognitive-normative tourist who is motivated by travel (Hvenegaard 2002). He criticises the use of categorising tourists into typologies because of the generalisations that are gleaned from the restrictive data that created the typology, the lack of consistent tourist categories and the complexities of tourist motivations and activities that make categorising problematic (Hvenegaard 2002).

Tapper (2006) says tourism is "highly responsive" to market demand and with increased interest has come an increasing diversification of wildlife tourism opportunities affecting a wider range of environments, species, types of activities as well as increased environmental awareness and use of interpretation. A tourist's main motivation for an animal watching trip may be entertainment but along with having fun they will have learned a lot about the wildlife viewed, contribute financially to the local community that depends on conservation for its survival (Tapper 2006, p. 11).

Previous studies have categorised tourists on their desire to interact with the environment ranging from hard-core and dedicated to mainstream and casual. The hard-core group are more likely to seek physically challenging activities and enjoy pursuits such as bushwalking and wildlife watching (Weaver 2002). Education and interpretation are crucial to ecotourism setting it apart from other tourism pursuits, the epitomy of which leads to changing the values and habits of the tourists towards issues of conservation and protection (Higham and Luck 2002).

3.5 Wildlife Tourism Experiences

Wildlife experiences are many and varied and dependent on the environment, management of facilities and emphasis on the animal encounters. In some instances, such as whale watching, animal safaris or bird-watching trips, the animals are the focus while in landscape-based tourism, wildlife watching may be part of the backdrop (Valentine and Birtles 2004, p. 16). Tourists may engage in unguided encounters, specialised tours, nature-based tours with a wildlife component or sightseeing with chance wildlife sighting. They may seek accommodation sited in wildlife locations such as beside a migratory pathway, or national park. Tourists may visit zoos to see wild animals in captivity, learn more about the animals or enjoy an entertaining day out with friends or family (Tribe 2004). Tourists may be interested in seeing particular species of animals, rare or endangered animals or they may want to see large numbers of animals. Bird watchers have their own criteria for wildlife watching which may include fulfilling bird species on a watchers life list or necessitate the creation of bird watching facilities (Higginbottom 2004, p. 21). Alternatively wildlife tours could focus on research, conservation and education where the primary role is not tourism (Valentine and Birtles 2004, p. 16).

3.6 Wildlife Tourism and Conservation

Conservation, according to the Oxford Dictionary, is the preservation, protection or restoration of something whether it be historic buildings, archeological sites, artefacts or, in this case, wildlife or undomesticated native animals and uncultivated native plants (Oxford Dictionaries, 2016). There is general consensus among international tourism and conservation organisations and governments that wildlife tourism should contribute positively to conservation. Wildlife tourism can benefit conservation by direct wildlife management and supporting research, educating visitors on conservation or using income derived from wildlife tourism to fund conservation initiatives. Sustainability is a wildlife tourism goal which is widely accepted and adopted by governments and businesses (Higginbottom and Tribe 2004). According to the World Tourism Organisation (UNWTO) 2012 annual report the protection of biodiversity and natural resources is a core element of sustainable tourism (UNWTO 2012).

While giving people close encounters with captive wild animals, zoos have the opportunity to educate and entertain their visitors about the animals and their natural habitats (WAZA 2015). The World Zoo Conservation Society (WZCS) encourages zoos to implement major and effective research projects which are mostly conducted ex situ. More than an incentive, zoos are subject to socio-political imperatives to contribute to conservation, reinforced by legislation to meet conservation requirements to be re-licensed (WAZA 2015).

In contrast, wildlife tourism management associated with wildlife watching in natural habitats is nearly always in situ and covers a range of activities including reintroduction of animals, control of exotic predators and tree planting. A major benefit to wildlife tourism operators is the sustainability of their wildlife for ongoing viewing and is the main incentive for their conservation investment. Some wildlife operators participate in direct wildlife management for conservation. Phillip Island Penguin Reserve, for example, run by a government-appointed board, hosts a popular Australian wildlife attraction, the Penguin Parade where visitors can gain up-close views of penguins making their daily walk from the sea to their burrows. The Reserve's Committee of Management has collaborated with researchers to oversee and help fund a large body of research and monitoring of the little penguin. In other instances wildlife tourism operators have lobbied for the conservation on which their businesses depend. Great Barrier Reef tourist operators lobbying to the Australian Government reportedly resulted in research on the crown of thorns starfish that was detrimentally affecting the reef. Raising public awareness of environmental issues is the primary mode of conservation for many wildlife tourism operators. This is also their main method of educating the public about conservation and anecdotal evidence shows wildlife tourism experiences are effective in influencing people's attitudes (Higginbottom and Tribe 2004).

Wildlife tourism raises direct revenue from charges including entrance fees, activity costs, accommodation and tourist commodities, which can be directed to conservation (Steven et al. 2013). The Seychelles in the Indian Ocean, for example, has introduced a US$90 tax on travellers entering the Seychelles with the revenue to be used to preserve the environment and improve tourism facilities (UNEP, report to the CSD 1999). Indirect contributions to conservation result when educating tourists participating in wildlife activities changes their behaviours and when wildlife activities benefit the local community financially resulting in a lessening of their dependence on the natural habitat and incentive to conserve the resource. As local communities benefit economically from employment and regional spenders its importance to the region increases and this translates into political votes (Buckley 2009). As politicians decide the future and funding of conservation, Buckley believes the real value of wildlife tourism is in its ability to attract political support, even despite describing protected areas as "life-support systems for humans and other species" and attributing a value on them at "beyond calculation" (Steven et al. 2013). Wearing and Neal (1999) agree that natural habitats are regarded in political terms by their value in meeting human needs, whether from tourism or for future human industry.

Tourism which combines engaging with animals and the experience of being in protected areas provides an economic rationale to preserve natural areas rather than developing them for alternative uses such as agriculture or forestry (Wearing and Neal 1999). Wilson believes giving landholders ownership of wildlife provides it with a value that would encourage owners to invest in wildlife conservation and sustainability. He suggests wildlife ownership would be an incentive to landowners to be more open to alternative land uses including wildlife tourism or captive breeding of wildlife for live selling and consumption and instead of the more common use of land for agriculture, forestry and urban areas (Wilson et al. 2016).

Enjoyment has been found to be a factor in the likelihood of tourists to learn and contribute to changing attitudes and behaviours. Research has found viewing alone is unlikely to have much impact on visitors' knowledge and wildlife conservation attitudes without effective interpretation. To be effective interpretation is believed to require participation, the inclusion of multi-sensory activities, easily understood interpretive material and the creation of personal connections (Moscardo et al. 2004). Widely recognised interpretive techniques include visitor centres, publications, guided tours and educational activities, displays, exhibits and signs

(Wearing and Neal 1999). However it is delivered, interpretation is a core element of the ecotourism experience and should convey meaning, stimulate a response in the visitor that impels that to consider their value base and behaviour in regard to conservation (Wearing and Neal 1999).

3.7 Negative Effects of Wildlife Tourism

The sustainability of wildlife tourism depends not only on its contribution to conservation but on the survival of the watched species, their habitats, benefits to local communities and its meeting of tourism demands (Tapper 2006). Wildlife tourism can have considerable effects on the environment such as erosion, noise and air pollution due to issues of access (Higham and Luck 2002). Activities such as horse riding and 4-wheel-driving in protected areas can cause significant environmental damage. The severity of the damage depends on the susceptibility of the environment, the behaviour and frequency of use and the management of use (Newsome et al. 2005). Increasing interest in wildlife tourism is likely to lead to increased pressure on wildlife watching sites, animal populations and habitats and monitoring by governments and conservation managers has been recommended (Tapper 2006).

Green and Higginbottom have identified three main categories of negative effects that wildlife tourism has on wildlife. These are the disruption of wildlife activity by the intrusion of tourists; direct killing or injury through hunting, fishing or collecting and unintentional killing or injury from road accidents, trampling or the burning of forest understorey to create firebreaks; and the alteration of habitat for tourist infrastructure (Green and Higginbottom 2001). Wildlife tourism can negatively affect the physiology and behaviour of wildlife ranging from increased mortality to reduced breeding. The presence in wildlife habitats of humans and their associated habitat clearing, artificial lights, noise can result in reduced resources, increased stress levels and disrupt normal behaviour. Similarly, Newsome recognises the negative impact to wildlife tourism relating to access and observation and also identifies close contact/feeding as a major factor (Newsome et al. 2005).

Orams agrees the feeding of wildlife, which is a popular practise used by tourist operators to facilitate tourist interaction with uncontained wildlife, show most cases have negative impacts on wildlife (Orams 2002). These practices can alter behaviour patterns and population levels and result in dependency of animals on human-provided food and their habituation to human contact. Supplementary feeding of wildlife may increase their probability of being viewed by tourists but may lack essential nutrients or cause them to become dependent on such food sources (Green and Giese 2004). The deliberate and long-term provision of food to wildlife has also been shown to lead to intra- and inter-species aggression where wildlife, in their efforts to obtain food, have harmed one another and tourists. Health implications have also been identified where injury and disease have resulted from the provision of artificial food sources (Orams 2002).The impact of wildlife tourism on wildlife is difficult to gauge particularly as the effects may vary from one species to another. Factors affecting wildlife include previous contact with humans, the type of transport used by visitors, the predictability of human activity, the habitat, the nature of wildlife activities and dependent offspring (Green and Higginbottom 2001).

Sustainable wildlife tourism requires management processes that identify negative effects and implement actions to correct them as well as ongoing monitoring, ideally from the outset of the project (Green and Higginbottom 2001). Across the scope of wildlife tourism there are various management approaches which deal with the various aspects of wildlife tourism access, observation and contact, and feeding (Newsome et al. 2005). Management of wildlife feeding, for instance, is varied and ranges from prohibition to promotion with little empirical research on the management merits of what is a controversial issue (Orams 2002). In addition to management, education and interpretation can be designed to contribute to wildlife conservation awareness through the delivery of messages which may influence visitors future behaviour (Newsome et al. 2005).

Despite the potential negative effects of wildlife tourism, its revenue is considered to be a growing method of conserving wildlife, particularly for those species threatened or endangered, and particularly in developing countries where government expenditure on conservation is low. With the burgeoning human population causing increasing pressures on land use for habitation and resources, protected areas of land are becoming both smaller and more significant for the survival of wildlife species (Wearing and Neal 1999). The contribution of wildlife tourism operators to conservation could be enhanced with captive breeding and predator management, something which has proved successful in improving the conservation status of birds in New Zealand (Wilson et al. 2016). However, the reliance on tourism revenue for conservation brings its own risks. Tourism is sensitive to market demand and socio-economic factors, and consequently revenues earned can vary considerably from year to year.

3.8 Case Study: The Hidden Vale Project

The Hidden Vale Retreat is one of eleven resorts in the Spicers group owned by Graham and Jude Turner and located across southeast Queensland and the Hunter Valley in New South Wales. Spicers Hidden Vale is a country retreat in the Scenic Rim, a picturesque rural area situated in the foothills of the Great Dividing Range surrounded by world heritage-listed national parks. The retreat boasts luxury accommodation, an award-winning restaurant and provides guests with a range of activities including trail-bicycle riding, spa treatments, farm animal experiences and environmental experiences including bushwalking and 4-wheel-driving (Spicers Hidden Vale website 2016).

The property includes more than 3000 ha of declared nature refuge, is part of a corridor of cooperative properties effectively forming a 100 km^2 conservation area, and is located just a half-hour drive from the University of Queensland Gatton campus. This nature refuge contains a wide variety of wildlife, but requires habitat restoration and effective predator and pest management as well as the integration of conservation with an existing livestock enterprise.

The Hidden Vale Project aims to use scientific research to establish endemic self-sustaining populations of fauna and flora on a privately-owned property in conjunction with existing cattle farming and ecotourism activities.

The central focus of the Hidden Vale Project is the Wildlife Centre, which was opened in March 2017. This purpose-built facility includes both wildlife-breeding aviaries and a research and teaching building to support research into best-practice breeding techniques for threatened native species, post-release monitoring, habitat restoration, and predator and pest control. Both post-graduate and undergraduate students from the University of Queensland will benefit from the Wildlife Centre through their involvement in research, practical sessions and work experience. Students, researchers, external groups such as Landcare and the RSPCA and members of the public will all be able to benefit from opportunities to learn, observe and participate in on-site conservation activities. Captive breeding agreements for a list of vertebrate species, including the vulnerable spotted-tailed quoll (*Dasyurus maculatus maculatus*), the endangered greater bilby (*Macrotis lagotis*) and the endangered eastern bristlebird (*Dasyornis bracbypterus*) have already received government support.

Research projects will benefit from findings of projects running concurrently within the project area. Research focused on the translocation of the greater bilby and spotted-tailed quoll, for instance, will be undertaken alongside research into the management of introduced predators (mainly feral cats and red foxes). A 17-month research project by Moseby & O'Donnell reported success in translocating greater bilbies into a reserve free of cats, foxes and rabbits with reproduction continuous during the research period, juveniles successfully recruited into the population and bilbies able to successfully able to recolonize parts of their former range (Moseby and O'Donnell 2003). Glen and Dickman said in their research an assessment of the dietary overlap between the spotted-tailed quoll, red foxes Vulpes vulpes and wild dogs Canus lupus indicated strong potential for competition between quolls and eutherian carnivores, showing control of introduced predators may be desirable, not only for the conservation of prey species but also for the protection of native carnivores (Glen and Dickman 2008).

In addition the Hidden Vale property will provide habitat for the release of rehabilitated wildlife from wildlife hospitals in South East Queensland. Habitat destruction through much of South East Queensland means that a large number of wild animals are rehabilitated and released each year. More than 35,000 animals are rehabilitated annually with the majority coming from the RSPCA (Qld) wildlife hospital at Wacol (Burton and Tribe 2016). Suitable habitat is needed for the release of some of these animals and the Hidden Vale property will provide an excellent site on which to conduct research into the post-release success of reintroduced rehabilitated wildlife.

3.9 The Hidden Vale UQ Wildlife Centre

The Hidden Vale UQ Wildlife Centre is a $5 million facility which includes a teaching space capable of seating up to 40 students, a postgraduate room for at least eight postgraduate students, office space for staff, three research laboratories, a veterinary clinic, large feed preparation and storage areas including a walk-in cold room, laundry facilities, toilets, a lunch room and an interpretation room with viewing to the clinic and first aviary. There will be six wildlife aviaries measuring 24 m long × 7 m high × 6 m wide with access corridors on the ground and upper levels, barriers to prevent animals escaping or entering and supplies of rainwater and bore water. A range of local native grasses, forbs and shrubs will be planted into soil distributed across the floor of the aviaries.

The architectural design of the Wildlife Centre takes advantage of its orientation. Winter sunlight penetrates the large aviaries as they face north and the building is situated on the top of a ridge to take advantage of summer breezes and water drainage. It is both energy and water self-sufficient.

Research projects began at the Wildlife Centre with the enlistment of a principal ecologist to focus on the restoration of habitat for native wildlife and a senior research officer engaged to work with postgraduates and industry in the development of multidisciplinary research projects focused on maximising the success of all elements of captive wildlife breeding and release.

Current research projects include:

- The description and evaluation of existing soil profiles, flora and fauna (vertebrate and invertebrate) in key locations for intensive long-term studies to determine the effect of climate change and changes in management.
- The control of introduced predators (primarily cats and foxes) using a range of techniques. The issue of feral cats is complex involving stakeholders who consider them a nuisance, the public at risk from zoonotic disease, people who are concerned about the welfare of feral cats, those concerned with wildlife impacts and the cats themselves (Robertson 2008). Methods of control have been controversial and have included culling, relocation, baiting, leg-hold traps, hunting, dogs and more recently trap neuter return (TNR) methods (Robertson 2008; Campbell et al. 2011). Robertson says because humans created the cat problem education on responsible pet ownership and the intrinsic value of animals is an important part of a solution (Robertson 2008) Cat eradication projects have reported to result in positive responses for populations of small mammals, reptiles and birds and helped make areas suitable for re-introduction of species that were extirpated. Cat eradication may also produce negative ecosystem impacts (Campbell et al. 2011).
- Habitat restoration including planting and weed removal. Despite significant expenditure of time and resources world-wide, restored habitats commonly fail to ameliorate the risks of habitat loss and provide the resources required by wildlife. Restorations are successful when animals assess them as functional habitats and respond to them adaptively (Hale and Swearer 2016).
- Identification and evaluation of thick timber regrowth area for potential timber thinning to encourage larger trees and the application of forestry techniques to encourage *Casuarina* growth for cockatoos, including Glossy Black cockatoo.
- Identification of large trees, species and presence, and dimensions of tree hollows for identification of potential sites for next boxes for a range of avian and mammal species.
- Capture and use existing data on rainfall and creek lines to understand water movement and its effect on erosion and the potential of flooding. Soil erosion by water is an important land degradation process in environments and is strongly linked to problems of flooding and channel management. Other erosion processes operating within catchments such as tillage erosion, land reshaping for land preparation or soil quarrying can have significant impacts on soil truncation and changes to land use affect the intensity of these processes. The conditions, position and connectivity of the runoff and sediment generating areas within catchments have a profound effect on flood characteristics within the main channels but the dynamics are not well understood (Poesen and Hooke 1997).
- Capture and use existing data on grazing and fire management to understand and implement appropriate management of flammable fuel (such as pasture and regrowth) and avoid impact on other projects such as restoration of creek lines and undergrowth.
- Rehabilitation and release of animals, including those from other sources such as RSPCA, re-establishing species missing from the landscape, and long term monitoring to determine survival and dispersal of animals released. Rehabilitating animals is a common practice that is rarely scientifically documented and monitored. An important aspect in ensuring an animals ability to be self-sustaining post-release is their behavioural development before release (Houser. et al. 2011). Critical to the conservation of the species is the health of the released animal, with the greatest risk in the release of rehabilitated animals being the transmission of disease from captivity to wild populations. The success rates of reintroduction has increased since the World Conservation Union (IUCN) in 1988 established the Reintroduction Specialist Group which formulated guidelines which place emphasis on the identification of release sites within the historic range of the species and acknowledge a need to ensure the previous causes of the decline have been addressed, both factors having strongly affected project success (Ewen. et al. 2012) One translocation project showing success has occurred through an ecological replacement in New Zealand. In 2008, 10 North Island kokako *(Callacas wilsoni)* were translocated from the central North Island to the southwestern corner of the South Island which was previously occupied by a southern kokako *(Callacas cinerea)*, declared extinct in 2004. Kokako were abundant on both islands before human settlement and their ecological roles included herbivory, fruit dispersal, pollination and perhaps seed dispersal. Monitoring of the birds showed they had settled in the area they were released and only one was deceased due to falcon (Falco noveaseelandiae) predation. The project is succeeding in re-establishing a keystone species in an ecosystem, restoring natural biodiversity, providing long-term economic benefits to the local economy and promoting conservation awareness (Ewen et al. 2012).
- Post-release monitoring of released animals using a range of technologies. New technologies have been developed in recent years to remotely track and study free-ranging animals. Combined with technologies for data collecting and recovery, satellite systems, and radio-frequency download systems, GPS can enable the collection of data such as temperature, activity, mortality and proximity of wildlife at determined locations (Tomkiewicz et al. 2010). Animal-attached remote sensing or bio-logging can enable scientists to examine wildlife in

the field with the same rigor normally reserved for the laboratory (Ropert-Coudert and Wilson 2005) Monitoring technology is a rapidly changing field, but integral to achieving the best results is to consider from the outset what is to be accomplished and how best to approach it.

The wildlife research and conservation will be integrated with Hidden Vale's cattle production (a strong feature of the property) to demonstrate the effective management of both shared land. The inclusion of agricultural land use with native conservation is practical and beneficial for teaching and research programs.

3.10 Linking Conservation and Wildlife Tourism

Under the partnership agreement the Wildlife Centre is a UQ research and teaching facility under the management of its School of Agriculture and Food Sciences. However, Spicers Retreat guests will also be able to visit the Centre, and to access information about the wildlife on the property and the work being conducted in the Centre. This will allow them to understand and observe the wildlife and their breeding, rehabilitation and release, as well as to attend wildlife activities and events to be held on the property. It is anticipated that more wildlife activities will be developed for the Retreat guests, including guided walks (day and night), wildlife talks in the Centre, and the opportunities to become involved in the radio tracking of wildlife released on the property. In this way, information can be spread to communities and the general public encouraging and inspiring them to become involved with and to support conservation. By encouraging guests to experience and understand natural flora and fauna and the importance of its preservation is likely to become more relevant and hence valuable to them. Research into visitor attitudes to wildlife conservation will also be undertaken.

3.11 Will Hidden Vale Retreat Visitors Be Interested in Wildlife and Conservation?

The importance of measuring visitor attitudes assists ecotourism operators to achieve their objectives by determining the factors which contribute to visitor satisfaction, guiding future planning to improve visitor satisfaction and raising the conservation awareness of visitors and enlisting their support (Moscardo and Saltzer 2004). A survey of 100 guests was conducted at the retreat in 2015 to measure visitor attitudes to potential wildlife interactions and the delivery of conservation messages. It found that wildlife interactions and the conservation of the environment were highly regarded.

Most Hidden Vale guests were interested in participating in wildlife activities if they were made available (78%) and 61% said they would be more likely to participate if they knew it was helping conservation. All the respondents came from Australia with 84% living within a two-hour drive of the resort. The majority of visitors (78%) were aged between 25 and 64 years and 36% visited with children, most (68%) aged 6–12 years. The survey found the majority of respondents visited Hidden Vale Retreat for the environment. Only 12% of visitors said their reason for visiting was the Australian animals, so it was somewhat surprising that the majority wanted to participate in wildlife activities if they were offered. More than 90% of respondents were interested in seeing wildlife in their natural environment (93%), about three quarters (73%) wanted to see nocturnal animals by guided night walks and 74% wanted to receive information from monitors on the whereabouts of wildlife. Respondents were also interested in gaining education along with their wildlife encounters, with most preferring their messages to be delivered by an expert guide (60%) while 24% preferred brochures and 32% interactive displays. However only a quarter to a third of respondents were interested in activities that could be considered more specialised such as attending an animal release (31%), viewing a rehabilitation area (27%) or watching wildlife on CCTV (23%).

The commercial viability of any wildlife tourism venture is linked not to its education and conservation role but to its ability to entertain its visitors. For zoos, a major challenge is to create exciting, interactive opportunities to attract visitors, while upholding their conservationist reputations and expanding their role in conservation (Tribe and Booth 2003). The way information will be relayed to visitors will be important at Hidden Vale for both its entertainment value and its ability to relay conservation messages. What the survey has not determined is the reality of the experience proposed, whether visitors do participate in wildlife activities once they are offered and whether these experiences deliver conservation messages to participants and affect their attitudes and behaviours.

References

Ballantyne R, Packer J (2013) International handbook on ecotourism. Edward Elgar Publishing, Cheltenham, UK

Buckley R (2009) Parks and tourism PLOS biology. http://journals.plos.org/plosbiology/article?id=10.1371/journal.pbio.1000143. Accessed 22 June 2015

Burton E, Tribe A (2016) The rescue and rehabilitation of Koalas (Phascolarctos cinereus) in Southeast Queensland. Animals (Basel) Set 15;6(9). https://www.ncbi.nlm.nih.gov/pubmed/27649249. Accessed 15 Nov 2016

Campbell KJ, Harper G, Algar D, Hanson CC, Keitt BS, Robinson S (2011) Review of feral cat eradications on islands. In: Veitch C, Clout MN, Towns DR (eds) Island invasives: eradication and management. pp 37–46.

Ecotourism Australia (2016) http://www.ecotourism.org.au/eco-experiences/why-choose-ecotourism/. Accessed 25 Oct 2016

Ewen J, Armstrong D, Parker K, Seddon P (eds) (2012) Reintroduction biology: integrating science and management. Conservation Science and Practice No. 9. Wiley-Blackwell, UK

Glen A, Dickman C (2008) Niche overlap between marsupial and eutherian carnivores: does competition threaten the endangered spotted-tailed quoll? J App Ecology 45(2). doi:10.1111/j.1365-2664.2007.01449.x

Green R, Giese M (2004) Negative effects of wildlife tourism on wildlife. In Higginbottom K (ed) Wildlife tourism CRC for sustainable tourism. Qld, pp 81–97

Green R, Higginbottom K (2001) Negative effects of wildlife tourism on wildlife. In: Wildlife tourism research report series 5, CRC for Sustainable Tourism. http://www.sustainabletourismonline.com/117/wildlife-tourism/negative-effects-of-wildlife-tourism-on-wildlife-wt-5. Accessed Dec 6 2016

Hale R, Swearer S (2016) When good animals love bad restored habitats: how maladaptive habitat selection can constrain restoration. In: J Appl Ecol. doi:10.1111/1365-2664.12829

Higham J, Luck M (2002) Urban ecotourism: a contradiction in terms. J Ecotourism 1(1):p36–p51. doi:10.1080/14724040208668111

Higginbottom K (ed) (2004) Wildlife tourism: impacts, management and planning. Common Ground Publishing, Victoria

Higginbottom K, Tribe A (2004) Contributions of wildlife tourism to conservation. In: Higginbottom K (ed) Wildlife tourism, CRC for Sustainable Tourism, Queensland, pp. 99–123

Houser A, Gusset M, Bragg C, Boast L, Somers M (2011) Pre-release hunting, training and post-release monitoring are key components in the rehabilitation of orphaned large felids. Sth Afr J Wildlife Res 41 (1):11–20. doi:10.3957/056.041.0111

Hvenegaard G (2002) Using tourist typologies for ecotourism research. J Ecotourism 1(1). http://dx.doi.org/10.1080/14724040208668109

Kutay K (1993) Brave new role: ecotour operators take centre stage in the era of green travel. In: Going green: the ecotourism research for travel agents. Supplement to Tour and Travel News 25(80)

Lyon T, Montgomery A (2015) The means and end of greenwash. Organ Environ 28(2):223–249

Moscardo G, Saltzer R (2004) Understanding wildlife tourism markets. In: Higginbottom K (ed) Wildlife tourism. CRC for Sustainable Tourism, Queensland, pp 167–185

Moscardo G, Woods B, Saltzer R (2004) The role of interpretation in wildlife tourism. In: Higginbottom K (ed) Wildlife tourism. CRC for Sustainable Tourism, Queensland, pp 231–251

Moseby K, O'Donnell E (2003) Reintroduction of the greater bilby, Macrotis lagotis (Reid) (Marsupialia: Thylacomyidae), to northern South Australia: survival, ecology and notes on reintroduction protocols. Wildlife Res 30(1):15–27. http://dx.doi.org/10.1071/WR02012

Newsome D, Dowling RK, Moore SA (2005) Wildlife tourism, vol 24. Channel View Publications.

Orams D (2002) Feeding wildlife as a tourism attraction: a review of issues and impacts. Tourism Manage 23:281–293. http://www.coastalmarineresearchgroup.com/PDFs/Orams%202002%20Feeding%20wildlife%20as%20a%20tourism%20attraction%20Issues%20and%20impacts.pdf. Accessed 7 December 2016

Poesen J, Hooke J (1997) Erosion, flooding and channel management. In: Mediterranean environments of Southern Europe. Prog Phys Geogr 21(2):157–199. doi:10.1177/030913339702100201

Robertson S (2008) A review of feral cat control. J Feline Med and Surg 10(4):366–375. doi:10.1016/j.jfs.2007.08.003

Ropert-Coudert Y, Wilson R (2005) Trends and perspectives in animal-attached remote sensing. Front Ecol Environ 3(8). doi:10.1890/1540-9295(2005)003[0437:TAPIAE]2.0.CO;2

Spicers Hidden Vale Website (2016) http://spicersretreats.com/spicers-hidden-vale/. Accessed 8 Oct 2016

Steven R, Castley J, Buckley R (2013) Tourism revenue as a conservation tool for threatened birds in protected areas. PLoS ONE 8(5):e62598. doi:10.1371/journal.pone.0062598

Stronza A, Pegas F (2008) Ecotourism and conservation: two cases from Brazil and Peru. Hum Dimensions Wildlife 13:263–279. https://www.researchgate.net/publication/232922489_Ecotourism_and_Conservation_Two_Cases_from_Brazil_and_Peru. Accessed 9 Oct 2016

Sustainable Tourism Cooperative Research Centre (STCRC) (2009) Wildlife tourism: challenges, opportunities and managing the future

Tapper R (2006) Wildlife watching and tourism: a study on the benefits and risks of a fast growing tourism activity and its impacts on species. United Nations Environment Programme (UNEP)/Convention on Migratory Species (CMS), Bonn, Germany. https://books.google.com.au/books?id=yaq2fm3_WYUC&pg=PA61&lpg=PA61&dq=wildlife+watching+tourism&source=bl&ots=lrRG-sXG1p&sig=ovI05ovTJbaook6EV2n9msqkOjM&hl=en&sa=X&sqi=2&ved=0CEEQ6AEwBWoVChMI9qOS7PKxxwIVBR-mCh1ZgAsJ#v=onepage&q=wildlife%20watching%20tourism&f=false. Accessed 9 Oct 2016

Tisdell C (2003) Economic aspects of ecotourism: wildlife-based tourism and its contribution to nature. Sri Lankan J Ag Econ 5(1). https://www.researchgate.net/publication/43447515_Economic_Aspects_of_Ecotourism_Wildlife-based_Tourism_and_Its_Contribution_to_Nature. Accessed 3 Dec 2016

Tomkiewicz S, Fuller M, Kie J, Bates K (2010) Global positioning system and associated technologies and ecological research. Phil Trans Royal Soc B 365(1550). doi:10.1098/retb.2010.0090

Tribe A (2004) Zoo tourism. Wildlife tourism: impacts, management and planning, pp 35–56

Tribe A, Booth R (2003) Assessing the Role of Zoos. Wildlife Conserv Hum Dimensions Wildlife Int J 8(1):65–74

United Nations Environment Programme (2012) Bird watching can help eco-tourism fly high in a green economy. http://www.unep.org/Documents.Multilingual/Default.asp?DocumentID=2683&ArticleID=9130&l=en. Accessed 20 July 2015

UNWTO Annual Report (2012) http://www2.unwto.org/publication/unwto-annual-report-2012. Accessed 10 Oct 2015

United Nation Environment Program Report to the CSD (1999) Tourism and environmental conservation http://www.unep.org/resourceefficiency/Business/SectoralActivities/Tourism/FactsandFiguresaboutTourism/ImpactsofTourism/EnvironmentalImpacts/TourismandEnvironmentalConservation/tabid/78779/Default.aspx. Accessed 8 Oct 2015

Valentine P, Birtles A (2004) Wildlife watching In: Higginbottom K (ed) Wildlife tourism. CRC for Sustainable Tourism, Queensland, pp 15–34

Wearing S, Neal J (1999) Ecotourism. Butterworth-Heinemann, Oxford

Weaver D (2002) Hard-core ecotourists in Lamington National Park. Aust J Ecotourism 1(1):19–35. doi:10.1080/14724040208778110

Weaver D, Lawton L (2007) Twenty years on: the state of contemporary ecotourism research. Griffith University. Gold Coast http://www98.griffith.edu.au/dspace/bitstream/handle/10072/26545/51660_1.pdf?sequence=1. Accessed 24 Oct 2015

Wilson G, Hayward M, Wilson C (2016) Market-based incentives and private ownership of wildlife to remedy shortfalls in government funding for conservation. Conserv Lett J Soc Conserv Biol. http://onlinelibrary.wiley.com/doi/10.1111/conl.12313/full. Accessed 24 Oct 2015

World Association of Zoos and Aquariums (WAZA) (2016) http://www2.unwto.org/publication/unwto-annual-report-2012p://www.waza.org/en/site/about-waza/what-we-do. Accessed 24 Oct 2015

World Conservation Union (IUCN) (2016) http://www.ecogo.org/tag/iucn-ecotourism-definition/. Accessed 23 Oct 2015

4 Birdwatching Tourism Sector in Rwanda: The Need of Self-learning and Self-improving for the Future of Wildlife Tourism Development?

Karel Werdler

Abstract

This chapter will focus on one, a rather specific but popular type of wildlife tourism, that is known as birdwatching tourism. Within Rwanda, the proposition of enjoying nature and more specific wildlife has traditionally focused on the observation of mountain gorillas in their natural habitat, but this resource has a certain scarcity and limitations, and only allows for a restricted amount of visitors. However, the composition of the bird fauna of more than 700 species (Van de Weghe and Van de weghe in Birds in Rwanda, an atlas and handbook, Rwanda development Board, Kigali, 2011) on an area of only 26.338 km^2 (10.169 sq. miles) seems to offer new possibilities for tourism product development and might even contribute to the creation of employment opportunities for the local population. But this possible development depends on certain factors ranging from political will and infrastructure and other facilities to the training of local guides, to name but a few and can only be accomplished when the hosts have a clear picture of demands and expectations of their bird watching guests. This chapter offers a first attempt at classifying these guests, based on both international examples and an overview of the Belgian-Dutch bird watching travel market and offers some recommendations on how to proceed with the marketing of these natural resources. For this chapter, a mixed method approach, among them desk research, interviews with Dutch and Rwandan stakeholders and participatory observation was used to collect the relevant data.

4.1 Introduction

"Bird-watching is either the most scientific of sports or the most sporting of sciences" (E.M. Nicholson in, *The art of Bird-watching, 1931*).

With this statement at the introduction of one of the first field guides for European birds published by Peterson in the Netherlands in 1954 (Peterson et al. 1969), the tone is set with a rather flattering description for all of those who would like to engage in this leisure activity. When you are a bird watcher you show a certain scientific interest and that makes you a different person from those that walk around in nature with no other objective than to enjoy themselves. Notwithstanding, there is not only a scientific layer to this activity, it also has a sports-like element implicitly referring to elements like endurance, training, maybe even a certain hardship, discipline and, last but not least—possible rewards. When travelling internationally the bird watcher therefore can be categorized as a representative of nature tourism, active tourism and maybe even adventure and sports tourism to name, but a few and this makes this large target group interesting for destinations that can offer combinations of all of these elements and enough bird species. As confirmed by the director of the Rwanda Development Board (RDB), the country has set out to cater to this type of tourism in the future and would like to position itself as one of the most interesting destinations on the continent for this target group. This chapter would like to make a contribution to that effort by taking a closer look at the Dutch bird watching

K. Werdler (✉)
Department of Tourism and Leisure Management Studies, Inholland University of Applied Sciences, Hoofddorp, The Netherlands
e-mail: Karel.Werdler@inholland.nl

I. Borges de Lima and R.J. Green (eds.), *Wildlife Tourism, Environmental Learning and Ethical Encounters*, Geoheritage, Geoparks and Geotourism, DOI 10.1007/978-3-319-55574-4_4

market, their organisations, need for information and possible differences in motivation and combine the outcome of this desk research with Rwanda's proposition and provide all stakeholders with some recommendations on how to proceed. The desk research was combined with non-structured interviews with relevant stakeholders representing specialised tour operators, local guides, birdwatchers in the Netherlands and in Rwanda and representatives of the Rwanda Development Board. Furthermore, the author participated in several guided birding tours in both Rwanda and other African countries and has been a dedicated birdwatcher for most of his life.

4.2 The Dutch Bird Watching Market

Bird watching is a popular (pseudo-scientific) pastime for many people in the Netherlands and there are quite a few organisations, local, regional and national that offer options for engaging in this activity. Some of them have a focus on leisure and regard bird watching as a social activity that can be engaged in a natural environment, others would like to contribute their findings in the field to science and organise inventory moments where birds in a given season or territory are counted. The findings of these activities are published by SOVON, a private organisation that cooperates closely with the Dutch Ornithological Society and the Dutch Society for the Protection of Birds (DSPB). Another related organisation is the Dutch Birding Association that stimulates the observation of birds in their natural environment and the documentation of the presence of special or rare species. One of their main activities is the maintenance of the national system Dutch Bird Alerts that informs participants of special observations. Most of these local and national organisations also take part in the Dutch Bird Fair, which is organised early autumn in one of the Netherlands' better known wetland areas, the *Oostvaarders-plassen*. This Bird Fair, that attracted some 4000 visitors during its last editions, offers a stage for all the organisations mentioned and many more, including such commercial partners as manufacturers of binoculars and cameras, but also representatives of destinations as the Spanish province of Navarra and individual tour operators. Apart from this, most organisations also have their own yearly congregations that include lectures and excursions.

For this research the magazine of the Dutch Society for the Protection of Birds (*Vogelbescherming Nederland*) was used as a first resource for possible related information on travel propositions for Dutch birdwatchers (Vogels 2013–2015). This magazine, aptly named "Vogels" (*Birds*) has a publishing history of more than 30 years and 5 yearly editions that are sent to its more than 150,000 subscribers. These subscribers receive the magazine as part of their membership of the DSPB. Within the magazine there are editorials related to the work of the society and specific articles that highlight national and international activities, sometimes in cooperation with international partner organisations such as Birdlife International, the Audubon Society and the Royal Society for the Protection of Birds (UK). One of the 2013 editions contained a small reference to the work of Rwanda's Association de la Nature au Rwanda that stimulated the local population around the Nyabarongo-wetlands to use and grow natural materials as papyrus that would help to rejuvenate the marsh and make it attractive to birds (Vogels 2013). Furthermore there are articles dedicated to typical Dutch birds and their environment, including some practical advice on where to spot specific birds in season, a photographic essay and summaries of recent scientific research in cooperation with organisations/publishers as Current Biology, Nature Communications and Proceedings of the National Academy of Sciences. Approximately 10–12 pages of the magazine of a total of 64 (18.75%) are reserved for advertisements and these can be divided in those that advertise specific products of the Society, such as bird feed and bird houses and those of commercial partners that sell cameras and binoculars, or offer photo-training and bird related travel propositions.

4.3 Africa as a Birdwatchers Destination

For this chapter, an inventory of travel related propositions was made of the editions of 2012–2015 of the Vogel-magazine and this yielded more than 10 general and specialised tour operators that cater to this market. With the help of internet research, further destinations could be added to the ones proposed in the magazines and a complete portfolio was established. Table 4.1 presents the tour operators on the Dutch birdwatchers market and their offer of destinations, divided in European, African and other destinations.

From this initial desk research, it became clear that some African destinations are more popular than others with Uganda topping the list with five references; or, rather, being offered by five different tour operators. However, when combinations are taken into account Gambia is offered three times as a single destination and four times in combination with Senegal by a total of seven tour operators. Interviews with the management of some of these companies confirmed the impression perceived during participation in birdwatching tours that Dutch clients, rather combine destinations and the opportunities to see more wildlife in general, than make birds the sole objective of their travel. More traditional destinations for wildlife viewing such as Kenya, Tanzania and South Africa are offered by three tour operators: Ethiopia, Madagascar and Namibia by two, and all the others —including Rwanda—by just one tour operator. Compared

Table 4.1 Tour operators on the Dutch market offering bird watching travel and destinations

Tour operator	European destinations	African destinations	Other destinations
Blue elephant (B)	Germany, Spain, Georgia, Greece, Finland, Romania, Scotland, Russia, Portugal, Azores	Uganda, Gambia	Bolivia, India, Brazil, Bhutan, Sri Lanka
SNP Natuurreizen/Vogelreizen	Germany, Greece, Portugal, Holland, Spain, France, UK, Bulgaria, Sweden, Hungary, Poland, Romania, Estonia, Georgia	Morocco, Kenya, South Africa	Ecuador, Turkey, Costa Rica, USA, Panama, India, Colombia, Sri Lanka, Mongolia, Indonesia
50+ Reizen	Poland, Georgia, Armenia	n.a.	n.a.
Birding breaks (B)	Poland, Cyprus, Georgia, Spain, Greece, Bulgaria, Romania, Germany, Holland, Russia, Finland, Iceland, Scotland, China, Tibet, Sweden, Portugal	Kenya, Uganda, Namibia, Madagascar, Gambia, Tanzania, Ghana, Ethiopia	Turkey, Kazakhstan, Papua New Guinea, Brazil, Costa Rica, USA, Ecuador, Peru, Oman, India, Colombia, Sri Lanka, Israel
Wild nature travel	Spain, Romania	Ethiopia, Gambia, South Africa, Kenya	Sri Lanka, Costa Rica, Indonesia
Beluga adventures (birding and photo travel)	Scotland, Spitsbergen (Norway), Greenland	n.a.	Asian Russia, Alaska, Antigua, Canada, South Georgia, Antarctica
Kleine Reiscollectief	Poland, Romania, Hungary, Sweden, Holland	n.a.	n.a.
Matoke tours	n.a.	Uganda, Tanzania	n.a.
Hamba Kahle natuurreizen in cooperation with birding Africa, outlook Safaris	Greece, Cyprus	Tanzania, Uganda, South Africa, **Rwanda**, Madagascar, Gambia/Senegal, Cameroon, Sao Tome, Namibia	Turkey, USA, Antilles, Brazil, Chile, Costa Rica, Australia, Indonesia
Globe Natuurreizen	Portugal, Iceland	Gambia/Senegal	Ecuador, Southern Atlantic
Inezia tours	Spain, Holland, Norway, Spitsbergen, UK, Romania, Hungary, Bulgaria, Poland, Greenland, Germany, Austria, Scotland, White Russia, Finland, Estonia/Latvia, Greece, Greenland	Gambia/Senegal, Uganda	Sri Lanka, Asian Russia, Turkey, Australia, Antarctica, Panama, Costa Rica
Vogelbescherming Nederland in cooperation with SNP/Inezia/Blue Elephant/Globe Natuurreizen	Greece, Madeira	Cabo Verde, Gambia/Senegal	Turkey, Taiwan, Bolivia, Indonesia

(B Belgian tour operator)

to the other continents, excluding Europe, African destinations are offered slightly more often (33 times) than the other regions (South and Meso-America 26 times, Asia 28 times). Although, it cannot be asserted at this moment since there has not been a follow-up of this research, the reason for this prominent position of the African continent might be related to lesser expenses for travel when departing from Europe, and/or the presence of other wildlife and natural scenery the continent is famous for.

4.4 Bird Literature and Field Guide Inventory

The next step in this research is taken from a "birdwatchers point of view". A birding destination becomes accessible when there are regular transport possibilities by air, as well as a variety of accommodations and other elements of tourism infrastructure, but even more so when field guides are available, allowing birdwatchers to compare their observations in the field with the existing literature. Field guides for ornithology as a hobby or a science that allow for determining different species have a long tradition in Europe, starting in the United Kingdom with such publications as: *Our country's birds, and how to know them* (Gordon 1894), and *Birds one should know, beneficial and mischievous* (Wood 1921). Similar guides were also available in France *Atlas de Poche des Oiseaux de France, Suisse et Belgique* (D'Hamonville 1908). In fact, the first description to be used as a field guide appeared in the county of Hampshire in 1789 and was written by a representative of the local clergy of Selborne, *The Natural History and Antiquities of Selborne* (Moeliker 2013). The publication, in 1904, of *Het Vogeljaar* (the Year in Birds), by J.P. Thijsse, was the first serious contribution that encouraged the Dutch to observe birds in the wild, but still lacked the structure of a taxonomic guide with clear references and illustrations. Such a field guide was first published in 1916 (Zwart 1921), but it took almost 40 years before Peterson's *Field Guide to the Birds of Britain and Europe* was translated and published (Peterson 1954, 10th ed. 1969). For many years, this remained the single available and often reprinted field guide.

During the next decades, however, more field guides—often translated from original English guides—became available. Some examples are Bruun and Singer's *Hamlyn guide to the birds of Britain and Europe* (1970); Heinzel et al.'s *The Birds of Britain and Europe with North Africa and the Middle East* (1973); and, Ferguson-Lees' *Shell guide to the birds of Britain and Ireland* (1983). Closer to the end of the twentieth century there is a remarkable change, and no longer translations from England are regarded as the best field guides, but those that were translated from original Swedish guides such as Jonsson's *Faglar in Europa*" (1993) and Svensson's *Fagelguiden Europas och Medelhavsomradets fagler I falt* (1999). Especially the last guide contributed to the growing popularity of bird watching as a leisure activity since it was published by the ANWB, the Dutch Automobile Association. This private organisation has over 4 million members and is not just a lobby representing car and bicycle-users, but it is also involved in publishing (on travel related subjects) and acting as a tour operator offering several different types of tourism from its portfolio, including the SNP (see Table 4.1).

It comes as no surprise that the first field guides for birds in Africa, south of the Sahara, focused on the traditional safari destinations in the eastern part of the continent. Williams' *Field guide to the birds of east and central Africa* was first published in 1963 and covered a region from the Red Sea and Ethiopia, in the North, to Madagascar and Mozambique, in the South; but it did not include Rwanda and Burundi (Williams 1977). Within a second field guide on the National Parks of East Africa, the same author included some plates to facilitate the observing of (rarer) birds, but most of the text and plates refer to the National Parks and the mammals one might observe and the region is restricted to Kenya, Tanzania and Uganda (Williams 1976). The *Helm field guide on Birds of East Africa,* published in 2002, was the first of its kind for this region that used the 'European' or 'British' system of full colour drawings of all species, showing the differences between male and female, flight patterns and possible dispersion over the region which now included Kenya, Tanzania, Uganda, Rwanda and Burundi (Stevenson and Fanshawe 2002). This was quite a difference with the previous guides that often showed a selection of species and not always in colour, but often in black-and-white drawings that did not facilitate the determination of the species. In the southern part of the continent, Newman (1983) was responsible for the first field guide for bird watching in Southern Africa and, for more than a decade, his work was a standard equipment for serious birders.

However, with the publication of the *Sasol guide,* in 1996, birdwatchers in this part of the continent had a newer and better, although heavier handbook available for the region that included Namibia, South Africa, Botswana, Zimbabwe and Mozambique (Sinclair and Hockey 1996). Field guides for the western part of Africa were harder to find. For more than 20 years, *The field guide to the Birds of West Africa,* by Serle et al. (1997), was the only one available and covered an area that ranged from Mauritania in the Northwest to Sierra Leone, Ghana, Nigeria, Cameroon, Congo and part of the C.A.R and Chad, and it also included the Cabo Verde islands and Sao Tome and Principe. As with many of the earlier field guides, many illustrations were in black & white and the authors admit it is near impossible to give a description of all the possible species in this enormous region.

At the end of the last century, Barlow and Wacher published the *Helm field guide to the Birds of The Gambia and Senegal* (1997) and provided the growing amount of birdwatchers to that region with a very complete handbook. The continent is catching up and has recently seen the publication of an up-to-date guide book for the birds of Ghana in 2010 (Borrow and Demey 2010) and one for the birds of the Horn of Africa (Ethiopia, Eritrea, Socotra, Djibouti and Somalia) in 2009 (Redman et al. 2009). Apart from these publications, there is a very extensive field guide that comprehends all birds in South of the Sahara (Sinclair and Ryan 2003), and there are several smaller field guides for specific regions or islands like Mauritius (Michel 1992).

One of the latest publications to date is the *Birds in Rwanda, atlas and handbook* that was published in 2011 by the Rwanda Development Board (RDB). This atlas and handbook differs in approach and presentation from those previously mentioned. Its focus is solely Rwanda; it includes a dispersion map for every species in the country, and it uses full colour photography instead of the current habit of explicit drawings. This presentation has both advantages and drawbacks. First of all, the photos show the different species in their natural environment and as all birdwatchers will confirm, 'the real bird is always a bit different from the drawings in the book'. But, the drawback is the fact that the individual pictures show each bird at a specific moment and under specific lighting and make it hard to compare them for size or flight details. As an atlas, the book is amazingly detailed giving the possible dispersion within 30 × 30 min grids that are subdivided in 167.5 × 7.5 min squares allowing for determining species based upon location. Another remarkable element of the guide is its attention for such topics as conservation and ecology and the different landscapes that make up Rwanda. Finally, it is remarkable and even unique that RDB as a government organisation has taken the trouble, energy and time to produce such a work; thereby, strongly illustrating its interest in promoting Rwanda as a destination for birdwatchers (Van de Weghe and Van de weghe 2011).

4.5 Who Are the Bird Watching Tourists and What Do They Want?

According to Moss (2009), quoted in Newsome (2015), exposure of birdlife on television documentaries and the rise of international bird watching tourism have led to a situation where hundreds of thousands of birdwatchers travel widely to see birds. So far there have been few studies dedicated to bird watching tourism either in Africa or abroad. Exceptions are the research by Green and Jones (2010) on the practices, needs and attitudes of bird-watching tourists in Australia; the case-study of Newsome in Papua New Guinea (2015); and, the work undertaken by Tisdell and Wilson (2003). The first study set out to refine and possibly expand the target of this specific tourism market, find out what bird-watching tourists want to see and to do, identify challenges and opportunities for the industry and the destinations regarding this type of tourism and explore the interest in and concern for conservation amongst these tourists. Although most of the respondents were Australian (164 of 201), the findings of this study are also relevant for other destinations and could be seen as an indication of the specific wants of this type of consumer.

One of the first findings was the fact that there were different types of birdwatchers, ranging from the 'occasional birders' who like to see birds both at home and during their holidays, to the 'dedicated birdwatchers', who were more likely than others to want to see endemic or rare birds and partly base their decision for a holiday destination upon these propositions. Finally, there is the category of the so-called 'thrill-seekers' whose main objective is spotting the more spectacular species such as big and colourful birds. For Rwanda, as for any other destination this would mean that the marketing of the country as a birdwatchers destination could be differentiated as well. When asked what they wanted to see most respondents answered: species not previously seen, or species unique to the country, or rare or vulnerable species.

The fact that Rwanda is home to 44 species that are endemic (RDB, no date given) gives the country a rather unique selling proposition for this last type of 'birder'. When asked how bird-watchers decide where to go birdwatching, it was found that travel agents figured very low and, with the exception of 'twitchers', the same could be said of the internet. Most of the respondents indicated that their visits were often combined with other activities and that the information from natural history organisations was regarded as inspiring and useful. It comes as no surprise that the information from friends and family was also regarded as influential. Watching birds in action and identifying as many species as possible were activities that rated high on the preference list of the respondents and in general most of them had a very positive attitude towards guided tours. It is worth mentioning that according to the research by Green and Jones (2010), the top three reasons for seeking a guided tour rather than undertaking it alone were for safety, reaching difficult places and finding difficult birds; all of which could be especially relevant for people making a first trip to Africa.

Another interesting result of the study was the finding that 'thrill-seekers' expressed less demands when referring to accommodation and could settle for budget-type hostels. Finally, more than 50% of the respondents indicated that they held a membership of a conservation organisation or joined meetings related to bird conservation (Green and Jones 2010). This seems to correspond with previous research in Australia where more than 50% of the respondents indicated to be strong or even extremely strong supporters of nature conservation (Tisdell and Wilson 2003). In her book on wildlife tourism, Shackley (1996) offers a first general view of the wildlife tourist. He, or just as often she, is slightly older (43.9 years vs. 40.8 years) than the leisure-oriented tourist, and is often making a first visit to a country and also more likely to travel in small groups rather than as an individual. These tourists are often motivated to enjoy cultural sites as well. Green and Jones found a similar percentage of male and female bird watchers, but did not enquire for any of the other topics (Green and Jones 2010). Personal communication with some of the tour operators mentioned in Table 4.1 yielded similar results when information was requested on the age and gender of the Dutch participants of the tours. However, when visiting the website of Dutch Birding and more specific its overview of members/contributors most portrait photo's showed men, 203 out of 206 (www.dutchbirdalerts.nl).

For Shackley, bird watching is the most popular nature watching activity and she also suggests that the enjoyment of unspoiled wilderness seems to have a considerable psychological benefit and can become an addictive pastime (Shackley 1996). The author of this chapter agrees with her when she further states that committed wildlife (e.g. bird watching) tourists are continually looking for new areas and new natural experiences, such as the observation of new species. But why would bird watching be more popular than, for instance, something like mammal—or reptile watching? The possible answer to this question is provided by Simon Barnes (2004), who states that the answer resides in the fact that birds are much easier to be found and observed that encountering a wild mammal in Europe is rather exceptional; birds are part of daily life and there are so many species that

Table 4.2 Composition Avifauna Rwanda (Van de Weghe and Van de weghe 2011)

Composition Avifauna Rwanda	Species	Breeding
Resident	491	447 with breeding records
Visitors	138	20 with breeding records
Occasional visitors	63	
Locally extinct	5	
Total	697	467 with breeding records
To be confirmed	4	
Total	701	

Table 4.3 Globally threatened bird species Rwanda (derived from Birdlife International 2015)

Species: Scientific name	Species: English name	Order
Aquila nipalensis	Steppe Eagle	Accipitridae (Hawks, Eagles)
Polemaetus bellicosus	Martial Eagle	Accipitridae (Hawks, Eagles)
Gyps rueppellii	Rüppell's Vulture	Accipitridae (Hawks, Eagles)
Necrosyrtes monachus	Hooded Vulture	Accipitridae (Hawks, Eagles)
Gyps africanus	White-backed Vulture	Accipitridae (Hawks, Eagles)
Trigonoceps occipitalis	White-headed Vulture	Accipitridae (Hawks, Eagles)
Torgos tracheliotos	Lappet-faced Vulture	Accipitridae (Hawks, Eagles)
Glaucidium albertinum	Albertine Owlet	Strigidae (Typical Owls)
Psittacus erithacus	Grey Parrot	Psittacidae (Parrots)
Balearica regulorum	Grey Crowned-crane	Gruidae (Cranes)
Balaeniceps rex	Shoebill	Balaenicipitidae (Shoebill)
Ardeola idae	Madagascar Pond-heron	Ardeidae (Herons)
Bucorvus leadbeateri	Southern Ground-hornbill	Bucerotidae (Hornbills)
Bradypterus graueri	Grauer's Swamp-warbler	Sylviidae (Old World warblers)
Chloropeta gracilirostris	Papyrus Yellow Warbler	Sylviidae (Old World warblers)
Nectarinia rockefelleri	Rockefeller's Sunbird	Nectariniidae (Sunbirds)
Cryptospiza shelleyi	Shelley's Crimson-wing	Estrildidae (Waxbills, grass finches, munias and allies)

'collecting them' might become a lifetime hobby. Also, the fact that they fly makes watching them attractive and interesting because that movement would appeal to the dream of every human being (Barnes 2004).

4.6 Bird Watching in Rwanda

According to RDB and Van de Weghe, Rwanda offers its visitors the opportunity to see almost 700 bird species of a total that is globally estimated to be close to 10,000. For a landlocked country of that size this is a remarkable amount and it can be partially explained by the wide variety of the landscape. The composition of the avifauna in Rwanda (Table 4.2) is as follows.

The amount of species registered by Birdlife International, one of the world's largest nature conservation partnerships, is 633 and divided into landbirds (520), seabirds (7), waterbirds (112), migratory (176). Of this total amount, 17 are to be found on the list of globally threatened species (Table 4.3); Figs. 4.1 and 4.2.

Fig. 4.1 Aquila Nipalensis, the English name Steppe eagle (*Credit* Kersti Nebelsiek/Fimbs-creative commons)

From an ornithologist's point of view, the Rwandese landscape can be divided in high altitude habitats (between 3000 and 4300 m); savannas with grass lands and thorn brush; forests and related habitats including bamboo, and secondary forests; swamps, habitats along lakes and rivers; non-vegetational habitats such as cliffs and rock outcrops; and, man-made habitats (Van de Weghe and Van de weghe 2011). Akagera National Park in the East, Nyungwe Forest National park, in the Southwest, and Volcanoes National Park that is mostly visited because of its gorilla population, are the most important bird watching destinations, but there are many other smaller locations throughout the country. For a more specific description of Rwanda's national parks and some selected locations, their species and the availability of guides, see Table 4.4.

The three National Parks are not only relevant for their avifauna and typical vegetation, but should also be considered as unique natural and ecological reserves in a country that mostly consists of man-made landscape (90%). This high percentage is due to the density of population which reaches almost 440/km^2 and is the highest in Sub-Saharan Africa. All three parks offer a landscape with specific vegetation that is related to the altitude and their micro-climate and the Table 4.5 offers a more detailed overview of these ecological elements Fig. 4.3.

The relation between the eco-climatic zone, the vegetation zone and the local birdlife is illustrated by some examples given by Van de Weghe and Van de weghe (2011) when he describes the presence of specific warblers, as the Brown Woodland Warbler and Cinnamon Bracken Warbler, as typical residents of the Afroalpine moorland of Vulcanoes National Park and the foraging of bushshrikes and boubou's amongst the thorny thickets. The great diversity of species in the tropical rainforest is generally accredited to such factors as long periods of climatic stability, geographical isolation and a great variety of ecological niches which allow symbiotic relationships between plant and animal (Park 1992). The grassland of Akagera is frequented by pipits, cisticola's and lapwings and sometimes the rare Denham's Bustard might be encountered, while the bright orange-colured flowers of the Lion's Ear (Leonotis sp.) attract sunbirds. Frazer's Eagle Owl and species as the Dusky Crimsonwing and Grey-chested Kakamega prefer the open canopy forest of Nyungwe National park and the malachite Sunbird can be found hovering near the Red-hot Poker flowers (Kniphofia granti) Figs. 4.4, 4.5, 4.6 and 4.7.

Apart from such infrastructural developments as roads, transportation, accommodation and other facilities one of the most important factors to take into account is the training of specialised guides. This is just one of the activities that has been undertaken over the last years by representatives of USAID in collaboration with RDB and the Wildlife

Fig. 4.2 Southern Ground Hornbill (Bucorvus leadbeateri) *Credit* Bernard Dupont

Table 4.4 Bird watching sites and selected species in Rwanda compiled by author

Location	Size	Species	Special species	Guides services provided (* based on 2011 data)
Volcanoes National Park	16,000 ha	165, incl. 17 endemics	Abyssinian owl, Ruwenzori double-collared sunbird	Several trained, one specialised in birds
Gishwati Forest	2000 ha	209, incl 20 endemics	Ruwenzori turaco, Ruwenzori nightjar	No birdwatching guides trained
Nyungwe Forest National Park	101,000 ha	310, incl. 27 endemics	Albertine owlet, Rockefellers sunbird, Noble francolin, Kivu groundtrush, Grauer's Rush Warbler, Red-collared Mountain Babbler	Several specialised birdwatching guides
Rugezi Swamp	8000	134	Grauers swamp warbler, Papyrus canary	Not provided
Mashoza Parike	16	151	Purple-crested turaco, Mountain Illiadopsis	One local RDB-trained guide
Akagera National Park	108,400 ha	482	Red-faced barbet, Shoebill stork, Denhams bustard, Cabanis's bunting	Biding guides available at southern entrance

Table 4.5 The eco-climatic and vegetation zones of Rwanda's National Parks

National Park	Eco-climate zone	Vegetation zone	Vegetation
Volcanoes N. P.	Afroalpine and humid zone	Afroalpine and high montane vegetation above 3200 m., Evergreen mountain forest	Ericaceous thickets, Tussock grassland, Bamboo
Nyungwe Forest N. P.	Humid zone	Evergreen mountain forest, both open and close canopy	Macaranga forest, Erica thickets, Forest Newtonia's, Guinea plumtree, Treefern, Lobelias
Akagera N. P.	Sub-arid zone	Acacia savannas (open grasslands) and dry sclerophyllous forest	Hyparrhenia grassland, various types of acacia, sclerophyllous and euphorbia

Fig. 4.3 Man-made habitat, tea plantation. *Credit* Karel Werdler

Conservation Society (Majanen 2012). While focusing on the Nyungwe Forest National Park this consortium had a much wider scope which included an ecotourism development for the benefit of local communities, leveraging private sector investment in the management, construction and maintenance of new and existing park infrastructure and in general working on a marketing strategy that would diversify the local economy, create jobs and increase family incomes, resulting in reduced poverty and threats (USAID no date).

Before taking a closer look at some of these larger objectives, it should be recognised that even a small-scale operation as guide training can contribute to these objectives, but because of the professional personal relationship between guides and their (bird watching) guests the

Fig. 4.4 Nyungwe National Park (*Credit* Karel Werdler)

emphasis of such an intensive training should not only be on specific knowledge of birds, other animals, plants and ecosystems, but also include communicative skills, foreign languages, a service attitude, customer care, a good physical condition and even basic knowledge of first-aid (Raadgers and Steenhuisen 2006). Furthermore, guides should be able to adjust their commentary and information to the characteristics of their audiences and engage in a dialogue which can contribute to more environmental awareness (Weaver 2006).

Bird watching tourism is generally regarded as one of the most responsible types of wildlife watching since its participants mostly travel in small groups, do not make use of motorised vehicles and in general are rather silent in their behaviour as to enlarge the chances of seeing and hearing more birds. The author does not want to embark on a discussion of more or less precise definitions, which are still subject to much debate, but agrees with the general opinion(s) that this type of tourism belongs to that group of (alternative) tourism forms that also includes sustainable tourism, ecotourism, environmental tourism, responsible tourism, wildlife tourism and/or low impact tourism. He also agrees with Barnes et al. (1992) and other authors that this type of wildlife tourism is a non-consumptive way of utilising wildlife resources to benefit human populations and that, if properly managed can offer a destination the opportunity to develop a high value added (tourism) industry. The example of Costa Rica's management strategy of focusing on bird watching tourism not only seems to confirm this opinion, but has also shown other positive developments such as increasing domestic tourism at a later stage and positive consequences for both wildlife and conservation issues, thereby indicating that there are educational and social values to gain as well (Shackley 1996; Brockington et al. 2008).

According to the last author wildlife tourism—and therefore bird watching tourism in Rwanda (*authors*

Fig. 4.5 Nyungwe National Park (*Credit* Karel Werdler)

addition)—can make a positive contribution to the economy and it can even become a focal point for development strategy and a focus of nation building (Brockington et al. 2008). In the future there will still be some questions to answer regarding such subjects as revenues, pricing and initial investments, but examples from the past have shown that this type of tourism can have a positive influence on both direct and indirect employment, stimulate domestic industries and diversify local economy in rural areas, stimulate improvements in local transport and communication and, last but not least produce foreign exchange. Research elsewhere has shown that the importance of a great variety of birds to a site can influence the decision to visit and additionally contribute to a larger revenue (Tisdell and Wilson 2003). However, as other studies have demonstrated caution should be taken in situations where critical behaviour takes place such as nest building, incubation, chick-rearing and moulting (Higham and Lusseau 2007).

4.7 The Future of Birdwatching in Rwanda: Conclusions and Recommendations

Nobody can tell what the future will be like and economic and other circumstances have often thwarted prophecies, expectations and even well-argued calculations. Yet, taking the suggestions of the authors mentioned earlier and the outcomes of the research, some ideas come to mind when the further development of Rwanda as a destination for birdwatching is taken into account. First and foremost is the supply of good guides and good guidebooks, which may be considered as part of the specific birdwatching infrastructure. Important steps have been taken, but to deal with a growing supply of visitors the responsible stakeholders should continue to invest in the training of guides.

Furthermore, a new handbook that is more like a traditional field guide, showing comparable drawings instead of photographs, might be considered. This should not

Fig. 4.6 Nyungwe National Park (*Credit* Karel Werdler)

replace the recent work by RDB and Van de Weghe, but form an additional work of reference. It might even be possible to enquire after cooperation with the publishers of existing field guides of the region. Another remarkable outcome of the research is the fact that there are different types of birding tourists that have different wishes and expectations, for instance, regarding accommodations. It was also found that they often relate to other sources than traditional travel agents and the internet when deciding upon their next birding destination. Combined with the fact that they are often members of natural history organisations, or refer to these as their sources of information, a low profile marketing approach directed at such organisations might be profitable for all stakeholders. This means that these organisations and their publications should not be targeted with advertisements, but supplying them with interesting information on birds, birdlife and conservation developments might find a very positive reception. Furthermore, intensified contacts with such organisations might open up other possibilities as well since they are often connected to other institutes, or offer their members specific opportunities to combine birdwatching travel with volunteer work on location.

Although not part of this research, some inspirational examples of this type of 'voluntourism' (or maybe visitor participation) might be found on the websites of the Audubon Society, the Earthwatch Institute and the Sierra Club (www.audubon.louisiana.org, www.earthwatch.org, www.earthwatch.org/expeditions,www.sierraclub.org/outings/volunteer). It is obvious that there is both political will and an entrepreneurial drive in Rwanda to 're-position' its resources as a tourist destination and the author firmly believes that all these activities could and should contribute to creating awareness of Rwanda as a birdwatching destination. Finally, a local educational program for the training of new birdwatching guides should be started, since the availability of these guides is rather modest at the moment and new generations should be involved, not just to provide services to future visitors, but also to create awareness of the country's unique resources.

Fig. 4.7 Nyungwe National Park (*Credit* Karel Werdler)

References

Audubon Louisiana (2015) Volunteer. http://la.audubon.org/get-involved/volunteer. Accessed 20 May 2015

Barlow C, Wacher T (1997) Helm field guide to the birds of the Gambia and Senegal. Helm, London

Barnes J, Burgess J, Pearce D (1992) Wildlife Tourism. In: Swanson TM, Barbier EB (eds) Economics for the wilds: wildlife, wild lands, diversity and development. Earth Scan Publications, London, pp 136–150

Barnes S (2004) How to be a bad bird watcher. Short Books, London

Birdlife International (2015) Country profiles. http://datazone.birdlife.org/country/rwanda. Accessed 10 July 2015

Borrow N, Demey R (2010) Field guide to the birds of Ghana. Helm, London

Brockington D, Duffy R, Igoe J (2008) Nature unbound, conservation, capitalism and the future of protected areas. Earthscan, London

Bruun B, Singer A (1970) Gids voor de vogels van Europa. Elsevier, Amsterdam

d'Hamonville BL (1908) Atlas de poche des oiseaux de France. Suisse et Belgique, Klincksieck, Paris

Dutch Birding (2012–2015) Dutch Birdalerts. Retrieved from http://www.dutchbirdalerts.nl. Accessed March 2015

Dutch Birding (2015) Smoelenboek. Retrieved from http://www.dutchbirdalerts.nl/smoelenboek. Accessed May 2015

Earthwatch (2015a) Volunteer opportunities. Retrieved from http://eu.earthwatch.org/. Accessed May 2015

Earthwatch (2015b) Volunteer abroad. Retrieved from http://eu.earthwatch.org/expeditions. Accessed May 2015

Ferguson-Lees J, Willis I (1987) Tirions Vogelgids. Tirion, Baarn

Gordon WJ (1894) Our country's birds and how to know them. Day & Son, London

Green RJ, Jones DN (2010) Practices, need and attitudes of bird-watching tourists in Australia. Sustainable Tourism Cooperative Research Centre, Queensland

Heinzel H, Fitter R, Parslow JF (1973) Elseviers gids van de Europese vogels. Elsevier, Amsterdam

Higham J, Lusseau D (2007) Defining critical habitats: the spatial-ecological approach to managing tourism-wildlife interactions. In: Higham J (ed) Critical issues in ecotourism. Butterworth-Heinemann, Oxford

Jonsson L (1994) Vogels van Europa, Noord-Afrika en het Midden-Oosten. Thieme, Baarn

Majanen T (2012) Natural resources management officer, personal communication, November 2012

Michel C (1992) Birds of Mauritius. Editions de l'Ocean Indien, Mauritius

Moeliker K (2013) Modern tijdverdrijf. In: Vogels, Summer 2013. Vogelbescherming, Zeist
Moss S (2009) Birding past, present and future – a global view. In: Del Hoyo J, Elliot A, Christie DA (eds) Handbook of the birds of the world, vol 14. Lynx Ediciones, Barcelona
Newmann KB (1983) Birds of Southern Africa. Southern Book Publishers, Capetown
Newsome D (2015) Conflicts between cultural attitudes, development and ecotourism: the case of bird watching tours in Papua New Guinea. In: Markwell K (ed) Animals and tourism: understanding diverse relationships. Channel View Publications, Bristol
Park C (1992) Tropical rainforests. Routledge, London
Peterson RT, Mountfort G, Hollom PAD (1969) Vogelgids. Elsevier, Amsterdam
Raadgers W, Steenhuisen E (2006) Reisleiding geven. Toerboek, Leiden
Redman N, Stevenson T, Fanshawe J (2009) Birds of the horn of Africa : ethiopoia, eritrea, djibouti, somalia and socotra. Princeton University Press, Princeton
Rwanda Development Board brochure (no date given). Rwanda Birding—Avitourism, a rare tweet. RDB, Kigali
Serle W, Morel GJ, Hartwig W (1997) Field guide to the birds of West Africa. Collins, London
Shackley M (1996) Wildlife tourism. International Thomson Business Press, London
Sierra Club (2014a) Volunteers. Retrieved from http://content.sierraclub.org/outings/volunteer. Accessed 8 June 2014
Sierra Club (2014b) Volunteers vacations. Retrieved from http://content.sierraclub.org/outings/volunteer. Accessed 12 June 2014
Sinclair I, Hockey P (1996) The larger illustrated guide to birds of Southern Africa. Struik Publishers, Capetown
Sinclair I, Ryan P (2003) A comprehensive illustrated field guide birds of Africa, south of the Sahara. Struik Publishers, Capetown
Stevenson T, Fanshawe J (2002) Birds of East Africa. Helm, London
Svensson L (2005) ANWB Vogelgids van Europa. ANWB media, Den Haag
Tisdell C, Wilson C (2003) Visitor profiles and environmental attributes, especially of birds, attracting visitors to Lamington National Park: tourist attitudes and economic issues. University of Queensland working paper no. 76. University of Queensland
USAID flyer "Nyungwe Nziza" Strengthening sustainable ecotourism in and around Nyungwe National Park (no date given)
USAID & WCS flyer Sustaining Biodiversity Conservation in and around Nyungwe National Park, Rwanda (no date given)
USAID (2013) Creating partners in conservation. Retrieved from https://www.usaid.gov/rwanda/newsroom/blog. Accessed 20 May 2014
Van de Weghe JP, Van de weghe GR (2011) Birds in Rwanda, an atlas and handbook. Rwanda development Board, Kigali
Vogels (Spring issue, 2013) Vogelbescherming Nederland, Zeist
Vogels (issues 2013–2015) Vogelbescherming Nederland, Zeist
Weaver D (2006) Sustainable tourism. Elsevier/Butterworth-Heinemann, Oxford
Williams JG (1976) A field guide to the National Parks of East Africa. Collins, London
Williams JG (1977) A field guide to the birds of East and Central Africa. Collins, London
Wood T (1921) Birds one should know, beneficial and mischievous. Gay and Hancock, London
Zwart K (1921) Vogelboek. Thieme & Co, Zutphen

Author Biography

Karel Werdler MA BA B.Sc is a senior lecturer and researcher at Inholland University of Applied Sciences, Department of Tourism and Leisure Management Studies. He has been a board member of Atlas Africa since 2003 and is pursuing his Ph.D. at London South Bank University. He has published on Tourism in Africa, Dark Tourism and City Marketing/Urban Tourism. He is guest lecturer at LSBU, CETT University of Barcelona and was project leader from Inholland for the Nuffic Niche project in Rwanda (2010–2014) in cooperation with Rwanda Tourism University College.

The Intersections Between Lacustrine Wildlife Tourism and Conservation: Scientific and Educational Opportunities in the Brazilian Southern Lakes

5

Rosane Maria Lanzer, Rita Gabriela Araujo Carvalho, and Aline Correa Mazzoni

Abstract

Taken as an activity on the rise in the contemporary world, tourism increasingly leans towards natural areas, especially coastal regions. Brazil stands out for its extensive coastline and attractions related to sun and beach. However, the interaction between tourism and these ecosystems is not much debated. In the state of Rio Grande do Sul, southern Brazil, a unique coastline can be found. This region has the most extensive coastal plain of Brazil, with 37,000 km^2 and about 100 freshwater shallow lakes along its coastline. Geological processes, combined with a specific biological system, have conditioned a differentiated landscape in the region, composed of an ecosystem mosaic of sand dunes, *restinga* forest formations, wetlands and shallow lakes, which harbor their own biodiversity, including human beings that depend on these ecosystems for survival. Such resources are constantly under pressure of human activities, tourism among them, since touristic activities in these environments do not take into account their fragility. As a result, several fauna and flora species are damaged by the pursuit of human leisure, as well as by mismanagement of water resources. Thus, this study aims to discuss the relations between tourism and the freshwater environments, seeking to quantify the lack of knowledge about the species inhabiting these ecosystems, which affects sustainable practices planning. This study shows that, even in the presence of unique landscapes in scenarios of great beauty and frail ecosystems, the rich biodiversity is impacted by urbanization. Human activities affect the lakes environmental quality, harming biodiversity and tourism development in the region. Management models need to undergo changes, through the implementation of environmental education programs and promoting new forms of tourism, like scientific tourism, and wildlife conservation.

5.1 Introduction

In many tourist destinations, the waters show great potential to attract tourists, either in its natural form, like lakes, waterfalls, rivers, or artificial, such as swimming pools, theme parks, dams, among others (ANA 2005). Lakes have a significant ecological, cultural, and economic value on supplying drinking water, fishery, regulation of local climates, and provide scenic landscapes for leisure and tourism (Jiang et al. 2016). Although the number of lakes on Earth is in the order of millions, the extension comprises only 2.5% of the planet's area (Dávid et al. 2012). Hall and Härkönen (2006) report that in many developed countries lakes are the main sites for developing tourism and leisure, as in England, Finland, Hungary, Scotland, Switzerland, Italy, New Zealand, Canada and United States. The authors also mention Andes lakes, as Lake Titicaca and destinations in Africa and Asia.

The International Union Conservation of Nature—IUCN stresses that 90% of the liquid freshwater on the surface is contained in lakes, providing habitat for many species and

R.M. Lanzer (✉) · R.G.A. Carvalho · A.C. Mazzoni
Laboratório de Toxicologia e Limnologia, Universidade de Caxias do Sul, Caxias do Sul, Brazil
e-mail: rlanzer@ucs.br

R.G.A. Carvalho
e-mail: ritageotourism@gmail.com

A.C. Mazzoni
e-mail: alinecmazzoni@gmail.com

I. Borges de Lima and R.J. Green (eds.), *Wildlife Tourism, Environmental Learning and Ethical Encounters*, Geoheritage, Geoparks and Geotourism, DOI 10.1007/978-3-319-55574-4_5

various services to mankind (Pittock 2015). Despite their biological importance, lakes are the least protected environments in the world (Hall and Härkönen 2006). These environments represent an important resource for touristic and recreational activities. Nevertheless, they require a high management level, due to the ecosystem fragility (Cooper 2006; Dávid et al. 2012).

Brazil is a country greatly rich in water resources, especially rivers, and has approximately 35,000 km of inland waterways and 9260 km freshwater reservoirs banks (Mi-nistério do Turismo 2010). Although less abundant, lakes are found throughout the territory, but it is along the coast that the shallow lakes call more attention.

Brazil's southernmost state, Rio Grande do Sul, presents around 100 coastal lakes inserted into a very heterogeneous mosaic of terrestrial ecosystems (Schäfer et al. 2011) with a highly complex and diversified structure that is not found anywhere else on the planet (Schäfer 1992), resulting in great biological diversity and representing a natural heritage poorly known and valued (Lanzer et al. 2013).

Wildlife tourism represents a significant proportion of a huge global market that is predicted to increase in the coming decades, with around 100,000 individuals involved (Moorhouse et al. 2015). Despite its great richness of fauna and flora, the use of biodiversity in Brazil is still incipient and mostly based on flagship species, which are often those considered charismatic (Lanzer et al. 2015). The wildlife touristic potential of Rio Grande do Sul coastal lakes is not well explored, being restricted to large, easily visible animals or economically important plants, while smaller inconspicuous species, which make up the majority of these ecosystems' biodiversity, are overlooked.

In the present study, the relations between tourism and the freshwater environments are discussed, identifying benefits and seeking to quantify the lack of knowledge about the species inhabiting these ecosystems, which affects sustainable practices planning.

5.2 Literature Review

Over the last decades, tourism has become a major economic, social and cultural activity in the modern world and continues to grow. According to the World Travel and Tourism Council—WTTC, which includes 184 countries, tourism in 2014 accounted for 9.8% of the global economy (WTTC 2015). Tropical destinations are among the most sought, mainly due to their warm weather all year round and Paradisiacal coastline landscapes (Paula et al. 2012), fostering the tourism of sun and beach, which has in environmental quality its most attractive features.

Brazil's continental dimensions extend over several climatic zones, as humid tropical north, semiarid regions and temperate areas in the south. This climate diversification forms a variety of biomes greatly wealthy in fauna and flora, causing Brazil to have the greatest biodiversity on the planet, housing more than 20% of the total number of species on Earth (MMA 2014). Its amazing natural heritage ranked on top of the 17 megadiverse countries list (MMA 2010). The country has a large water supply, with an average of 260,000 m^3/s running through Brazilian territory, of which 205,000 m^3/s are in the Amazon River basin, leaving the rest of the territory with an average flow of 55,000 m^3/s (ANA 2015).

Wetlands play important roles that benefit social-ecological systems. Brazilian inland waters are of enormous global significance for Algae (25% of the world's species), Porifera (Demospongiae, 33%), Rotifera (25%), Cladocera (Branchiopoda, 20%), Annelida (12%), Decapoda (10%) and fishes (21%). Threatened freshwater species include 44 species of invertebrates (the majority Porifera) and 134 fishes (mostly Cyprinodontiformes, Rivulidae), primarily distributed in south and southeastern Brazil (Agostinho et al. 2005). Nonetheless, diversity in freshwater environments has been poorly studied and is much less known than in other ecosystems (Rocha 2002). Consequently, studies about the relationships between biodiversity and tourism in Brazil are not as efficient as they could be (Lanzer et al. 2015). In addition to this, freshwater habitats show alarming extinction rate trends (Ricciardi and Rasmussen 1999). In Brazil, the main causes of direct loss of biodiversity in these environments are pollution, eutrophication, siltation, construction of dams, fishery, species introduction, contaminants by heavy metals and pesticides.

Moreover, unplanned touristic activities may also pose a threat to aquatic biodiversity in these coastal ecosystems, since they are extremely important natural resources for tourism. Such threat is heightened when aquatic environments are located in the coastal region, with several social and environmental issues arising due to its popularization and massification (Coriolano 2014). However, tourism developed in a responsible way can be a major ally in aquatic ecosystems conservation, enhancing environments such as lakes and rivers, through segments like scientific tourism and environmental education activities.

Coastal lakes are physiographically diversified systems (Schäfer 1992; Esteves et al. 2008) ranging in a variety of sizes, morphometries, water chemistry characteristics, landscape positions and trophic status (Caliman et al. 2010), a set of conditions that offers a large number of habitats and niches for many animal, plant and microorganism species, whether inside the aquatic environment or in surrounding associated areas. These ecosystems are usually located in densely populated areas, in which human activities are often an integral part of their ecology (Berkes and Seixas 2005). When this human use of aquatic resources is not properly managed, they can pose a serious threat to the environment. Hence, coastal lakes are among the most impacted systems in the world, along with their biodiversity and ecosystemic services (IOM 2014).

The geomorphologic province "Coastal Plain of Rio Grande do Sul" is 540 km long, from Torres in the North to La Coronilha, Uruguay, in the South. It has an area of approximately 37,000 km^2, in which 22,740 km^2 (61%) are terrestrial area and 14,260 km^2 (39%) are covered by water bodies (Schäfer 1992). The region comprises about 100 freshwater lakes, distant between 0.5 and 20 km from the Atlantic Ocean (Schäfer 1992).

According to the Brazilian Forest Code, natural lakes are considered Permanent Preservation Areas (APPs) and any human activity should respect a band with a minimum width of 30 meters from the water body (Brasil 2012). Notwithstanding, the occupation and modification of areas surrounding the lakes are much smaller.

The northern portion of Rio Grande do Sul littoral is one of the regions with the highest urban growth rates in 2000–2010 decade, when the last population census was taken, ranging from around 3 to 4.25% per year, while population growth for the entire state was 0.49% (Seplan-RS 2016). The development of several towns in this region has led to a large occupancy and increasing demand of natural resources (Rocha and Hartz 2013). With constant growth of these urban zones, the lakes, which are tourist attractions, along with the littoral zone, have been increasingly exploited for aquatic sports practice and construction of summer residences (Pedrozo and Rocha 2007) without any restrictions. In many cases, they also serve as receivers of domestic sewage and agricultural effluents, in spite of being drinking water intake source. Furthermore, the irrigation channels system facilitates alien species entry (e.g. *Corbicula fluminea*, *Limnoperna fortunei*), that have already invaded these lakes, leading to competition with native species (Schäfer 1992; Lanzer 2001).

From 2007 to 2016, the Coastal Lakes project (Lagoas Costeiras), supported by Petrobras, conducted studies about the ecological conditions of shallow lakes in the coastal plain of Rio Grande do Sul. The studies showed the devastation of terrestrial ecosystems and the misuse of water resources, stressing the lakes pollution, reduced water volume during rice irrigation season, disorderly and improper use of wells and high contamination risk of surface water and groundwater, particularly by pesticides and fertilizer application in rural areas, facilitated by the sandy soils high permeability (Schäfer et al. 2011, 2013). In its third edition, Lagoas Costeiras 3 project was developed in Osório, from 2014 until 2016 and aimed the ecological characterization of shallow lakes, along with the identification of their direct and indirect uses and as a tool for environmental education. Studies were conducted by a multidisciplinary team that collected data in 16 lakes. During this period, tourism situation in lakes was also diagnosed, showing that, despite their touristic potential, the way activities have been carried out does not reflect the ecosystem fragility.

Therefore, new tourism alternatives must be devised, which contribute to the lakes conservation, encouraging understanding of their intrinsic, cultural, social, educational, scientific and economic values. Scientific tourism would be an excellent tool to develop in coastal lakes.

Bourlon and Mao (2011) describe scientific tourism as a niche that considers learning and encourages creativity in travels and visits, being an activity focused more on experience than on consumption. In this way, tourism can be an economic and social benefit and, at the same time, contribute to the coastal lakes preservation. According to Pichlerová (2007), the basis of scientific tourism is satisfying educational, cultural and relaxing needs of a group of people interested in the same thing. The main parts of scientific tourism are excursions led by experts in a specific field, but seminars and various audio—visual media can be included as well (Molokácová and Molokác 2011).

Osório's coastal lakes have a rich wildlife biodiversity, making them great attractions for the practice of scientific tourism. In this sense, it is believed that scientific tourism, coupled with awareness-raising and environmental interpretation actions may revalue lake ecosystems, approaching and strengthening the relationship between humanity and nature.

5.3 Methodology

Osório is located on the north coast of Rio Grande do Sul (29°53′12″ S; 50°16′11″ W), at an altitude of 16 meters, and has an estimated population of 43,897 inhabitants distributed in 664 km^2 (IBGE 2015). The region's climate was classified as temperate humid with hot summer (Cfa), with rain all year round. Temperatures in the warmest month are above 22 °C and in the coldest month vary between −3 and 18 °C. A sequence of longitudinal formations, parallel to the coast, runs up to Serra Geral mountains, followed by Maquiné and Três Forquilhas Valleys until the mountain slopes, over which an Atlantic Forest permanent preservation area is found. Apart from the larger lakes, the area has a considerable number of smaller water bodies, some located along the coast and some inland (Fig. 5.1).

The relation between tourism and coastal lakes was described from morphological and ecological characteristics, uses and biodiversity of five shallow lakes, which already have tourist activities, and its surroundings, (Barros,

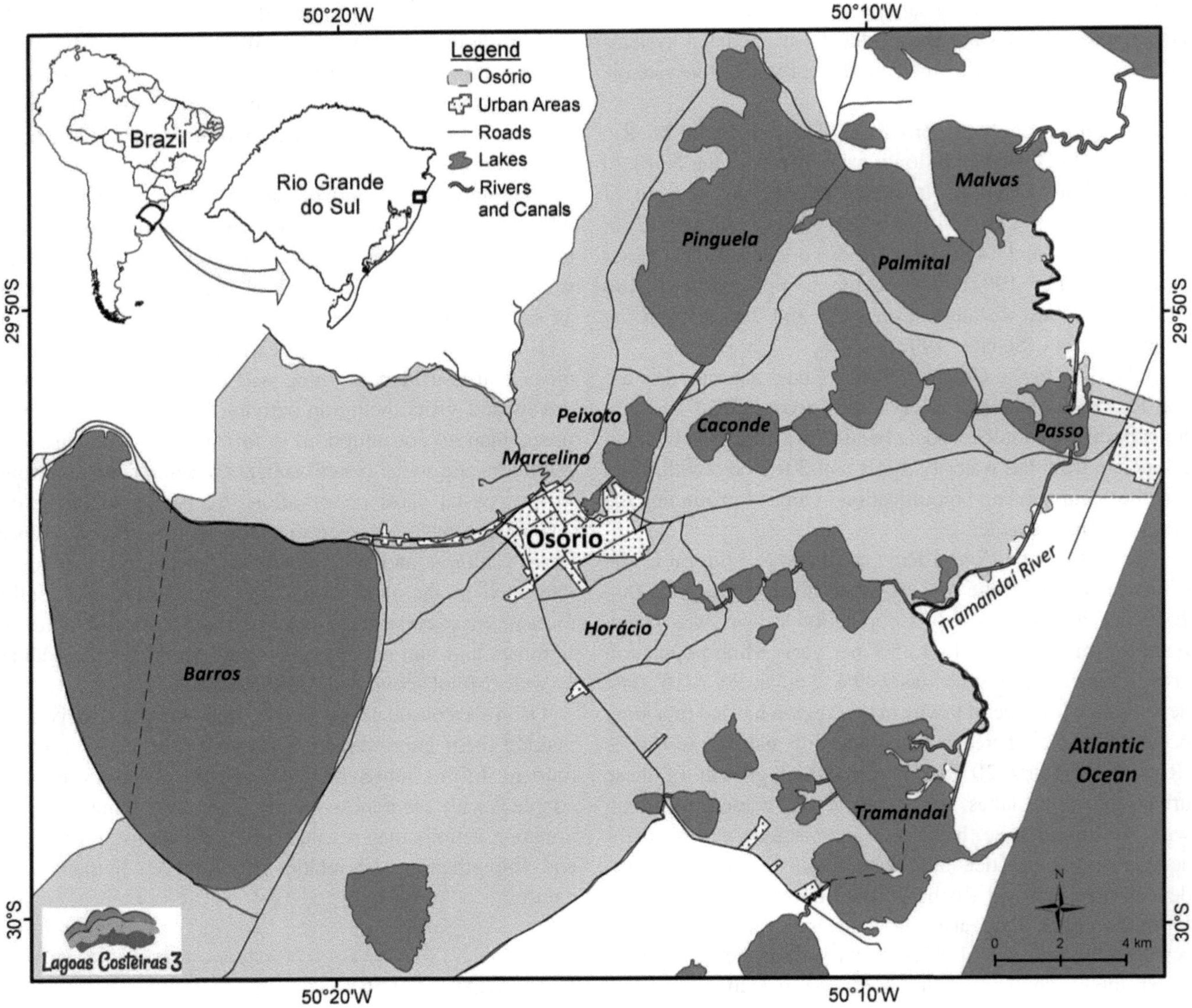

Fig. 5.1 Shallow lakes studied in the city of Osório, north coast of Rio Grande do Sul state, Brazil. *Source* Projeto Lagoas Costeiras 3

Marcelino, Peixoto, Pinguela-Palmital-Malvas, Horácio). Lakes were classified according to their eutrophication status using Schäfer's adaptation (1992) of Carlson Trophic State Index (TSI) (1977). The lakes morphology was defined with measurements of surface, maximum, average and relative depth, perimeter and Fetch. Depth measurements were performed with Echo Sounder 500 Fish Elite Eagle (Lawrence Electronics Inc.), which relates the depth data to geographic coordinates, through associated GPS receiver (average accuracy of 5 m). Spatial parameters were collected from satellite images (Landsat 8). The area and shoreline (perimeter) were settled in ArcGIS software. Aquatic biodiversity was assessed by surveying macroinvertebrate and macrophyte communities. Fish occurrences were based on Malabarba et al. (2007) data. Vegetation types and diversity of amphibians, reptiles, birds and mammals were surveyed for the municipal area. The data about vertebrate were obtained from Teia—Projetos Ambientais and will be later published in an Environmental Atlas organized by Lagoas Costeiras 3 project. Additional data were provided by the municipal administration, document analysis and literature review.

Knowledge about local biodiversity importance for the community was analyzed through interviews conducted in January 2016, where 117 questionnaires were applied to area residents aged over 18 years.

This set of data was used to evaluate nature and scientific tourism development potential, as a way to relate their benefits with the lakes management, aiming their conservation. Considering this situation, Lagoas Costeiras 3 project held, during the years 2015 and 2016, a series of environmental educational and interpretation activities, such as workshops, lectures and courses, with the goal of filling this gap on wildlife knowledge in the region. The activities had as a theme the natural resources of Rio Grande do Sul coastal plain, its uses and impacts, covering the genesis, morphology, ecological condition, aquatic biological communities and food webs of the coastal lakes, as well as the surrounding vegetation types, vertebrate fauna and current and potential use of tourist attractions.

5.4 Results

Coastal plains are among the most striking features of the Brazilian coastline, especially its south and southeast portions (Fig. 5.2). Osório's terrain is formed by plains and plateaus between sea and mountain with Atlantic Forest cover. Its location is not only favored by geography, but also by the road network, allowing a connection to several localities in addition to the proximity of the state capital (95 km). Another contributing factor is the position near BR 101, the highway bordering from the southern coast until the north of the country, interconnecting several Brazilian states.

Thus, Osório is a passage for tourists seeking further north beaches, many from Argentina and Uruguay. Despite its prime location and landscape richness, low importance is given to the coastal lakes regarding the actual service these ecosystems provide to the regional community and visitors.

5.4.1 Lakes and Tourism

Lakes provide important resources to sustain human livelihoods and economic development. Their water is used for different economic activities, among which stands out the use for supply, agriculture, especially for rice irrigation, and as urban waste receiver. Fishing, tourism and leisure are also benefits provided by these ecosystems. Lakes morphological features can have an influence on the type of activities carried out in the water and surrounding areas. Data about the studied lakes morphology are shown in Table 5.1.

Lake Barros is considered the biggest in the region, being divided between Osório and Santo Antônio da Patrulha cities (Table 5.1). Its main economic importance lies on the lake water removal for rice cultivation (November and February). In drier summers, the low water level exposes a wide margin area, damaging aquatic vegetation, mainly composed by

Fig. 5.2 Aerial view of a part of the lakes complex located in Osório, Rio Grande do Sul, Brazil. *Source* Projeto Lagoas Costeiras 3

Table 5.1 Morphological characteristics and trophic state of six lakes in Osório, Rio Grande do Sul, Brazil

Lake	Area (km^2)	Perimeter (km)	Fetch (km)	Z_{av} (m)	Z_{max} (m)	Z_{rel} (%)	TSI
Barros	91.78	39.5	10.94	4.7	6.1	0.06	Eutrophic
Marcelino	0.27	2.7	0.71	0.7	1.2	0.21	Hypereutrophic
Peixoto	2.99	8.1	2.37	1.6	2.7	0.14	Hypereutrophic
Pinguela-Palmital-Malvas	51.18	59.9	7.62	2.5	4.2	0.05	Eutrophic
Horácio	0.74	3.4	1.11	1.4	2.7	0.28	Eutrophic

Z_{av} average depth; Z_{max} maximum depth; Z_{rel} relative depth; *TSI* trophic state index
Source The authors

Schoenoplectus californicus. The littoral zone reduction has drastic consequences to animals inhabiting margins. On the other hand, there is an increase in the area available for bathers. In high rainfall summers, as occurred in 2015 due to El Niño, the higher water level caused destruction of pathways and led residents to abandon their houses. This is due to the closing of natural outflow to ensure the water volume for irrigation. But when that is not so required, flooding the banks brings damage to other users, also affecting tourism activities. Lake Barros extensive east margin (Table 5.1) offers good conditions for water sports practice, such as kite surf, windsurf and sailing. The road connecting Osório to Santo Antônio da Patrulha borders the lake southern side, facilitating open access for bathers in the summer. However, there is no touristic infrastructure, lifeguards or water quality monitoring. Visitors camp freely, making fires, removing vegetation, improperly disposing garbage and placing their vehicles inside the Permanent Protection Area.

Lake Marcelino, near the urban area, receives Osório's untreated domestic sewage, which has consequences on biodiversity. The leisure public complex, installed along the southern shore, has a pier, skate park, outdoor gym and playground, and is regarded as an excellent recreational space for the community. Fishing is observed in this lake, even though not recommended, but other water activities are nonexistent. From 1916, canals were built interconnecting lakes, in order to allow navigation towards Torres at the state northern border and facilitate products flow to capital Porto Alegre, in addition to rail and animal transport. In 2015, the Waters Memorial (*Memorial das Águas*) was inaugurated where the old harbor was located, which gathers some antique objects and a photo collection. Besides its historic relevance, the place's scenic beauty is valued by the community, which recognizes a recovery need for this water source.

Lake Peixoto also suffers pollution impacts, due to an artificial connection with Lake Marcelino. Nevertheless, its waters are used for public supply. This lake also undergoes urban pressure and margins occupation by private condos. The Municipal Camping is open all year round for users to spend the day or camp, but tourism infrastructure, use zoning and balneability control are only seen during the summer. Although classified as hypereutrophic, swimming, fishing and various water sports are held. Stand-up-paddle, jet ski and other motor vehicles use was also observed near the connection with Lake Pinguela. Balneability, during the observation period, varied in dependence of cyanobacterial blooms. The lack of knowledge by visitors about a possible toxins presence and health risks of using this water needs to integrate environmental education actions.

Lakes Pinguela, Palmital and Malvas, although known by their own names, form one single water body. The group has around 51 km surface area (Table 5.1) and is connected to the south with Lake Peixoto, north to Lake Quadros and east with Tramandaí river, which flows into the Atlantic Ocean. Despite the large extent, public access is restricted. There are numerous private tourism enterprises installed along the banks, with complete infrastructure, hosting and various leisure activities, such as horseback riding, swimming pools, dock, sport courts and space for events. Opened in 1929, Santa Martha factory produced sugar, rum and ethanol, operating at Lake Pinguela margin and is part of lacustrine navigation history. The second largest fetch is seen in Lake Pinguela, favoring the practice of wind related activities, which are explored by a kite surf school near Lake Palmital. The almost 60 km perimeter offers several areas of great scenic beauty and rich vegetation, which can only be accessed using boats.

Valuing of lands for urban expansion modifies the occupation of margins, once in rural properties and now replaced by private condominiums. This can be observed in Lakes Marcelino, Peixoto and Pinguela-Palmital-Malvas group. Privatization of margins has expanded in recent years with more visitors opting for freshwater activities instead of the sea. Most owners are vacationers who come from other cities and use the lakes for swimming, fishing and water sports. With the increasing privatization of margins, public open access to some lakes will only be possible by waterway. Condominiums built around lakes often bring as a benefit a greater control of water protection legal aspects, which are not seen in places freely accessible. However, the public nature of water use becomes compromised.

Lake Horácio, the second smaller studied lake (Table 5.1), has a campground at the northeastern margin, in a small area of native vegetation. Despite open access, use is sporadic and only during the summer a temporary touristic infrastructure is available, comprising lifeguards and water quality control. Navigation is rarely seen in the lake and camping and bathing are, virtually, the only observed activities. The ban of vehicle access along the shores contributes to nature preservation.

5.4.2 Wildlife Biodiversity and Environmental Education

Osório geographical features, coupled with Atlantic Forest and Pampa biomes elements, results in a great diversity of environments, represented by various terrestrial and aquatic ecosystems, like wetlands, shallow lakes, grasslands, dunes, hillside forests, *restingas* (sandbanks) and swamp forests. This environmental variety houses an important species biodiversity (Fig. 5.3).

Macrophytes are aquatic plants that constitute a very diverse group in the region, with an estimated number of 127 species. Some occur in almost every lake, such as *Echinodorus grandiflorus*, *Eichhornia azurea*, *Cabomba caroliniana*, *Paspalidium geminatum*, *Schoenoplectus californicus* and *Potamogeton ferrugineus*, but they widely vary in density and population size (Fig. 5.4a). Families with the highest representativeness in floristic richness were Cyperaceae (24) and Poaceae (11) comprising 27.5% of the species found. Many Brazilian endemic species are found in the lakes surroundings. These species are of great importance because, due to restricted geographic distribution, they are more susceptible to environmental degradation, with a few already classified as endangered, such as *Baccharis penningtonii*, *Syngonanthus caulescens* and *Limnobium laevigatum*. A total of 87 species endemic to Brazil was registered along the lake margins. Moreover, 26 species are in some extinction risk category for Rio Grande do Sul state,

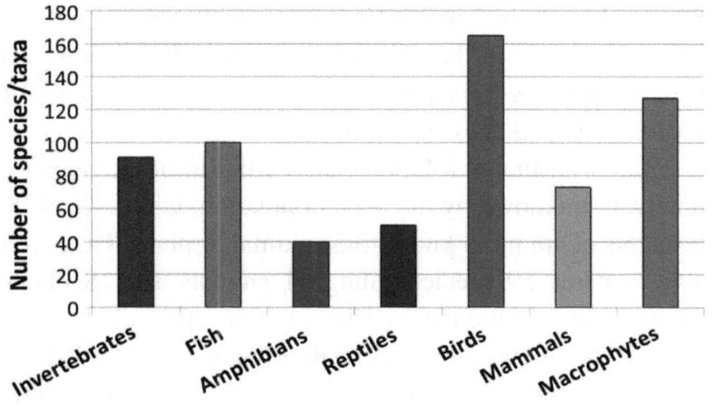

Fig. 5.3 Species richness of aquatic invertebrates, vertebrates and macrophytes in Osório region, Rio Grande do Sul, Brazil. *Source* Lagoas Costeiras 3 Project; fishes from Malabarba et al. (2007)

highlighting the importance of these environments preservation.

The lakes' aquatic invertebrate assemblages (Fig. 5.4b), known as macroinvertebrates, were composed of 49 families and 42 genera of insects, mollusks, crustaceans, annelids, flatworms and mites. Higher taxonomic richness was observed in Lakes Caconde (49), Peixoto (48) and Horácio (40). Lake Marcelino had the lowest taxa richness (24) and a dominance of non-biting midges larvae, regarded as being resistant to pollution, which represented 89% of the macroinvertebrate community. Lake Peixoto also showed this dipteran predominance, despite its higher diversity. In the other lakes, a more homogeneous taxa distribution was observed, with great diversities of caddisflies, water beetles, dipteran larvae (Insects) and seed shrimps (Crustacea-Ostracoda). Noteworthy is the presence of invasive alien species *Limnoperna fortunei* and *Corbicula fluminea*, in the studied lakes. Golden mussel colonies can injure bathers or water sports practitioners' feet.

Regarding the ichthyofauna, about 100 species of freshwater fish are known for the region and some have yet to be described, corresponding to about a quarter of the species found throughout Rio Grande do Sul (Malabarba et al. 2007). Among them *Oligosarcus jenynsii*, *Rhamdia quelen*, *Odontesthes bicudo*, *Odontesthes bonariensis* and *Odontesthes piquava* are used in sport fishing and for human consumption, along with several species of *Astyanax*. The highest fish diversity was registered in Lake Pinguela-Palmital-Malvas.

In the Herpetology field, there is an estimated 40 amphibian species (Fig. 5.4c) and 50 reptile species within Osório limits. This richness for both groups is quite significant and represents about 40% of amphibians and reptiles in Rio Grande do Sul. Among amphibians, 39 species are in the order Anura and one belongs to Gymnophiona. Nearly 80% of Rio Grande do Sul amphibian richness can be found in the study area. Reptiles are distributed in 35 species of snakes, nine lizards, three pond turtles, two amphisbaenians and one alligator. Rare species were also recorded associated with Pinguela-Palmital-Malvas group.

In relation to birds (Fig. 5.4d), the coastal region is home to about 50% of the existing species in Rio Grande do Sul. Among these, 34% are dependent on aquatic habitats for their survival, such as coastal lakes, marshes, flooded fields and beaches. Around 20% of bird communities in the region are migratory and Lakes Peixoto and Pinguela-Palmital-Malvas holds their higher incidence, due to sandy shores that favor avifauna occupation. Many neartic water birds are found along the banks, like the solitary sandpiper (*Tringa solitaria*) and lesser yellowlegs (*Tringa flavipes*), also austral migratory birds, such as rufous-chested plover (*Charadrius modestus*) and spectacled tyrant (*Hymenops perspicillatus*) can be seen. Furthermore, about 5% of the

Fig. 5.4 Examples of biodiversity in the region of Osório, Rio Grande do Sul, Brazil. **a** *Eichhornia azurea*, *Credits* Lagoas Costeiras 3; **b** *Oecetis* sp., *Credits* Aline Correa Mazzoni; **c** *Dendropsophus microps*, *Credits* Marcelo Duarte Freire; **d** *Calidris melanotos*, *Credits* Cristian Marcelo Joenck

species in the region are considered relevant to conservation, as curve-billed reedhaunter (*Limnornis curvirostris*), red-necked tanager (*Tangara cyanocephala*) and rusty-collared seedeater (*Sporophila collaris*), which are categorized as "Near Threatened". Besides these, several bird species found in the area are classified into some threat category.

Around 73 mammals species are found in the region wetlands (marshes, temporary ponds, rice plantations, irrigation canals/drainage and lake margins), including Coypus (*Myocastor coypus*), Brazilian guinea pig (*Cavia aperea*), Capybara (*Hydrochoerus hydrochaeris*) and Neotropical otter (*Lontra longicaudis*). In the lakes surroundings, ten "Endangered" species are found. Among these are the small mammals Tuco-tuco of the dunes (*Ctenomys flamarioni*) and Tate's woolly mouse opossum (*Marmosa paraguayana*), as well as medium and large size mammals, like Southern tamandua (*Tamandua tetradactyla*), Southern brown howler (*Alouatta guariba clamitans*), Margay (*Leopardus wiedii*), Jaguarundi (*Puma yagouaroundi*) and South American coati (*Nasua nasua*). Some invasive mammals found in the region live in the wild, as European hare (*Lepus europaeus*) and rodents (*Rattus rattus*, *R. norvergicus* and *Mus musculus*).

Even with this rich biodiversity, wildlife in the region is relatively unknown by the local population, with 88% of the residents claiming to know some animal or plant. They were able to name 59 species, citing 41 animals and 18 plants. Among the mentioned species, reptiles prevailed (21%) (Fig. 5.5), followed by mammals (20%), birds (20%) and fishes (19%). Regarding the flora, 13% said they knew some kind of aquatic plant (see Fig. 5.5). Amphibians and aquatic invertebrates were among the lesser known species, despite being groups closely related to the lakes.

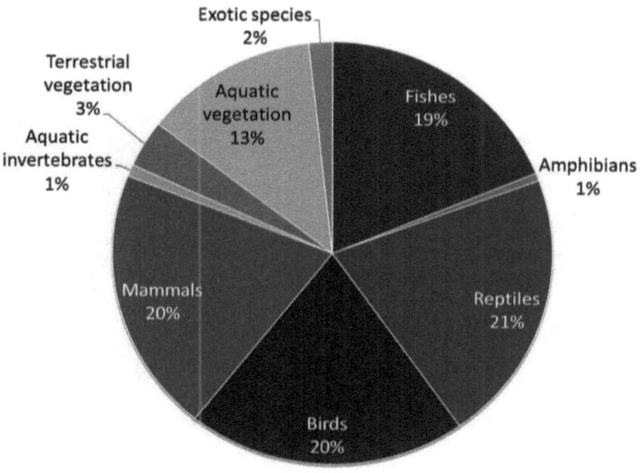

Fig. 5.5 Fauna and flora identified by residents of Osório, Rio Grande do Sul, Brazil. *Source* The authors

Residents cited mainly those species most frequently observed or that have economic value. In this sense, activities that arouse people's interest in knowing and understanding the ecosystem importance, like environmental interpretation and scientific tourism, are extremely relevant for lakes biodiversity conservation and local tourism development.

In order to promote and contribute to the knowledge about biodiversity in the region, Lagoas Costeiras 3 project team conducted a series of environmental activities, both in-person and at distance, with educational material available in virtual learning platforms, social networks and video-sharing websites. In practical lessons, the participants were taken to the margins of lakes Caconde and Peixoto, where they could meet macroinvertebrates and birds' fauna *in situ*, as well as macrophytes and vegetation surrounding the lakes. Though open to general public, the events were attended mainly by teachers and students from Osório's public schools.

Among the proposed activities, there was a "macroinvertebrates hunt" (Fig. 5.6), with people receiving tweezers, sieves and magnifying glasses to search for aquatic invertebrates in the sand and macrophytes at Lake Peixoto's margins. Once the animals were found, an explanation was given about their identification and ecological importance, so that people could actually see and know more about these small organisms, which often go unnoticed.

Nature trails were held at Lake Caconde surroundings, in which participants were able to observe both fauna and flora communities, and learn about terrestrial vegetation types and how they are integrated with other ecosystem elements, like the birds that promote seed dispersal and pollinating agents, such as wind and insects.

In one of the courses, people also entered Lake Caconde in a boat and received explanations about the complex interrelationships between physical and biotic environments, as well as on the interdependence of terrestrial and aquatic systems, which determine the ecosystem dynamics. They learned, for example, about the role of macrophytes, like the carnivorous bladderworts *(Utricularia* sp.), or how to recognize traces of animals, such as otter (*Lontra longicaudis*) excrements.

5.5 Discussion

Osório is popularly known by two nicknames, "City of Lakes", in reference to its 23 lakes, many of them interconnected, and "City of Good Winds" due to a strong wind incidence in the region (Prefeitura de Osório 2015). Because of its location, it is considered a transitional zone, and influenced just as much by marine elements as by countryside and forestry components, resulting in a high biodiversity. Lakes comprise about 41% of the municipal area, while marshes and other wetlands occupy around 17% (Fujimoto et al. 2006).

UNWTO (2013) refers to water as a major tourist attraction and an essential element for the activity, generating jobs and income for many communities in the world. Lakes provide important resources to sustain human livelihoods as they supply drinking water, fishery, irrigation of crops, supply of raw materials for handicrafts, scenic landscapes, regulate local climates, biodiversity, ecological processes, leisure and tourism, and environmental integrity.

The shallow lakes system along Rio Grande do Sul coastal plain represents a unique ecological complex on the planet by gathering a large amount of fresh water bodies near the coast (Schäfer 1992; Schäfer et al. 2009). Rio Grande do Sul northern coast concentrates the most populated beaches during summer. This population increase, from December until March, brings significant environmental changes affecting water supply and quality, beaches balneability, sewage and solid waste disposal, and resulting in a higher demand for health services, conflicts of use on the beach strip, more irregular houses, water pollution, noise and visual pollution and disregard for the environment, among others problems (Fujimoto et al. 2006).

Many lakes worldwide, especially the small ones, are experiencing great changes caused by eutrophication. Consequently, species composition goes through changes, toxic algal blooms come up and drinking water supplies are affected. The exploration of the lakes for tourism purposes requires a high level of management due to the ecosystem frailty (Cooper 2006; Dávid et al. 2012). The lack of infrastructure and control of the activities generate impacts on different levels, jeopardizing these ecosystems conservation and accelerating natural aging processes and

Fig. 5.6 Participants collecting aquatic invertebrates in environmental education activities at Lake Peixoto. *Source* Projeto Lagoas Costeiras 3

degradation. Absence of management in lakes results in a lack of water quality control. Lake Marcelino provides to residents and visitors a landscape of scenic beauty and recreation site along its margins, but its water are heavily polluted. Lakes Peixoto and Pinguela-Palmital-Malvas which concentrate touristic activities, are quite eutrophic, with frequent cyanobacterial blooms, which should restrict its use for tourism. In all of these lakes, users are at risk of accidental water ingestion when practicing water sports, swimming or even bathing. Mota (2008) emphasizes the water aesthetic use for recreational purposes and contemplation. Thus, elements interfering in landscape, as waste, stench or different water color can spoil people's experience in these environments.

Contributions of nature-based tourism are being increasingly studied and measured (Carlsen and Wood 2004; Hughes and Carlsen 2009). The valuation of wildlife for tourism purposes has the potential to demonstrate tangible economic benefits attributable to wildlife and thus, to present a business case for wildlife conservation. Consequently, arguments for conservation frequently make reference to tourism revenue (Catlin et al. 2013).

The ecological features found in the studied region enable its classification among the priority areas for animal conservation, labeled "high value" and "very high value" for biodiversity (MMA 2000). With regard to vertebrate animals, most species found in Rio Grande do Sul are represented in the study area, especially those related to aquatic environments. Many of these plant and animal species are under some protection status and a few are considered rare.

Some species become better known among residents by establishing a closer relationship with human beings, as in the case of fishes, whose scales are used for handicraft (Coelho-de-Souza et al. 2013). Aquatic vegetation also provides raw material for craftsmen, with extractive activities acting as an additional income for some families (Silveira et al. 2011). Other species, on the other hand, are not as much valued simply because they are not well known.

According to Abbot et al. (2001), the perceptions of biodiversity held by local communities play a key role in determining whether conservation efforts succeed or fail. Tourism can play an important role in local residents' perception of fauna and flora values and help maximize conservation (Broadbent et al. 2012). Local community has mostly recognized species of vertebrates in their biodiversity. This is possibly due to ease of viewing at Lake Marcelino, which, contradictorily, is inhabited by several species, due to the great organic enrichment and adjacent marshes. Among the most mentioned animals were capybara, coypus (mammals), pond turtles, alligators (reptiles), along with several bird species. Macrophytes, amphibians and macroinvertebrates, despite being well represented in lakes, were less recognized by local residents. This can be justified because these animals are not considered charismatic species. In the case of invertebrates, it adds up their small size and absence on the biodiversity boards placed at Lake Marcelino.

The awareness-raising activities conducted by Lagoas Costeiras 3 team contributed to change people's perception about the lakes. Previously seen as simply water reservoirs for consumption and leisure, the lakes came to be perceived as environments full of life and ecologically complex. Participants showed a great interest in non-charismatic species, especially benthic macroinvertebrates, which used to be completely unknown by most people. From the moment these animals were introduced, people were actually able to connect with them and recognize their importance. Wildlife tourism often relies on charismatic megafauna to bolster conservation outcomes, using the inherent charisma of a species to rally public awareness and support (Skibins et al. 2016). However, there is strong evidence to suggest that tourists can form an emotional connection to a wide array of species, which leads to adoption of pro-conservation behaviours extended to species beyond those in the limelight (Skibins 2015). More research is now required for in-depth analysis of alternative markets to those created by charismatic biodiversity, in order to create new opportunities to protect less charismatic biodiversity (Di Minin et al. 2013). Thus, the creation of environmental education programs and initiatives to develop scientific tourism becomes necessary, in order to value ecosystems and their biodiversity.

Bourlon and Mao (2011) report that scientific tourism involves a host of tourist activities, some of which work on adventure and exploration aspects, others under a cultural or educational perspective, or even in the scientific strict sense. Many types of tourism encourage the conservation of large, charismatic species, while other life forms and ecosystems are neglected. Local communities' knowledge on ecosystems and biodiversity has a positive influence on tourists' conservation and pro-conservation behaviors. Thereby, it is assumed that tourism can be an excellent way to promote the knowledge about regional biodiversity and brings a new approach in the use of these natural attractions for both tourists and the local community.

However, Osório's municipal administration does not have an environmental education program contemplating the coastal lakes wildlife at the moment. Based on the experience with the environmental events conducted by Lagoas Costeiras project, it was noticed that there is a demand for these kind of activities, which could be carried out continuously with the local community. Thus, seminars, training courses, awareness-raising workshops focused on wildlife conservation will help prepare the community to work with activities involving scientific tourism. Such activities may involve, in addition to practices with benthic macroinvertebrates, bird watching at selected sites, self-interpretive nature trails encompassing the terrestrial biodiversity surrounding the lakes, boat rides stopping at strategic learning points. Besides these, other activities addressing the historical-cultural heritage associated with the lakes could help diversify the tourist offer.

Specifically, future choice experiments could include a wider range of smaller species, ranging from those that are well-known to little-known, or species which are under different levels of threat. The results of such studies may help decision makers market currently less popular conservation areas, which lack charismatic megafauna. It is also important to start educating the next generation of visitors to conservation areas about broader aspects of biodiversity than simply of charismatic species (Di Minin et al. 2013).

Besides taking effective actions to increase the local communities' knowledge on wildlife, improvements in the lakes touristic infrastructure are also needed, as well as the involvement of various stakeholder and adoption of measures to avoid or minimize impacts resulting from tourism.

5.6 Conclusion

This study shows that, even in the presence of unique landscapes in scenarios of great beauty and frail ecosystems, the rich biodiversity is impacted by urbanization. Human activities affect the lakes environmental quality, harming biodiversity and tourism development in the region. Management models need to undergo changes, through the implementation of environmental education programs and promoting new forms of tourism, like scientific tourism. Community's role is extremely important for ecosystems conservation. In order to inform the public more broadly about protected areas, we need to educate them about other aspects of biodiversity, not just mammals and birds, and about ecosystem functioning, ecosystem services and topography.

Acknowledgements This research was funded by PETROBRAS. The authors would like to thank the whole team of Lagoas Costeiras 3 project, researchers, technicians and students for their support in data survey and field work. We also thank the Teia—Projetos Ambientais Ltda team, for providing the data about vertebrates.

References

Abbot JIO, Thomas DHL, Gardner AA et al (2001) Understanding the links between conservation and development in the Bamenda Highlands, Cameroon. World Dev 29:1115–1136

Agostinho AA, Thomaz SM, Gomez LC (2005) Conservação da biodiversidade em águas continentais do Brasil. Megabiodiversidade 1(1):70–78

ANA Agência Nacional de Águas (2005) Cadernos de recursos hídricos: turismo e lazer e sua interface com o setor de recursos hídricos. http://www.dominiopublico.gov.br. Accessed 8 July

ANA Agência Nacional de Águas (2015) Conjuntura dos recursos hídricos. http://conjuntura.ana.gov.br/docs/crisehidrica.pdf. Accessed 8 July

Berkes F, Seixas CS (2005) Building Resilience in Lagoon social-ecological systems: a local-level perspective. Ecosystems 8 (8):967–974

Bourlon F, Mao P (2011) Las formas del turismo científico en aysén, chile. Gest Tur Chile 15:74–98

Brasil (2012) Código Florestal. Law n° 12.651. http://www.planalto.gov.br/ccivil_03/_ato2011-2014/2012/lei/l12651.html. Accessed 8 July 2016

Broadbent EN, Zambrano AMA, Dirzo R, Durham WH, Driscoll L, Gallagher P, Salters R, Schultz J, Colmenares A, Randolph SG (2012) The effect of land use change and ecotourism on biodiversity: a case study of Manuel Antonio, Costa Rica, from 1985 to 2008. Landscape Ecol 27:731–744

Caliman A, Carneiro LS, Santangelo JM et al (2010) Temporal coherence among tropical coastal lagoons: a search for patterns and mechanisms. Braz J Biol 70(3):803–814

Carlsen J, Wood D (2004) Assessment of the economic value of recreation and tourism in western Australia's national parks, marine parks and forests. http://citeseerx.ist.psu.edu/viewdoc/download?doi=10.1.1.471.3580&rep=rep1&type=pdf. Accessed 8 July 2016

Catlin J, Hughes M, Jones T, Josnec R, Campbell R (2013) Valuing individual animals through tourism: science or speculation? Biol Conserv 157:93–98

Coelho-de-Souza G, Perucchi LC, Kubo RR (2013) Patrimônio socioambiental da bacia hidrográfica do rio Tramandaí. Via Sapiens, Porto Alegre

Cooper C (2006) Lakes as tourism destination resources. In: Hall MC, Härkönen T (eds) Lake tourism: an integrated approach to lacustrine tourism systems, aspects of tourism, vol 32. England, p 235

Coriolano LN (2014) Turismo e meio ambiente interfaces e perspectivas. In: Coriolano LN, Vasconcelos FP (eds) O turismo e a relação sociedade-natureza: realidades, conflitos e resistências EdUEC, Fortaleza, p 444

Dávid L et al (2012) Lake tourism and global climate change: an integrative approach based on Finnish and Hungarian case-studies. Carpath J Earth Env 7(1):121–136

Di Minin E, Fraser I, Slotow R, MacMillan DC (2013) Conservation marketing and education for less charismatic biodiversity and conservation businesses for sustainable development. Anim Conserv 16:263–264

Esteves FA, Caliman A, Santangelo JM et al (2008) Neotropical coastal lagoons: an appraisal of their biodiversity, functioning, threats and conservation management. Braz J Biol 68(4):967–981

Fujimoto NSVM et al (2006) Litoral norte do estado do Rio Grande do Sul: indicadores socioeconômicos e principais problemas ambientais. Desenvolv Meio Ambiente 13:99–124

Hall CM, Härkönen T (2006) Lake tourism: An Introduction to lacustrine tourism systems. In: Lake tourism: an integrated approach to lacustrine tourism, aspects of tourism, vol 32. London, UK, p 235

Hughes M, Carlsen J (2009) Developing a knowledge platform on value of parks for tourism. In: Hergesell A, Liburd J (eds) BEST EN Think Tank IX: the importance of values in sustainable tourism. James Cook University Campus, Singapore, pp 260–275. http://espace.library.curtin.edu.au/R?func=dbin-jump-full&local_base=gen01-era02&object_id=121028. Accessed 8 July 2016

IBGE—Instituto Brasileiro de Geografia e Estatística (2015) Estimativas da População Residente para os Municípios e para as Unidades da Federação Brasileiros com data de Referência em 1° de Julho de 2015. http://www.ibge.gov.br. Accessed 20 Jan 2015

IOM Institute of Medicine (2014) Understanding the connections between coastal waters and ocean ecosystem services and human health: workshop summary. The National Academies Press, Washington, DC, p 145

Jiang B, Wong CP, Ouyang ZY (2016) Beneficiary analysis and ecological production function to measure lake ecosystem services for decision-making in China. Acta Ecol Sin 36(8):2422–2430

Lanzer RM (2001) Distribuição, fatores históricos e dispersão de moluscos límnicos em lagoas do sul do Brasil. R Bras Bioci 9(2):63–84

Lanzer RM, Reichert L, Pinto L, Souza D (2015) The use of biodiversity in the Brazilian tourism promotion: analysis of state tourism websites. In: Mirosława Czerny; Wojciech Doroszewicz. (Org.) Sustainable development in peripheral regions. WUW, Varsóvia, pp 181–194

Lanzer RM, Ramos BVC, Marchett CA (2013) Impactos ambientais do turismo em lagoas costeiras do Rio Grande do Sul. Caderno Virtual de Turismo. Rio de Janeiro, 13(1):134–149

Malabarba LR et al (2007) Guia de identificação dos peixes da bacia do rio Tramandaí. Via Sapiens, Porto Alegre http://www.onganama.org.br/pesquisas/Livros/Guia_Peixes_Bacia_Rio_Tramandai_marco_2013.pdf. Accessed 8 July 2016

Ministério do Turismo (2010) Turismo Náutico: Orientações básicas. http://www.turismo.gov.br/sites/default/turismo/o_ministerio/publicacoes/downloads_publicacoes/Turismo_Nxutico_Versxo_Final_IMPRESSxO_.pdf. Accessed 8 July 2016

MMA—Ministério do Meio Ambiente (2000) Avaliação e ações prioritárias para a conservação da biodiversidade da biodiversidade da Mata Atlântica e Campos Sulinos. MMA/SBF, Brasília. http://www.mma.gov.br/estruturas/sbf_chm_rbbio/_arquivos/Sumario%20Mata%20Atlantica.pdf. Accessed 8 July 2016

MMA—Ministério do Meio Ambiente (2014) Biodiversidade Brasileira. http://www.mma.gov.br/biodiversidade/biodiversidade-brasileira. Accessed 8 July 2016

Molokácovác L, Molokác S (2011) Scientific tourism—tourism in science or science in tourism? Acta Geotur 2(1):41–45

Moorhouse TP, Dahlsjö CAL, Baker SE, D'Cruze NC, MacDonald DW (2015) The customer isn't always right—conservation and animal welfare implications of the increasing demand for wildlife tourism. PlosOne 10(10):1–15

Mota S (2008) Gestão ambiental de recursos hídricos. ABES, Rio de Janeiro

Paula DP, Morais JO, Dias JMA et al (2012) A importância da Praia do Futuro para o desenvolvimento do turismo de sol e praia em Fortaleza, Ceará, Brasil. Braz Geogr J Geosci Hum Res Medium Ituiutaba 3(2):299–316

Pedrozo C, Rocha O (2007) Environmental quality evaluation of lakes in the Rio Grande do Sul Coastal Plain. Braz Arch Biol Technol 50 (4):673–685

Pichlerová M (2007) Scientific tourism development in the Slovak Carpathians. In: Hamor F et al (eds) Ecotourism and sustainable development in the Carpathians. Proceedings from the international conference, Carpathian Biosphere Reserve, Rakhiv, Ukraine, pp 251–255

Pittock J (2015) Managing freshwater, river, wetland and estuarine protected areas. In: Worboys GL, Lockwood M, Kothari A et al (eds) Protected area governance and management. ANU Press, Australia, pp 569–609

Prefeitura de Osório (2015) http://www.osorio.rs.gov.br/. Acessed 20 Jan 2015

Ricciardi A, Rasmussen JB (1999) Extinction rates of North American freshwater fauna. Conserv Biol 13(5):1220–1222

Rocha O (2002) Perfil do conhecimento de biodiversidade em águas doces no Brasil. In: Lewinsohn TM, Prado PI (eds) Biodiversidade brasileira: síntese do estado atual do conhecimento. Editora Contexto, São Paulo, pp 165–169

Rocha EA, Hartz SM (2013) Estabilidade e persistência de uma comunidade de peixes em lagoa costeira neotropical. Rev Bras Biocienc 11(2):149–156

Schäfer A (1992) Die Küstenseen Südbrasiliens. Ein Beitrag zur Biogeographischen Raumanalyse der Restinga. Erdwissenschaftliche Forschung 26, Sttuttgart, Franz Steiner

Schäfer A, Lanzer R, Pereira R (2009) Atlas socioambiental: municípios de Mostardas, Tavares, São José do Norte e Santa Vitória do Palmar. EDUCS, Caxias do Sul

Schäfer A, Lanzer R, Sbersi F et al (2011) Influence of land use on freshwater lakes in the middle part of the Coastal Plain of Rio Grande do Sul. In: Biblio C, Hensel O, Selbach JF (eds) Sustainable water management in the tropics and subtropics—and case studies in Brazil, vol 1. Fundação Universidade Federal do Pampa, UNIKASSEL, PGCult-UFMA, Jaguarão/RS, pp 607–627

Schäfer A, Lanzer R, Scur L (2013) Atlas socioambiental: municípios de Cidreira, Balneário Pinhal e Palmares do Sul. EDUCS, Caxias do Sul

SEPLAN Rio Grande do Sul, Secretaria do Planejamento, Mobilidade e Desenvolvimento Regional (2016) Atlas Socioeconômico do Rio Grande do Sul. http://www.atlassocioeconomico.rs.gov.br. Acessed June 2016

Silveira TCL, Rodrigues GG, Coelho-de-Souza GP et al (2011) Effects of cutting disturbance in *Schoenoplectus californicus* (C.A. Mey.) Soják on the benthic macroinvertebrates. Acta Sci Biol Sci 33 (1):31–39

Skibins J (2015) Ambassadors or attractions? Disentangling the role of flagship species in wildlife tourism. In: Markwell K (ed) Animals and tourism—Understanding diverse relationships. Channel View, Clevedon, pp 256–268

Skibins J, Powell R, Hallo J (2016) Lucky 13: conservation implications of broadening "Big 5" flagship species recognition in East Africa. J of Sustain Tour 24(7):1024–1040

UNWTO Tourism and Water: Protecting our common future, World Tourism Organization (2013) http://www.toinitiative.org/fileadmin/docs/ActivityReports/press_rel/Background_paper_Tourism___Water.pdf. Accessed Jan 2015

WTTC World Travel and Tourism Council (2015) Travel and tourism: economic impact. World. http://www.wttc.org//media/files/reports/economic%20impact%20research/regional%202015/world2015.pdf. Accessed 8 July 2016

Wildlife Tourism Safaris, Vehicle Decongestion Routes and Impact Mitigation at Chobe National Park, Botswana

6

Naomi Moswete, Kenosi Nkape, and Mpho Tseme

Abstract

The Chobe National Park River Front (CNPRF) is renowned for a high population and variety of wildlife species in Botswana. The park has become popular for nature-based tourism and wildlife safaris. With increased numbers of wildlife tourists there have been reports on problems of overuse and vehicle congestion in some parts of the Chobe National Park. In order to mitigate crowding and vehicle congestion on the popular Chobe River Front route, the DWNP introduced and implemented Upper and Nogatshaa routes. The purpose of the study is to assess wildlife tourists' frequency of use and potential environmental impacts on the Chobe River Front, Nogatshaa and Upper routes of the Chobe National Park. Data were collected in June 2013. A semi-structured questionnaire and face-to-face interviews were employed to elicit information from guides operating from fixed lodges, guides from mobile tour safaris and wildlife officials based at Sedudu gate. In addition, participant observation was also used to collect additional data for this study. The results revealed that the Chobe River Front of the CNP was heavily utilized by wildlife tourists, followed by the Upper route and the least used was Nogatshaa. The Chobe River Front route was the most preferred, while Nogatshaa is the least preferred route. The study revealed that there are benefits associated with the newly created vehicle decongestion routes at the CNP. Observations have been made to indicate that the two new routes have relatively relieved the Chobe River Front from tourist vehicle pressure; lessened the congestion of tourist vehicles particularly at animal sightings or encounters of predators (leopards, lions), have relatively relieved the Chobe River Front from tourist vehicle pressure; lessened the congestion of tourist vehicles particularly at animal sightings or encounters specifically predators (leopard, wild dogs, lions) and also creation of a few waterpoints along the Upper and Nogatshaa routes appears to have contributed towards spreading of wild animals over a large area thereby alleviating competition for foraging and water and thereby reducing grassing pressure at the CRF. However, there are still issues of congestion during game drives particularly along the River bank route and at the CRF

N. Moswete (✉)
Department of Environmental Science, University of Botswana, 4775 Notwane Road, Gaborone, Botswana
e-mail: moatshen@mopipi.ub.bw; nomsamoatshe@yahoo.com

K. Nkape
Department of Tourism & Recreation, Botswana Wildlife Training Institute, Maun, Botswana
e-mail: knkape@gmail.com

M. Tseme
Department of Wildlife and National Parks, Wildlife Warden, Kasane, Botswana
e-mail: tmpho14@gmail com

I. Borges de Lima and R.J. Green (eds.), *Wildlife Tourism, Environmental Learning and Ethical Encounters*, Geoheritage, Geoparks and Geotourism, DOI 10.1007/978-3-319-55574-4_6

viewing site. Hence, we still can make a general statement that the decongestion strategy that was meant to alleviate tour operators and tourists' traffic pressure from the Chobe River Front has possibly not achieved the intended purpose as yet. Managerial implications include improving the use of Upper and Nogatshaa routes by providing better facilities and service to all types of visitors and tourists to make it appealing. It is recommended that the park management should consider devising a strategy to attempt to demarket the Chobe River Front route to reduce visitor pressure, vehicle congestion and alleviate negative impact on animals and associated resources of the CNP.

6.1 Introduction

Wildlife-based tourism plays an important role in economies of many developing countries. Until recently, the benefits of wildlife tourism included generation of employment, hides and skin, game meat and other animal-related products to urban and rural areas (Botswana Review Board 2014; Higginbottom 2004; Lepper and Goebel 2010; Mbaiwa 2003; Shackley 1996). Important aspects of this form of tourism are socioeconomic; and include creation of business opportunities for gateway communities, lodging facilities for tourists, such as safari lodges and upmarket wildlife and wilderness built campsites (Akama and Kieti 2003; Barnes 1996, 2001; Higginbottom et al. 2001; Lepp 2007; Mbaiwa 2003; Moswete and Dube 2013; Tisdell and Wilson 2004). In addition, wildlife-based tourism has led to conservation of natural resources (Green and Higginbottom 2000; Hemson et al. 2009; Owen-smith 1996; Parry and Campbell 1992) that includes, but is not limited to endangered and rare species of wild animals and birds and the associated habitats (GoB 2001; Higginbottom and Tribe 2004; WTO 2014), forest resources that include forest tea, herbs and medicinal plants and trees (Garekae et al. 2016; GoB 2001; Lepetu 2007). However, over the years, wildlife-based tourism has been found to cause adverse impacts on the natural resource base from which travel and tourism industry is dependent on (Eagles and McCool 2000; Higginbottom et al. 2001; Manning and Valliere 2009; Mbaiwa 2005b; Moswete and Mavondo 2003; Owen-Smith 1996).

Visitor numbers to national parks and game reserves of countries offering safari/wildlife based tourism have increased over the years (Akama and Kieti 2003; DWNP 2007, 2010; WTTC 2007, 2013). Subsequently, the numbers have led to overcrowding and congestion of vehicles by nature tourists and wilderness campers in or near protected areas (Mbaiwa 2005b). There have been negative impacts of nature and safari tourism on wildlife which include interference with feeding and breeding (Ewert et al. 1999; Green and Giese 2004; Higginbottom 2004). These challenges have to an extent compelled the World Health Organization (WHO) and World Travel and Tourism Council (WTTC) to call for initiatives aimed at supporting the concept of sustainable tourism, in an attempt to minimize negative impacts on the environment and wildlife resources (Holloway 1998; Lindsey et al. 2006, 2007; Mbaiwa 2004b). As discovered, overcrowding wildlife tourists in national parks, game reserves and other protected areas in many countries can have adverse impacts on the environment and the natural resource-based tourism product (Ewert et al. 1999; Finnessey 2012; Manning 1999), hence reducing visitor satisfaction (Akama and Kieti 2003; Kalisch and Klaphane 2007; Mabunda and Wilson 2009; Magole and Gojamang 2005; Mbaiwa 2004a, 2005a, b). Similarly, literature on wildlife shows that safari and nature based tourism have caused crowding and congestion in popular protected areas including Amboseli National Park and many other conservation areas (Akama and Kieti 2003; Munyi 1992).

Several studies on protected area based tourism have uncovered negative impacts of wildlife tourism in some developed countries (Finnessey 2012; Green and Giese 2004; Kalisch and Klaphane 2007) and in many developing countries, for example; Botswana (Mbaiwa 2004a, b; 2012; Moswete and Mavondo 2003; Omphile and Powell 2002), South Africa (Mabunda and Wilson 2009), Namibia (Richardson 1998); Tanzania (Mgonja et al. 2015); Nepal (Nyaupane and Thapa 2006) and Suriname (Westin 2007). Close viewing and interaction with wild animals and birds can impact negatively on popular charismatic species especially the big five (leopard, buffalo, lion, elephant, rhinoceros). Nature based and/or safari tourism is more rampant in some wildlife parks and reserves in east and southern Africa (Kaltenborn et al. 2011; Mgonja et al. 2015; Moswete and Dube 2013; Munyi 1992; Silva and Mosimane 2012).

In protected area, tourists take more of their time on viewing big cats, thus are likely to disrupt the animals during resting moments and thereby affecting their behavior (Mbaiwa 2004a; Valentine and Birtles 2004). A case in point is the Kenya's Amboseli National park where thousands of local and international tourists enter the park each year have experienced negative impact of wildlife watching on some species such as lions and leopards (UNWTO 2014). Similarly, in South Africa's Kruger National park and Botswana's

Okavango Delta studies have revealed that tourists disturb hippos during resting time especially that they are known to sleep during the day and forage at night (Mbaiwa 2005b; Mabunda and Wilson 2009). Weaver (1998) argues that if such tourism behaviors are not controlled and monitored that sensitive species such as cheetahs and hippos (Botswana & South Africa) behaviors (e.g., hunting patterns) have been altered especially in habitats that tend to be frequented by tourists (Guesset et al. 2009; Hemson et al. 2009; Lindsey et al. 2007; Mbaiwa 2005b; Westin 2007). In some situations, increased death and disappearance of some species of wild animals and birds have been reported due to continuous disruptions during nestling and foraging (Green and Higginbottom 2000; Hachileka 2003; Moswete and Mavondo 2003). In the case of the Okavango Delta in Botswana, soaring visitor numbers has led to increased production of garbage (solid and liquid waste) from some popular tourism built campsites causing death and injury of some animals (Mbaiwa 2012; Moswete and Mavondo 2003).

6.2 Wildlife and Safari Tourism

Wildlife tourism is described as a form of tourism that involves encounters with non-domesticated animals either in their natural environment or in captivity (STCRC 2009), and includes a wide range of activities such as bird watching, general wildlife watching, non-consumptive hunting, photography (nature including wild animals, birds), and recreational fishing activities (STCRC 2009; UNWTO 2014). Safari tourism refers to activities in which a group of visitors or tourists are conveyed into a protected area (national parks, game reserves) by a professional guide, usually for a period ranging from a day or more (see Valentine and Birtles 2004: 27). The vehicles used are typical as they are four-wheel drive, off-road vehicles with open—tops and high sites for quality viewing by tourists (Valentine and Birtles 2004). According to the UNWTO (2014, p. 12), safari is a term that is generally used as a synonym for game viewing or wildlife watching (Weaver 2001) and is also associated with wildlife tourism in east and southern Africa (Akama and Kieti 2003).

6.2.1 Wildlife Tourism Safaris in Botswana

Botswana's tourism industry is built on wildlife and wilderness resources that abound the country. For example biodiversity hotspots and wealth of natural resources including wildlife are found mainly in and around the Okavango Delta and the surrounding protected conservation areas. Of particular mention are the Chobe National Park, Makgadikgadi and Nxai Pans national park, Moremi Game Reserve in the north and the greater Central Kalahari Game Reserve, Kgalagadi Transfrontier Park Kutse Game reserve in the Kalahari region. Most of wildlife tourism safaris take place in protected areas where hunting or consumptive tourism is not permitted. In the last few decades, hunting safaris or consumptive tourism activities were permitted only in designated buffer zones or wildlife management areas (WMAs) in the country.

The WMAs are explained in the Botswana National Atlas (GoB 2001, p. 197) as areas reserved primarily as zones in which the major land use permitted is consumptive and non-consumptive wildlife utilization. These areas constitute 20% of conserved land areas in Botswana and were intended to serve as buffer areas between conflicting land uses and as wildlife migratory corridors (p. 197). It is worthy of note that the WMAs are further subdivided into smaller units referred to as Controlled Hunting Areas (CHAs) which are scattered throughout the country and their main function is to facilitate the controlled use of wildlife (see GoB 2001). Thus, the WMAs and CHAs are multiple resources use areas in which sustainable utilization of wildlife and wilderness resources for nature tourism are promoted (Moswete 2009, p. 63–64).

In these areas individual (international tourists and citizens) and tourism safaris operators were only allowed to hunt specific species in particular Controlled Hunting Areas (CHAs) throughout the country using a quota system (DWNP 2001; Lovelock 2007). Since safari hunting was tightly controlled, a safari operator or an individual hunter was required by law to have a permit to hunt a specified number and species of wild animal at the time (GoB 2001). According to the Botswana tourism policy of 1990 (GoB 1990, 2007), wildlife conservation regulation of 2000 and the national ecotourism strategy of 2003 (GoB 2001) wildlife and safari tourism activities were to be practiced with due diligence to ensure long term sustainability of natural resources. Subsequently, the government devised and adopted a marketing strategy through which sustainable wildlife and safari tourism be maintained. This, "High cost, Low volume" (HCLV) tourism strategy became a yard stick through which safari tourism activities were practiced in all parks and reserves in Botswana (DWNP 2001; Magole and Magole 2011). Subsequently, the HCLV strategy was employed so as to minimise negative environmental impacts, improve conservation activities of natural resources and above all to maximise ecological and socio-economic benefits of park based tourism (Department of Tourism (DOT) 2009; Magole and Magole 2011).

It is worthy of note that over the last few decades, wildlife and safari activities have been a profitable tourism-related businesses benefitting many citizen and foreign owned enterprises (Barnes 2001; STCRC 2009; Massyn 2010; Mbaiwa 2005a; Moswete et al. 2009). In the most recent past, critics begun to blame wildlife-related tourism activities, particularly trophy hunting as the cause for the decline of some wildlife species in some parts of the country (Lovelock 2007). This somewhat led to Botswana government's decision to ban consumptive hunting safaris and begun promoting of photographic tourism in 2014 (Botswana Review 2015).

6.2.2 Tourism Policy

The tourism policy of Botswana encourages sustainable tourism (GoB 1990). Wildlife based tourism is presumed to be sustainable as it is regarded as non-consumptive. Thus, the tourism policy encourages responsible tourism that promotes low negative impacts on the ecological resources that the industry depends on. The government encourages high-cost, low volume tourism practice in all protected areas (PA). This type of tourism practice discourages casual camping, but promotes wildlife tourists to stay in permanent accommodation such as safari lodges and built or developed campsites (BTDP 2000; DWNP 2001, 2007; GoB 1990, 2007).

Furthermore, the policy restricts high numbers of tourists in some ecological and cultural sensitive areas within protected conservations areas (BTDP 2003). The high cost low volume strategy encourages sustainable tourism and environmental protection. Above all, the policy emphasizes the need to ensure that wildlife tourist activities are carried out sustainably. Based on Botswana's tourism policies and conservation tourism-related strategies, a 3-tier pricing system is used to manage tourists' entrance fees in all state controlled and managed protected areas categorized as citizens, residents and non-residents (DWNP 2016; Sebopeng 2010). In a pursuit to monitor use and discourage mass tourism in all PAs the government through the DWNP has established carrying capacity of visitors and recreators (DWNP 2001; Mbaiwa et al. 2008; Sebopeng 2010), lodging facilities are licensed based on the number of beds, and the number of vehicles permitted in a national park is set at 25 per day (DOT 2009). The question of interest is whether this is implemented and adhered to in all Botswana's PAs including the CNP.

Recently, there have been signs of unsustainable practice. Although the tourism policy has positive socio-economic and environmental impacts in general, problems have emerged. Research has revealed that the policy arrangement fails to address issues of sustainable use of environmental resources in some areas including the Chobe National Park and Okavango Delta (Mmopelwa et al. 2007; Mbaiwa et al. 2008). Additional challenges associated with weak tourism policies include situations when tour operators do not observe and/or adhere to regulations but to maximize profit (Mbaiwa 2005). There is need to ensure ecological sustainability of an important resource such as the Chobe National Park in northern Botswana.

The CNP is popular for its scenic natural landscape and abundance of wildlife (especially the African elephants) that attract a large number of domestic and international nature tourists alike. A Study by Magole and Gojamang (2005) found that nearly 90% of nature tourists visit parks and reserves in northern Botswana to see wild animals (Mmopelwa et al. 2007). Until recently, there has been an increase of wildlife and nature tourists to Botswana's Chobe National Park (Table 6.2), resulting in instances of crowding and congestion at some specific sites such as the Chobe River Front Route (DWNP 2010; 2012; Magole and Gojamang 2005; Mbaiwa 2003, 2004a). The Department of Wildlife and National Parks (2009–2015) has recorded the highest number of visitors and safari—based tourists to northern parks (Table 6.1) with the highest numbers of wildlife tourists entering and visiting the CNP more than the Moremi Game Reserve (MGR) and Nxai Pan National Park and Makgadikagdi Pan Game Reserve (MPGR) (Table 6.2).

As illustrated in Table 6.2, in a six year period (2009–2014) the number of private or self-drivers[1] and mobile tour operator[2] clients entering the CNP almost doubled. It is interesting to note that the total number of all types of visitors/tourists increased from 184,677 in 2009 to 267,274 in 2014 with a slight drop in 2015 (Table 6.2).

According to Mbaiwa (2004b) if large number of safari vehicles and nature-based tourists visit one place of interest at the same time they cause problems and disruptions such as noise pollution, trampling, littering and disturbance of animal during feeding and resting (Moswete and Mavondo 2003; Mbaiwa 2004a; Mogende 2013).

There has been concern of tourist vehicles congestion (Fig. 6.1) in the CNP particularly the Chobe River Front route (Mafa and Habala 2011a, b). Consequently, increase in visitor and tourists numbers to the CNP especially during peak season and certain months in a year (DOT 2009; Table 6.1) has prompted the Department of Wildlife and National Parks (DWNP) to introduce the Upper and Nogatshaa routes within the northern part of the CNP. This was a strategy to disperse (spread) safari/wildlife tourists within the park so as to reduce vehicle congestion along the "Chobe river front". Artificial water-points were constructed at certain locations along the Upper route and Nogatshaa route as a strategy to attract animals to these routes and increase attractiveness of the routes to the tourists. A colour-coded

[1]Refers to visitors using their own vehicles to enter the park, e.g., (i) Visitors to Botswana who have entered the country using their own vehicles (ii) Residents and citizens of Botswana using their own vehicles to enter the park.

[2]This category is made up of those operators who convey visitors into the Chobe National Park, but are not necessarily linked to a fixed camp, lodge or hotel in the Chobe and Kasane area. It comprises of visitors staying at lodges and hotels who are conveyed by mobile operators to whom game activities have been outsourced by these lodges and hotels.

Table 6.1 Wildlife and safari tourism visitor numbers to Botswana's northern Parks (2007–2015)

Year	CNP[a]	MGR[a]	NPNP/MPGR[a]	Total
2007	19,1254	28,684	7462	227,400
2008	100,540	24,598	11,775	136,913
2009	0	0	6004	6004
2010	76,160	22,232	1506	99,898
2011	191,830	35,475	11,817	239,122
2012	100,280	35,741	32,022	168,043
2013	241,650	40,976	23,710	306,336
2014	131,731	33,224	13,593	178,548
2015	235,832	8556	4059	248,447

0—No data
[a]Chobe National Park (CNP); Moremi Game Reserve (MRM), and Nxai Pan National Park (NPNP) & Makgadikgadi Pan Game Reserve (MPGR)
Source Department of Wildlife and National Parks (2015)

Table 6.2 Chobe National Park wildlife tourism statistics for 2009–2015

Type of visitors	2009	2010	2011	2012	2013	2014	2015
Private visitors	18,089	25,373	25,577	29,227	32,462	36,563	34,854
Mobile tour operator clients	47,277	49,408	53,438	63,179	78,257	90,515	78,128
Inside fixed camps/lodge clients	18,559	14,655	17,885	18,934	19,570	20,298	18,303
Outside fixed camps/lodge clients	86,982	96,975	96,900	100,195	111,361	118,763	104,547
Non-fee paying tourists	13,770	10,480	11,091	2132	1788	1135	1803
Total	184,677	196,891	204,891	213,667	243,438	267,274	237,635

Source Department of Wildlife and National Parks (2015)

token of 10 cm in diameter was used to identify the routes and was hung on the inside rear view mirror for ease of view by wildlife authorities who undertook law enforcement patrols in the park. A booking system for tourists and tour operators (groups) utilizing the three routes was introduced in order to protect the park environment and ensure nature based tourism sustainability (DTMC 1993; DWNP 2000; GoB 1990; Mafa and Habala 2011a, b). The booking was to be accepted for a period not less than a week in advance, and confirmation was to be sent back to the client two days before the booking period expires. The booking system was used as a strategy to manage, monitor and regulate the number of safari tourism vehicles entering the CNP, and for people visiting the Chobe River Front in particular (Sebopeng 2010).

Hence, the main aims of the study were to determine the frequency of use of the Chobe River Front, Nogatshaa and Upper routes by tour operators and tourists and to determine the effectiveness of the decongestion strategy introduced by the Department of Wildlife and National Parks as a management tool for the Chobe National Park. The strategy was introduced to ease off wildlife/safari tour operators and tourists on the existing River Front route. Using the case study of the CNP, the following issues and questions are explored: (i) What is the frequency of use of the Riverfront, Nogatshaa and Upper routes? (ii) Are the CNP River Front, Nogatshaa and Upper routes utilized equitably by wildlife/safari tour operators and tourists? (iii) What are possible challenges encountered by operators and tourists? And (iv) How effective is the CNP's decongestion strategy on the newly created routes?

6.3 Background and Study Area

The study took place at the Chobe National Park (CNP), which is located in the Chobe district in northern Botswana (Fig. 6.2). It is the first national park in Botswana and covers an area of 10,566 sq/km (GoB 2001). The CNP is renowned

Fig. 6.1 Sedudu main gate tourists waiting to enter the Chobe National Park, Botswana. *Photo* K. Nkape

for the greatest concentrations of African elephants (Fig. 6.9) estimated at 120,000 (Botswana Review 2015; DWNP 2010); charismatic species of game (known as the Big 5[3]), and high populations of birds found in the southern African region. Its uniqueness is the abundance and variety of wildlife (Fig. 6.3); the waters of the Chobe River and the rich biodiversity of this well-kept and preserved national park (Botswana Review 2005; GoB 2001; Mafa and Habala 2011a, b). The park is comprised of four distinct ecosystems: the Chobe River Front with floodplains and thick teak forest, Savuti marsh in the west; the Linyanti swamps in the North West and the hot dry winter land in between (Deloitte & Touche Consultants [DTMC] 1993; Ecosurv 2000; GoB 2001). The Serondela and CNP River front zones are the popular wildlife and nature tourists' destinations within the CNP (Chobe National Park Management Plan 2009–2015; Magole and Gojamang 2005).

The focus for this study is on the CNP's River Front (Fig. 6.2). This part of the park runs along the Chobe river bank, and covers about 17 km from Sedudu main gate to Serondela campsite (Fig. 6.3) (DWNP 2010). The Chobe River Front finds itself as a centre of attraction for game viewing due to its scenic beauty and diversity and abundance of wildlife. It is a small area of the CNP which acts as a magnet for wild animals due to its year-long availability of surface water provided by the perennial Chobe River. A wide diversity of wild animals converges in large numbers, especially during the dry season, along the Chobe River to drink. These animals include the African elephant (*Loxodonta africana*), Cape buffalo (*Syncerus caffer*), Giraffe (*Giraffa camelopardalis*), African Lion (*Panthera leo*), Hippopotamus (Hippopotamus amphibious) and many African ungulates, primates and water birds (GoB 2001; Mafa and Habala 2011a, b). It is this natural beauty of lush green and plenty free roaming wild animals that come to the river bank which attract safari tour operators; hence, they bring their clients to this route for quality nature experience.

The Nogatshaa route starts from the Sedudu-Ngoma gate and meanders along the southern boundary of the CNP towards Nogatshaa for about 60 km. Similar to the River Front, this route is a centre of attraction for game viewing and safari tourism due to its aesthetic beauty of the natural landscape and animal population. There are three boreholes along this route and some animals such as elephant, eland, zebra and buffalo are usually seen along the route and at the boreholes (DWNP 2000; Mafa and Habala 2011a, b). There is however, limited access that connects to the existing boreholes leading to long distant travel along this route (Ecosurv Pty 2000).

The Upper route is about 23 km and starts at Sedudu gate and runs left at the Sedudu valley. The route runs parallel to the Chobe River Front and intersects the fire break and winds back through the Ngoma-Sedudu tarred road

[3]The BIG 5 refers to the big 5 animals—Buffalo, elephant, leopard, lion, rhinoceros that are found in relatively high numbers in Botswana and southern and eastern Africa, hence the Big 5 is marketing slogan of these animals that tend to excite tourists (see GoB 2001 p. 195–210).

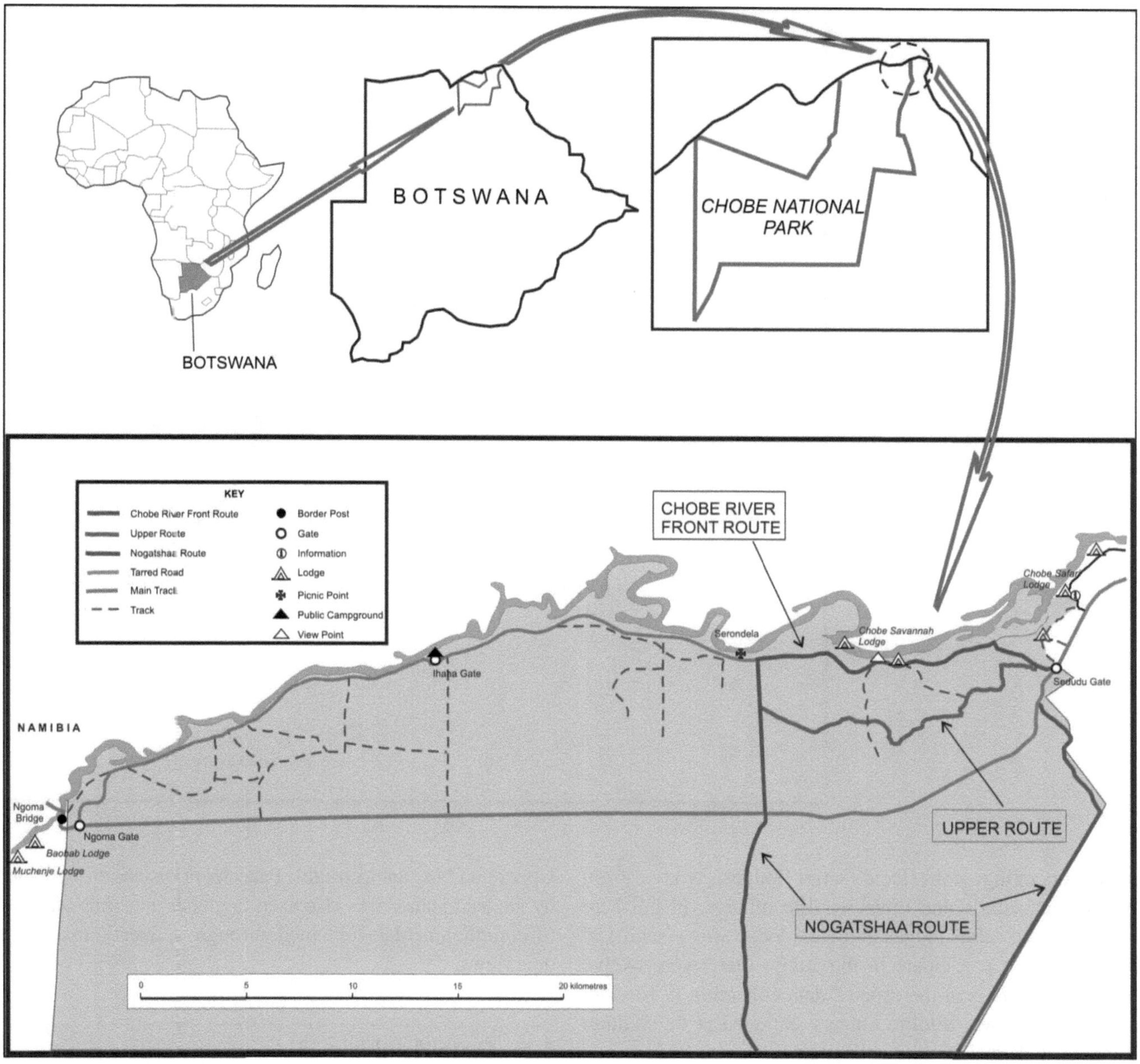

Fig. 6.2 Study map and location of the Chobe National Park and routes (*Source* by G. Koorutwe)

(Fig. 6.4). The vegetation is dominated by Baikea woodland (GoB 2001), and wild animals foraging in this area include elephant, tshessebe, zebra and the rare sable antelope (Mafa and Habala 2011a, b).

6.4 Research Methods

Data for this study was collected by means of a semi-structured questionnaire or survey with open and close-ended questions. The field work was conducted during the month of June in 2013. Local guides from fixed safari lodges and mobile tour safaris were contacted in Kasane and asked to participate in the study. All the twenty-five individual professional guides contacted agreed to take part in the study. Questionnaires included open and-close-ended questions about the frequency of use, potential tourism impacts upon the three routes of the Chobe River Front (River Bank), Nogatshaa, and Upper Routes. A face-to-face interview was also used to solicit information from wildlife officials (guides and patrols in the CNP) of the Department of Wildlife and National parks. Respondents were informed about the purpose of study and thereby assured of the importance of their views and opinions about the use of the nature tourists' within the routes at the CNP. Convenience sampling of 25 professional tour guides in total, with ten

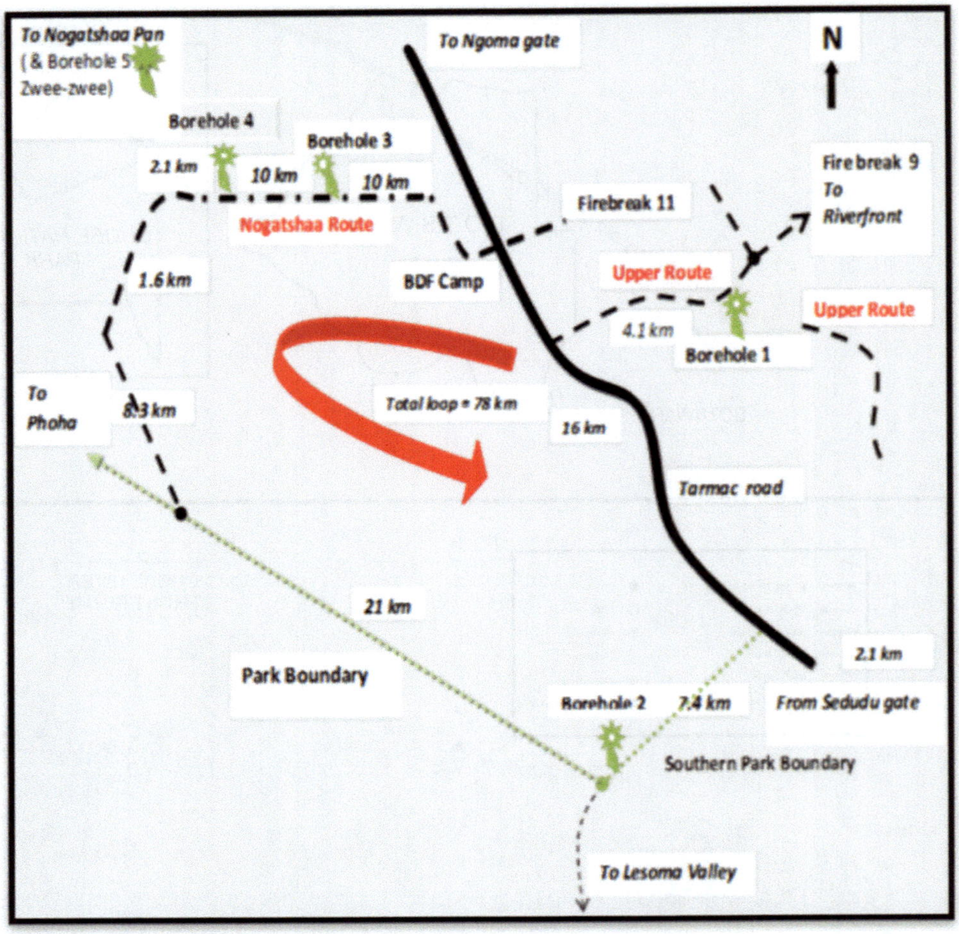

Fig. 6.3 Sketch showing tourists' vehicle routes within the Chobe National Park (Mafa and Habala 2011a, b)

guides operating from fixed safari lodges, twelve from mobile tour safaris and three wildlife officers. In all there were ten fixed safari lodges and each lodge was visited and guide asked to participate in the study; there were twelve mobile tour safaris at the time of data collection in Kasane; and there were ten wildlife officers stationed at the Sedudu gate (Fig. 6.4) during data collection.

All professional tour guides were targeted for this study, requested to participate in the survey and were interviewed because they were regarded as experienced and knowledgeable, and were the ones conducting game drives and sightseeing rides for nature-based tourists in the park. They are also familiar and conversant with the use of routes and trails within inside the park. Purposive sampling method was used to select wildlife officials stationed at Sedudu gate for interview.

The Sedudu gate (Fig. 6.4) forms the main entrance into the Chobe National Park. In all, the park manager and two experienced officers were interviewed. The wildlife officers were interviewed because they patrol the park and know the frequency of the utilization of the Chobe river front, the Upper and Nogatshaa routes. Participant observation method by the lead author was also used to gather data that was used to compliment data obtained through a questionnaire and interviews.

6.5 Data Analysis

Quantitative data was analyzed using Excel software and statistical package for the social science (SPSS) to generate descriptive statistics as the study was exploratory in nature. Pivot tables were used in the presentation of results. In analyzing the face-to-face interviews, content analysis was employed for analysis of qualitative data. Content analysis helps to identify meaning in the words as were said by the interviewee. The transcribed interview data was reorganized, similar words, phrases and the associated meanings were grouped and set aside. Then similar words, phrases, and sentences were matched and contrasted to each other. This procedure helped us to identify emergent themes in the interview data.

Fig. 6.4 Sedudu gate to the Chobe National Park information for visitors. *Photo* K. Nkape

6.6 Results

6.6.1 Demographic Profile of Respondents

The findings of the study present opinions and views of local guides from fixed safari lodges and mobile tour safaris in Kasane and wildlife personnel with specific reference to how they perceive frequency of use of the CNP river front, Nogatshaa and the Upper routes as well as the effectiveness of the decongestion strategy introduced by the DWNP to spread tourists and visitors in the CNP and reduce vehicle congestion at the popular CNP river front. A total of 25 professional guides with fixed safari lodges and mobile tour safaris were interviewed. There were more males (92%) than females (8%). Most of the respondents were younger with age ranging from 25 to 35 (60%) and older persons were fewer with age 36 and more (40%). In order to establish length of guiding experience, the respondents were asked about how long they have been involved in wildlife/safari guiding. A significant number of the guides (76%) had guiding experience ranging from 6 to 22 years, while a sizable proportion (24%) have guiding experience ranging between 1 and 5 years at the Chobe National Park River Front (Fig. 6.5).

In order to establish whether the three routes of the Chobe River Front, Nogatshaa, and Upper routes at the CNP were utilized equitably. The respondents were asked if the Chobe River Front, Nogatshaa and Upper routes were equitably used for wildlife based tourism purposes. The 'use' variable was measured by a binary question of yes/no. Almost all the respondents stated that the three routes were not used equally by mobile tour operators and fixed safari lodge or nature tourists. All the local guides observed that the Chobe River Front was the most popular, preferred and utilized when

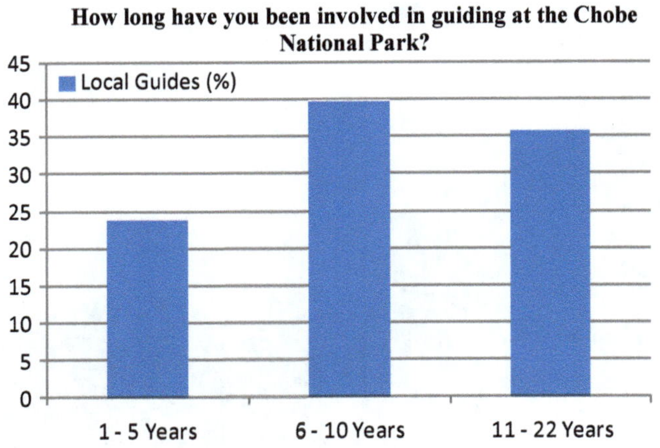

Fig. 6.5 Number of years respondents worked as professional guides at the CNP. *Source* the authors

compared to the Nogatshaa and Upper routes. Further, an open-ended question was asked to seek the reasons behind differences in use of the three routes.

The overwhelming majority of respondents from fixed safari lodges and mobile tour safari demonstrated their knowledge and experience as regards why the Chobe River Front was the most popular and highly utilized route in the CNP. The Chobe River Front route was described as:

> Mobile tour Safaris:
> The route is very attractive because of its abundance of wild animals, the natural landscape and the ever green Sedudu island. . . because animals are concentrated to the river front, . . . the route is also very short [distance] and it is not strenuous to transverse.

> Wildlife officer:
> Most game is found near to the Chobe river - … concentrations of animals or game are found within the river bank – with that given time frame for a game drive . . . for the game drive it is easy; it saves much time since time given for one game drive is short. . .

> Mobile tour Safaris:
> Most of the wild animals are water dependent, so almost all animals are seen at this part of the park most of the time. . . one may think that it is because Chobe river front is not as far as Nogatshaa

> Fixed lodge local guide:
> . . . opportunity to encounter many wild animals when they come to drink. Also to see the big cats (leopards, lion, cheetah) along the Chobe river as they follow their prey. . .

> Fixed Safari lodge:
> [Chobe River Front] Time allocated for game viewing is too short so when using the Chobe River front route we can manage to cover it in three (3) hours and have clients encounter a lot of game.

In addition, the respondents were asked as to which of the three routes (River Front, Nogatshaa and Upper routes) were the least preferred for wildlife/safari tourism. Of all the local guides, 79% indicated that Nogatshaa was the less popular, followed by Upper route with 17% and the Chobe River Route at 4%. A follow up open-ended question was asked to gather reasons why such a route was the least preferred. A large number of the respondents said that Nogatshaa was least preferred because it is a long route and too far to cover in a 3 h period as prescribed by the park management.

> Fixed lodge:
> [Nogatshaa] We are given three (3 h) in the park as per park authority and to use this we cannot reach Nogatshaa in 3 h: . . . the route is far and cannot be covered in a short time: it needs 6 or 7 h to complete.

> [Upper] Not many wild animals that can be encountered or seen in that route. Time allocated for one game drive is not sufficient to cover the whole route.
> [Nogatshaa] The route is long and difficult to cover and enjoy wildlife encounters in 3 h – as per the policy at the CNP . . . The Nogatshaa route is long and not well maintained.

> Mobile tour safaris:
> [Nogatshaa] There are less animal spotting and encounters; the route is not as attractive as the other because there is no water and the road is not well maintained . . . no waterhole closer to the road. There is only water in the pan at Botswana Defense Force camp which is far from the road.

> [Nogatshaa] Not many game or wild animals to see because the area is dry, the route too long and far and there are no stretch points.

> Wildlife manager:
> [Upper] . . . also natural waterholes dry during winter . . . less game is seen or spotted during dry season.

The respondents were asked an open-ended question to state challenges likely to encounter when using the three routes. Table 6.3 shows overall issues observed by the respondents. As indicated, challenges included traffic and/or too many tourist vehicles (76%) on the River Front Route (Fig. 6.1), 48% said the Nogatshaa route was too long and/far. while others mentioned that there was dense vegetation that obstruct good sighting and encounter of animals at the Upper route.

Upon the number of challenges that the respondents outlined above (Table 6.3), a question was asked to establish what they (guides) would recommend as per use of the routes. Slightly more than half of the respondents recommended that the DWNP must create more routes in the park that will link Nogatshaa and the upper route as well as some that will connect the Upper and Chobe River Front. Subsequently, a sizeable number of the respondents made a suggestion that extra artificial waterholes be introduced to attract animals especially Nogatshaa and the Upper route. Others indicated and recommended the need for mobile tour safaris and private self-drives going on Nogatshaa to be allocated more time (>3 h) because they have more time than those from fixed lodges.

Table 6.3 Challenges likely to be encountered when using the three routes in the CNP

Statements/Challenges	Chobe River Front route	Nogatshaa route
Dense vegetation, not easy to see animals	0[a]	0
Sandy, uneven road & not well maintained	1 (4%)[b]	3 (12%)
Traffic/too many vehicles, congestion	**19 (76%)**	0
Route is too long/far to cover in 3 h	0	**12 (48%)**
Less animals to encounter or spot	0	**6 (24%)**
Few feeder routes	0	2 (8%)
Less attractive natural landscape	0	0
No stretch point/Rest	0	1 (4%)
Insufficient water/few boreholes	0	1 (4%)

Source The authors
0[a]means 'not a challenge for that route'; [b]percentages show the proportion of total encounters for the three routes for each item/statement

Fig. 6.6 Elephants with calves at the Chobe National Park River Front, Botswana. *Photo* N. Moswete

6.7 Discussion

The data supported our questions, observations, and predictions that the Chobe River Front was more frequently used by private or self-drive, fixed safari lodge and mobile tour safaris visitors and tourists. In case of whether the three routes: the Chobe River Front, Nogatshaa and Upper routes were utilized equitably for wildlife tourism safaris; the study uncovered that the River Front route was in constant high demand, and thereby heavily utilized compared to Nogatshaa and Upper routes. Consequently, the Chobe River Front was the most preferred route as compared to the Nogatshaa and Upper routes. The reasons advanced by the respondents both long term and professional guides and wildlife officials were that the river front is endowed with a high number and variety of wild animals. Also, both aquatic and terrestrial species are attracted to the Chobe River by water. It is therefore easy to locate groups of trophy animals such as lions, leopards, elephants and buffaloes as well as many other predators as they tend to follow their prey to the river. It is because the animals as well are attracted by the availability of all year round evergreen forage and water of the Sedudu Island (Figs. 6.6 and 6.7) (GoB 2001).

According to the new DWNP decongestion strategy, each game viewing tour in the park is allocated a maximum of 3 h at a time regardless of the route chosen as a mitigation tool.

Fig. 6.7 Chobe National Park, some species found: Giraffe (Giraffa camelopardalis), kudu (Tragelaphus strepsiceros) and warthog (Phacochoerus aethiopicus). *Photo* K. Nkape

Thus, all the respondents observed that the time allocated for game drives is adequate for the Chobe River Front, but not for the other two routes. The local guides for example, reasoned that it is because scores of animals (number and variety) can be viewed and more encounters be enjoyed within the three hour slot as stipulated in the DWNP decongestion strategy document (Mafa and Habana 2011a, b). Many regional and long-haul wildlife tourists who choose to visit Botswana's northern parks come to see wildlife that tends to be found in greater numbers at the CNP (Botswana Review 2015/16; Magole and Gojamang 2005).

The Chobe River route is observed as the best part of the CNP because tourists also enjoy viewing sunset; (Figs. 6.8 and 6.9), the beautiful landscape of the Chobe River and the open grassland of this part of the park. In addition, the River Front route is relatively short in terms of length (and time), and can be covered within the 3 h allocated for game drives and sightseeing in the park. The respondents also observed that the route is well serviced compared to the Nogatshaa and Upper routes. As alluded by some of the respondents there are ample stretch points and sun-downer spots at Serondela (Mafa and Habala 2011a, b; Mogende 2013), where tourists and visitors have the chance to relax and stretch their legs when using the River bank route. Above all, the combination of water, evergreen forage and many ungulates make the River route more interesting and experiential to wildlife- based and safari tourists than the Upper and Nogatshaa. At most, the natural attractiveness and aesthetic beauty of the Chobe River Front route has led to problems of vehicle congestion and crowding during wet and dry season respectively (Mogende 2013).

With reference to the question on possible challenges encountered when using the three routes within the CNP there were varying views and opinions. For the Chobe River Front, respondents indicated that there were some challenges associated with its utilization for wildlife tourism safaris. The wealth of wild animals encountered particularly when a predator has killed its prey and a group of lions are found devouring the carcass, or lions and wild dogs found fighting for the meat has contributed to illegal behavior where vehicles are veered off-road to see and photograph. In addition, many animals freely foraging or resting and the lush green of the Sedudu Island and its serenity have all contributed to vehicle and tourist congestion on the River bank route especially if some lions or cheetahs are encountered. Respondents indicated that during peak season the tourist vehicle congestion becomes high and uncontrollable especially around one group of animals (e.g. leopards, elephants). Also, the route becomes too congested such that even the tourists complain about dust especially during peak or 'open' season (Mogende 2013).

Among the three wildlife routes in the CNP, the least preferred and utilized was the Nogatshaa when compared to Upper and River Front routes. The respondents observed that

Fig. 6.8 Sunset and tourist driving along the River bank at the Chobe National Park. *Photo* N. Moswete

Nogatshaa route was too long and far (distance to be covered in one tour) thereby compelling the guides to drive with less and brief stops to cover the 3 h allocated to one game drive per session (DWNP 2001). Based on their views, many guides thought the route is less attractive as it has less wild animals to see compared to the other routes. It was also observed by many that at the available water holes, there are no observation posts and the route is not well maintained. A portion between Nogatshaa wildlife camp and nearby Nantanga camp becomes too muddy and impassable during wet season. However so, most interesting parts of the park including some spots within the River bank route also become difficult to transverse during wet season but becomes part of the fun as tourists are not looking for the state of the art routes when in a game watching expeditions in national parks and game reserves.

The majority of the respondents chose the Upper route as the second most liked, preferred and utilized after the River Front. Its scenic natural beauty, lots of wild animals to spot particularly when they come to drink at the existing natural water holes during dry season. The route is popular and preferred compared to Nogatshaa due to the fact that during game drives it eventually loops onto the Chobe River Front route (Fig. 6.3) towards its terminal point. This has made the wildlife and safari tourists categorizing it as their second best choice after the River Front. However, the respondents' highlighted some challenges they face when using the Upper route at the CNP. The route is too long for the three hour period given for a single game drive. Almost all the respondents mentioned that the Upper route has dense vegetation that makes wild animal sighting, encounter and viewing difficult. The route has no stretch points and observation posts at water points.

Regardless of the views and opinions from the interviews of professional guides at the CNP, there are signs of benefits of the decongestions strategy as espoused by some

Fig. 6.9 Tourists watching elephants during boat cruise experiditions on the Chobe River Bank. *Photo* N. Moswete

professional guides and wildlife officials of the Chobe National park. However, this park generally received thousands of day visitors from tour operators coming from nearby Zambia and Namibia (Table 6.1), and the numbers have been increasing in each year. Since the decongestion strategy was introduced as a mitigation tool in 2011 there has been a relatively sensible spread of wildlife and safari enthusiasts within the CNP. Overall, this has eased off the number of visitors not at the CNP but only at the Chobe River Front route and at the viewing spot. Additional benefits include the DWNP mandate of providing additional water holes along Nogatshaa and Upper routes of the CNP to avail drinking water to animals especially during dry seasons.

Recurring droughts in Botswana have led to increased mortalities of some water dependent species in the park, hence the need to provide artificial waterpoints in the CNP. Provision of water for animals is not necessarily for tourism, (though they add value to wildlife or game viewing) but as it gets very dry and all natural water holes dry up. In Botswana's PAs (wildlife and nature) there is less provision of water for animals. It is so far known that having fewer to no artificial water sources is supported by some reactionary pro-conservationists. However drawn from Botswana internationally recognized conservation practice only few waterpoints are availed at the CNP animals for drinking and mud swimming buy high numbers of elephants and other water dependent species. This study reveals that only a small number of tour guides from fixed lodges and mobile safari tours argued for provision of additional waterpoints on the Nogatshaa and Upper routes to attract animals because their clients pay a lot of money (e.g., for lodging and game watching trips). This could be because they lack knowledge on the effects of excessive waterholes in a conservation area such as the CNP (Owen-Smith 1996).

To avoid repetition of benefits of wildlife tourist safaris (Akama and Kieti 2003; Humavindu and Barnes 2003; Silva and Mosmane 2012; UNWTO 2014), we cannot forgo the financial contributions to conservation that accrue from user fees at the CNP. Monetary benefits are generated from gate takings by visitors, camping at DWNP wilderness campsites, use of wilderness safari trails and more. Research and filming at the CNP generate more as for example, filming that include advertising, feature films, documentaries per person is BWP5000 (USD$500) per week for a non-resident, and (DWNP 2016). For conservation alone, there is increased number of some animal species in the park and the routes have provided some visibility of some species that do not come to the river.

6.8 Conclusion

To reiterate, the main aim of the study was that the Upper and Nogatshaa routes were created specifically to alleviate tourists' traffic congestion from the Chobe River bank routes and at the wildlife watching area. The research sought to find out whether these two routes were utilized equitably by fixed safari lodges and mobile tour safaris as compared to the Chobe River Front. Above all, to assess whether the decongestion strategy was effective as was proposed and implemented.

We conclude that the respondents had vast experience (3–22 years) and thus the local tour guides interviewed were knowledgeable and experienced about the utilization of the routes at the CNP: as such their perceptions and views can be trusted. The study revealed that the Chobe River Front, Nogatshaa and the Upper routes were not equitably utilized. It appeared, however, that despite the introduction of the Upper and Nogatshaa routes a large number of wildlife and safari tour operators and self-drive tourists still chose the Chobe River Front far more than the other two routes. Also, the Chobe River Front route was the most preferred; seconded by the Upper route, while Nogatshaa was the least preferred route. As such, there were still challenges of soaring visitor numbers using the River bank route because of proximity to the river, and short time to cover it in a day.

The results indicate that the decongestion strategy that was meant to alleviate tour operators and tourist's traffic pressure from the Chobe River Front as well as spreading tourists' within the CNP has possibly not served the intended purpose. There could be problems resulting from decongestion mitigation strategy implementation loopholes or flaws by those DWNP official stationed at the Sedudu gate. Hence, the need for further interrogation of the congestions and crowding that are still observed at the Chobe River route.

Overall, the mitigation mechanism of the three routes appears to be somewhat working because the DWNP uses a booking system (online and telephone) which has so far contributed considerably in that when tour operators and other visitors call in they have to indicate the route they wish to use and pay for, indicate number of clients as well as the number and type of vehicles to use in the park a day. However, the findings from this study is contradictory, leaving only questions of why there are still congestion of vehicles, crowding of tourists vehicles around the same group of animals during game drives in the park. Who is not adhering to the CNP code of conduct? Possibly there is an urgent need for further enquiry into the whole system of management at the CNP and authority.

Indeed we see an urgent need for further research into the CNP so as to unearth why there are still crowding issues on the River route and why change is slow. The CNP is a critical resource for Botswana and adjacent communities because wildlife watching and safari tourism contribute to the local economy of Chobe, Kasane and the Okavango Delta through jobs creation, and also to conservation of resources (wildlife and other natural resources). However, if the natural environment that supports wildlife-based tourism industry is not carefully managed, resources protected and tourism activities monitored to ensure sustainable practice conservation based tourism for the CNP will cease to exist.

The study revealed that there are benefits associated with the newly created vehicle decongestion routes at the CNP. There has been indication that the two new routes have relatively relieved the Chobe River Front from tourist vehicle pressure; lessened the congestion of tourist vehicles particularly at animal sightings or encounters of predators (leopard, wild dogs, lions), and also creation of a few waterpoints along the Upper and Nogatshaa routes appears to have contributed towards spreading of wild animals over a large area thereby alleviating competition for foraging and water and thereby reducing grassing pressure at the CRF. However, there are still issues of congestion during game drives particularly along the River bank route and at the CRF viewing site. Hence, we still can make a general statement that the decongestion strategy that was meant to alleviate tour operators and tourists' traffic pressure from the Chobe River Front has possibly not achieved the intended purpose as yet. Managerial implications include improving the use of Upper and Nogatshaa routes by providing better facilities and service to all types of visitors and tourists to make it appealing. It is recommended that the park management should consider devising a strategy to attempt to demarket the Chobe River Front route to reduce visitor pressure, vehicle congestion and alleviate negative impact on animals and associated resources of the CNP.

We conclude that our study could be used as a baseline to do further surveys on nature based tourists or wildlife visitors at the CNP to solicit their views and enquire if they too feel or visualize crowding during game drives, and if that affects their satisfaction and experience levels of wildlife watching and other related wilderness activities in and adjacent to the park.

6.8.1 Recommendations

For Nogatshaa and Upper Routes: (i) in order to alleviate Chobe River front from tourist congestion, the Department of Wildlife and National Parks (DWNP) should improve the utilization of the Nogatshaa and Upper route by providing other incentives for utilizing the routes. Such could carry out a study to find out if there is need for providing water points along the routes to attract some animals from the Chobe River Front but this should be thoroughly checked for sustainability; (ii) DWNP must construct safe stretch points (rest stop) and sun-downer spots along Nogatshaa route where tourists could briefly relax and stretch their legs than one long drive to the end; (iii) DWNP should consider lightly increasing game viewing time for the Nogatshaa and Upper routes especially for fixed safari lodges.

References

Akama JS, Kieti DM (2003) Measuring tourist satisfaction with Kenyan's wildlife safari: a case study of Tsavo West National Park. Tour Manag 24(1):73–81

Barnes JI (1996) Economic characteristics of the demand for wildlife viewing tourism in Botswana. Dev South Afr 13(3):377–397

Barnes JI (2001) Economic returns allocation of revenues in the wildlife sector of Botswana. S Afr J Wildl Res 31(3–4):141–153

BRB (Botswana Review Board) (2014) Botswana review of commerce and industry, 33rd edn. R and T Directories (pty), Limited, Gaborone

BTDP (Botswana Tourism Development Programme) (2000) Botswana tourism master plan. Department of Tourism, Commission of the European Union, ARCA Consulting, Gaborone

BTDP (Botswana Tourism Development Programme) (2003) Botswana national ecotourism strategy. Department of Tourism, Gaborone. (Government Printer)

Department of Wildlife and National Parks (DWNP) (2010) Revised management plan. Chobe National plan, DWNP, Gaborone, Botswana

DOT (Department of Tourism) (2009) Tourism facts and figures. Ministry of Environment, Wildlife and Tourism, Gaborone, Botswana

DTMC (Deloitte and Touche management consultants) (1993) Chobe National Park management plan volume 2: management and development programmes. Commissioned by the Kalahari Conservation Society with Funding from the United States Agency for International Development, USAID; project # 690-0251-33

DWNP (Department of Wildlife and National Parks) (2000) Chobe national park draft management plan. DWNP, Gaborone, Botswana

DWNP (Department of Wildlife and National Parks) (2001) Chobe national park final draft management plan. DWNP, Gaborone, Botswana

DWNP (Department of Wildlife and National Parks) (2007) Parks and reserves tourism statistics annual report 2007. DWNP, Gaborone, Botswana

DWNP (Department of Wildlife and National Parks) (2016) Fee structure for national parks and game reserves. DWNP, Kasane, Botswana

Eagles PFJ, McCool SF (2000) Tourism in national parks and protected areas: planning and management. Wallingford, CABI Pub, pp 97–149

Ecosurv (Pty) (2000) Chobe River Front Management Plan (CRFMP). Gaborone Government Printer, Botswana

Ewert AW, Dieser RB, Voight A (1999) Conflict and the recreational experience. In: Jackson EL, Burton, TL (eds) Leisure studies: prospects for the twenty-first century. Venture pub, Pennsylvania, pp 335–345

Finnessey L (2012) The negative effects of tourism on national parks in the United States. Bachelor of Science Honors Thesis, Johnson & Wales University [Online] http://www.scholararchive.jwu.edu/cgi?article. Accessed July 2016

Garekae H, Thakadu OT, Lepetu J (2016) Attitudes of local communities towards forest conservation in Botswana: a case study of Chobe forest reserve. Int Forest Rev 18(2):180–191

GoB (Government of Botswana) (1990) The national conservation strategy, Government Paper No. 2 of 1990. Government Printer, Gaborone

GoB (Government of Botswana) (2001) Botswana national atlas. Department of surveys and mapping, Government Printer, Gaborone

Green R, Giese M (2004) Negative effects of wildlife tourism on wildlife. In: Higginbottom K (ed) Wildlife tourism: impacts, management and planning. CRC sustainable tourism. [Online] available, http://www.sustainabletourism.publisher-site.com (August 2016), pp 81–97

Green R, Higginbottom K (2000) The effects of non-consumptive wildlife tourism on free-ranging wildlife: a review. Pac Conserv Biol 6:183–197

Hachileka E (2003) Sustainability of wildlife utilization in the Chobe District, Botswana. S Afr Geogr J 85(1):50–57

Hemson G, Mills MGL, Johnson PJ (2009) Community, lions, livestock and money: a spatial and social analysis of attitudes to wildlife and the conservation of value of tourism in a human-carnivore conflict in Botswana. Biol Conserv 142:2718–2725

Higginbottom K (ed) (2004) Wildlife tourism: impacts, management and planning. Common Ground Publishing, Altonia

Higginbottom K, Tribe A (2004) Contribution of wildlife tourism to conservation. In: Higginbottom K (ed) Wildlife tourism: impacts, management and planning. Common Ground Publishing, Altonia, pp 99–124

Higginbottom K, Northrope C, Green R (2001) Positive effects of wildlife tourism on wildlife. Wildlife tourism research report series: No 6. Status assessment of wildlife tourism in Australia series. CRC for Sustainable Tourism Pty Ltd [Online]. Available: www.crctourism.com.au. Accessed 24 Aug 2016

Holloway JC (1998) The business of tourism, 5th edn. Addison Wesley Longman Ltd, New York

Humavindu MN, Barnes JI (2003) Trophy hunting in the Namibian economy: an assessment. S Afr J Wildl Res 33(2):65–70

Kalisch D, Klaphane A (2007) Visitors' satisfaction and perception of crowding in a German National Park: a case study on the Island of Hallig Hooge. For Snow Landscape Res 81(1&2):109–122

Kaltenborn BP, Nyahongo JW, Kideghesho JR (2011) The attitudes of tourists towards the environmental, social and managerial attributes of Serengeti National Park, Tanzania. Trop Conserv Sci 4(2):132–148

Lepetu J (2007) Socio-economic impact and stakeholder preference to conservation of forest reserves: a case study of Kasane Forest Reserve, Botswana. Unpublished PhD dissertation, University of Florida, USA

Lepp A (2007) Residents attitudes' towards tourism in Bigodi village, Uganda. Tourism Manag 28:876–885

Lepper CM, Goebel JS (2010) Community-based natural resource management, poverty alleviation and livelihood diversification's case study from Northern Botswana. 27(5):725–739

Lindsey P (2010) The future of wildlife based land use in Botswana. Available: http://www.africanwildlifeconservationfund.org/wp-content/uploads/2011/09/Lindsey-2010-The-future-of-WBLUs-in-Botswana.pdf. Accessed 4 Sep 2016

Lindsey PA, Alexander R, Frank LG, Mathieson A, Romanach SS (2006) Potential of trophy hunting to create incentives for wildlife conservation in Africa where alternative wildlife based land uses may not be viable. Anim Conserv 9:283–291

Lindsey PA, Roulet PA, Romanach SS (2007) Economic and conservation significance of the trophy hunting industry in sub-Saharan Africa. Biol Conserv 134(4):455–469

Lovelock B (ed) (2007) Tourism and the consumption of wildlife: hunting, shooting and sport fishing. Routledge, USA

Mabunda DM, Wilson D (2009) Commercialisation of national parks: South Africa's Kruger National Park as an example. In: Saarinen J, Becker F, Manwa H, Wilson D (eds) Sustainable tourism in southern Africa: local communities and natural resources in Transition. Channel View pub, Bristol, pp 116–133

Mafa M, Habala L (2011a) Strategy to address Chobe National Park river front congestion. Prepared for the Department of Wildlife and National Parks, Chobe District, Botswana

Mafa M, Habala L (2011b) Chobe National Park decongestion strategy. Department of Wildlife and National parks, Government print, Botswana

Magole LI, Gojamang O (2005) The dynamics of tourist visitation to National Parks and game reserves in Botswana. Botswana Notes Rec 37:80–96

Magole L, Magole L (2011) Revisiting Botswana's high-value, Low-volume tourism. Tourism Anal 16(2):203–210

Manning R (1999) Crowding and carrying capacity in outdoor recreation: from normative standards to standards of quality. In: Jackson EL, Burton TL (eds) Leisure studies: prospects for the twenty-first century. Venture pub, Pennsylvania, pp 323–369

Manning R, Valliere W (2009) Crowding, conflict, and coping in outdoor recreation. In: Manning RE (ed) Parks and people: managing outdoor recreation at Acadia National Park. University of Vermont Press, Burlington, pp 109–121

Massyn JP (2010) Citizen participation in the lodge sector of the Okavango Delta. In: Spenceley A (ed) Responsible and critical issues for conservation and development. Routledge, pp 225–238

Mbaiwa JE (2003) The socio-economic and environmental impacts of tourism development on the Okavango Delta, north-western Botswana. J Arid Environ 54:447–467

Mbaiwa JE (2004a) Wildlife resource utilisation at moremi game reserve and Khwai community area in the Okavango Delta, Botswana. J Environ Manag 77(2):144–156

Mbaiwa JE (2004b) The socio-economic impacts and hallenges of a community based safari hunting tourism in the Okavango Delta, Botswana. J Tourism Stud 15(2):37–50

Mbaiwa JE (2005a) Enclave tourism and its socio-economic impacts in the Okavango Delta, Botswana. Tourism Manag 26(2):157–172

Mbaiwa J (2005b) The problem and prospects of sustainable tourism development in the Okavango Delta. J Sustain Tourism 13(3):203–227

Mbaiwa J, Bernard F, Orford C (2008) Limits of acceptable change for tourism in Okavango Delta. Botswana Notes Rec 38:92–112

Mgonja JT, Sirima A, Mkumbo PJ (2015) A review of ecotourism in Tanzania: magnitude, challenges and prospects for sustainability. J Ecotourism 14(2& 3):264–277

Mmopelwa G, Kgathi DL, Molefhe L (2007) Tourists perceptions and their willingness to pay for park fees a case study of self-drive tourist and clients for mobile tour operators in moremi game reserve, Botswana. Tourism Manag 28(4):1044–1056

Mogende E (2013) Perceived ecological impacts of the development of wildlife based tourism: a case study of the Chobe National Park River Front, Botswana. Unpublished Thesis, University of Botsana, Gaborone

Moswete N (2009) Stakeholder perspectives on the potential for community-based ecotourism development and support for the Kgalagadi Transfrontier Park in Botswana. Unpublished PhD dissertation, University of Florida, Gainesville, FL, p 292

Moswete N, Dube OP (2013) Wildlife-based tourism and climate: potential opportunities and challenges for Botswana. In: D'Amore L, Kalifungwa P (eds) Meeting the challenges of climate change to tourism: case studies of best practice. UK, Cambridge Scholars Pub, pp 395–416

Moswete N, Mavondo F (2003) Problem facing the tourism industry of Botswana. Botswana Notes Rec 35:69–77

Moswete N, Toteng E, Mbaiwa J (2009) Resident involvement and participation in urban tourism development: a comparative study in Maun and Gaborone, Botswana. Urban Forum 19(4):381–394

Munyi SW (1992) Kenya ecotourism workshop. Contours (Bangkok) 5 (8):30–32

Nyaupane G, Thapa B (2006) Perceptions of environmental impacts of tourism: a case study at ACAP, Nepal. Int J Sustain Dev World Ecol 13(1):51–61

Omphile U, Powell J (2002) Large ungulate habitat preference in Chobe National Park, Botswana. J Range Manag 55(4):341–349

Owen-Smith N (1996) Ecological guidelines for water points in extensive protected areas'. S Afr J Wildl Res 26(4):107–112

Parry D, Campbell B (1992) Attitudes of rural communities to animal wildlife and its utilization in Chobe Enclave and Mababe Depression, Botswana. Environ Conserv 19(3):245–252

Richardson J (1998) Wildlife utilization and biodiversity conservation in Namibia: conflicting or complementary objectives? Biodivers Conserv 7:549–559

Sebopeng R (2010) Management of protected areas by department of wildlife and national parks: Perfomance audit report No. 8 of 2010, Government printer, Gaborone

Shackley M (1996) Wildlife tourism. International Thompson Business Press, London

Silva JA, Mosimane AW (2012) Conservation-based rural development in Namibia: a mixed-methods assessment of economic benefits. J Environ Dev 22(1):25–50

STCRC (Sustainable Tourism Cooperative Research Center) (2009) Wildlife tourism: challenges, opportunities and managing the future. CRC for Sustainable Tourism, Gold Coast, QLD, 68 p., Retrieved 10 Aug 2016 from, http://www.sustainabletourismonline.com/awms/Upload/Resource/Wildlife%20Tourism%20Snapshot_LoRes.pdf

Tisdell C, Wilson C (2004) Economics of wildlife tourism. In: Higginbottom K (ed) Wildlife tourism: impacts, management and planning. CDC Sustainable Tourism. Victoria [Online]. Available: http://www.sustainable tourism, publisher-site.com. Accessed 16 June 2016

Valentine P, Birtles A (2004) Wildlife watching. In Higginbottom K (ed) Wildlife tourism: impacts, management and planning, pp 15–33. CRC Sustainable Tourism [Online]. Available: http://www.sustainable tourism, publisher-Site.com. Accessed 9 Sep 2016

Weaver D (1998) Ecotourism in the less developed world. CABI Press, New York

Weaver D (2001) Ecotourism. Wiley, Australia

Westin JL (2007) Effects of ecotourism on the behavior and health of red howler monkeys (alouatta seniculus) in Suriname. Unpublished PhD dissertation, University of Michigan, USA [Online]. Available: https://deepblue.lib.umich.edu/handle/2027.42/57667?show=full. Accessed 3 Aug 2016

WTO (World Tourism Organisation) (2014) Towards measuring the economic value of wildlife watching tourism in Africa. Briefing paper [Online]. Available www.unwto.org. Accessed 18 Sep 2016

WTTC (World Travel and Tourism Council) (2007) Botswana: the impact of travel and tourism on jobs and the economy. World Travel and Tourism Council, London

WTTC (World Travel and Tourism Council) (2013) Botswana: travel and tourism economic Impact. World Travel and Tourism Council, London

Westin JL (2007) Effects of ecotourism on the behavior and health of red howler monkeys (alouatta seniculus) in Suriname. Unpublished PhD dissertation, University of Michigan, USA [Online]. Available: https://deepblue.lib.umich.edu/handle/2027.42/57667?show=full. Accessed 3 Aug 2016

Part II

Ecological Learning in Wildlife Encounters

7 Let the Oceans Speak: The Synergistic Interaction Between *Intensity* and *Interpretation* During Wild Dolphin Experiences

Gayle Mayes

Abstract

The desirable outcomes of interpretation in a wildlife setting include the stimulation of visitor intent to minimise impacts and maximise support for the marine environment and its wildlife. Existing frameworks and models of wildlife tourism focus on interpretation and the cognitive domain. However, wildlife tourism encounters, especially with whales and dolphins, can be intense, emotional, powerful and profound affective experiences. Theory and models from experiential education were more relevant and applicable for exploring synergistic relationships between the affective and cognitive domains that resulted in changes in attitudes, beliefs, behaviours and actions. This paper investigates the interactive effects between the *intensity of wild dolphin tourism experiences* in the presence of *education/interpretation* commentaries on: overall satisfaction levels; changes in knowledge; and changes in pro-environmental attitudes, beliefs, intended behaviours and intended actions of participants. A multiple case study approach was used to collect data on six matched and paired dolphin tourism activities which varied in intensity and educational and interpretive input. Chi-square, one and two-way ANOVAs and MANOVA analyses (with a reduced number of factors) were used on data from 600 structured self-administered questionnaires. Results suggested that high intensity wild dolphin experiences are highly satisfying, and that the overall satisfaction of participants is not enhanced by the inclusion of a high quality interpretation component. High intensity wildlife experiences in themselves appear to have significant educative impact or power which can impede positive effects of high quality interpretation commentaries on increases of knowledge of participants and intended pro-environmental behaviour change. Minimal information in these intense instances appears to gain maximum impacts—so interpretive staff should let the oceans speak! However, as the intensity of wildlife experiences decrease, the quality and presence of interpretation in wild dolphin encounters play a greater part in enhancing overall satisfaction, knowledge of participants, pro-environmental attitudes and beliefs, and intended behaviours and actions. Therefore, as the intensity level of wild dolphin encounters decrease, guides need to deliver increasingly high quality commentaries as essential inputs of the tourist experience.

G. Mayes (✉)
Tourism, Leisure and Events, Sustainability Research Centre, University of the Sunshine Coast, Maroochydore, DC 4558, Australia
e-mail: gmayes@usc.edu.au

I. Borges de Lima and R.J. Green (eds.), *Wildlife Tourism, Environmental Learning and Ethical Encounters*, Geoheritage, Geoparks and Geotourism, DOI 10.1007/978-3-319-55574-4_7

7.1 Introduction

Wildlife-related tourism accounts for some 20–40% of international tourism, attracting millions of people around the world and contributing hundreds of billions of dollars to the global economy on an annual basis (Higginbottom 2004). Various forms of wildlife tourism have grown rapidly in recent years (Ethos Consulting 1991; Shackley 1996; Roe et al. 1997), especially marine wildlife tourism, with approximately 87 countries and overseas territories offering whale watching tours to more than nine million people (Hoyt 2001). The rapidly increasing numbers of wildlife and Cetacean tourists are mainly seeking opportunities to see wild and unique animals up close, behaving naturally in their natural environments (Moscardo and Saltzer 2005). Cetacean and dolphin tourists also seek a recreational experience in their leisure time (Driver and Tocher 1979; Dierking 1998; Muloin 1998). Dolphin based tourism activities meet all of these requirements and offer a diverse choice of opportunities for encounters that are unequalled by any other species as a whole in wildlife tourism.

Since Australia stopped hunting whales in 1979 the whale watching industry has grown rapidly, with a rapid acceleration of growth occurring in the last decade (IFAW 2004). In 2004, approximately 1.6 million people swam with, fed and/or watched Cetaceans, contributing close to A $300 million to Australia's tourism industry and injecting a major contribution to the nation's economy (IFAW 2004).

As a part of the whale watching industry, the dolphin tourism industry in Australia is also growing rapidly (IFAW 2004). In 2005, the Federal Government released guidelines for dolphin and porpoise tourism encounters which differed from the previous whale watching guidelines (DEH 2005). Until then, the encounter guidelines were the same for all cetacean tourism businesses in Australia. Each Australian state was also able to develop and adapt their interpretation of the Cetacean tourism permitting and regulatory systems, the encounter guidelines and management practices. This situation has given rise to a range of differing dolphin encounter opportunities across the Australian states, along with a lack of reliable figures on: how many operators are involved in which dolphin tourism activities; where or when they are operating; or who the operators are. In several cases, groups of dolphin watch and dolphin swim operators developed their own regulations in the absence of guidance and leadership from government and marine wildlife conservation management organisations.

With the rapid increase in demand for whale and dolphin tourism opportunities, an increasing supply of operators are offering interesting and varied opportunities to encounter an increasing range of marine wildlife species. In some instances, self-regulated operators developed best practices in managing the wildlife encounters and associated high quality education/interpretation programs. In other instances poor, and at times harmful, practices may occur. Hoyt (2001) mentioned that Australia's whale watching industry performed poorly in delivering quality education and/or interpretation components in comparison to the other 87 whale watching countries. This means that numerous valuable opportunities to positively affect the pro-environmental attitudes, beliefs, behaviours and actions of marine wildlife tourists are potentially lost annually.

Education and/or interpretation are regarded as integral tools for ecotourism and sustainable tourism practices as they foster conservation and sustainability for wildlife and the environment (Ham 1992; Gray 1993; IFAW 1995, 1997; Eagles 2001; Newsome et al. 2002; Garrod and Wilson 2003; Moscardo 2003; Higginbottom 2004). The desirable outcomes of interpretation in a wildlife setting are the stimulation of visitor intent to minimise impacts and maximise support for the marine environment and its wildlife. The framework for cetacean watching (IFAW 1997) recommended a set of practices to assist in developing a more sustainable industry. Exposing tourists to on-site interpretive experiences may have a positive effect on pro-environmental attitudes and actions (Knudsen et al. 1995; Beaumont 1999; Knapp and Benton 2005). Limited research exists on wildlife encounters and integrated interpretive messages that act as effective tools in eliciting pro-environmental behaviour change in participants (Orams 1995a, 1997b; Orams and Hill 1998; Higham and Carr 2002).

A pool of information, frameworks and models of wildlife tourism exist (Ham 1992; Orams 1997b; Moscardo 1999; Moscardo and Saltzer 2005; Ham and Weiler 2004), however, these interpretive and behaviour change practices are based predominantly on cognitive dissonance, communication, and education theories. Such theories neglect to recognise or include the affective domain and the intensity of the experience—an approach which does not align with the recommended Cetacean watching interpretive framework (IFAW 1997) that promotes an intuitive, affective and 'at the moment' approach.

More recently, interpretive researchers have explored the integration of the affective and cognitive domains and the effect of the intensity of experiences on learning and education (Eagly and Chaiken 1993; Esses et al. 1993; Pooley and O'Connor 2000; Markwell 2004; Packer and Ballantyne 2004). Wildlife tourism encounters, especially whale and dolphin tourism activities, are experiential (Maher 2005), with potential for emotional, powerful and profound experiences for participants (Ham 2004b). This set of affective components has the potential to lead to psychological and behavioural change (Tarssanen and Kylanen 2005). Herein lies the gap in current interpretive theory and practice. This gap is also identified by the review of available literature which focuses on establishing and measuring the role of the intensity of experiences in education/interpretation in natural marine environments where humans have participated in commercial encounters, specifically in the case of wild dolphins.

This paper investigates the interaction effects between the *intensity of wild dolphin tourism experiences* in the presence of *education/interpretation* commentaries on levels of overall satisfaction, knowledge, pro-environmental attitudes, beliefs, intended behaviours and intended actions of participants.

7.2 Literature Review

The literature review addresses five areas which firstly established the direction for this study and led to development of the SWIM model which proposes a synergistic interaction between the intensity of wildlife experiences and the education/interpretation component. Wildlife ecotourism is an interpretive experience and adventure. Therefore, literature on the theory and practice of adventure and experiential education formed an integral part of the search for information where guides, facilitators and educators use experiences to influence participants' attitudes, beliefs, behaviours and/or actions. The five reviewed areas of literature are:

1. Models and frameworks of wildlife and dolphin tourism experiences
2. Wild dolphin tourism interpretation and application to wildlife encounters
3. Evaluation of wildlife, marine and Cetacean interpretation
4. Foundations and underlying theories of education and interpretation
5. Theories and practices of tourism as experiences, experiential learning
6. Education, environmental, adventure-based and outdoor education

7.2.1 Models and Frameworks of Wildlife Tourism

In Australia, participants can choose from feeding, swimming-with and observing dolphins from land and/or a range of boat-based platforms of various sizes and in many locations. These activities, especially dolphin watch and swim-with experiences from aboard a vessel, parallel safari type adventures (IFAW 1997; Müller and Cleaver 2000). Dolphin tourism offers visitors varying types of encounters and differing levels of excitement, challenge, risk, adventure, and intensity ranging from very profound to every day experiences.

Neither of the two popular but differing models of wildlife encounters by Orams (1996) and Higginbottom (2004) give details on the nature, role or effect of the intensity of the marine wildlife experience. Higginbottom's (2004) central component called The Visitor Wildlife Encounter/Wildlife Tourism Product is a blank box, whereas Orams' (1996) model includes the spectrum of wildlife encounter activities and education is regarded as an integral part of the wildlife tourism experience.

Reynolds' and Braithwaite's (2001) Wildlife Experience framework and Tradeoff Matrix (Fig. 7.1), is key to this study because it deals with the contributing factors to visitor satisfaction and the relationship between the intensity of wildlife experiences and impacts on visitors. The framework serves as a basis for describing, comparing and evaluating commercial wildlife experiences, and was therefore used for plotting the three intensities of watching (B/C), feeding (B) and swimming with dolphin (A) activities.

The three diagonal lines separate four broad categories of a complex range of features. The greater the trade-off for short-term goals, including negative impacts on the animals, the greater the need for sustainable management practices, the higher the payment by the users and the greater the need for special consideration by the providers (Reynolds and Braithwaite 2001). The six Quality Wildlife Tourism Factors "intrinsic to the situation that capture the essence of quality and richness (intensity) of the wildlife tourism encounter for the person experiencing it" (Reynolds and Braithwaite 2001, p. 9), constitute a set of ambient and intangible factors. Combined, these factors create an overall level of profundity and strength of the wildlife experience for the visitors (see Table 7.1).

The Wildlife Experience framework also suggests that wildlife tourists may experience increased satisfaction from the ten aspects listed in Table 7.2.

The model by Moscardo and Saltzer (2004), which applied Mindfulness Theory (Langer 1992) to wildlife experiences, proposing that in any given situation a person can be "mindless" or "mindful", the latter being "a state of active cognitive or mental processing where mindful people pay attention to the information available in the environment" (Moscardo and Saltzer 2004, p. 180). While the Mindfulness Model recognises the visitor attentiveness component, it doesn't offer a framework that includes the factors of quality, richness or intensity. The Reynolds and Braithwaite (2001) Experience Tradeoff Matrix and detailed framework for wildlife encounters are therefore more detailed, relevant and applicable for this study.

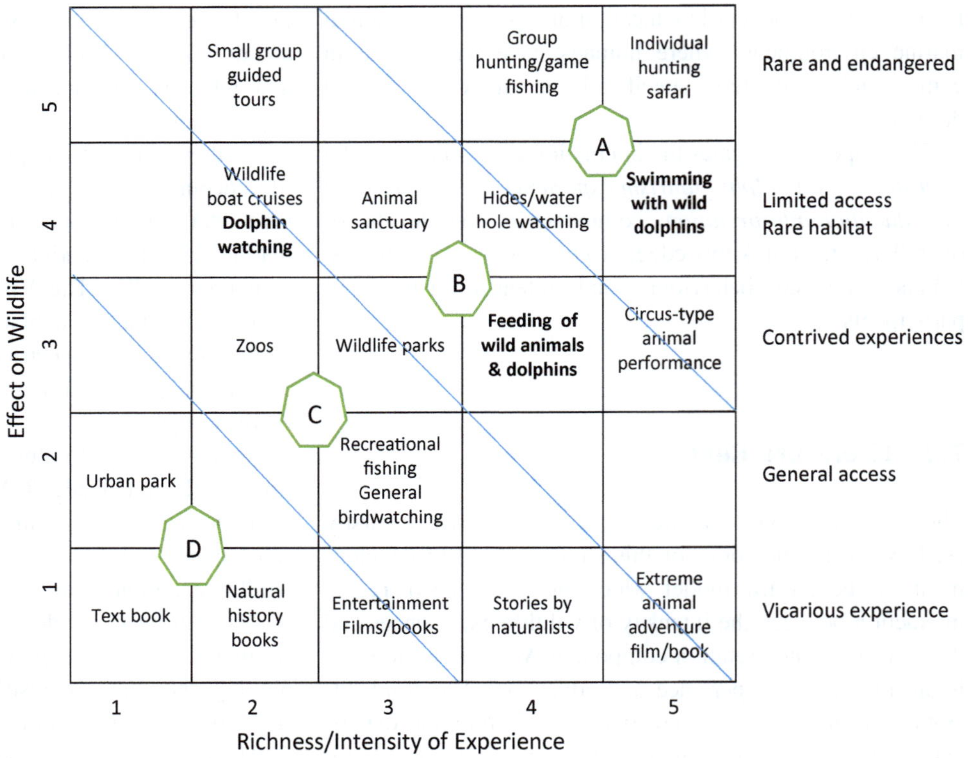

Fig. 7.1 Wild dolphin tourism encounters plotted on the Wildlife Experience Tradeoff Matrix (*Source* Adapted from Reynolds and Braithwaite 2001, p. 39)

Table 7.1 Intrinsic quality and richness (intensity) factors of wildlife tourism

	Factors	Factor description
1	Authenticity	Not contrived—and exhibiting natural behaviour
2	Richness/Intensity	Level of excitement, exhilaration, enthrallment
3	Uniqueness	Special, unusual animal, feeling of privilege by visitors
4	Duration	Critical length of time of interaction and experience
5	Species popularity	Physical attractiveness, size, danger, drama associated with the species
6	Species status	Rarity of the animal/species—endangered

Source Reynolds and Braithwaite (2001, pp. 35–36)

Table 7.2 Factors increasing satisfaction with wildlife interactions

	Satisfaction factors
1	Greater understanding and high levels of expert knowledge of an exhibit or animal
2	The presence of educational resources, tools, and educational aids
3	Delivery of interpretation through guides
4	Level of knowledge of guide/s (if applicable)
5	Communication skills of guide/s
6	Personal guide-observer rapport
7	Motivation levels of guide and observer
8	A variety of communication strategies and on-site interpretation aids
9	Communication with previous visitors
10	Pre-reading by observers

Source Reynolds and Braithwaite (2001, pp. 37–39)

7.2.2 Interpretation and Application to Wildlife Encounters

> Interpretation is communication which influences how humans think, feel, and behave… and success in interpretation is measured by what it is intended to achieve (e.g. provocation to thought, an enjoyable experience, acquisition of a belief or attitude, or perhaps a behaviour like staying on a trail, buying a souvenir or a repeat visit) (Ham 2004a: 1)

Interpretation practices have been in use by park and protected area managers since advocated by Tilden (1957). Interpretation is therefore not a new concept or term, even though little specific research is available and only very basic theoretical discussion of how to design effective wildlife interpretation has occurred (Moscardo 2003). Interpretation used in the context of marine wildlife tourism as a part of dolphin encounters can assist in reducing the negative impacts of marine tourism (Orams 1993), and assist in turning tourists into 'greenies' (Orams 1997b) in the long-term.

Ecotourism practices support conservation and sustainable resource management by deliberately setting out to increase the awareness and knowledge of tourists and to "influence them in such a way that the experience will convert them into somebody keenly involved in conservation issues" (Ceballos-Lascurain 1991, p. 25). The intention of marine ecotourism contributing to conservation applies to coastal and marine wildlife ecotourism, development and management (McIntyre 1995).

Well managed and high quality experiences have the capacity to support, conserve and promote the health of coastal and marine ecosystems (Twynam et al. 1998; TIES 2000). Marine wildlife ecotourism includes an interpretation and learning component and aims to contribute positively to the conservation of natural marine environments and the inhabitants (Epler-Wood 2002). Education and interpretation are therefore regarded as cornerstone components of both sustainable marine wildlife management and ecotourism.

In the case of whale watching encounters, the IFAW (1997) developed and presented a generic framework to assist guides in conducting education and interpretation programs for commercial Cetacean boat-based experiences. Cetacean encounters offer challenging and creative opportunities for guides. The foci in the framework are the sighting, behaviour, intensity and quality of the encounters with the whales and/or dolphins. The interpretive program in the framework is presented as a multi-staged process, with each stage having a purpose for contributing to the cumulative effect on visitors' pro-conservation attitudes, beliefs, intended behaviours and actions. Visitors will experience and meaningfully connect with Cetaceans in their own individual way and may need space, quiet and their own time in which to do so. Aware interpreters can also take advantage of 'Effective Learning Moments' (ELMs), 'Sensitivities', 'Emotional Moments' (EMs) and 'Interpretive Instants'. The framework also approaches the encounters from the visitor's perspective, with the emphasis on the affective domain, integrating the cognitive inputs and an intuitive commentary with the visitor experience. This approach offers enriching strategies for cognitive domain inputs and potential for enhancing the power of the impacts on conservation attitudes and intended behaviours of the humans.

The key competency in the IFAW (1997) approach is the ability to work intuitively and at the affective moments with the variations in intensities of wildlife encounter or recreational activity, in order to create the maximum impact on participants. The Cetaceans are the attention-grabbing focus while the intensity of the experiences is the catalyst for the intuitive approach. By intentionally and meaningfully communicating with visitors, the desired outcomes of enhanced support for conservation and adoption of minimal impact pro-environmental behaviours and actions may be achieved. The adapted IFAW framework acknowledges and emphasises the importance and need to intuitively and purposefully time the affective highs with the imparting of the cognitive inputs (education and interpretation) as integral parts of the these encounters.

Research on wildlife tourism products and visitor experiences is scarce (Green et al. 2001; Moscardo and Saltzer 2005), and even more so in the marine wildlife niche products and market. Wallace and Pierce (1996), Markwell (1998), Muloin (1998), Ryan et al. (2000) and Smith et al. (2005). Muloin (1998), suggested that the high levels of satisfaction amongst whale watch cruise passengers may result from high level intensity whale encounters, characterised by energetic displays of natural behaviour by a large number of whales surfacing close to the cruise vessel and remaining on the surface for an extended period of time. Muloin (1998) also recommended that encountering whales was a critical component to a successful experience.

7.2.3 Evaluation of Wildlife, Marine and Cetacean Interpretation

Although education and interpretation are fundamental elements of the principles of sustainable tourism (Moscardo 2003), finding a substantial body of research and/or evaluation on impacts of interpretation on participants of marine tourism encounters and/or visitor behaviour proved challenging, especially in relation to conservation behaviour. Even fewer studies exist in the specific field of interpretation impacts in Cetacean tourism contexts.

The predominant foci of marine tourism evaluative studies focused on increases in knowledge (Cottrell and Graefe 1994), visitor perceptions and awareness of degradation and impacts resulting from marine recreation activities (Priskin 2003), visitor management and a major emphasis on protection of marine wildlife species such as stingrays (Shackley 1996; Newsome et al. 2004), Minke whales (Valentine et al. 2004), turtles (Tisdell and Wilson 2000, 2004; Wilson and Tisdell 2003), whale sharks/basking sharks (Davis et al. 1997; Norman 2000, 2001; Speedie 2003), coral reefs (Moscardo 1999; Townsend 2003), sea lions (Ingram 2001), manatees (Sorice 2001; Sorice et al. 2003), sub-Antarctic and Antarctic wildlife (Enzenbacher

1992, 1994; Patterson et al. 1996; Bennet and Kriwoken 2001; Stewart et al. 2005), and Yellow Eyed Penguins as a tourist attraction (Schänzel and McIntosh 2000).

Studies reporting specifically on the positive impacts of interpretation on conservation attitudes and behaviour and visitors' appreciation of marine wildlife appear to be few and results appear disappointing. For example, Schänzel and McIntosh (2000) reported that only two of the 40 interviewed visitors to a penguin conservation centre and tourist attraction in New Zealand intended becoming more involved in conservation activities.

In Chesapeake Bay, Maryland, Cottrell and Graefe (1993) found a relationship between attitudes and pro-environmental behaviour of boaters. They established that as self-perceived general knowledge of ecology increased, subsequent increases in proactive behaviour occurred. Cottrell and Graefe (1994) further established that knowledge of specific issues had a direct positive effect on responsible marine environment behaviour. Both findings are relevant and important for the design and presentation of interpretive commentaries and resources.

The case study of the Avoca Beach rock platform (Garrod and Wilson 2003) is an example of the successful use of a strategically planned and comprehensive coastal and marine education and interpretation behaviour management program which aimed to reduce high impacts on the marine life and reduce rubbish left from numbers of tourists and fisher-folk. In Kaikoura, New Zealand, one of the world's premier marine mammal tourism destinations, tourists' awareness and knowledge about marine mammals was significantly modified, thus the experience had educational value, but no significant improvement was detected in the conservation values of participants (Beasley 1992).

More recent and pertinent research (Higham and Carr 2002) involved 12 operators, eight of whom were involved with coastal and marine ecotourism, and three offered Cetacean encounters as the primary wildlife experience. Results indicated "that ecotourism visitor experiences may be an effective means of influencing visitor environmental values and behaviours" (Higham and Carr 2002, p. 291). Results also suggested that interpretation programs can foster "behavioural change relating to domestic lifestyle that may contribute to the long-term benefit of the environment" (Higham and Carr 2002, p. 279). However, the focus of the study was to examine the components, content and process of the visitor ecotourism experience and not the impacts or influence of the experience on pro-environmental attitudes, beliefs, behaviours and actions, as per this thesis. Exploration of relationships between visitor ecotourism experiences and environmental values and behaviours were also flagged as a part of Higham's and Carr's recommendations for continuing research.

The existing visitor attitudes towards perceptions of the impacts of Cetacean tourism on the Cetaceans were the focus of Finkler's and Higham's study (2004). Once again, they did not measure changes in, or influences on, visitor attitudes. Their results have relevance to this study in that the boat-based visitors were aware of, and concerned about the Cetacean tourism management practices such as manoeuvring and noise of powerboats in the path of killer whales causing disturbance to the whales. The land-based visitors also expressed concerns about powerboats compromising the safety of whales. Both groups reported visual disturbance from the number and activity of recreational powerboats amongst the wildlife encounter.

A study by Parsons et al. (2003) reported higher levels of knowledge and greater awareness of whale watching opportunities amongst Cetacean watchers compared to general tourists in western Scotland. The whale watching tourists were also reported as more environmentally aware and conservation-active compared to other tourists, but the research did not extend past this point to consider the direct influences of the wildlife encounter on the participants or the influences on their intended pro-conservation activity.

Participants in the only place in Australia and possibly the world for swimming with Dwarf Minke whales were invited to comment on the management practices of the encounter (Valentine et al. 2004). However, no published research was available on the actual experience and/or evaluation of the impacts of the Dwarf Minke whale swim encounter on visitors' pro-conservation attitudes or intended behaviours.

A considerable research effort by Orams and associates have addressed the following goals of interpretation: (1) enhance visitor experiences (Orams 1997c); (2) protect resources at sites (Orams 1996); (3) protect visitors (Orams 1995b; Orams et al. 1996); (4) increase public support for an agency and its management policies (Orams 1995c, 2002; see also Neil et al. 1996); (5) add to or broaden visitors' perspectives about a place or idea; enhance their knowledge (Neil et al. 1996; Orams 1997a); and foster positive attitudes and behaviours (Orams 1995c, 1997b; Orams and Hill 1998) with respect to the natural and cultural environments (Orams and Forestell 1998).

Although Orams (1995b) was unable to establish significant enhancement of pro-environmental attitudes in the short-term, "the interpretive program prompted a significantly higher level of action across all the other indicators measured" which "caused participants to become more environmentally responsible in the long-term" (Orams 1995a, p. 256). These long-term changes in pro-environmental behaviour ultimately "benefit the marine environment and, through this, the animals that live in it" (Orams 1995a, p. 256). Orams' results suggest that time may be a significant contributing factor to enhancing pro-environmental behaviour as a desirable outcome of wildlife encounters. This aligns with experiential education theory where reflection after the experience enhances transfer from the encounter to the home environment.

Strong support was given to the existence, importance and place of intensity in tourism by this comment which was part of an address to a conference in Tasmania by Ham (2004b): "Numinous experiences are typically characterised by three psychological qualities: intense engagement (or focus); a loss of the sense of time passing; and a transcendence of self…meaning that lots of dopamines and endorphins are flowing during a numinous moment" (Ham 2004b, p. 15). Ham also suggested that "we have all been moved at one time or another by a place, a view, an object, or even by an idea that was so profound it transcended us, or provoked us to contemplate a higher meaning of things" (Ham 2004b, p. 16). Interpretive programs should activate the affective domain by seeking to include or emphasise an emotional component as a means of communicating its message/s (Orams 1995c).

7.2.4 Foundations and Underlying Theories of Education and Interpretation

Subsequent to the models and approaches to wildlife and interpretation by Orams (1995c), Woods and Moscardo (2003), and Moscardo and Saltzer (2004), three key psychological theories based on Tilden's (1977) approach to interpretation gained wide acceptance as theoretical foundations of interpretation (Ham and Weiler 2004). These three key theories are: the Schema Theory, the Theory of Planned Behaviour (Ajzen 1985, 1991), which is an extension of the Theory of Reasoned Action (Fishbein and Ajzen 1975), and the Elaboration Likelihood Model (Petty and Cacioppo 1984; Petty et al. 1992).

The Elaboration Likelihood Model (ELM) (Petty and Cacioppo 1984) links the amount of thinking or elaboration we undertake on issues to the influence on our attitudes and behaviour. In essence, the more we think or elaborate on a theme or in the case of wildlife tourism, a pro-conservation message, the more influence the issue or message will have on our attitudes and behaviour (Ajzen 1992). For a review of empirical studies on the effectiveness of persuasive communication as a management tool in reducing visitor impacts and conflicts, see Manfredo (1992), and for more recent research in the area: Beeton et al. (2005).

These foundations have led to the growth of approaching interpretation from an interpreter-centred and cognitive-based perspective. The emphasis has been on the interpreter creating the motivation which potentially leads to changes in attitudes and feelings. Orams (1995c) included the affective domain in his model of effective interpretation, suggesting that the encounter or experience was a part of the affective component. Orams importantly acknowledged that emotional involvement was a likely short-cut to inducing behaviour change (Orams 1995c). Even so, Orams focussed on the cognitive domain as the point of reference and delivery, neglecting the place or priority of the affective domain in the interpretation process.

The neumen-seeking or profound and transcending tourist wildlife experiences discussed by Ham (2004b) are obviously highly emotional, engaging the affective domain. The neumen moments appear to involve the affective domain to the exclusion of everything at the moment. Ham (2004b) did not expand upon or introduce any theoretical discussion about the connections or interactions between high intensity experiences and interpretation or the affective and cognitive domains. Nor did he discuss flow-on effects of neumen moments on pro-conservation behaviours or attitudes. Nor did he allude to a relationship between the profound experiences and the current cognitive-based learning frameworks (Ham 2004b).

7.2.5 Experiences, Experiential Learning and Education

The premise that the tourism industry is "in the business of selling experiences" (Ooi 2004, p. 72), is widely supported (Mannell and Iso-Ahola 1987; Li 2000) and therefore accepted for the purposes of this paper. However, "…understanding tourism experiences is difficult because of the existential, multi-faceted and highly personal nature of experiences which arise from the activities, the environment and the social context" (Ooi 2004, p. 72).

Six approaches appear to be used by researchers to capture tourism experiences. The first approach concentrates on the cognitive psychology of experiences and how participants' perceptions affect experiences (Mannell and Iso-Ahola 1987; McIntosh and Prentice 1999. The second approach argues "… that tourism activities enable tourists to gain experiences that are regarded as beneficial to them personally … improving their psychological mood and wellbeing, and allowing self-identity and learning about places and cultures" (Ooi 2004, p. 72). The third approach focuses on mind states and depth of experiential engagement (Ellis 1973).

The concepts of "optimal experiences" or special, meaningful and out-of-the-ordinary experiences (Walker et al. 1998) and "flow", where participants experience a feeling of intense emotion, concentration, focus or "transcendence" (Csikszentmihalyi 1991), are accepted aspects of the experiential approach. The fourth or phenomenological approach (Cohen 1979; Li 2000; Lengkeek 2001) attempts to capture ranges of personal experiences that are "not just intense and optimal … such as the restoration of well-being, escape from boredom, search for aesthetic meaning, and quest for alternative lifestyles" (Ooi 2004, p. 73). The fifth approach pays attention to the gap and relationship between locals and tourists (Urry 1990; Hannabuss 1999; Ooi 2002). The sixth and final experiential perspective regards tourists as a diverse group with behaviours that are not easily predictable and who have diverse perceptions of their tourism experiences.

Nature and adventure-based activities have become playgrounds for experience-seeking tourists (Gyimothy and Mykletun 2005) where two parallel trends have occurred: the growth of environmental consciousness; and a trend towards more educative and challenging vacations and activities (Urry 1995; Lindberg et al. 1998). These trends combined may contribute towards explaining the rapid growth in nature, adventure and ecotourism services and products where companies and providers are seeking to sell powerful and long-lasting experiences (Lindberg et al. 1998; di Castri and Balaji 2002).

The marketing of ecotourism as environmentally and culturally friendly for aware consumers has moved towards the selling of unique experiences with conservation as a sub-theme (Gössling 2005). The recent emphasis is on marketing ecotourism as a powerful experience, often containing adventure tourism elements and reflecting the changes in the tourism production system (Gössling 2005). Experience products and services, including wildlife encounters, are very successful on the market, with the niche of wild, new, extraordinary and extreme activities proving very popular for a demographic of "particularly wealthy or willing older people" (Gössling 2005, p. 35). For example, Maher (2005) conceptualised the nature of extreme experiences of visitors to the Ross Sea Region, a remote and until recently unexplored part of Antarctica, concluding that the process "is best described in experiential education literature" (Maher 2005, p. 71).

Experiential learning theories tend to be holistic, incorporating cognition and integrating behaviour with conscious perceptions and reflections on experience (Boud et al. 1985; Weil et al. 1989; Kraft and Sakofs 1991). Experiential education is defined as "a process through which a learner constructs knowledge, skill, and value from direct experiences" (Luckmann 1996, p. 7). Experiential learning put simply is "learning by doing with reflection" and is "based on the belief that people learn best by direct and purposeful contact with their learning experiences" (Priest and Gass 1997, p. 17).

Reflection by participants is regarded as an important aspect in enhancing learning (Boud et al. 1985). It is these theories of experiential education and learning that recognise, incorporate and explain the synergistic relationship between the cognitive and affective domains, thus presenting relevant theoretical frameworks for experiences in wildlife encounters. The process of experiential education has generally been represented in the form of a triangle (Dewey 1938); cycle (Kolb 1984), spiral (Joplin 1981) or wave (Schoel et al. 1988), where both experience and reflection are key components and enhance each other (Martin 2001). Dewey's (1938) model of experiential education involved observation of surrounding conditions, knowledge from past experiences and judgment combining knowledge and observation.

Experiential learning is distinguished from experiential education in that experiential learning is a process of change involving reflection on individual experience, whereas experiential education is a "transactive process between an educator and a student…considering the socio-economic, political and environmental elements in the learning environment" (Itin 1999, pp. 91, 92). Similarly, guided wildlife ecotourism experiences could be regarded as transactive processes between an interpretive guide and wildlife tourists.

7.2.6 Adventure-Based Learning and Education

Three inter-related fields exist under the umbrella of experiential education: adventure, environmental and outdoor education. Figure 7.2 demonstrates the inter-relatedness of those three fields which are defined, briefly explained and applied to commercial wildlife tourism encounters and especially to interpretation. Swimming with wild dolphins, feeding and watching wild dolphins from boat-based viewing platforms are identified as adventure-based activities, and can therefore be included as part of experiential learning and/or education activities, depending on whether or not the experience is guided and/or contains interpretive elements.

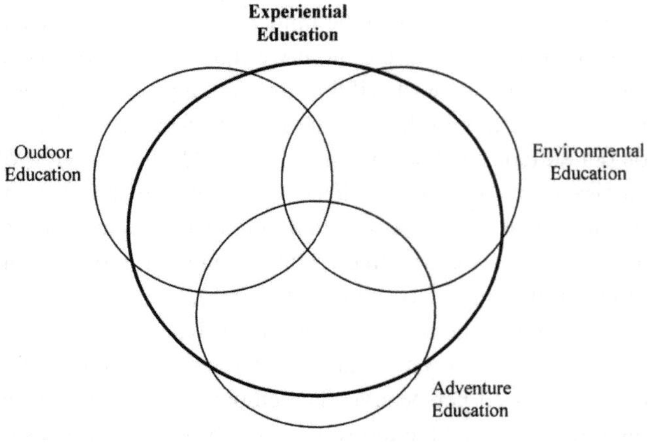

Fig. 7.2 Three inter-related fields of experiential education (*Source* Martin 2001, p. 23)

Adventure education involves purposeful planning, educational processes and some risk, which may differ from environmental and/or outdoor education (Mortlock 1984; Zook 1986; Miles and Priest 1990). Real life adventures can also increase self-understanding and develop individual capabilities and a greater understanding of nature (Zook 1986). Adventure can be defined as "an experience that involves uncertainty of outcome … and because of the holistic nature of the experience, there is personal growth" (Hopkins and Putnam 1993, p. 227). Adventure education therefore includes challenge, high adventure and new growth experiences. The Adventure Experience Paradigm (Miles and Priest 1990) suggests that adventure-based experiences should offer high challenge and high perceived risk but low real risk. Physical, social, psychological, and financial risks are all included as varying forms of risk and pivotal (Liddle 1998) or integral aspects of adventure education (Dickson et al. 2000). Interestingly, adventure has also been described as adult play where humans behave in such a way as to avoid boredom or over-stimulation.

The major portion of the behaviour serving this drive for stimulation or arousal is concerned with sensation and/or stimulus seeking (Ellis 1973; Carpenter and Priest 1989). This concept was extended by Dodson and Yerkes (1908) and Hanin (2000) who proposed that a relationship existed between optimal arousal and optimal performance (Fig. 7.3). Individuals have also been portrayed as adventure-seeking and striving to find opportunities for challenges and attainment of individual optimal arousal by Csikszentmihalyi (1991) who coined the phrases "states of flow" and "peak experiences" (Fig. 7.3), which are closely connected to states of optimal arousal.

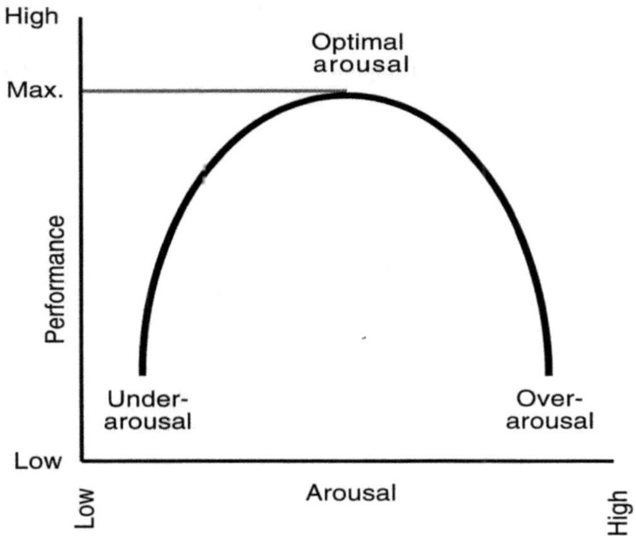

Fig. 7.3 A graphic representation of optimal arousal theory (*Source* Dodson and Yerkes 1908, pp. 459–482)

Flow is a state of "being"—when totally engrossed, absorbed or involved in an activity. Flow is also a theory of how activities produce an optimal experience based on individuals' perceptions of skill and challenge (Fig. 7.4). Theoretically, when participants are in a state of flow, they experience intrinsic feelings of enjoyment, wellbeing, and personal competence. This affective "high" in turn, encourages a desire to return to the activity in an attempt to recapture the feelings (Priest and Gass 1997). Ham's (2004b) "neumen" states resembles these affective "highs".

The current range of commercial wild dolphin encounters offers adventures with varying levels of intensity and/or arousal. In the context of this project, jumping off a boat into open ocean and swimming with wild dolphins meets the criteria of a high intensity (hard) adventure activity with a combined set of circumstances that allow or create optimal arousal levels and potentially: states of 'flow'. In comparison, the risk, challenge and arousal level associated with swimming with wild dolphins is greater than the arousal level associated with feeding or watching wild dolphins, which are regarded as moderate and low intensity (soft) adventures.

The Adventure Wave model (Schoel et al. 1988) compares the pattern of experiences in adventure and experiential education programs with a series of waves with peaks and valleys and periods of turbulence, excitement, activity and calm (Fig. 7.5). The model includes a preparation phase at the beginning and an analysing phase at the end, with briefing and debriefing sessions before and after each wave

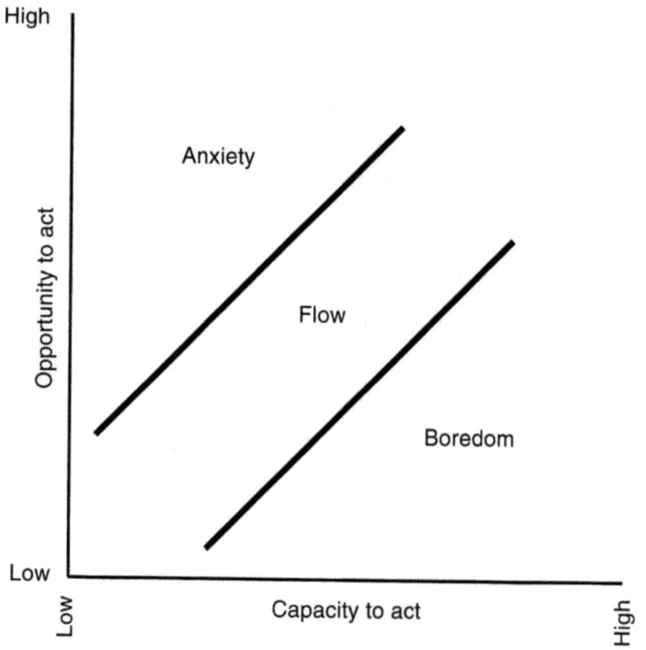

Fig. 7.4 A graphic presentation of flow theory (*Source* Csikszentmihalyi 1975, p. 49)

of experience. The Adventure Wave model (Schoel et al. 1988) emphasises the role of the facilitator in achieving effective and long-lasting change. The interpretation-based framework for Cetacean watching suggested that tour guides should capture the interpretive instant and take advantage of any situations that arise, being especially sensitive to the first Cetacean encounter of the cruise. At these heightened emotional/affective moments (EMs) of the encounter, participants may be more receptive to cognitive inputs resulting from an interaction between the intensity of the experience and the impact/effectiveness of the education/interpretation message (ELMs). This in turn, may result in changing/enhancing pro-environmental behaviours (IFAW 1997).

The Adventure Wave experiential education model fits well with the suggested Framework for Cetacean Education and Interpretation Programs (IFAW 1997), where the 'Activity' represents when humans have the wild dolphin tourism encounters (Fig. 7.5).

The IFAW (1997) framework included a pre-journey briefing and another briefing on the way to the Cetaceans. These activities parallel the preparation and negotiation phase in the wildlife encounters, and the wave crests represent the times of maximum focus, connection, intensity and individualisation in Fig. 7.5. The crests are followed by calm periods where the interpretive guide can introduce real time information specific to the Effective Learning Moments (ELMs) and the activity that occurred at the preceding crest.

The Adventure Wave Model recommended formal planning and preparation (Schoel et al. 1988). Cetacean behaviour cannot be planned for as it happens without prior warning, prediction or control. Therefore, guides must be ready to take advantage of the high intensity, opportune moments (EMs) for effective interpretation and learning to occur. Cetacean tourism experiences and encounters are characterised by unplanned sightings (represented by the crests and waves of adventure), followed by periods of quiet or opportune moments for debriefing (represented by the valleys). On the return trip, the guides should summarise the experience, ask participants to reflect on the intensity of the experience, develop transfer from the natural environment to home, impart, repeat and reinforce the take-home or call to action conservation message/s. This format fits the analysing and accounting phases of the Adventure Wave Model and aligns with the experiential education practice of including the debriefing and reflection phases in calm or "Valley" stages of the wildlife encounters.

By combining funnelling (Priest and Naismith 1993) as a debriefing method (Fig. 7.6) with effective use of the adapted Adventure Wave Model for wildlife encounters, the process of meaningful connections between the pro-conservation message with the call to action assist and motivate participants to transfer and apply new/enhanced pro-conservation intentions to actual behaviours and actions in their home environment (Priest and Gass 1997). The funnelling approach "guides the group through a series of steps that funnel clients' attention from the experience toward making beneficial changes in their lives" (Priest and Gass 1997, p. 194). The facilitator intentionally begins by asking participants to recall and remember the affect and effect, thus creating an awareness of the learning that took place. The facilitator—in the case of wildlife tourism: interpretive guides—then create connections between the participants' adventure experiences and their home environments through metaphors, showing how the skills learned in the outdoor environment can be transferred to the home, work or other environments. The facilitator then gains a commitment from participants to change and sustain those behaviours and actions in their lives.

Participants of Adventure Experiences finally declare who, in their lives will act as a support system for their new behaviours. The participant and the experience are the foci of the facilitation process which involves a forward commitment from the participants. Each person is encouraged to

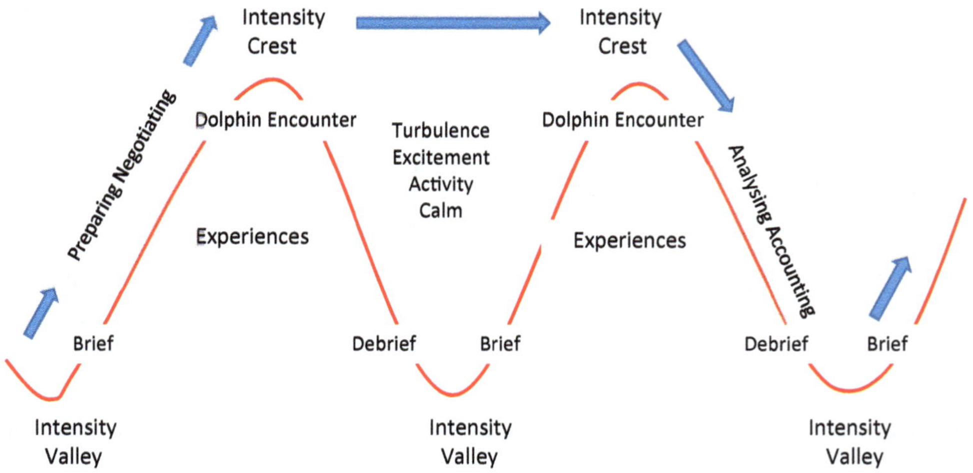

Fig. 7.5 The Adventure Wave Model and Dolphin Encounters (*Source* Adapted from Schoel et al. 1988, p. 29)

find their own individual level of connection and meaning in the experience. This debriefing and facilitation process can be transferred and used for developing high impact, effective interpretation practices in the case of wildlife tourism encounters.

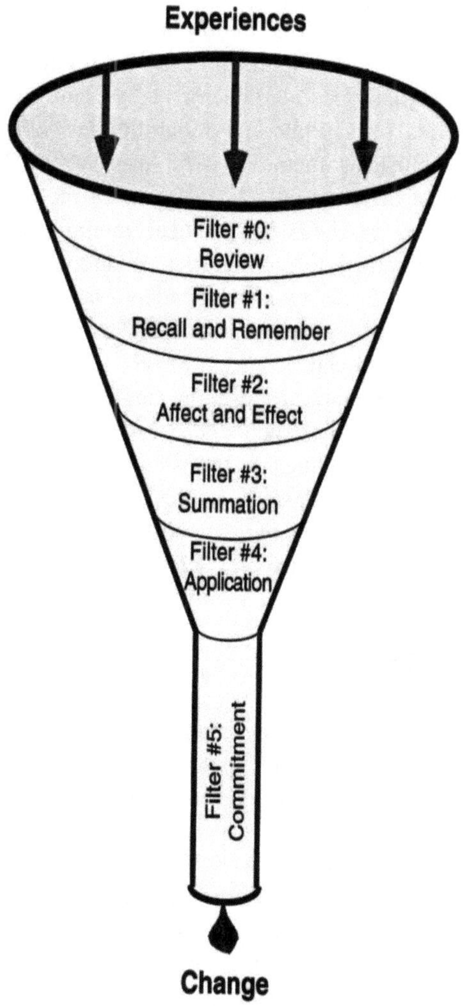

Fig. 7.6 The debriefing funnel (*Source* Priest and Gass 1997, p. 196)

7.3 Methodology

The overall goal of this study is to assess the role that the *intensity of the experience* has on *education/interpretation* in marine environments where humans have encounters with wild dolphins. This investigation is a quasi-experimental design even though a relationship between at least two variables is under investigation. Although participating groups were purposely chosen, the setting up and use of control groups was not possible. Additionally, a single post-test design was chosen in preference to a pre-test/post-test format. The terms *dependent variables* and *independent variables* are usually applied to experimental research (Sproull 2002) and they will also be used for the purposes of this quasi-experimental research design (Table 7.3).

Figure 7.7 presents the Synergistic Wildlife Interpretation Model (SWIM) including details of the independent and dependent variables used to address the research aims. In this paper, pro-environmental attitudes are regarded as a cluster of single or specific pro-environmental beliefs (Ajen and Fishbein 1980; Mayes et al. 2004). Additionally, behaviours and actions are also distinct constructs where pro-environmental behaviours are clusters of pro-environmental actions, with an action being "a specific pro-environmental behaviour performed by an individual" (Ajzen and Fishbein 1980, p. 31).

The approach of this study includes: quantitative evidence gathered through descriptive observations; an education/interpretation direct observation and recording sheet; plus a questionnaire, under the umbrella of a comparative multiple case study method and protocol (Yin

Table 7.3 The independent variables and brief definitions

Independent variables	Definitions
Intensity	Intrinsic quality and/or richness of a wildlife tourism experience = level of profundity. Intensity is a combination of authenticity, level of excitement, exhilaration, enthrallment, uniqueness, duration, species popularity and rarity status
Interpretation	Interpretation is more than education and includes education, recreation and conservation. It enhances knowledge, understanding and awareness. An identified outcome of interpretation is the stimulation of the visitor to adopt minimal impact behaviours and increase their support for the conservation of the environment, the fauna and the flora
Environmental education	Education is a part of interpretation. The focus of education is to increase knowledge and/or skills through teaching, learning and instruction, but does not necessarily aim to stimulate minimal impact behaviours and/or support for the environment

Source The author

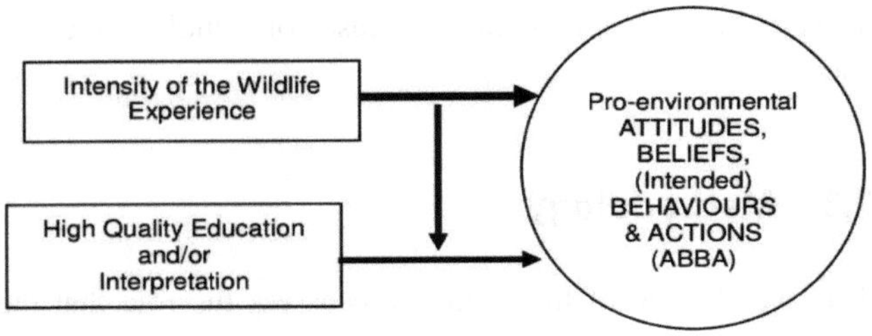

Fig. 7.7 The proposed synergistic wildlife interaction model (SWIM) (*Source* The author)

2003). Qualitative and quantitative evidence and data were gathered through primary and secondary research.

The population for this study is the adult participants of three pairs of wild dolphin-based tourism experiences in Australia, namely two low intensity watching encounters using boat-based platforms; two moderate intensity shore-based feeding encounters, plus two high intensity boat-based swimming-with dolphin encounters. One case of each matched pair of watching, feeding and swimming-with dolphin operators had a high quality interpretive component as part of the experience. The other case in each pair offered a matched mode of wild dolphin activity focussed on the encounter and/or the experience and lacked the high quality interpretation components.

A self-administered questionnaire, written in English, was the main data gathering instrument. The questionnaire consisted predominantly of Likert attitude scales plus three open-ended questions which asked participants to state what they liked least and best about their experience, and if they had any other comments about the encounter.

A structured observation sheet was designed for gathering data on the content of the interpretation component which visitors experienced as part of the dolphin encounters. This instrument provided a structured and systematic process for recording the type and number of times specific items and/or topics were mentioned by the information giver/guide as part of the tourist experience. The observation sheet was not designed to assess the methodology or strategy used by the deliverer of the information, nor did it measure or evaluate the competency level or skill level of the person in delivering the education/interpretation component. Each time a particular educational or interpretive item, message or piece of relevant information was given verbally, a tick was recorded in the appropriate place on the content instrument.

This final phase of a multi-phase study undertook a complex cross-case comparison of all six dolphin encounter cases and initially required factor analysis in combination with a series of validity and reliability tests (Table 7.4). ANOVA statistical methods were used to establish the magnitude of effect or influence of each of the independent variables, and MANOVA statistical methods established whether an interaction or synergistic relationship existed at the intercept of the two independent variables and the dependent variables. Data were analysed using the Statistical Package for the Social Sciences (SPSS).

Table 7.5 summarises the eight single item factors representing the eight dependent variables/constructs and includes the names of the constructs and the number of items which contributed to those constructs. High construct reliability occurred for the eight new constructs with Cronbach's alpha scores ranging between 0.82 and 0.94.

7.4 Results

The primary analysis in this study tested the relationships between the two key independent variables, *intensity of experience* and *education/interpretation*, and a number of factor-analysed dependent scales. The analysis comprised one-way ANOVAs and MANOVAs. The MANOVA provided an opportunity to examine the interactions between *intensity of experience* and *education/interpretation* whereas the ANOVA could not. The summary of these results follows. The ANOVAs (Table 7.6) revealed five significant effects:

1. *Intensity of experience* had a positive effect on DV1: Overall_satisfaction (2.8%)

Table 7.4 Summary of the statistical methods used for cross-case comparison

Three combined matched pair cross-case comparisons 1. Watching dolphins from boat-based platforms 2. Feeding dolphins from shore 3. Swimming with dolphins from boats	**Two dependent variables Intensity and education/interpretation**	Factor analysis Varimax rotation Normality of distribution Test for multivariate outliers Chronbach's alpha Levene's test for homogeneity Data transformation ANOVA and MANOVA

Source The author

Table 7.5 The dependent variables, constructs and Chronbach's alpha

The new single-item dependent variables	New constructs	Cronbach's alpha scores	No of items	Origin of constructs
1. Level of overall satisfaction	Overall_satisfaction	N/A	1	Q 6
2. Increase in knowledge about the dolphins and their marine environment	Knowledge_change	N/A	1	Q 3
3. Satisfaction with information given about dolphins generally	Information_general	0.86	3	Q 5
4. Satisfaction with information given about how to help conserve dolphins and their marine environment	Information_conservation	0.95	2	Q 5
5. Change in level of motivation to be pro-environmentally active at an organisational level	Motivation_general	0.92	5	Q 9
6. Change in level of motivation to be pro-environmentally active at an individual level	Motivation_individual	0.93	5	Q 9
7. Support for conservation of dolphins and the marine environment	Support_conservation	0.94	3	Q 10
8. Support for assisting in marine conservation	Assist_conservation	0.82	2	Q 10

Source The author

2. *Intensity of experience* had a negative effect on DV4: Information_conservation (1.9%)
3. *Education/interpretation* had a positive effect on DV4: Information_conservation (1.4%)
4. *Education/interpretation* had a negative effect on DV2: Knowledge_change (0.06%)
5. *Education/interpretation* had a positive effect on DV7: Support_conservation (0.03%)

The advanced analysis employed a MANOVA across each of the eight dependent variables and the two independent variables *intensity* and *education/interpretation,* and included the interactive term *intensity of experience* x *education/interpretation*.

Five significant MANOVA results (Table 7.6) were attributed to the interaction effect of *intensity of experience* x *education/interpretation* on the following dependent variables:

1. DV1 Overall_satisfaction
2. DV2 Knowledge_change
3. DV3 Information_general
4. DV4 Information_conservation
5. DV7 Support_conservation

Table 7.6 presents a summary of the statistically significant ANOVA and MANOVA results in order to facilitate an understanding of the complex nature of these two sets of results associated with the synergistic interaction between the *intensity of the experience* and *education/interpretation* and the dependent variables.

The overall results in Table 7.6 indicate a basic, yet complex situation represented by two key sentences which summarise the outcomes of the investigation: (1) *In the presence of high intensity experiences, high quality education/interpretation commentaries do not have a role;* and (2) *In the absence of high intensity experiences, high quality education/interpretation commentaries do have a role.*

7.5 Limitations

This study focussed on commercial human-dolphin encounters, so differing results may be obtained by studying other wildlife species encounters, especially when contrasting the differing intensities of the encounters. Wild dolphin encounters with commercial operators were researched in four of the seven states of Australia and differences may occur in the states not included in the project. Pro-environmental attitudes, beliefs, intended behaviours and intended actions can only be studied in the immediate short-term with this research design. Intentions do not necessarily lead to actions (Ajzen and Fishbein 1980). Therefore, a long-term study is essential as a follow-up to this investigation.

Table 7.6 Summary of key results for the ANOVA and MANOVA analyses

FACTOR	Overall Satisfaction (DV1)	Knowledge Change (DV2)	Information General (DV3)	Information Conservation (DV4)	Support Marine Conservation (DV7)
ONE-WAY INTERACTION —ANOVA RESULTS					
INTENSITY OF EXPERIENCE F Sig	* * *			* * *	
Strength of Effect: One-way eta^2	2.8%			1.9%	
Direction of effect	+			—	
EDUCATION/INTERPRETATION F Sig		* *		* * *	*
Strength of Effect: One-way eta^2		0.6%		1.4%	0.3%
Direction of effect		—		+	+
TWO-WAY INTERACTION —MANOVA RESULTS					
INTENSITY OF EXPERIENCE X EDUCATION/INTERPRETATION F Sig	* * *	* *	* * *	* *	*
F value Strength of effect	9.54	6.51	11.41	3.75	2.91
Direction of effect Experience only	+	High *in* <u>*high*</u> *Intensity*	+	Lower *in* <u>*moderate*</u> *intensity*	Lower *in* <u>*moderate*</u> *intensity*
Education/Interpretation	Partially	Low *in* <u>*high*</u> *intensity*	—	Higher *in* <u>*moderate*</u> *intensity*	Higher *in* <u>*moderate*</u> *intensity*

Source The author

Key: *<0.1, **<0.05, ***<0.001

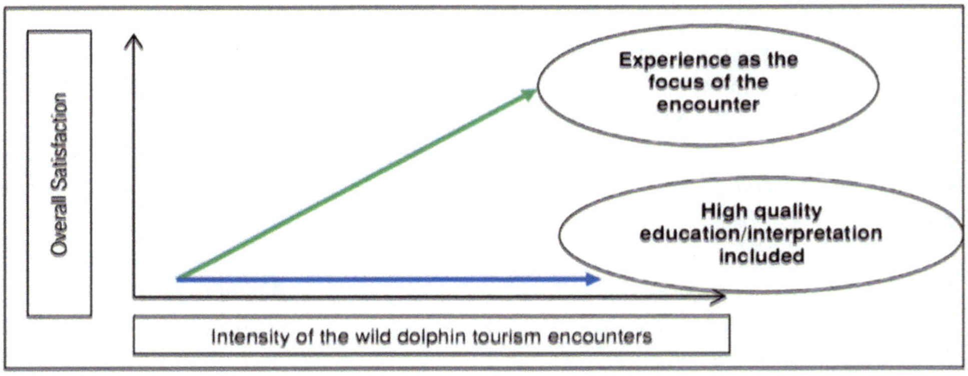

Fig. 7.8 Relationship between *overall satisfaction* and *intensity of the experience* (*Source* The author)

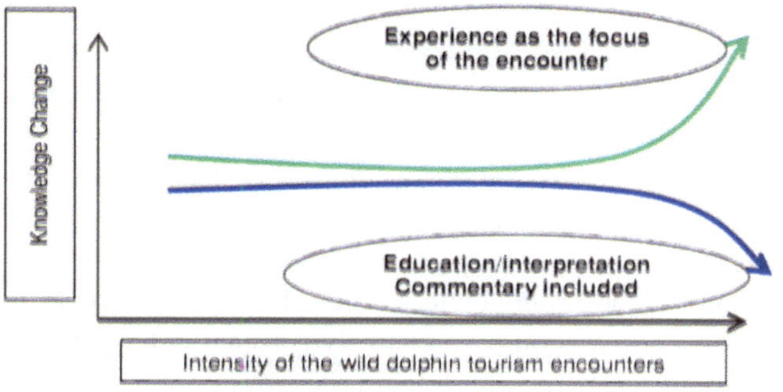

Fig. 7.9 Relationship between knowledge change and intensity of the experience (*Source* The author)

7.6 Discussion

This paper investigates the interaction effects between the *intensity of wild dolphin tourism experiences* in the absence and presence of *education/interpretation* commentaries on the levels of Overall Satisfaction, Knowledge, Pro-environmental attitudes, beliefs, intended behaviours and intended actions of participants. Each factor and effect is discussed in turn.

Overall_satisfaction: A very strong effect was detected for the interaction between *intensity x education/interpretation* ($p < 0.001$) for *Overall satisfaction*. Participants appreciated the inclusion of the education/interpretation programs as an augmented and value-added product, creating a higher quality tourism service/product. However, the inclusion of an interpretation component did not affect participants' feelings of *Overall_satisfaction* with the experience (Fig. 7.8).

As the *intensity of experience* decreased, the excitement, exhilaration and enthrallment similarly decreased in the absence of a high quality education/interpretation component. In response to the second set of research questions, it appears that *intensity of the experience* has a direct effect on *Overall_satisfaction* and an interaction effect occurs between *intensity x education/interpretation* for *Overall_satisfaction*, but this does not hold true for *education/interpretation* alone (Fig. 7.8).

Knowledge_change: The MANOVA revealed a significant and strong effect for *intensity of experience x education/interpretation.* The results suggest that the highest increases in knowledge occurred with the highest level of intensity swimming-with dolphins encounter in the absence of a high quality education/interpretation component. Increases in knowledge are a logical product of education and learning in the presence of the *education/interpretation* components, however, *education/interpretation* had a counter-intuitive and negative effect on *Knowledge_change* in the Level A encounter (Fig. 7.9).

The interaction between *intensity of experience x education/interpretation* is significantly related to increases in knowledge, indicating a complex effect between the two independent variables. For both moderate and low intensity experiences, the changes in knowledge were similar. In the case of high intensity experiences, *education/interpretation* had a dramatic but unexpected effect. The participants of the high intensity experience-based encounter reported the greatest changes in knowledge while participants of the high intensity experience with a high quality education/interpretation component reported the second lowest result.

A possible explanation for this result may be that *education/interpretation* becomes a liability for increases in knowledge in high intensity experiences. The level of *intensity of experience* is not the significant factor, but none the less is important as it is a principal element in a complex interaction with *education/interpretation.* The results indicate that *intensity of experience* on its own did not have a significant relationship with increases in knowledge, but had a significant effect in combination with *education/interpretation* as the main contributing factor. This is further supported by the strength of the effect of *education/interpretation* alone, on *Knowledge_change*, statistically the second strongest effect on the dependent variables.

The signs and symptoms of high intensity swimming encounters parallel numen (Ham 2004b), peak (Ellis 1973) and flow experiences (Csikszentmihalyi 1975). High intensity, risky, adventurous and physically challenging activities demand high levels of focus and concentration. For example, jumping off the back of a boat and swimming with wild dolphins offers a novel, challenging, high perceived risk, exciting, high arousal, 'hard' adventure experience. The demand for concentration and focus and a preoccupation with the profound and enthralling experience of being close to wild dolphins in their own environment may be met at the cost of the cognitive inputs of education and interpretation messages. Whereas, watching dolphins is a low intensity 'soft' adventure, with less challenge, risk and fewer physical demands but high levels of enthrallment and profundity. Watching dolphins may demand less concentration and focus, allowing participants to focus on other (augmented) aspects of the experience such as processing an education/interpretation commentary which engages the cognitive domain.

The cognitive domain can accept and process more of the incoming information while participants are involved in lower intensity activities such as watching and feeding wild dolphins. Therefore *education/interpretation* is significantly associated with greater increases in knowledge, but the high intensity wildlife experiences block or impede receptivity to the positive effects of education on knowledge.

Information_general: The results for *Information_general* were unique compared to the other five samples as no significant effects were found for the ANOVA, yet a very strong significant effect was recorded for the interaction effect between *intensity x education/interpretation* and *Information_general* (level of satisfaction with information given about dolphins generally). In the three cases that offered high quality education/interpretation commentaries as a part of the experience, self-reported levels of visitor satisfaction with the information given about dolphins decreased as the level of *intensity of experience* increased. Once again, this complex situation appears to support the two notions that, in the presence of high intensity experiences, high quality education/interpretation inputs have a very limited role, whereas, in the absence of high intensity experiences, high quality education/interpretation inputs have a major role.

Information_conservation: ANOVA results for this construct were different from the previous one-way analyses. Highly significant effects, although in opposite directions, occurred for both independent variables, *intensity of experience* and *education/interpretation*, for level of visitor satisfaction with information given about how to help conserve the dolphins and their marine environment. For the first time across all results, the highest mean for the MANOVA analysis occurred in one of the moderate intensity samples for this construct. This result appears to support the notion of a synergistic relationship between *intensity of experience* and *education/interpretation*, but only for moderate level intensity activities. The combination of the moderate level of *intensity of experience* with the *education/interpretation* component created the highest level of satisfaction with *information about conserving dolphins* and *information about the marine environment*. It appears that a moderate level of *intensity of experience*, combined with a high quality *education/interpretation* commentary, creates the best results for maximum effects and enhancing this important outcome. A high *intensity of experience* creates a negative or nullifying effect, and low *intensity of experience* appears ineffective in arousing the affective domain which creates an optimal state of cognitive readiness for processing inputs and connecting with messages given in the oral presentation or accompanying commentary.

Support_conservation: The impact of the two independent variables, *intensity of experience* and *education/interpretation*, on *Support_conservation* is of particular interest as the results indicated that the level of support for marine conservation was relatively constant in both the low and high intensity wild dolphin tourism experiences. However, results differed markedly for the moderate intensity experiences. Specifically, the role of high quality *education/interpretation* components served to promote significantly increased support for marine conservation in moderate intensity experiences. It appears, then, that combining a moderate level intensity wildlife encounter activity with a high quality education/interpretation component creates the most successful and effective circumstances for strengthening participants' motivation to support conservation of dolphins and the marine environment (Fig. 7.10).

As noted previously, moderate level intensity dolphin tourism experiences provide optimal arousal from which information is internalised. The main reason for the negative effect of high intensity experiences on *Support_conservation* may be due to the total focus required by participants to participate in high intensity dolphin tourism activities. Further to this, the heightened level of arousal, referred to as a state of 'flow' or 'peak' experience (Csikszentmihalyi 1991) engages the affective domain to such an extent that additional or cognitive inputs cannot be processed. The highly affective core experience may fully eclipse the cognitive inputs of high quality education/interpretation components. The lower responses and results for the low intensity dolphin encounter experiences suggest that lower intensity experiences do not eclipse the cognitive domain, thus allowing other inputs and involvement of the cognitive domain to process information such as conservation messages.

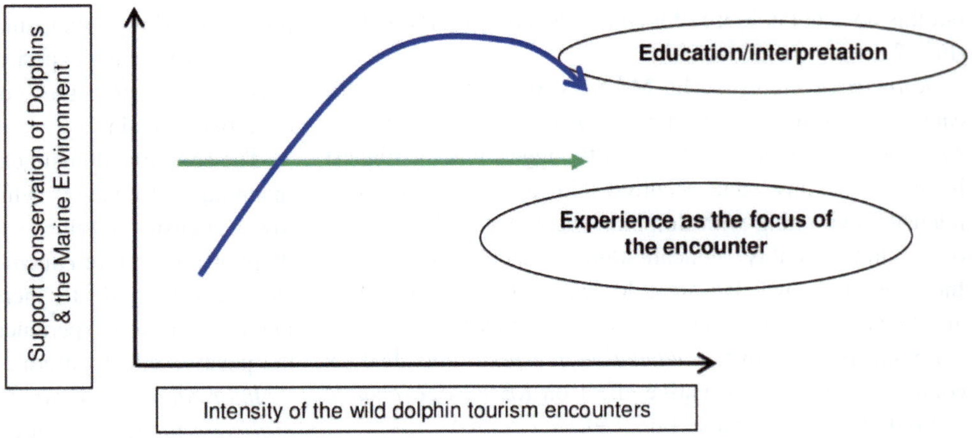

Fig. 7.10 Relationship between support for conservation and intensity of the experience (*Source* The author)

This evidence also appears to support the notion that moderate level intensity experiences in combination with high quality education/interpretation commentaries are the key to developing and delivering an effective and successful sustainable dolphin tourism experience that has a significant positive effect on the pro-environmental attitudes of participants. These optimal effective learning moments (ELMs) and emotional moments (EMs) (IFAW 1997) therefore require a moderate state of intensity or arousal plus pertinent and high quality educative/interpretive information for effective and significant enhancement of support for marine conservation (Fig. 7.10). The evidence also supports the notion that high intensity wildlife encounters are a powerful affective experience or emotional moment (EM) and have a significant effect on their own in enhancing pro-environmental attitudes and intentions of participants as effective learning moments (ELMs). Finally, low intensity experiences do not appear to create an optimal level of arousal or ELMs, and therefore require input from high quality educational and/or interpretation commentaries that have the potential to stimulate or arouse participants' affective levels.

7.7 Conclusion

High intensity experiences have the potential to interfere with, and even exclude, cognitive inputs such as education, learning, interpretation and pro-environmental effects on attitudes, beliefs, intended behaviours and intended actions. This comment needs further clarification. For general information about dolphins and the marine environment, *intensity of experience* and *education/interpretation* tend to work in different directions. Among lower intensity experiences, education/interpretation becomes an important driver of higher satisfaction with general information. With high intensity experiences, the role of education works against satisfaction with general information. The corollary of this is that intensity once again eclipses education through multi-sensory impact. In the absence of such high intensity sensory experiences, education/interpretation is more effective.

The overall results of this paper suggest the following facts about the role of *intensity of experience* in dolphin tourism activities where a high quality education/interpretation component is present:

1. *in the presence of high intensity encounter experiences, high quality education/interpretation components have a lesser role;*
2. *in the presence of moderate intensity encounter experiences, high quality education/interpretation components have a leading role; and,*
3. *in the presence of low intensity encounter experiences, high quality education/interpretation components must have a role.*

The MANOVA suggests that *education/interpretation* plays a role in increasing *Overall_satisfaction* in the absence of a high-intensity encounter experience, specifically among moderate and low intensity encounters. Alternatively, *education/interpretation* has a negative effect on *Overall_satisfaction* in high intensity dolphin encounters. In either explanation, *education/interpretation* is only displaced by high intensity experiences, playing a contributory role towards satisfaction in moderate and low intensity dolphin encounters.

Therefore, educational/interpretive commentaries do enhance knowledge, but it appears that this only occurs in the lower intensity wildlife encounter experiences. The participants of the B and C (lower) intensity wildlife situations may experience lower demands for their focus and concentration. They may experience lower to minimal levels of challenge and little need to master new skills. Lower intensity experiences allow participation in the activity and the ability to focus on other cognitive inputs.

In the case of the two C (lowest intensity) level activities where the *Overall_satisfaction* mean was higher in the presence of an *education/interpretation* component, participants may experience the lowest demands on their levels of concentration and focus, allowing the greatest opportunities for focusing on, and processing cognitive inputs. The *educative/interpretive* cognitive component may be enhanced by the lower intensity experience because the participants are able to accept and process the cognitive inputs in the form of education/interpretation information or the value added experience.

7.8 Recommendations

The Outward Bound approach to personal development and mastery used by some instructors is based on a belief in the intensity and profound experience of challenging and affective-based activities in remote and mountain environments. Intuitive instructors and guides refer to this as 'Letting the mountains speak'. This phrase parallels Tilden's (1977) recommendations for guides to be aware of the impact of the environment and to let the resource speak for itself. Tilden cautioned against saying too much, promoting instead an approach that gives minimal information with maximal impact. It appears that in the same context and with the same underlying theoretical foundations, intuitive marine wildlife guides operating with Effective Learning Moments, Heightened Emotional Moments, and peak experiences must recognise and utilise the high intensity experiences that have maximal effect and should consider 'Letting the oceans speak'. The highest intensity experience, in the absence of an education component, created the highest levels of overall satisfaction and resulted in the greatest mean increase in levels of knowledge about dolphins. High intensity wild dolphin experiences therefore appear to develop understanding and knowledge without the input of too much information.

When all the factors are taken into account, the intense and adventure-based process used to expose tourists to wildlife experiences appears to be a highly effective and educationally powerful experiential-based method of helping adults to learn about dolphins, the marine environment and the conservation of both in their leisure time. The peak experiences of high intensity dolphin encounters and the interactive and potentially synergistic relationship between moderate level intensity experiences and high quality education/interpretation components appears to offer valuable experiential education opportunities. This study also shows that different approaches are needed by interpretive guides for the high, moderate, and low intensity wild dolphin encounters.

Heightened emotional or affective states called Emotional Learning Moments (ELMs) were identified as states of arousal that guides should watch for and anticipate as an outcome of high intensity dolphin encounters and use strategically as an important affective input and essential synergistic component of the interpretation method. Wildlife guides need to be aware of: the role of intensity in wildlife encounters; the complex components of affect; and how to harness the affective domain in order to increase the effect of the interpretive component and take home messages on the pro-environmental attitudes, beliefs, intended behaviours and intended actions of participants.

The level of intensity of marine wildlife tourism encounters is an important aspect of the commercial tourism experience when a high quality education/interpretation commentary and more comprehensive education/ interpretation program may be included. The way in which the level of intensity of the experience affects the reception, processing and resulting attitudinal and behavioural outcomes of the education/interpretation commentary is complex. Intuitive and experienced guides, managers and facilitators who employ experiential education methods and strategies will be better equipped by understanding this complex and sometimes synergistic role of *intensity of experience* in wildlife tourism encounters. Wildlife guides will be able to develop more effective strategies and methods and have more effective tools for maximising the impact of the various levels of intensities of wildlife experiences on the participants. To this end, the Adventure Wave Model was adapted for marine wildlife encounters to assist in understanding how to approach an education/interpretation commentary for high, moderate, and low intensity wild dolphin encounter experiences.

The crests are the high, moderate or low intensity moments, which serve as the effective learning moments (ELMs). For high intensity encounters that fully involve the affective domain, the guides should wait until the peaks of the experiences have passed and then introduce non-complex cognitive inputs and/or interpretive messages. In moderate intensity encounters, guides can use the peaks of the experiences more rapidly in the commentary, building a high quality commentary around the encounters or peaks of the experience. Low intensity experiences have much lower wildlife adventure wave peaks. Therefore interpretive guides need to have a greater affective input in these low arousal sessions, delivering a more descriptive, detailed and comprehensive high quality interpretive commentary designed to arouse the participants, involve their affective domains, thus creating optimal circumstances for reception of pro-conservation messages.

Encounters with whales and dolphins can be high intensity experiences and skilled guides should work with the intensity and 'let the oceans speak'. Conservation information can be interspersed throughout the trip and/or confined to the return journey. Further to this, if participants are involved in high intensity dolphin encounters which fully involve the affective domain, the results suggest that the education/interpretation commentary should be kept to a minimum. The relevant information and conservation messages can be left until later or be introduced during the return trip to the place of disembarkation.

The return journeys represent the reflection and debrief stages of an experiential education cycle, where reflection and sharing of experiences builds on and reinforces what participants experienced and learning takes place (Joplin 1995; Priest and Gass 1997; Boud et al. 1993). This is the time when participants are able to reflect on their experience, be more receptive to information, and process any new or introduced cognitive inputs. This is the time to summarise the experience, reflect on the trip encounters, deliver the take home conservation messages and increase participants' pro-environmental motivation and intentions and ensure that transfer from the experience to the home environment occurs.

Understanding the effects of the differing levels of intensity of experiences on participants' information processing abilities will better equip interpretation guides for maximising the impacts of wildlife tourism encounters on participants' pro-environmental attitudes, beliefs, intended behaviours and intended actions and for minimising the impacts on the dolphins and the marine environment. The results of this study can also form the basis of recommendations for the dolphin tourism industry and marine wildlife operators, guides and teachers involved with wildlife tourism activities, thus contributing to the sustainability of the marine wildlife tourism industry. The results also contribute to the small pool of knowledge and research addressing optimal conditions for the synergy between the affective and cognitive domains or in this investigation: the intensity of experiences and education/learning.

References

Ajzen I (1985) From intentions to actions: a theory of planned behaviour. In: Beckman JKJ (ed) Action-control: from cognition to behaviour. Springer, Heidelberg, pp 11–39

Ajzen I (1991) The theory of planned behaviour. Organ Behav Hum Decis Process 50:179–211

Ajzen I (1992) Persuasive communication theory in social psychology: a historical perspective. Influencing human behaviour. Sagamore Publishing Inc., pp 1–27

Ajzen I, Fishbein M (1980) Understanding attitudes and predicting social behaviour. Prentice Hall, Englewood Cliffs

Beasley I (1992) Marine mammal tourism: educational implications and legislation. University of Otago, Dunedin

Beaumont N (1999) Ecotourism: the contribution of educational nature experiences to environmental knowledge, attitudes and behaviours. Doctor of philosophy thesis, Griffith University, Brisbane

Beeton S, Weiler B, Ham S (2005) Applying persuasive communication theory to managing visitor behaviour: a scoping study at Port Campbell national park. CRC sustainable tourism monograph series. Gold Coast, QLD, Australia

Bennett M, Kriwoken LK (2001) Managing wildlife on Australia's sub-antarctic Macquarie Island. Paper presented to first national conference on sustainable wildlife tourism, Hobart, Tasmania, 28–30 Oct 2001

Boud D, Cohen R, Walker D (eds) (1993) Using experience for learning. SRHE and Open University Press, Buckingham, UK

Boud D, Keogh R, Walker D (1985) Reflection: turning experience into learning. Kogan Page, London

Carpenter G, Priest S (1989) The adventure experience paradigm. Leisure Stud 8(1):65–75

Ceballos-Lascurain H (1991) Tourism, ecotourism, and protected areas. In Kusler JA (ed) Ecotourism and resource conservation, a collection of papers, vol 1(1). Omnipress, Madison, pp 24–30

Cohen E (1979) A phenomenology of tourist experiences. Sociology 13:201

Cottrell SP, Graefe AR (1993) General responsible environmental behaviour among boaters on Chesapeake Bay. In: Proceedings of the 1993 northeastern recreation research symposium, Park and Recreation Resources Department, Michigan State University, Michigan

Cottrell SP, Graefe AR (1994) Specific responsible environmental behaviour among boaters on Chesapeake Bay: a predictive model. In: Proceedings of the 1994 northeastern recreation research symposium, Park and Recreation Resources Department, Michigan State University, Michigan

Csikszentmihalyi M (1975) Play and intrinsic rewards. J Humanist Psychol 15(3):41–63

Csikszentmihalyi M (1991) Flow: the psychology of optimal experience. Harper and Rowe, New York

Davis D, Birtles RA, Valentine P, Cuthill M, Banks S (1997) Whale sharks in Ningaloo Marine Park. Tourism Manag 18(5):259–271

DEH (2005) Australian national guidelines for whale and dolphin watching. Department of Environment and Heritage, Canberra. http://www.deh.gov.au/coasts/publications/pubs/whale-watching-guidelines-2005.pdf. Accessed 25 June 2005

Dewey J (1938) Experience and education. MacMillan, New York

di Castri F, Balaji V (eds) (2002) Tourism, biodiversity and information. Backhuys Publishers, Leiden

Dickson TJ, Chapman J, Murrell M (2000) Risk in the outdoors: the perception, the appeal, the reality. Aust J Outdoor Educ 4(2): 10–17

Dierking LD (1998) Interpretation as a social experience. In: Uzzell D, Ballantyne R (eds) Contemporary issues in heritage and environmental interpretation. The Stationary Office, London, pp 56–76

Dodson RM, Yerkes JM (1908) The relationship of strength stimulus to rapidity of habit formation. J Comp Neurol Psychol 18:459–482

Driver B, Tocher S (1979) Toward a behavioural interpretation of recreational engagements with implications for planning. In: Van Doren CS, Priddle GB, Lewis JE (eds) Land and leisure: concepts and methods in outdoor recreation, 2nd edn. Maaroufra Press, Chicago, pp 86–104

Eagles PFJ (2001) Planning and management of wildlife: meeting the sustainability challenge. Paper presented to the first national sustainable wildlife tourism convention, Hobart, Tasmania, 28–29 Oct 2001

Eagly AH, Chaiken S (1993) The psychology of attitudes. Harcourt Brace Jovanovich, Fort Worth

Ellis MJ (1973) Why people play. Prentice Hall Inc., Englewood Cliffs

Enzenbacher D (1992) Tourists in Antarctica: numbers and trends. Polar Rec 28(164):17–22

Enzenbacher D (1994) Antarctic tourism: an overview of the 1992/93 season activity, recent developments, and emerging issues. Polar Rec 30:105–116

Epler-Wood M (2002) Ecotourism: principles & policies for sustainability. United Nations Environment Program, Paris

Esses VM, Haddock G, Zanna MP (1993) Values, stereotypes, and emotions as determinants of intergroup attitudes. In: Mackie DL (ed) Affect, cognition, and stereotyping: interactive processes in group perception. Academic Press, San Diego

Ethos Consulting (1991) The commercial wildlife viewing report: a discussion of market and development potential. Report prepared for the Ministry of Tourism, BC, Canada

Finkler W, Higham J (2004) The human dimensions of whale watching: an analysis based on viewing platforms. Hum Dimensions Wildl 9(2) Summer:103–117

Fishbein M, Ajzen I (1975) Belief, attitude, intention and behaviour: an introduction to theory and research. Addison-Wesley, Reading

Garrod B, Wilson JC (2003) Aspects of tourism 7: marine ecotourism—issues and experiences. Aspects of tourism series. Footprint Books, Church Point

Gössling S (2005) Ecotourism as an experience industry: nature-based tourism and ecotourism in Scandinavia. In: Kylanen M (ed) Articles on experiences, vol 2(2). University of Lapland Press, Rovaniemi, pp 28–39

Gray GG (1993) Wildlife and people. University of Illinois Press, Urbana

Green R, Higginbottom K, Northrope C (2001) A tourism classification of Australian wildlife. Wildlife tourism research report series no 7. Status assessment of wildlife tourism in Australia series. CRC Sustainable Tourism, Pty Ltd, Gold Coast

Gyimothy S, Mykletun RJ (2005) Play in adventure tourism. The case of Arctic trekking. Ann Tourism Res 31(4):855–878

Ham S (1992) Environmental interpretation: a practical guide for people with big ideas and small budgets. Fulcrum/North American Press, Golden

Ham S (2004a) Interpretation isn't magic—an academic's view on the role of research in teaching interpretation. J Interpretation Res 9 (1):65–68

Ham S (2004b) Creating unforgettable event experiences - making a difference by making meaning. Paper presented to 2004 events Tasmania exchange conference, Launceston, Tasmania, 19 May 2004
Ham S, Weiler B (2004) Introduction to interpretation. Paper presented to a workshop on interpretation at tertiary level: beyond visitor satisfaction. QUT, Brisbane
Hanin YL (2000) Emotions in sport. Human Kinetics, Champaign
Hannabus S (1999) Postmodernism and the heritage experience. Library Manag 20:295–302
Higginbottom K (2004) Wildlife tourism: impacts, management and planning. Common Ground Publishing Pty Ltd, Altona
Higham JES, Carr A (2002) Ecotourism visitor experiences in Aotearoa/New Zealand: challenging the environmental values of visitors in pursuit of pro-environmental behaviour. J Sustain Tourism 10(4):277–294
Hopkins D, Putnam R (1993) Personal growth through adventure. David Fulton Publishers, London
Hoyt E (2001) Whale Watching 2001 Worldwide tourism numbers, expenditures, and expanding socioeconomic benefits. International Fund for Animal Welfare, Yarmouth Port
IFAW (International Fund for Animal Welfare) (1995) Report on the workshop on the scientific aspects of managing whale watching. IFAW, Monte Castello di Vibo
IFAW (International Fund for Animal Welfare) (1997) Report of the workshop on the socioeconomic aspects of whale watching. IFAW, Kaikoura
IFAW (International Fund for Animal Welfare) (2004) The growth of whale watching tourism in Australia. IFAW Asia Pacific, Surry Hills
IFAW, IWWF, WDCS (International Fund for Animal Welfare, International World Wildlife Fund, Whale & Dolphin Conservation Society) (1997) Report of the international workshop on the educational values of whale watching, Provincetown, Massachusetts
Ingram C (2001) An assessment of the possible impacts of recreation and tourism on the Australian sea-lion at Carnac Island Nature Reserve. Paper presented to the first national sustainable wildlife tourism conference, Hobart, 28–30 Oct 2001
Itin CM (1999) Reasserting the philosophy of experiential education as a vehicle for change in the 21st century. J Exp Educ 22(2):91–98
Joplin L (1981) On defining experiential education. J Exp Educ 4 (1):17–20
Joplin L (1995) On defining experiential education. In: Warren K, Sakofs M, Hunt JJS (eds) The theory of experiential education. Kendall/Hunt Publishing Company, Boulder, pp 15–22
Knapp D, Benton G (2005) Long-term recollections of an environmental interpretive program. J Interpretation 10(2):51–53
Knudsen DM, Cable TT, Beck L (1995) Interpretation of cultural and natural resources. Venture Publishing Inc., State College
Kolb DA (1984) Experiential learning. Prentice Hall, Englewood Cliffs
Kraft R, Sakofs M (eds) (1991) The theory of experiential education. Association for Experiential Education, Boulder
Langer EJ (1992) Matters of mind: mindfulness/mindlessness in perspective. Conscious Cogn 1(3):289–305
Lengkeek J (2001) Leisure experience and imagination: rethinking Cohen's modes of tourist experience. Int Sociol 16:173–184
Liddle J (1998) Risk management: walking the tightrope. J Exp Educ 21(2):61–62
Lindberg K, Furze B, Staff M, Black R (eds) (1998) Ecotourism in the Asia Pacific region: issues and outlook. Routledge, London
Li Y (2000) Geographical consciousness and tourism experience. Ann Tourism Res 27:863–883
Luckmann C (1996) Defining experiential education. J Exp Educ 19 (1):6–7
Maher PT (2005) The nature of the sea: a framework for exploring visitor experiences in the Ross Sea Region, Antarctica. In: Kylanen M (ed) Articles on experiences 2. University of Lapland Press, Rovaniemi
Manfredo MJ (1992) Influencing human behaviour. Theory and applications in recreation, tourism, and natural resources management. Sagamore Publishing Inc., Champaign
Mannell RC, Iso-Ahola SE (1987) Psychological nature of leisure and tourism experience. Ann Tourism Res 14:314–331
Markwell K (1998) Taming the 'Chaos of Nature': Cultural constructions and lived experience in nature-based tourism. Unpublished PhD dissertation thesis, University of Newcastle, Newcastle, New South Wales
Markwell K (2004) Constructing, presenting and interpreting nature: a case study of a nature-based tour to Borneo. Ann Leisure Res 7 (1):19–33
Martin AJ (2001) Towards the next generation of experiential education programmes: a case study of Outward Bound. Thesis presented in partial fulfilment of the requirements for the degree of doctor of philosophy thesis, Massey University, New Zealand
Mayes G, Dyer P, Richins H (2004) Dolphin-human interaction: changing pro-environmental attitudes, beliefs, behaviours and intended actions of participants through management and interpretation programs. Ann Leisure Res 7(1):34–53
McIntosh AJ, Prentice RC (1999) Affirming authenticity: consuming cultural heritage. Ann Tourism Res 26:589–612
McIntyre AD (1995) Human impact on the oceans: the 1990s and beyond. Mar Pollut Bull 31(4–12):147–151
Miles JC, Priest S (1990) Adventure education. Venture Publishing, State College
Mortlock C (1984) The adventure alternative, 2 edn. Cicerone Press, Amazon International
Moscardo G (1999) Making visitors mindful: principles for creating sustainable visitor experiences through effective communication, vol 2. Sagamore Publishing, Champaign
Moscardo G (2003) Interpretation and sustainable tourism: functions, examples and principles. J Tourism Stud 14(1):112–123
Moscardo G, Saltzer R (2004) Understanding wildlife tourism markets. In: Higginbottom K (ed) Wildlife tourism. Common Ground Publishing Pty Ltd, Altona, pp 166–185
Moscardo G, Saltzer R (2005) Understanding tourism wildlife interactions. CRC for Sustainable Tourism, Pty Ltd, Gold Coast
Muller TE, Cleaver M (2000) Targeting the CANZUS baby boomer explorer and adventurer segments. J Vacation Mark 6(2):154–169
Muloin S (1998) Wildlife tourism: the psychological benefits of whale watching. Pac Tourism Rev 2:199–213
Neil DT, Orams MB, Baglioni AJ (1996) Effect of previous whale watching experience on participants' knowledge of, and response to, whales and whale watching. Encounters with Whales '95, Australian Nature Conservation Authority, Canberra
Newsome D, Moore SA, Dowling R (2002) Aspects of tourism 4: natural area tourism—ecology, impacts and management. Footprint Books, Mona Vale
Newsome D, Lewis A, Moncrieff D (2004) Impacts and risks associated with developing unsupervised stingray tourism at Hamelin Bay, Western Australia. Int J Tourism Res 6:305–323
Norman B (2000) The whale shark: current conservation initiatives. Australian Marine Conservation Society, Perth
Norman B (2001) Whale shark tourism = whale shark conservation. Paper presented to the first national sustainable wildlife tourism convention, Hobart, Oct 2001
Ooi CS (2002) Cultural tourism and tourism cultures: the business of mediating experiences in Copenhagen and Singapore. Copenhagen Business School Press DK

Ooi CS (2004) Tourism experiences as attention products. Lapland Centre of Expertise for the Experience Industry & The Experience Institute, Rovaniemi, Lapland

Orams MB (1993) The role of education in managing marine wildlife—tourist interaction. Paper presented to the 7th annual marine education society of Australasia conference, Brisbane

Orams MB (1995a) Development and management of a feeding program for wild bottlenose dolphins at Tangalooma, Australia. Aquat Mammals 21(2):137–147

Orams MB (1995b) Managing interaction between wild dolphins and tourists at a dolphin feeding program, Tangalooma, Australia. Doctor of philosophy thesis, University of QLD Brisbane

Orams MB (1995c) Using interpretation to manage nature-based tourism. J Sustain Tourism 4(2):81–93

Orams MB (1996) A conceptual model of tourist-wildlife interaction: the case for education as a management strategy. Aust Geogr 27 (1):39–51

Orams MB (1997a) Historical accounts of human-dolphin interaction and recent developments in wild dolphin based tourism in Australasia. Tourism Manag 18(5):317–326

Orams MB (1997b) The effectiveness of environmental education: can we turn tourists into 'Greenies'? Prog Tourism Hospitality Res 3:295–306

Orams MB (1997c) Wild dolphin based tourism: minimising the risks and maximising the benefits. Paper presented to pacific congress on marine science and technology, Honolulu, Hawaii

Orams MB (2002) Humpback whales in Tonga: an economic resource for tourism. Coastal Manag 30:361–380

Orams MB, Forestell PH (1998) From whale harvesting to whale watching: Tangalooma 30 years on. Paper presented to recent advances in marine science and technology '94, Townsville

Orams MB, Hill JG (1998) Controlling the ecotourist in a wild dolphin feeding program: is education the answer? J Environ Educ 29 (3):33–38

Orams MB, Hill GJE, Baglioni AJJ (1996) 'Pushy' behaviour in a wild dolphin feeding program at Tangalooma, Australia. Mar Mammal Sci 12(1) January:107

Packer J, Ballantyne R (2004) Is educational leisure a contradiction in terms? Exploring the synergy of education and entertainment. Ann Leisure Res 7(1):54–71

Parsons ECM, Warburton CA, Woods-Ballard A, Hughes A, Johnston P, Bates H, Leück M (2003) Whale-watching tourists in West Scotland. J Ecotourism 2(2):93–113

Patterson DL, Holm EJ, Carney KM, Fraser WR (1996) Effects of tourism on the reproductive success of Adélie penguins at Palmer Station: preliminary findings. Antarct J Unit States 31(1)

Petty R, Cacioppo JT (1984) The effects of involvement on responses to argument quantity and quality: central and peripheral routes to persuasion. J Pers Soc Psychol 46:69–81

Petty R, McMichael S, Brannon L (1992) The elaboration likelihood model of persuasion: applications in recreation and tourism. In: Manfredo MJ (ed) Influencing human behaviour: theory and applications in recreation, tourism and natural resources management. Sagamore Press, Champaign, pp 77–101

Pooley JA, O'Connor M (2000) Environmental education and attitudes: emotions and beliefs are what is needed. J Environ Behav 32 (5):711–723

Priest S, Gass MA (1997) Effective leadership in adventure programming. Human Kinetics, University of New Hampshire, UK

Priest S, Naismith M (1993) The debriefing funnel. J Adventure Educ Outdoor Leadersh 10(3):20–22

Priskin J (2003) Tourist perceptions of degradation caused by coastal nature-based recreation. Environ Manage 32(2):189–204

Reynolds PC, Braithwaite D (2001) Towards a conceptual framework for wildlife tourism. Tourism Manag 22(1):31–42

Roe D, Leader-Williams N, Dalal-Clayton B (1997) Take only photographs, leave only footprints: the environmental impacts of wildlife tourism. International Institute for Environmental Development, London

Ryan C, Hughes K, Chirgwin S (2000) The gaze, spectacle and ecotourism. Ann Tourism Res 27(1):148–163

Schanzel HA, McIntosh AJ (2000) An insight into the personal and emotive context of wildlife viewing at the Penguin Place, Otago Peninsula, New Zealand. J Sustain Tourism 8(1):36–52

Schoel J, Prouty D, Radcliffe P (1988) Islands of healing: a guide to adventure based counselling. Project Adventure Inc., Hamilton

Shackley M (1996) Wildlife tourism. International Thomson Business Press, London

Smith L, Weiler B, Ham S (2005) How tourists react to profound wildlife experiences. In: TRU stories: insights from tourism research, Tourism Research Unit, Monash University, Melbourne, Victoria

Sorice MG (2001) A multistakeholder perspective on human interactions with the West Indian Manatee (*Trichechus manatus*) in Crystal River, Florida. Master's thesis, Texas A&M

Sorice MG, Shafer CS, Scott D (2003) Managing endangered species within the use/preservation paradox: understanding and defining harassment of the West Indian Manatee. Coastal Manag 1(31):319–338

Speedie C (2003) Marine ecotourism potential in the waters of South Devon and Cornwall. In: Garrod B, Wilson JC (eds) Marine ecotourism issues and experiences. Channel View Publications, Clevedon, pp 204–214

Sproull NL (2002) Handbook of research methods: a guide for practitioners and students in the social sciences, 2nd edn. Scarecrow Press, Lanham

Stewart EJ, Draper D, Johnston ME (2005) A review of tourism research in the Polar Regions. Arctic 58(4):383–394

Tarssanen S, Kylanen M (2005) A theoretical model for producing experiences—a tourist's perspective. In: Kylanen M (ed) Articles on experiences 2, vol 2. Lapland Centre of Expertise for the Experience Industry, Rovaniemi, Lapland, pp 132–151

Tilden F (1957) Interpreting our heritage. University of North Carolina Press, Chapel Hill

Tilden F (1977) Interpreting our heritage, 3rd edn. University of North Carolina Press, Chapel Hill

Tisdell C, Wilson C (2000) Wildlife-based tourism and increased support for nature conservation financially and otherwise: evidence from sea turtle ecotourism at Mon Repos. University of Queensland Press, University of Queensland, Working paper on economics, ecology and the environment

Tisdell C, Wilson C (2004) Economics, wildlife tourism and conservation: three case studies. CRC Sustainable Tourism Pty. Ltd., Gold Coast

TIES (The International Ecotourism Society) (2000) Marine ecotourism guidelines project, draft for review. The International Ecotourism Society, North Bennington, Vermont

Townsend C (2003) Marine ecotourism through education: a case study of divers in the British Virgin Islands. In: Garrod B, Wilson JC (eds) Marine ecotourism. Channel View Publications, Clevedon

Twynam D, Johnston M, Payne B, Kingston S (1998) Ecotourism and sustainable tourism guidelines: an annotated bibliography. The International Ecotourism Society, North Bennington

Urry J (1990) The tourist gaze. Leisure and travel in contemporary societies. Collection Theory, culture & society. London, Sage Publications

Urry J (1995) Consuming places. Routledge, London

Valentine PS, Birtles AM, Curnock M, Arnold P, Dunstan A (2004) Getting closer to whales—passenger expectations and experiences, and the management of swim with dwarf minke whale interactions in the Great Barrier Reef. Tourism Manag 25(6):647–655

Walker GJ, Hull RB, Roggenbuck JW (1998) On-site optimal experiences and their relationship to off-site benefits. J Leisure Res 30(4):453–471

Wallace GN, Pierce SM (1996) An evaluation of ecotourism in Amazonas, Brazil. Ann Tourism Res 23(4):843–873

Warner Weil S, McGill I (eds) (1989) Making sense of experiential learning. Diversity in theory and practice. Open University Press, Milton Keynes

Wilson C, Tisdell C (2003) Conservation and economic benefits of wildlife-based marine tourism: sea turtles and whales as case studies. Hum Dimensions Wildl 8:49–58

Woods B, Moscardo G (2003) Enhancing wildlife education through mindfulness. Aust J Environ Educ 19:97–108

Yin RK (2003) Case study research design and methods, 3rd edn. Sage Publications, Thousand Oaks

Zook LR (1986) High adventure pursuits. Publishing Horizons, Columbus

8 Wildlife Resources, Habitats and Ecosystems for Visitors' Experiential Learning: Educative Wildlife Tourism in the Australian Context

Ismar Borges de Lima

Tilden's fundamental thesis:
"Through interpretation, understanding; through understanding, appreciation; through appreciation, protection."
National Park Service, U.S. Department of the Interior Administrative Manual (1953, p. 38)

Abstract

The aim of this chapter on Australian wildlife and tourism is manifold. It discusses the major existing elements necessary for consolidating an educative wildlife tourism within an experiential learning perspective for the visitors by considering some aspects of Kolb's theory. The chapter is concerned with ecological and biological resources, and related phenomena, that are relevant for a meaningful environmental interpretation and education; one of the foundations for an educative tourism together with conservation. The chapter begins by presenting the current protected areas in Australia and their relevance as natural settings and habitats for wild animals and tourism. The discussion continues by critically appraising the role of rangers in managing protected areas, natural resources and visitors. The role of rangers and guides in Parks is fundamental for enhancing visitors' experiences and understanding of natural and cultural settings, landscapes, wildlife, and ecosystems. Rangers also play an important role in promoting visitor education as a way of mitigating possible negative impacts in sensitive natural areas. Yet, the chapter outlines the most popular wild animals by providing a comprehensive description of koalas, kangaroos and Tasmanian devils. The biofacts, physical characteristics, behaviour and pertinent ecological aspects are presented to demonstrate how rich and important wildlife is for tourism, especially for an educative learning tourism that can contribute to connect humans to nature in many ways. The chapter was written based on the outcomes of post-doctoral research qualitatively oriented, based on the pertinent literature, active and observant participation, and on the analysis of websites and documents. Considering a relative paucity of publication on educative wildlife tourism, the chapter seeks to fill some gaps in the literature and to advance the debates on the importance of conservation and protection of wildlife resources within an environmental science perspective.

I. Borges de Lima (✉)
Southern Cross University (SCU), Gold Coast, Australia
e-mail: ismarborgeslima@gmail.com; ismarlima@yahoo.com.br

I. Borges de Lima
Recinatur Foundation and MultiAmazon Lab, UNIFESSPA, State University of Roraima (UERR), Boa Vista, Brazil

I. Borges de Lima and R.J. Green (eds.), *Wildlife Tourism, Environmental Learning and Ethical Encounters*, Geoheritage, Geoparks and Geotourism, DOI 10.1007/978-3-319-55574-4_8

8.1 Introduction

About 1800 years ago, Ptolemy, an ancient Greek geographer, included Australia on one of the earliest maps of the world. It was named *Terra Australis Incognita*, an island continent, which means "unknown southern land", the smallest of the seven continents, but the only one to hold just one country…"Australia's geography and wildlife are unlike anywhere else in the world…[throughout its eight ecoregions], animals such as kangaroos, koalas, kookaburras…[and platypuses]" (Banting 2003, p. 4). Australian unique wildlife and varied landscapes are both of great appeal for tourism, and they demand conservation and protection. With a rich biodiversity, Australia becomes an ideal place for outdoor environmental learning experiences for both domestic and foreign visitors. In her handbook on wildlife tourism, Ronda Green mentions why Australia has been so different compared to the rest of the world with regards to its ecosystems and, especially to its wildlife resources,

> Australia was once part of the great southern super-continent called Gondwana […] Australia left Gondwana before any hoofed animals, cats, bears, monkeys, rodents or other placental mammals reached it, but it did have monotremes (egg-laying mammals) and marsupials. As it drifted northwards [it became] isolated from other continents for many millions of years…and developed many unique species, even whole families – koalas, numbats, lyrebirds and many many more. Much more recently (about 3,000 years ago) humans introduced dogs (dingo), and may have introduced a few other species (e.g. bladey grass, bracken) that are widespread in southeast Asia and have been in Australia a long time […] white explorers and settlers have introduced a great variety of animals and plants that have gone wild – rabbits, buffaloes, foxes, blackberries, lantana… the list is very extensive (Green 2014, pp. 31–32).

This brief, but extraordinary citation, reveals how especial is wildlife for tourism and people in Australia. It is advocated in this chapter that the ecological and biological resources and related phenomena are critical for a meaningful environmental interpretation and education; one of the foundations for an educative tourism in line with conservationist management, practices and attitudes in benefit of the natural world and humankind. Protected areas in Australia, the role of rangers and guides, wildlife resources, experiential learning, environmental interpretation and education, and facts and biographic distribution of kangaroos, wallabies, platypuses, dingos, wombats, flying foxes, cassowaries, greater bilby, echidnas, koalas, crocodiles, and Tasmanian devils are part of the discussion in the chapter. A great collection of figures with pictures and diagrams, and tables, helps to illustrate the content. Biofacts, physical characteristics, wild animal behaviour and pertinent ecological aspects are thoroughly presented. What is necessary for developing a meaningful educative wildlife tourism through experiential learning? This is the leading question to be answered.

8.2 Australian Protected Areas, Conservation and Visitors: Natural Settings for Environmental Interpretation and Education

Australia has over 9000 protected areas which cover roughly 95 million hectares, which makes it one of the nations with the greatest proportions of protected areas in terms of land-mass in the world (TTF 2013, p. 5). National Park is one of the categories as a protected area. The notions and definitions of a park vary institutionally and geographically, and the term eludes various approaches regarding its many possible uses in different regions and countries; a natural setting that is known as a park in one place may be perceived as a recreational area in another. A historical record shows that a park has been "diverse things as a place to bathe, a hunting preserver, a formal garden…a common space for tethering livestock prior to bartering…in some countries, a place for exercising, walking and nature viewing" (Lankford et al. 2011, p. 4). The term 'national park' is something of a misnomer in Australia, and that most of so-called 'national parks' (of which there are hundreds) are actually state-run.

National Parks usually accommodate an array of outdoor recreation and adventure activities from organised sports, such as mountain-biking, canoeing, rock climbing, abseiling, whitewater rafting to bushwalking, wildlife watching, and nature contemplation (Bell 2005). Often parks are classified based on the types of activities they support and by their use, and a management system can have a holistic approach or a narrow one. Two common categories of recreation areas are activity-oriented structured recreation, with developed structures—thus, more anthropocentric oriented; or resource-oriented non-structured recreation, which gives to this type of park a more biocentric orientation (Frawley 1989; Hu 2002; Lankford et al. 2011; Cocks and Simpson 2015).

In terms of management systems for parks and protected areas, Australia has one of the oldest systems in the world (Frawley 1989; Parkin 2006). The Australian system for park management, for example, is spread at different jurisdictional levels, a matter for Territory, State, and local governments (Baird 1986). Parkin (2006) identified that each "state and territory has its own conservation-focused legislation for the creation and management of protected areas and, or, other natural resource legislation for the protection of flora and fauna" (p. 8). This type of arrangement has led to ten different systems to manage protected areas in Australia

(Worboys et al. 2001). Management can be defined as a set of activities related to decision-making, leadership, planning, and controlling in regard to the various resources of an organisation, such as infrastructural, structural, informational, human and financial resources. The main objective of management is to produce satisfactory managerial and operational outcomes (Davidson et al. 2006).

According to Merriam-Webster Dictionary, management refers to "the act or skill of controlling and making decisions about a business, department, sports team; etc.; the act or process of deciding how to use something". As applied to a park context, management can be understood as any decision-making aiming to promote the effective operationalisation of the Parks, including staff, visitors, finances, conservation, protection, enforcement of law, monitoring, landscapes, water catchment and waterways, and the well-desired protected state of natural areas, that is, decisions over biotic, abiotic elements, and the ecosystems, in the protected areas (Cunningham et al. 2005; Odum 2006). Howard (2013) defines 'managers' as individuals in charge of coordinating efficiently and effectively many resources in their duty areas, for example, the National Parks. The great load of responsibilities in Park management is given to field-related government agencies and departments.

Australian protected area agencies manage a very significant proportion of Australia's natural and cultural assets at local, state, territory, and national level (refer to Table 8.1). In Queensland, protected areas management is under two main Departments: the Department of Natural Resources and Mines; and the Environmental Protection Authority (EPA), which has a specific agency for park management, the Queensland Parks and Wildlife Service (QPWS). In Northern Territory, New South Wales and ACT, parks services have specific policies, plans and manuals which guide them through distinct aspects of natural, cultural, and heritage assets management.

The Australian States do not have a sole integrated park management framework, and "the management of protected areas in Australia involves elements of the multiple use and ecosystem models of land management" (Lawrence 1996). Though this leads to levels of autonomy in terms of management, on the other hand, it may create a more complex park management system with each State ruling and interpreting laws and policies in a distinct and particular view. But, by assessing the main aspects of parks management frameworks it is noted they have several intersections and approaches. Some protected area agencies manage huge territorial land, for example, Parks Victoria is in charge of managing 16% of that State (Stone 2001). Buckley et al. (2003) explain that "each national park under the administration of Parks Australia includes guidelines for asset management in its individual park-management plans" (pp. 56–57).

Australia has 17.88% of its landmass protected in the National Reserve System, NRS, totalling 10,339 units of protected areas over eight states and territories with 137,501,551 ha. The Australian IUCN Reserve Types in the National Reserve System (NRS), there are seven major categories of protected areas: Strict Nature Reserve (IA), Wilderness Area (IB), National Park (II), Natural Monument (III), Habitat/Species Management Area (IV), Protected Landscape/Seascape (V), and Managed Resource Protected Area (VI) (refer to Fig. 8.1). Under the NRS, it was identified 1,086 National Parks covering an area of 38,053,578 ha (Department of Environment, Australia Government). This gives a notion of the terrestrial extension of protected areas in the country not including the marine reserves.

According to the Department of Environment, Australian Government, the vast majority of land belonging to the NRS is open for public access, and visit is controlled by each management plan of the protected area to minimise possible negative impacts and disturbance to sensitive fauna and flora. Restrictions also apply to Indigenous sacred sites in respect to ethnic and cultural issues. Apart from the government protected land, there is private land under the status of protected areas totalling 1,223 units covering 1% of Australia; in general, this type of property belongs to private landholders, community groups, organisations, trustees, and most of them are also open for the public and have a pivotal role in protecting biodiversity in peri-urban or rural areas; they also run volunteer programs, and some properties are equipped and have facilities such as camping sites and walking trails to host independent visitors and tour groups (Department of Environment, Australia Government).

Despite of the existing management plans for running the protected areas, the National Parks Australia Council, NPAC, a non-governmental organisation which represents the views of State and Territory NGOs in the country since its creation in 1975, in a public communication, alerted that the national parks across Australia have faced critical managerial and operational threats such as over-development, including commercial one. In response to these issues, NPAC has promoted awareness campaigns to encourage the various government spheres to implement strategies which can strengthen and reinvigorate the national parks system seeking to guarantee it as a legacy for all Australians and visitors.

In the 1950s and 1960s, the national parks as protected areas with special flora and fauna to be used as recreational resources by the public brought concerns over the adequate level of its use (Manning 2002), and in the beginning of the 1960s the notions of 'carrying capacity' started to pervade debates and the literature. At that time, it became a common sense that natural settings and resources on earth would have their limits for use, making both wilderness management and

Table 8.1 Protected Area Management Agencies in Australia at national, state, territory, local level in the National Reserve System (NRS)

At national level	Commonwealth	• **Environment Australia, including**: – **Parks Australia**: Six Commonwealth National Parks, the Australian National Botanic Gardens, and 58 Commonwealth Marine Reserves – Great Barrier Reef Marine Park Authority – Wet Tropics Management Authority		
			State area (ha)	**Number protected areas, hectares, and land percentage**
At state and territory level	Australian Capital Territory	• Australian Capital Territory Parks and Conservation Service	235,813	**46** 130,214 ha **(55.22%)**
	New South Wales	• Department of Environment and Conservation's Division of: Parks and Wildlife • State Forests	80,121,268	**925** 7,293,630 ha **(9.10%)**
	Northern Territory	• Parks and Wildlife Commission	134,778,762	**81** 25,129,386 ha **(18.64%)**
	Queensland	• **Environmental Protection Authority (EPA)**'s Division of: **Queensland Parks and Wildlife Service (QPWS)** • Department of Natural Resources and Mines	172,973,671	**1086** 14,108,222 ha **(10.26%)**
	South Australia	• Department for Environment and Heritage	98,422,137	**1995** 29,394,607 ha **(29.87%)**
	Tasmania	• Tasmanian Parks and Wildlife Service	6,840,133	**1524** 3,015,707 ha **(44.09%)**
	Victoria	• Department of Natural Resources and Environment • Parks Victoria	22,754,364	**3056** 4,012,124 ha **(17.63%)**
	Western Australia	• Department of Conservation and Land Management	252,700,808	**1607** 54,375,439 ha **(21.52%)**
At local level	Municipalities/Districts	Management of specific district protected areas: wetlands, river corridors, and bushland reserves by local government agencies (e.g. Councils) which are usually directed/guided by pertinent state or territory legislation and, or, local law		

Source Adapted from Parkin (2006), and Worboys et al. (2001), with additional information obtained online in the government agencies and reports, among them Department of Environment, Australia Government, CAPAD 2014

visitors' management as crucial to park management. How much is too much for parks to accommodate visitors? As mentioned by Manning (2002), "the working hypothesis was that increasing numbers of visitors causes greater environmental impact as measured by soil compaction, destruction of vegetation, and related variables" (p. 307). A CRC Report prepared by Higginbottom and Buckley (2003) has a thorough study and data on terrestrial wildlife viewing in Australia. Fredline (2007) also makes her contributions by assessing the domestic market for wildlife tourism in Australia; the study was also concerned with the wildlife tourism behaviour and visitors' attitudes toward animals.

The notions of national parks have been the mainstay of nature conservation (Hockings 2000, 2003). It was only three decades ago, in the 1980s, that the idea of protected areas became evocative as a system for Parks (McNeely and Miller 1984): that is, Parks being perceived as a place of special attributes and assets to be preserved and to be used for recreational, educational, and scientific purposes (Parkin 2006, p. 6; Hockings 2000, 2003). According to the Guidelines for Protected Area Management Categories (IUCN 1994), a 'National Park' is defined as a "protected area managed mainly for ecosystem conservation and recreation [which can] provide a foundation for spiritual, scientific, educational, recreational and visitor opportunities, all of which must be environmentally and culturally compatible". A study report produced by Higginbottom et al. (2001a, b, p. ii; iii) identified at that specific year some of the direct

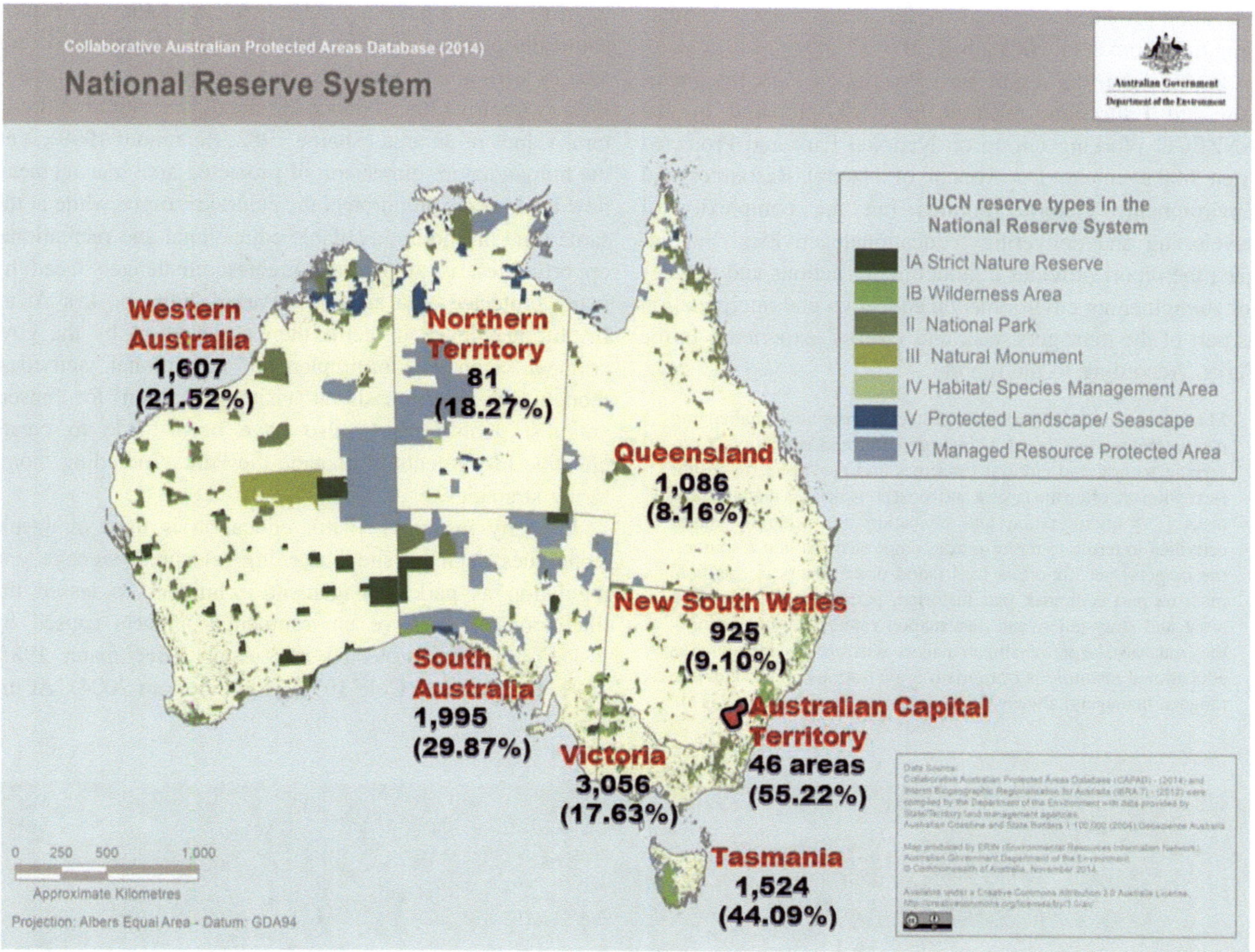

Fig. 8.1 Map of protected areas under the Australian IUCN classification in the National Reserve System (NRS) and the percentage of them compared to an overall State or Territory area. *Source* Adapted from the, Collaborative Australian Protected Areas Database (2014), Department of Environment, Australian Government. The original image was produced by ERIN (Environmental Resources Information Network), and is available under a Creative Commons Attribution 3.0 Australia License, November 2014. To note: For effects of citation, check for official updated percentages and figures directly on official gov. reports, websites, etc

positive impacts of wildlife tourism on wildlife regarding the Australian context:

- Government-owned wildlife tourism attractions and activities in Australia […] provide significant financial input into conservation in a few instances;
- Wildlife tourism appears to have led to some small-scale shifts towards more conservation-oriented land-use and wildlife management practices outside of protected areas;
- Wildlife tourism is associated with significant practical contributions to conservation;
- The nature and magnitude of costs and benefits of wildlife tourism to wildlife will vary according to many factors such as type of tourism activity, vulnerability of the wildlife population, effectiveness of interpretation, and conservation ethic of the operator.

In the literature, the role of park rangers for visitors' environmental learning is still scarce; somehow neglected, not fully developed, and it demands a more comprehensive investigation particularly with regards the relevance of including 'educational activities' and 'pro-conservationist messages' as part of the environmental-oriented attractions (Fig. 8.2). Mostly, the problem lies in the complexity of delivering 'environmental interpretation' at a managerial level in the parks, and the set of competences required for a successful and interactive delivery with visitors. In Parks, it is common that environmental interpretation and education is delivered by guides, educators, and volunteers, rather than the rangers themselves, who are more committed to conservation work and park maintenance and monitoring, rather than playing an educational role presenting natural assets to visitors. Notwithstanding, in Australia "the magnitude of benefits to wildlife associated with education provided as part of wildlife tourism are unknown" (Higginbottom et al. 2001a, b, p. iii).

In Australia, the report Best Practice in Park Interpretation and Education, released in 1999, prepared by the ANZECC Working Group on National Park and Protected Area Management, Department of Natural Resources and Environment, Victoria, pointed out the complexity of developing and delivering 'educational activities' by rangers; the report provides a guideline for actions and policies for strengthening environmental education and interpretation as part of the managers' task and visitors' experience in the Parks. According to the report,

> Managers of national parks and protected areas have challenging responsibilities in regard to interpretation and education. Conserving natural and cultural resources and providing for visitor recreation are often the largest and most conspicuous management tasks [...] interpretation and education are generally minor activities in terms of the resources employed [human resources, the rangers] yet important [...] Good practitioners in this field must be part ecologist, part historian, part anthropologist, part artist and story-teller, and part market researcher. Increasingly they must also be partnership managers assisting providers such as educational institutions or tourism organisations rather than [the rangers, managers] always delivering services direct (pp. 9–11).

Most visitors have a fundamental need for information about the places they visit, and while most visitors do not visit to learn about conservation per se, it is clear that many seek to improve their knowledge about the natural and cultural values of an area (Sharpe 1982; Beaumont 1999). Yet, the human/nature dimension of protected area management, how to conserve and protect the natural resource while at the same time promoting available educational and recreational opportunities, is among the greatest challenges faced by many protected area agencies (Parkin 2006, p. 45). At an Institutional level, it seems there is an effort by the government agencies to implement educational activities, though with an emphasis on 'visitor education' for conservationist goals. Efforts also have been made to create effective instruments to manage the Parks, including monitoring strategies.

In many instances visitor education is used alongside techniques such as site hardening, closures, signage and regulation as park management techniques to lessen the likelihood of negative environmental impacts caused by visitation to the protected area estate (Beckmann 1988, 1991; Hammit and Cole 1998; Higginbottom 2004). At the

Fig. 8.2 A ranger beginning a guided tour at the Rainforest Nature Park, Kuranda, Cairns Region, Queensland, Australia. *Source* Authors' own work, 2015

same time, the traditional role of visitor education has been to provide information to increase public awareness and appreciation of natural resources (Carter 2001; Sharpe 1982; Anderson et al. 1998; O'Neill et al. 2004) used effectively, it (visitor education) can enhance the quality off the visitor experience and address management issues such as:

- Protecting fragile resources (by directing visitors to other areas);
- Reducing intentional and unintentional vandalism;
- Reducing accidents by explaining unusual dangers;
- Increasing understanding of, and compliance with, management activities;
- Increasing knowledge of land management objectives (reservation, conservation);
  (Adapted from Beckmann 1991, p. 41, and Moscardo 1999, pp. 8–14, apud Parkin, 2006, p. 46).

In Queensland, the management planning process for protected areas, like many other states and territories in Australia, is based on classification (NCAct 1992, p. 14) and prescribed management principles (NCAct 1992, pp. 15–27). The Queensland Parks and Wildlife Service (QPWS), a Division of the Queensland Government's Environmental Protection Agency, is the State government agency responsible for the administration and management of protected areas under the Nature Conservation Act (1992), which is the principal piece of legislation that guides the administration and management of protected areas in Queensland (Parkin 2006, p. 82), Marine Parks Act (1984), Recreation Areas Management Act (1988), Brisbane Forest Park Act (1970) and Forestry Act (1959). The QPWS's primary purpose is to implement the Government's environmental objectives to ensure the protection, conservation and proper management of Queensland's natural and cultural values (Qld Govt 2001).

The QPWS emboldens people to visit and enjoy the protected areas through active nature-based outdoor recreational activities as long as they do not conflict with conservation and preservation of the sites, including sacred, cultural, heritage settings (Batt 2004, as cited in Parkin (2006)). In order to guarantee that visitors will behave in a proper sustainable way in the protected areas, the QPWS has run visitor education programmes to raise awareness, direct and influence visitors' behaviour, attitudes, and perceptions aiming to minimise the negative recreational impacts in those sensitive areas (Bauchop and Parkin 2000; Higginbottom 2004), and the framework used to this purpose is provided by the organisation's interpretation and education strategy (I&E Strategy) (QPWS 2000). The goal of the I&E strategy is to guide the visitor education activities performed by the QPWS, its interpreters and park rangers in the Queensland's parks and reserves.

8.3 The Role of Rangers and Guides in Promoting Environmental Interpretation and Education in National Parks in Australia

Park rangers play a key role in conservation and visitor management in protected areas. They manage Australia's network of parks and reserves making efforts to conserve the nation's biodiversity. For Howard (2013), "an understanding of the role of park rangers and the professional skills they require is therefore also important for future workforce capability" (242). But, how to better define a park ranger? Howard (2013) explains that "park rangers are middle managers who are primarily responsible for implementing the policies and plans developed by the main office" (p. 243).

The Park rangers and managers are responsible for looking after 1086 protected areas in Queensland, covering 14,108,222 ha which represents 8.16% of the overall land of the State (refer to Fig. 8.1; Table 8.1) (Department of Environment, Australian Government). The rangers need to ensure conservation, protection, resource and visitor management in the 213 National and Regional Parks in Queensland, which covers a landmass of 6,661,888 ha, they are also in charge of managing seven National Parks Scientific with an overall area of 52,181 ha which represents 0.03% of the State.

By reviewing the literature it was noted that the research on the skills and role of park rangers is ample, and this career tracks a set of specific abilities (Burns and McInermey 2010). One of the most recent publications is the research of Howard (2013) on the role of park rangers and the skills they need for managing the natural environment. Day (1995), for example, investigated the needed skills and training for conservation staff, and pointed out 50 skills necessary for becoming a park ranger. McGahan and Bassett (1999) identified the need of training and improved skills for managers, including rangers, in five main areas, as a way of "improving ranger knowledge and understanding of geography, climate, natural ecosystems, wildlife and plants; nature interpretation and the production and use of communications materials and outreach equipment; organisation of training, workshops, youth camps and other environmental education activities; public relations and public speaking; and knowledge of local languages and cultures, community relations and community participation" (p. 72).

To date, there is still little research on the interpretative and educational role of park rangers in National Parks in

Australia, particularly Queensland region. Most existing information in this field can be found in the documents and reports of Australian government agencies, for example, the Interpretation and Education Strategy 2000–2002, internal document, Queensland Parks and Wildlife Service, Brisbane, Qld.; Statewide Interpretation Workshop (5–8 March 2001) Report, internal document; Queensland Parks and Wildlife Service, Brisbane, Qld.; Interpretation and Community Education Situation Report (1999–2001), internal document, Queensland Parks and Wildlife Service, Brisbane, Qld.; QPWS interpretative Planning Handbook, internal document, Queensland Parks and Wildlife Service, Brisbane, Qld.; QPWS Community and Education Manual, internal document, Queensland Parks and Wildlife Service, Brisbane, Qld.; Master Plan for Queensland's Parks Systems 2001, The State of Queensland, Environmental Protection Agency, Brisbane, Qld.; and the Queensland Department of Environment and Heritage, 1998a, Public Contact Manual—A Guide to Effective Community Education, Heritage Interpretation and Extension, unpublished document.

In the Northern Territory, the Parks and Wildlife Commission classifies the rangers into two main groups: Park rangers and Wildlife rangers, and emphasises that their work is "highly rewarding", never the same, never a routine by dealing with outdoor issues, for example, wildlife protection. Being a ranger implies dealing with challenges. According to the Northern Territory Commission, an ideal ranger is expected to have a set of skills: a specific qualification (e.g. a tertiary education in Natural Resource and, or, Park Management, Conservation, Land Management, or, related fields), pertinent management experience and empathy traits, such as high levels of motivation, disposition to handle wildlife, and communication skills to manage human resources, including visitors and staff, etc.

Katz (1974) argued that people need certain skills to perform as managers, in which the rangers' role fits in, and suggested three encompassing categories of skills regarding a managerial work: technical, human, and conceptual. The technical skills are those necessary to accomplish or understand the specific kind of work being done in an organisation; the human skills are related to the ability to communicate with and understand other people; and the conceptual skills are those abilities to think abstractly and logically as part of the process of innovating and integrating work (as cited in Howard 2013, pp. 243–244).

The Parks and Wildlife Commission Northern Territory sets the role of rangers into four dimensions: environmental management and protection, visitor management and services, wildlife management, and law enforcement. As for visitors' management and services, some of the main tasks are: the delivery of "face to face interpretative activities such as guided walks and talks, slide shows and junior ranger activities to promote understanding and appreciation of natural and cultural aspects of...Parks", and the representation of Parks and Wildlife on a daily basis with interaction with park visitors to let them know about the park rules and regulations, as well as conducting "law enforcement duties". As for wildlife management in the Northern Territory, the rangers have the following duties:

- Taking part in problem wildlife control programs and providing advice to others.
- Providing technical assistance to other departments in regards to feral animal management.
- Monitoring the snake removal hot-line, providing the appropriate advice, removal and relocation of the animals as required.
- Monitoring the crocodile sightings hot-line and taking appropriate action when sightings or other information is reported.
- Trapping, capturing and removing crocodiles.
- Maintaining crocodile traps and equipment.
- Assisting with sample collections from crocodiles and other wildlife for research.
- Surveying crocodile and waterfowl populations by boat and plane.
- Assisting with scientific surveys and the protection of threatened and endangered species populations.
- Working with other organisations in relation to mistreated or problem animals.
- Community engagement.
- Providing information to people, businesses and school groups in relation to native wildlife, pest animals, permits and wildlife crime.

A publication in the website of Parks Victoria has a section on the role and responsibilities of a park ranger. The main responsibilities are divided into two major duty groups: conservation and recreation. These aspects are reorganised into subcategories to facilitate the understanding of how complex is the role of a ranger; and the multiple tasks and duties the rangers are engaged on a daily basis (refer to Table 8.2).

8.4 National Parks, Rangers and Visitors Management Tools: Visitor Education

Page (2011) lists a number of techniques of how to manage visitors' impacts through regulation and restrictions. According to him, the most salient techniques cited in the literature are: regulating access by area (sacred sites, indigenous lands), by transport (vehicle-free environments), by visitor numbers and group size, by types of visitors permitted (discouraging specific groups trough segmented marketing), regulating visitor behaviour and equipment use,

Table 8.2 Conservationist and recreational responsibilities of rangers in park management

Rangers' Role	Conservation (Protection and maintenance of natural & cultural assets in Parks)
General conservationist role of wildlife and flora	• Protection, enhancement and management of natural assets • Identify and protect populations of threatened or endangered animals • Identifying weeds and pest animals and eradicate or control those that hold the most threat to native plants and animals • Work with volunteers on projects such as weed control, maintaining tracks and other infrastructure
Burning prevention	• Organise and conduct prescribed burning operations Assisting with fire suppression (fire-fighting) for fires on public land throughout the state
Heritage & cultural protection & maintenance	• Protection, enhancement and management of cultural assets • Promote and maintain historic assets, such as gold mining sites and historic huts • Develop co-operative relationships with local indigenous groups
Scientific and business activities management	• Issue permits and oversee researchers studying within the park • Monitor and issue permits to businesses operating within parks, such as, tour operators, ski resort operators, hydroelectricity, cafés and beekeeping
Rangers' Role	**Recreation (Helping visitors to enjoy and understand Parks)**
General role in Recreation	• Implementation and management of projects, programs and contracts • On a broader level, the role of a park ranger involves extensive planning, researching, strategic thinking and people management to effectively balance conservation and recreational values of each asset.
Delivery of educational activities for visitors	• Delivery of interpretation and education services including guided tours, demonstrations and talks
Developing visitors' facilities and recreational/ adventure settings	• Maintain and develop visitor facilities such as picnic areas, camping areas and toilets • Create and maintain trails for hikers, mountain bike riders, four-wheel driving, etc.
Visitors' safety	• Respond to emergency situations such as 'Search and Rescue'
Supporting human resources management	• Management and development of staff, volunteers, contractors & work experience students

Source Ismar Lima (2015), adapted from Parks Victoria, Role of a Park ranger. Available online at http://parkweb.vic.gov.au/learn/informationfor-students/managing-our-parks/role-of-a-park-ranger and by promoting preventive modifications of sites such as pathways, boardwalks, that can direct visitors in natural settings; and the provision of interpretation and education schemes for visitors (Page 2011, p. 320).

Visitor education has been regarded as an important park management tool by Australian Park agencies (Parkin 2006; Marion and Reid 2007; Brown et al. 2010). Visitor education seeks to open opportunities for enhanced visitors' experiences while seeking to minimise related negative impacts to natural settings in protected areas (Green and Higginbottom 2001; Higginbottom 2004; Weiler and Black 2015). Some educational programmes have been created in an attempt to influence visitors' behaviour leading them into more pro-conservationist and pro-environmental attitudes while strengthening conservation actions in parks. These educational programmes, usually managed by park rangers, employ interpretation, talks, story-telling, and demonstration techniques as the means to address visitor-related damages or impacts on cultural and natural assets (de Lima 2016a). According to Marion and Reid (2007), "findings reveal that most of the visitor education efforts evaluated did effectively alter visitor knowledge, behaviour and/or resource and social conditions in the intended direction" (p. 5).

The use of visitor education as a park management tool is part of the "sustaining recreational and tourism opportunities" element, a meaningful method to spur on people's awareness and engagement in conservation, while providing "visitors with facilities…constructed and maintained to meet safety standards…with information…of the hazards in parks" (EPA 2001). According to EPA (1999), the use of environmental education and interpretative services to serve to put in evidence the values of Parks and of other protected areas in Queensland in terms of community awareness and conservation outcomes.

In general, 'visitor education' programmes centred on visitors' safety and awareness are worldwide designed as the main tools for nature protection by pertinent environmental agencies, and these educational programmes differ from environmental education programmes that are much broader in their targets (Parkin 2006, p. 11). Such programmes not only raise visitor awareness about the natural and cultural settings and resources, but also aim at developing a meaningful understanding about nature, its biomes, flora and fauna, and its ecosystems for the visitors. The content approaches and emphases may hold a great distinction between both visitor educational programmes; they have distinct goals, but employ similar interpretative and mediatory techniques. For Morgan and Soucy (2006), non-formal environmental education oftentimes implies natural resource communication at park locations, and both terms 'non-formal environmental education' and 'environmental interpretation' are close in meanings and effects; thus, in the literature sometimes they are used interchangeably (p. 596).

Even more complex is the interpretative and educational role of the rangers in visitor management. This is the main issue to be discussed in the paper. Within the tourism literature, terms used are 'tour guide', 'tourist guide', 'tour leader', 'tour manager', 'tour escort', and 'courier' (Weiler and Black 2015, p. 2), and even 'tour conductor'. There is a sort of consensus among researchers about the instrumental (leadership) role of the guide in order to keep a tour running successfully for the visitors in terms of safety, logistics, and certainly as individuals in charge of the mediation and interpretation of content and sites, "this in turn has drawn attention to the importance of the communicative competency of guides, including the application of best practice principles in interpretation and intercultural communication" (Weiler and Black 2015, p. 2), which, are also elements of a successful tourism operation.

8.5 Environmental Interpretation and Education: A Challenging Task for Rangers and Guides

As explained by Beck and Cable (2002), interpretation is a communicational process which helps to interconnect the visitors to the [cultural, nature] resource [or place]: thus it is visitor centred. For example, interpretation is habitually perceived as effective in terms of managing the interactional processes between 'visitors' and 'wildlife' because it can result in levels of environmental awareness with an augmented view of a conservation ethic (Beckmann 1988, 1991; Moscardo 1998; Howard 2013), and Orams (1996) emphatically states that interpretation [in guiding] is the most effective strategy for managing wildlife encounters. Interpretation should provoke visitors to reflect and to connect with cultural and natural elements of visited sites, to local people, culture, artefacts, and to historical events to the extent it can fill them with information which can lead to thoughtfulness about care and of stewardship (Weiler and Black 2015, p. 18). There is thus a distinctive difference between interpretation and information; the latter refers just to communication of facts; it gives plain facts; conversely, "interpretation can provoke ideas, perhaps even jolt people into a completely new understanding of what they have come to see" (Carter 2001).

As McIntyre et al. (2014) explain it, "interpretation is a mission-based communication process that forges emotional and intellectual connections between the interests of the audience and the meanings inherent in the resource", and in order to effectively approach ecological themes and deliver the content to visitors, it is essential to understand who the audience is and what they are looking for, so interpretation can become meaningful and fulfil expectations. Sometimes, the visitors only wish to contemplate and observe nature.

A significant number of publications on interpretation and communication processes are based on Freeman Tilden six interpretative principles of 1957; the foundations and gifts of interpretation. Environmental interpretation and environmental education: what is the difference? According to Veverka (2001, 2014),

> Environmental Education (either the formal education process, or the hopeful result of a program or exhibit), can be presented in either an informational "instructional" approach or using an interpretative approach. Remember, interpretation is a communication process. If the process works in presenting and translating the information about the environment in a way that is meaningful for the audience, then environmental "education" occurs…The interpretative communication process can be used for interpreting anything, any subject. If the interpretative communication is effective, then "education" can occur about that subject. Interpretation is an objective driven, and…audience focused process that looks for results (the accomplishment of stated objectives).

For Ward and Wilkinson (2006), it is highly relevant to distinguish the essence of interpretation from education in terms of values and purposes. According to them, the main aspect that separates interpretation from education, including environmental education, is the available time frame for delivering a content to the audience. "In education, there is typically a longer time frame and repeated exposure through which to build knowledge and learning. With interpretation [there is] one opportunity to achieve [this] goal…of short time…but instead should serve as a catalyst for learning" (p. 21). By taking this understanding into account, it is possible to assert that 'environmental interpretation' is the main tool to promote and achieve environmental education (de Lima 2011, 2016a, b); and the latter is dependent on an effective interpretation; they are disconnected for educational purposes. Tilden (1957) mentioned that environmental interpretation is "an educational activity which aims to reveal meanings and relationships through the use of original objects, by first-hand experience, and by illustrative media, rather than simply to communicate factual information" (p. 8). Ham (1992), in his book entitled, Environmental Interpretation: A Practical Guide for People with Big Ideas and Small Budgets, explains 'environmental interpretation' as the use of techniques to communicate wonders and complexities of nature science to common people, that is, "translating the technical language of nature science or related fields into terms and ideas that people who aren't scientists can readily understand. And it involves doing it in a way that's entertaining and interesting to these people", and interpreters—nature guides—play a pivotal role for achieving it.

A guide is a type of individual in charge of using environmental interpretation to achieve levels of environmental education through a interwork which demands skills and knowledge to escort groups of visitors in venues, places and sites of touristic interest such as natural areas, historic buildings, zoos, sanctuaries, parks, museums; thus, they are expected to provide interpretation of natural and cultural assets "in an inspiring and entertaining manner" (Weiler and Black 2015, p. 3), and this also applies to rangers in parks in charge of guiding visitors. As part of the process of presenting natural and cultural settings, Mediation is thus critical for touching one's perception and feeling in regard to specific themes and topics in hosting places, particularly in terms of post-visit postures, "the strategic use of tour guides to influence on-site behaviour and change post-visit attitudes and behaviours might also be considered as mediation" (Weiler 2015, p. 35). Tour guiding (or group guiding by rangers) commonly demands eclectical skill, abilities and training in introducing and mediating culture, places, ecosystems, landscapes, and local people attributes. Rangers involved in guiding in Parks are expected to hold the same attributes and skills as those of outsourced tour guides, and also they are expected to have a set of knowledge and skills specific for working in protected areas as already presented in Table 8.1. For acting in the parks, usually contracted skilled rangers are allowed and, or, authorised guides and tour companies. But, in general, for guiding and delivering meaningful biofacts to visitors, a skilled, knowledgeable and trained person in this field can carry on the guiding tasks; this person can be an environmentalist, biologist, a school teacher, a instruction, a tutor, etc. A skilled guide usually gathers not only effective communicational abilities, but at least some major ecological and biological knowledge of the field, areas, wildlife the person is in charge of providing environmental interpretation and education (de Lima 2016a, b).

Concerned with the role of guides and the benefits and enhancement they could provide for visitors, local stakeholders, and destination sites, Cohen (1985) presented two conceptual spheres with course of action for the guides: tour management in which guides have an instrumental (leadership) role in organizing and managing group(s); and the experience management in which guides have as a role to facilitate visitors' engagement and learning (mediation) (Refer to Table 8.2). In 1993, Weiler and Davis (1993) advanced the discussion by adding a third sphere to Cohen's model with a focus on the role of the guide (or of the ranger) in a site/resource management. Cohen's (1985) and Weiler and Davis's (1993) "conceptual frameworks have stood the test of time in drawing attention to both the diversity of guiding roles that are common to all contexts and types of tour guiding, and the specialist roles that ecotour/nature guides are required to perform" (p. 25).

By taking into account the three spheres, a framework is proposed in this paper aiming to examine the roles of guides (or of rangers in guided tours) and the relevance of guiding, that is, instrumental (tour management), mediatory (experience management), and interpretative (resource management)

Table 8.3 The three key spheres of tour guiding and the roles of a contemporary tour guide, including the rangers' role in guiding, visitor management, and interpretation

Sphere 1: Group management	**Instrumental (leadership) roles** focused on organising and managing the group
Sphere 2: Experience management	**Mediatory roles** focused on facilitating individual's engagement and learning
Sphere 3: Resource/Site management	**Interpretative and role-modelling roles** focused on the sustainability of host environments, communities and destinations

Source Adapted from Weiler and Black (2015, p. 28)

(Table 8.3). In regard to 'resource management' it can refer to both cultural/heritage and natural resources. Rangers in charge of guided tours or of visitor education can use interpretation or mediation to explain or connect visitors to some aspects of an Indigenous community, or, Indigenous lifestyle and traditional knowledge (de Lima 2016a, b).

Within the perspective of these three spheres, guides can add value to a visitor experience and to a local site, or destination, contributing to the conservation process. That is, "nature-based tour guides also encourage participants to reduce their impacts on-site, and they facilitate a change in values towards long-term conservation" (Weiler and Black 2015). The guides can also introduce outsiders to a specific culture providing specific information, raising awareness of and respect for Indigenous peoples. As an example, bush tucker or a wildlife encounter guided (led and mediated) by Indigenous people can be a fascinating experience in getting to know about a local forested area by using the senses. Guiding implies a multitude of ways for acquiring knowledge.

Jennings and Weiler (2006) explain that guides can mediate a visitors' connection to localities and local issues to the extent that they can enhance or detract them from their experience, either facilitating or inhibiting outcomes, because the guides perform both an instrumental and mediatory role. Weiler and Black (2015) provide four domains in a framework to examine the mediatory role of guides, and they make a distinction between mediation and interpretation, in that 'interpretation' is a role in itself with a collection of techniques necessary for mediation by using interpretative strategies such as analogies, anecdotes, narratives, storytelling, metaphors, and even non-verbal communication such as artefacts and experiencing through the senses (touching, listening, tasting, smelling, seeing) (Cohen 1985; Moscardo 1998; Colquhoun 2005; Jennings and Weiler 2006; Weiler and Davis 1993).

Put simply, there is no mediation without interpretation, because the techniques used in the interpretation can help "visitors to understand and feel empathy towards objects, persons, sites or environments" (Weiler and Black 2015, p. 35), it is the guide's role to get the visitors "under the skin of visited destinations" (McGrath 2007, p. 376), and the mediation role is all-encompassing in regard to enhancing a visitor's experience as pointed out by Weiler and Black (2015): mediating/brokering physical access; mediating/brokering encounters (interactions); mediating/brokering understanding (intellectual access); and mediating/brokering empathy (emotional access).

In order to satisfactorily act as a guide, a set of competences are necessary in guiding, particularly in dealing with heterogeneous, multicultural visitors. Such competences are: fluency in the visitors' language; a local culturally knowledgeable person; social-interpersonal skills; expression and demonstration of cultural pride; discernment in what is culturally appropriate to share; and engaging in two-way communication (Weiler and Black 2015, p. 65). For Indigenous guides, culturally sensitive issues can be better approached and shared with visitors because the guides have a local cultural upbringing which can position them as genuine knowledge mediators of their own culture.

In regard to the role of guides and natural resource management, the cases examined in the literature reveal that the guides face restrictions in achieving wide-ranging conservation outcomes. Most of their roles in terms of nature management rest on reducing on-site impacts by delivering 'conservation messages' to visitors while putting emphasis on their conduct at the moment of having contact with natural assets either a forest or a reef (de Lima 2016a, b). In their studies, Medio et al. (1997) bring up the role of guides in mitigating impacts on coral reefs by divers or snorkellers.

The guides can play an interventionist role in guiding visitors on the trails by working with them in order to avoid excessive noise, off-track walks, collection or removal of natural elements, including those of cultural value such as sacred rocks, petrified wood, etc. (Littlefair and Buckley 2008), and certainly "guided tours and roving interpretation rangers [can]… convey important conservation messages to visitors, helping them to enjoy, connect with and value our significant and special places" (Colquhoun 2005, p. 7). In their literature review, Zeppel and Muloin (2008) stated that visitors who are exposed to environmental messages are reported to have higher levels of pro-conservation behaviour, and are more environmentally cognizant.

Some evidence shows that a tour guide who makes himself/herself authoritatively respected can lead visitors into more responsible behaviour during their stay in natural areas (Littlefair and Buckley 2008). By taking into account

Fig. 8.3 A ranger in action at the Rainforest Nature Park, Kuranda, Cairns Region, Australia: Enhancing visitors' experiences and learning through interpretative and demonstration techniques (Dimension I). *Source* Ismar Lima, field work in Cairns, Queensland, Australia, 2015

these facts, Indigenous tourism operators and Indigenous guides can contribute to reducing impacts in the visited areas: controlling visitor access to sites, using licensing, law enforcement, and observing regulations restricting the use of renewable and non-renewable resources by the tourism industry (Weiler and Black 2015; de Lima 2016a, b). The three dimensions in which guides can get involved in helping to encourage sustainability (Weiler and Black 2015, pp. 72–75), are as follows:

- *Dimension 1*: *Enhancing* visitors' understanding and valuing of a site, communities, cultures and environments.
- *Dimension 2*: *Influencing and monitoring* visitors' behaviours, en route, on-site and at destinations.
- *Dimension 3*: *Fostering* visitors' *post-visit*, pro-environmental and pro-heritage conservation attitudes and behaviours.

A conservationist role of guides as mediators has limitations, and it happens because sustainability outcomes in terms of conservation and nature/heritage protection are largely under the responsibility of protected area managers and of local/regional government agencies. The creation of pro-conservationist policies is something out of the scope of a guide role, "the guiding profession, let alone an individual guide, may thus feel relatively powerless to make a difference in contributing to the sustainability of a particular activity, tour, business, community, industry or environment" (Weiler 2013, pp. 14–15). There are however several ways through which the park rangers can contribute to address sustainability targets through guiding and interpretative talks, and the list includes the enhancement of the visitors' understanding and valuing of communities, cultures and environments, as well as purposeful actions aiming at influencing visitors' behaviour on-site (Fig. 8.3).

8.6 The Need of Interpretative Planning and Strategies: Addressing Messages on Wildlife

Interpretative planning is a first step in a planning and design process for supporting Institutions in their informal learning-based programmes and actions where interpretation has a critical role to deliver biofacts, messages, and experiences to visitors, such as in zoos, nature centres, heritage sites, parks and wildlife sanctuaries and reserves, etc.; it is above all a decision-making process that binds the most effective ways to deliver a content to a targeted audience; the planning consists of integrating the available nature resources (fauna, flora, etc.), the management demands and the visitor informative (learning) expectations. For Veverka (1998, 2001, 2014), interpretative planning is a process that identifies and describes significant visitor experiences in a resource-based recreation area, and recommends ways to provide, encourage, sustain, facilitate or otherwise assist those experiences (Veverka 1998, 2001, 2014). Interpretation of informal learning institutions focuses particularly on relating content in a meaningful manner to a visitor's self experience, and for achieving this goal, usually it is sought to provoke emotion, thought or further inquiries into a subject, getting the attention and engagement of visitors for information transfer and knowledge building, and most interpretative plans are based on a thematic approach to interpretation…to communicate to various audiences (Brochu 2003; Coghland and Kim 2012; Veverka 2014).

Interpretative planning helps Institutions, organisations and companies to organise environmental interpretation and education opportunities for visitors, so they can explore the nooks and crannies of natural settings and their wildlife; to learn key information, biofacts and details of a natural world through interpreters, guides and rangers mediation. An interpretative plan sets a communicational process, through which valuable information—meanings and relationships of the natural world—, are disclosed to visitors through experiences by combining techniques and strategies, which include the use of objects, artefacts, props (Fa et al. 2011), etc., so they can take the most as a learning moment in natural settings, e.g., zoos, parks, etc. (Veverka 2001; AldrichPears Associates). "The Experience" is what visitors take from a park. The provision of opportunities for visitors to interact with park resources in a manner that is both safe for visitors while leaving the resources unimpaired is unimpaired what has been termed "visitor experience planning" (Dave Dame, cited in Harpers Ferry Center 1998, p. 2), and this is the core of park planning and development.

For elaborating any interpretative plan, it is utterly necessary a familiarisation with the pertinent site or natural setting; it is worth noting that just an occasional visit hardly provides the means to gain knowledge enough to outline the basics of an interpretative planning; conversely, it usually requires a detailed physical exploration as well as contacts and discussions with key stakeholders. For example, the Interpretation Master Plan for the Angel Island State Park of 2012, an area managed by California State Parks, the largest island in San Francisco Bay with 740 acres and six miles of shoreline, considers the geology, climate, hydrology, and biology—the island's natural resources—as the foundation for interpretative services at the park with a focus on ecological knowledge transfer. Also it is relevant to cite the guidelines and reports used in Queensland, Australia, for visitor education and Park interpretation, among them: the QNPWS Interpretation Manual (1984), a first step towards documenting the Service's interpretation philosophy and activities; the QPWS interpretative Planning Handbook: Connecting people with nature through interpretation, extension and community education (2001). The Handbook provides a step-by-step guide to interpretative planning from individual to state-wide strategic planning, and was developed to assist interpreters and educators, e.g. the Park rangers and managers, to write and implement strategic plans appropriate for the demands of a specific area, community, conservation or resource issues (Parkin 2006, pp. 103–104).

For a consistent plan, it is necessary to gather all information and data; by knowing a site better it is possible to have a thorough mental image of the area, its layout, the arrangement of its physical features, its natural resources; the wildlife; the local ecosystem with its flora and fauna, and related phenomena, and how visitors access and use it. This will help planners to comprehensively understand who uses it, why they use it, what they like about it, and the type of improvements that should take place as priority for improving interpretation and educational outcomes, for example, in a park, wildlife sanctuary, or zoo.

An interpretative plan furnishes planners, interpreters, guides, and other field-related people with instructions and suggestions on important elements to be considered in a planning process; consequently, they can develop content and strategies to deliver biofacts, for example. The elaboration of an interpretative plan has many different stages and phases, and the person needs to take into account the scales of its application and use, for example, from a macro perspective (a whole region) to a specific setting (a single display). Within a National Park scope, there are usually a park-wide interpretation strategies; local interpretative plans for hot-spot areas, as well as individual interpretative plans for each visitor centre (Carter 2001). By dealing with visitors of all ages, particularly children, school-visitors, McIntyre et al. (2014) alert that interpretation should be enjoyable and entertaining as an essential quality. In order to have it, they suggest the use of a conversational tone; to avoid reading from notes; to incorporate humour, music, sounds, two-way communication; incorporate objects (biofacts); use comparisons, analogies, and metaphors.

The visitors are interested in the wildlife, but not in overly-serious lectures. The role of interpreters is to convey information in ways that allow visitors to have fun while they are learning; a recreational learning experience is one where the visitor attends or participates in a program through which the person can gain both scientific and entertaining knowledge (McIntyre et al. 2014; de Lima 2016a, b). In a nutshell, the basic principles of interpretation are: to provoke, to relate, to reveal, to address the whole; and, to strive for message unity (Veverka 2001, 2014). In order to achieve it, it is necessary to consider combining multiple sources of interpretation to repeat the interpretative message in nature-based tourism; thus, the interpreters should consider interpretative layerings at an attraction by using a variety of interpretative sources on visitors' understanding of the attraction (Coghlan and Kim 2012), and of the wildlife. In 2011, Education Scotland, an Executive Agency of the Scottish Government, released a practical guide for outdoor learning, in which some general benefits from taking learning outdoors within and across curriculum areas are cited (Education Scotland 2011, p. 7); it highlights that,

- connections made experientially with the real world help to develop skills, knowledge and understanding in a meaningful context;
- the outdoor environments and surroundings act as a rich stimulus for creative thinking and learning. This affords opportunities for challenge, enquiry, critical thinking and reflection;
- the multi-sensory experience outdoors helps children and young people to retain knowledge more effectively;
- learning in a less structured environment can provide a different learning experience; being outdoors can be a more relaxing learning experience for many learners.

With an increasing emphasis and opportunities for learning outdoors about ecological and biological aspects of the wildlife and ecosystem, an educational nature-based tourism takes shape and can advance public understanding on the human and nature relations and interactions. The 'learning component' in tourism activities, either visiting a Park or visiting a zoo, adds great value to people's experiences and to tourism itself; visitors as learners have an opportunity to make their visits and stay a more meaningful self-experience. This can include, for example, the participation in interactive and sensory activities mediated by guides or interpreters. The visitors can also choose to participate in a more hands-on and open-air learning tourism with bush tucker and bush medicine, by combining it with Indigenous tourism, wildlife tourism and geotourism. In geoheritage areas, visitors have a chance to learn and understand about the natural landscapes and the character of a geopark (Newsome and Dowing 2010), and they also have a chance to better understand the fragile ecosystem and wildlife that are usually present in savannah and deserts. Geotourism can be an experience associated with wildlife tourism, by "integrating fun and geosciences through geotourism…as a strategy to attract more visitors" (Farsani et al. 2012). For example, to engage visitors in a bush tucker at Alice Springs Desert Park, in Australia, conducted by Aboriginal guides or rangers, to harvest and taste native bush foods, while enjoying the uniqueness of the largest sandstone rock, the Uluru, also known as Rock Ayers, and listening to native dream-time stories of the place and of the culture.

Table 8.4 shows some aspects that should be taking into account for planning environmental interpretation and education in the context of an outdoor recreation and educational nature-based tourism. The framework includes information on natural resources and guidance on topics suitable to different age ranges.

8.6.1 David Fleay's Wildlife Park: An Overview

David Fleay's is a wildlife park nestled just west of the Burleigh Heads, on Gold Coast, that allows visitor to "stroll through the tranquil surroundings to experience some of Queensland's most iconic natural habitats and meet the resident wildlife" […] (NPRSR, Queensland, online). The park was built from 1952 to 1983 and has played an important role in demonstrating the conservation initiatives of David Fleay, who established the property in 1951: a naturalist who became the first to breed platypus in captivity, and his concept was that rescued and threatened birds and other animals should be kept in conditions similar to their natural environment, if they cannot be in open ranges. Figure 8.4 shows an early picture of Fleay with a rescued floodwater baby platypus in hands for a very close kids' appreciation and getting the news with the creation of a "platypusary". The platypuses (*Ornithorhynchus anatinus*) are monotremes; some people regard them odd mammals; the females lay up to three eggs a time; the platypus has a flat bill like a duck, feet like an otter, a paddle-shaped tail like a beaver, and a furry body. The platypus is considered as one of the Australian wild species visitors should see during their visit to the country because of its uniqueness, however it is not an intention in this chapter to create a sort of hierarchy ranking the wildlife, but some species may have more tourist appeal than others; notwithstanding, all wild animals have their values, peculiarities and tourism attractiveness. Higginbottom and Buckley (2003) based on their study on terrestrial wildlife of Australia, recommended to increase the demand from domestic and international tourists to see a wide range of Australian wild animals in their natural settings,

Table 8.4 Approaches for environmental interpretation and education across ages in the context of outdoor recreation and educational nature-based tourism

Age range	Desirable topics and approaches	Not recommended topics	Interpreting topics across ages	
			Fauna (wild animals), flora, and ecosystems	Natural outdoor Settings / sites, zoos, sanctuaries (*Forests, Savannah, Deserts; Marine, River, Lake Environments, etc.*)
Birth to 3 years-old ***Sensory***	* Animals are cool * Sensory experiences * Surrounding animals * Animals affections as family, moms, dads, babies	* Ecosystems (too abstract) * Life cycles (birth, death) * Endangered species * Environmental problems	* Imitation, mimesis: pretending some animal basic behaviours	* Listen to sounds of nature or of a site you are in and reproduce them (birds, frogs, water sounds, etc.)
4–7 years-old * *Sensory* ***Empathy***	* Animal homes * Farm/domestic animals * Predators/prey * compare/contrast animals to self * Animal groups * Life cycles * Desirable environmental attitudes (recycling, reusing, turning off lights, etc.)	* Ecosystems (too abstract) * Endangered species * Environmental issues * Consequences of not behaving ecologically friendly (habitat loss, pollution, endangered species, etc.)	* Role play of animal's life and behaviour (hatch, stretch, chirp, eat, snuggles against mom, sleep, defense)	* Comparisons of humans to forest animals by using facial expressions, hands, body size, etc. * Nature-based play games * Nature discovery activities according to their age
8–11 years-old * *Sensory* * *Empathy* ***Exploration***	* All of the above * Good environmental manners (tree-planting, habitat cleaning-up, etc.) * Ecosystems * Physical adaptations * Animal habitats and needs * Site-specific investigations and, or, observations * Cycles (life, water, etc.) * Basic notions on good and desirable environmental manners	* Dire consequences of not choosing and practising good ecological manners (human impacts on nature), e.g. avoid anything too depressing, frightening or gory	* Discuss animals' habitat and life cycle; to make comparisons among the animals or to people as a way to illustrate an issue/animal	* Build nature, biome, ecosystem models on the sand, on a paper, board, and present the various layers and animals that live in a site/place, e.g., forest * Nature-based play games * Nature discovery activities
12 and up *(Heterogeneous audience—young people and adults)* * *Sensory* * *Empathy* * *Exploration* ***Action***	* Behavioural adaptations * Consequences of not being ecologically friendly, not not using good environmental manners * Ecosystem investigation with concrete experiences * Endangered species	* Most topics are appropriate, if presented in a sensitive manner; they should be preferably presented as a way of building a sense of affection and care for nature and its dwellers, and presenting problems and what we can do about them, but avoiding a sense of hopelessness	* Discuss wild animal habitats, ecosystem and flora, and try to engage the participants/visitors in any appropriate hands-on activity or game that is ecologically beneficial to nature	* Build nature, biome, ecosystem models * Nature-based play games * Nature discovery activities * Tree species learning and tree-planting; reforestation
School-Visitors *(Curriculum-based)* * *Sensory; Empathy; Exploration; Action;* ***Experimenting***	The interpretative and educative sessions can be tailored to accommodate the school-visitor group(s) according to their curriculum-based demands and interests, present the	* All topics are appropriate, and may present a challenging level. Basic and some in-depth content is part of the learning process in and	* All the above, and experiments and other learning tools can be used, such as working sheets, etc. * Learn about different types of	* Build nature, biome, ecosystem models * Use comparisons tables * Flora and fauna list elaboration * Biological and ecological tests and observations

(continued)

Table 8.4 (continued)

Age range	Desirable topics and approaches	Not recommended topics	Interpreting topics across ages	
			Fauna (wild animals), flora, and ecosystems	Natural outdoor Settings / sites, zoos, sanctuaries (*Forests, Savannah, Deserts; Marine, River, Lake Environments, etc.*)
	challenge of finding the right questions to ask: this is so often neglected, and an important part of both creative and critical thinking	with a natural setting and wildlife	biomes/ecosystems and resources	* Nature-based play games * Nature discovery activities * Tree species learning and tree-planting; reforestation
Specific Interest Groups (Adults) *(College, University students; researchers; professional wildlife watchers; etc.)* * *Sensory; Empathy; Exploration; Action; Experimenting,* ***Testing, Finding, and Developing new understandings***	The interpretative, educative and sessions can be tailored to accommodate the groups according to their demands, interests and focus	* All topics are appropriate, and in-depth content is expected as part of the sessions	* All the above, and experiments, and other learning tools can be used * Learn about different types of biomes/ecosystems and resources	* Build nature, biome, ecosystem models * Use comparisons tables * Flora and fauna list elaboration * Biological and ecological tests and observations * Experiments * Scientifically focussed activities

Source The author. This table was built adapted from multiple sources, among them Veverka (1998, 2001, 2014), and based on the author's research and self-experience on the ground in Australia, Gold Coast, Cairns, and Darwin, in 2015, and particularly in New Zealand, between 2004 and 2008, following the environmental interpretation, education and conservation work of Kuaka New Zealand with visitors on the Bay of Plenty as part of his doctoral research

Fig. 8.4 David Fleay, "Platypusary" and school visitors on Gold Coast, Australia. *Source* Author own work. Picture taken from an informative outdoor sign at David Fleay Wildlife Park, Gold Coast, Australia, 2015 (*original *black* and *white photograph* from the David Fleay Natural history collection)

> For international visitors, it may be possible to create an Australian equivalent to Africa's 'Big Five' (e.g. 'You've seen Africa's Big Five, what about the Seven Wonders from Down Under'?, in Australia). Suitable species might be the koala, kangaroo (red or eastern grey), saltwater crocodile, platypus, bilby, and wombat. This could be mirrored on a regional/state scale, emphasising species of particular local interest (p. 39).

Currently, the David Fleay Wildlife Park continues Fleay's work by gathering different threatened native animals in one location for public education and for breeding with ultimate release back into the wild. The platypus is an attraction of great appeal for the visitors with them crowding at the David Fleay's Nocturnal House during the feeding and educational sessions managed by the rangers.

At present days, the Park is managed by Queensland Parks and Wildlife Service, under the Australian Environmental Protection Agency and Department of National Parks, Sport and Racing (NPRSR) legislation and norms, and it aims to raise community awareness about the need to protect native animals, particularly endangered and threatened ones. Cassowaries, emus, platypuses, possums, crocodiles, greater bilby (Macrotis lagotis), and koalas are some of the animals in the Park. At the Park, visitors of all ages and adults have an opportunity to attend educational sessions with rangers several times a day; there is the session at the amphitheatre, another in an indoor theatre, plus crocodiles and platypus feeding, etc. Curriculum-based visitors have used David Fleay's as an outdoor learning complementary to their school subjects on biology and ecology, among other scientific disciplines and topics. During the sessions, rangers and school teachers work in association to explain in an enticing and educative way the major aspects of the wildlife and its habitats as well as related ecological phenomena (see Fig. 8.5).

Apart from the educational sessions and ranger's interpretative mediation, the visitors have also an opportunity for a self-guided learning through several signs on wild animals and ecosystems, and by observing and appreciating the wildlife in natural semi-captive settings (Fig. 8.4). At David Fleay Park, visitors are not allowed to touch, handle, cuddle, hold, or feed the animals. This is strictly prohibited at the Park whose major mission is wildlife conservation and breeding. Generally only the rangers manage and handle the animals, and visitors are passive in this process that assembles elements for an experiential learning, although they do sometimes allow visitors to touch, but not to hold or feed the animals.

8.6.2 Currumbin Wildlife Sanctuary: An Overview

The Sanctuary was established in 1947 by beekeeper and flower grower Alex Griffiths, who started feeding wild lorikeets of the region as the means to prevent them from causing damage to his blooms. "The feeding of the colourful lorikeets soon developed from a local curiosity to a popular tourist attraction […], and" in 1976, the sanctuary was donated to the National Trust of Queensland—a like-minded organisation dedicated to preserving the state's natural and cultural heritage. The Trust continues to operate the sanctuary on a not-for-profit basis, "with all revenue reinvested back into the park, in conservation-based research, caring for sick and injured wildlife and public education" (CWS Organisation, Alex Griffiths and our history, online). On 1st of July 2014, the National Trust of Queensland changed the Sanctuary's name to 'National Trust of Australia' (Queensland), and this way it became independent of government. It has hundreds of wild Australian animals on display, as well as in natural bush land and rainforest settings; it has been intensely visited by domestic and foreign visitors.

At Currumbin's, visitors can have direct physical contact with some species such as koalas, kangaroos, emus, wallabies, snakes, and birds during the flying shows. Kangaroos and lorikeets feeding is part of the visitors' experience (Fig. 8.6). Holding a koala or young crocodile for pictures and in other attractions is a paid possibility at the place, so the Sanctuary can also gain extra finance for maintaining the place. Currumbin Sanctuary has many entertaining and educational options for the visitors, including sessions with their own staff (rangers) and self-learning informative signs spread throughout the property (see Fig. 8.6). It defines itself as being a wildlife conservation and educational business, and as such has become competitive with David Fleay Wildlife Park,—located approximately eight kilometres (10 min drive) away.

The following statement was posted by Long (2015), sharing his opinion as a visitor at Currumbin Wildlife Sanctuary, and it summarises a visitor feeling of being part of an experiential learning and wild animal encounter in Australia,

> I've evolved over the years on the issue of zoos and aquariums and for the most part, I don't like them. I know the counter-arguments, that the research they conduct actually helps preserve species, but I can't help but feeling that it just doesn't seem right. Sanctuaries and refuges are different, usually, and it's their focus on aiding and protecting local wildlife that draws me to them time and time again, just as I was when I visited the Currumbin Wildlife Sanctuary along the Gold Coast of Australia. A heritage listed establishment, the Sanctuary has been helping animals and educating visitors for almost 70 years. Today it's home to a wide variety of native Australian animal species as well as a hospital where they take in thousands of sick or injured animals every year. Yes, animals are in enclosures and thousands of visitors crowd around them each year. But it's an establishment that has, since the very beginning, been all about protecting native wildlife and trying to find ways to integrate them into the human population explosion found around the state. It's also a way for people to learn more about them, because once we see and even touch an unfamiliar animal, we can't help but feel responsible for it. The Sanctuary has a lot of educational programming options for all ages and a personal favourite experience was to hang out with kangaroos for the afternoon, feeding and even petting them as they slept under leafy trees. (Matt Long, Destinations Landlopers blog, July 2015, online).

Orams (2002, p. 289), based on the available literature, explained some aspects of wildlife's and visitors' management at the Sanctuary,

Fig. 8.5 School visitors at the David Fleay Wildlife Park: experiential learning on wildlife and conservation through environmental interpretation and education and signs. *Source* Author's own work and pictures collection, Gold Coast, Australia, 2015

…[Currumbin] has been feeding wild Rainbow Lorikeets (a small parrot) since the 1940s. These feeding sessions are closely controlled by trained staff. The food provided is a mixture developed to prevent dietary deficiency (Cannon 1979) and food receptacles are disinfected prior to and after use. In addition, an interpretation programme is delivered to tourists during the feeding sessions. Staff at the sanctuary also monitor the numbers and health of the birds as well as support research (Burger 1997).

Fig. 8.6 Interactive and educational opportunities for visitors: Kangaroo feeding and ecological signs. *Source* Author own collection of pictures, field work on Gold Coast, in 2015

8.7 Australian Wildlife, Interpretation and Education: Ecological and Biological Elements

Wildlife interpretation and, consequently environmental education on wildlife, particularly through interpretative tourism, functions to propitiate distinct levels of connections between visitors and the world of science. Through the art of interpretation, which involves communicative strategies and tools, the natural world, landscapes and aesthetic aspects are presented to people. According to Ward and Wilkinson (2006), some of the most used presentation strategies are: characterisation, demonstration, storytelling, puppets, guided imagery and guest speaker. All these strategies can be combined with interactive nature-visitor play activities, for example, sensory ones such textures, colours, shapes, smell, nature stuff collection, etc. One of the strengths of environmental interpretation and education in Tourism is the potentiality of a segmented learning with a focus on ecological and biological aspects of species, and on their habitats, animal behaviour, but with less stress on in-depth scientific data and more attention on observation and interrelationships among and between visitors, ecosystems and species with the assimilation of it into human affairs. Wildlife interpretation enriches individual and group experiences in a site or destination by revealing meanings about the natural, historical and recreational resources mostly in an interdisciplinary and holistic way.

Apart from the biological characteristics of wild species, of habitats and of aesthetic components, other elements are nuclei to interpretation in wildlife tourism, such as wild species conservation and management, pest control, control of human disturbances (Green and Higginbottom 2001; Higginbottom 2004), carrying capacity, as well as ecological restoration, reforestation, re-wilding to improve habitats, issues of hydric resources and of soil components and erosion, etc. With regards to environmental science (Nebel and Wright 1993; Odum 2006; Asthana and Asthanba 2006), the following disciplines and issues have been object of basic presentation and interpretation to visitors, such as natural history of species; flora ecology; aquatic ecology; fisheries; oceanography which includes marine life and ecology; avian ecology; insect biology; cycles and influences of weather and climate; and geology. Wildlife tourism, as a sub-set of nature-based tourism, is defined by Higginbottom (2004), as a type of "tourism based on encounters with non-domesticated (non-human) animals… [that] can occur in either the animals' natural environment or in captivity" (p. 2). Newsome et al. (2005) define wildlife tourism as a type of tourism based on the observations of wildlife and human-wild animal interactions, it is a form of tourism "undertaken to view and/or encounter wildlife. It can take place in a range of settings, from captive, semi-captive, to in the wild, and it encompasses a variety of interactions from passive observation to feeding and/or touching the species viewed" (Preface, ix), but some studies draw attention to negative impacts caused by human-wildlife interactions (Hughes and Carlsen 2008), and Green and Higginbottom (2001) explored potential impacts on wildlife in Australia.

In a guided tour, the delivery of such content is purposeful to an individual or group interest which can be either superficial with very basic messages and data or it can be much more in-depth and scientific grounded, for example, if the group of visitors is curriculum-based seeking outdoors complementary knowledge for their disciplines. The targeted audience matters significantly at the moment of tailoring and delivering ecological and biological information on wildlife. Age, background, country of origin, command of English with satisfactory listening skills, individual or group purposes are some factors that should be taken into account at the moment to meaningfully present and mediate the 'natural world' to 'visitors'.

8.8 Kolb's Experiential Learning Theory, and Interpretative and Educational Wildlife Tourism

Kolb's Experiential Learning Theory (ELT) considers an ideal learning spiral based on the dialectics of conceptualising and experiencing, and of acting and reflecting as responsive outcomes of a learning situation. Kolb (1984) perceives learning as 'the process whereby knowledge is created through the transformation of experience. Knowledge results from the combination of grasping and transforming experience' (p. 41). His proposed learning model is dialectically related to four experiential modes: *Concrete Experience* (CE); *Reflective Observation* (RO), *Abstract Conceptualisation* (AC), and *Active Experimentation* (AE). It is a cyclical process through which concrete experiences lead to reflective observations, which is followed by an 'abstract conceptualisation'—a stage in which new concepts are thus created. These three former stages enable the development of implications for actions called as 'active experimentation', which, by its turn, leads to a 'concrete experience'. As Packer and Ballantyne (2013) put it in simpler terms, "this is a cycle of experiencing, reflecting, thinking and acting" (p. 170).

Kolb's experiential learning proposal can be fully applied to explain a learning process which involves visitors-nature-guides/ranchers as mediators with environmental interpretation and education in wildlife tourism being the means and tools for achieving it. Figure 8.7 shows the whole learning process by presenting the biotic, abiotic elements and factors, and associated phenomena, interlinked to ecological and biological scientific approaches, which serve as a foundation for guides and ranchers, to holistically mediate the ecosystems with its specific fauna and flora to visitors. The role of guides and rangers as mediators can be active by bridging 'visitors' to nature by using a series of strategies and techniques, such as demonstration, characterisation, which are mostly guide-centred, that is, the guides play a major role as protagonists for presenting an ecosystem to visitors. Conversely, nature interactive activities, such as sensory and nature modelling ones, are mostly visitor-centred, and the visitors themselves become the main protagonists in their contact with nature, occasionally under the guidance or supervision of a person (guide, rancher, tutor, instructor, fellow, colleague, teacher, etc.) on duty for it. Passive mediation in environmental interpretation and education refers to those tools and environments in which guides and rangers have a minor role in presenting nature themselves due to the use of technological and audiovisual

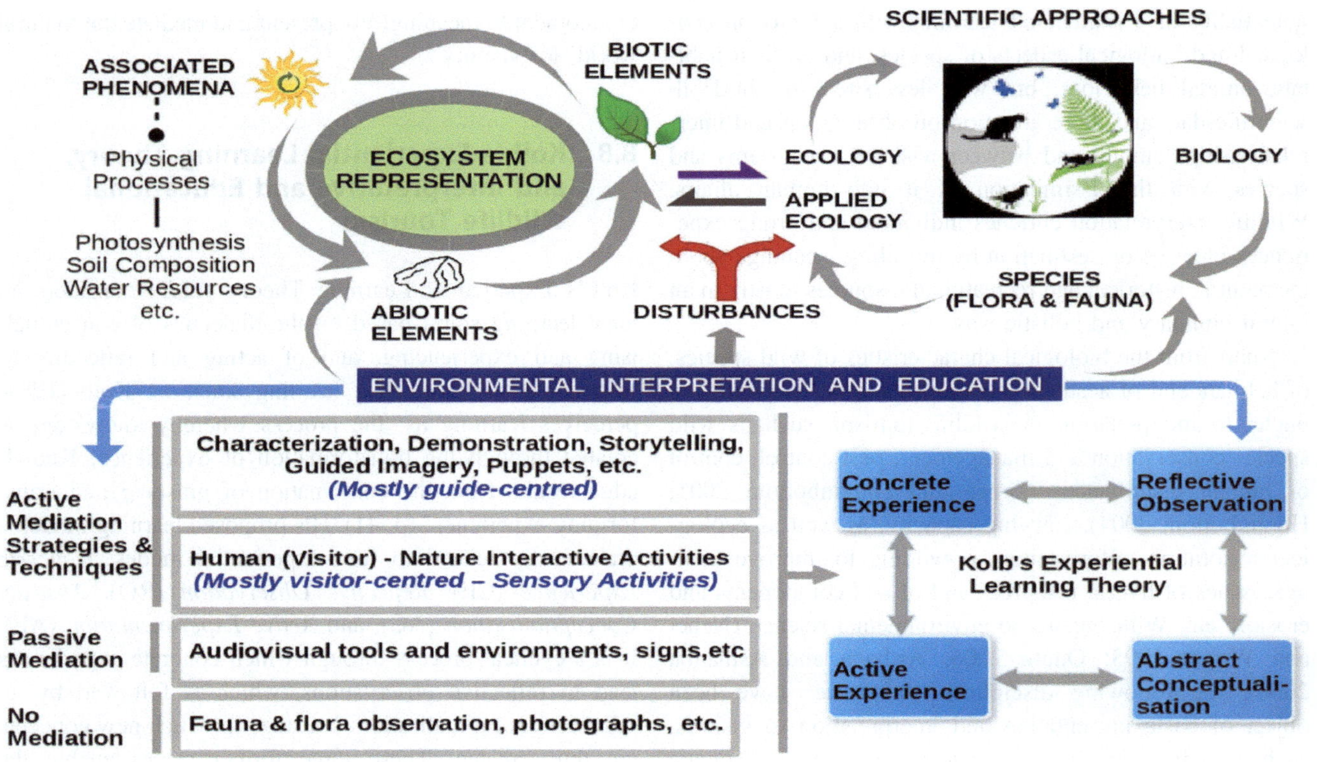

Fig. 8.7 Conceptual diagram for all biotic and abiotic elements, and ecosystem associated phenomena to be part of a holistic environmental interpretation and education for visitors' experiential learning. *Source* The author

instruments, such as video, documentary, slideshow, mobile applications, signs, etc.

The techniques, activities and audiovisual equipments for environmental interpretation and education can be developed to propitiate edu-recreational, conservationist, discovery, educational curriculum-based experiences, etc. On the other hand, ecological and biological learning and experiences can be a self-achievement without any mediation of guides and rangers. For example, visitors can choose to discover, contemplate, experience and learn about an ecosystem by watching, viewing, photographing the fauna and flora, by observing, recording, and exploring the abiotic elements and the physical processes (or the results of them). The point is that ecological and biological factors are inherently part of the process of being in contact with nature. For all them, experiential learning is a multileveled factor in environmental interpretation and education either being a mediated or self-conducted learning. It involves at some point 'Concrete Experience' (CE), 'Reflective Observation' (RO), Abstract Conceptualisation (AC), and Active Experience (AE) in a spiral learning cycle. For example, nature contemplation can be 'reflective observation', thus, 'concrete experience', and it leads to 'abstract concepts' on nature, and this process can result in an 'active experience'. This cycle (CE, RO, AC, and AE) can be even more noticeable as mediation plays a critical role in promoting experiential learning, lets say, in wildlife demonstration and characterisation (Fig. 8.7). In terms of sustainability, an experiential learning in tourism, mediated by a guide or ranger, can for example contribute to enhance visitors' understanding of a site and of environments and foster visitors' post-visit, pro-nature conservation decisions and behaviours (Weiler and Black 2015). As noted, the analysis is mostly conceptual for illustrating the case; this Chapter does not aim to present and crisscross all the variables, situation, content, and elements to outline a visitor-nature-guide/rancher learning experience in light of Kolb's theory. This is an aspect to be investigated in a future research.

In Fig. 8.7, the 'ecosystem', and its components and factors, are taken as the main '*arena*' and 'focus' for an environmental interpretation and education in wildlife tourism; it is represented by biotic and abiotic elements, and associated phenomena. The ecosystem is "composed of a biological community and its physical environment. The environment includes abiotic factors (nonliving components) …as well as biotic factors" (Cunningham et al. 2005, p. 57).

In the field of biology and ecology, abiotic elements are widely understood as all non-living things and resources and physical conditions that can affect living organisms, for example, degradation of a substance by hydrolysis, and abiotic factors can include water resources, and their state and conditions, light, temperature, atmosphere, and soil and its components, such as rock, sand, minerals, etc. The waves can also be regarded as an abiotic factor in a marine context (Sadava et al. 2014; Chapin et al. 2011; Hogan 2010). In this chapter, the non-living things—individual object or groups of objects—are classified as abiotic components. The physical, chemical and geochemical processes are called named 'associated phenomena', which include the effects of temperature, rain, radiation on an ecosystem, flood-related erosions, micro-climate; that is, abiotic phenomena or processes (Cunningham et al. 2005). The biotic elements or factors are related to all living beings, and in a very simplistic explanation: the green plants are classified as '*producers*' because of the photosynthesis they do; the domestic and wild animals as '*consumers*'; the microorganisms as '*decomposers*' (Zahran 2010; Krebs 2007; Cunningham et al. 2005; Nebel and Wright 1993). Photosynthesis involves biotic and abiotic components in a process whereby sunlight (abiotic) is captured by green plants, algae and some bacteria with synthesis of sugars and proteins in tiny membranous organelles called chloroplasts that reside within plant cells (so, it is also biotic) (Cunningham et al. 2005, pp. 55–56).

8.9 Natural and Anthropogenic Disturbances on Wildlife

Disturbances are caused by all natural and non-natural factors that have an impact on nature, on the ecosystems, which directly or indirectly affect the biotic and abiotic elements, and any associated phenomena, and Walker (2012) also classifies 'disturbance' as allogenic and autogenic; and, disturbance by addition; its main characteristics can be typified by frequency, intensity, severity, extent, and interactions. Impact is understood as any endogenous and exogenous interference or intervention on a natural order and, or, state of the Earth to an extent that changes it positive or negatively (Elmqvist et al. 2003). Natural disturbances (extreme disturbances) can be caused by physical, geophysical, and, or chemical interferences on the natural state of the world such as earthquakes, volcanoes, floods (Zahran 2010; Prestemon et al. 2008), erosion, tsunamis, landslides, etc. Prestemon et al. (2008) define it as a "process that results in significant changes in ecosystem structure, leading to alterations in function and the goods and services that humans derive from nature", for example, natural disturbances in forested areas can take place "by physical and biological processes. Large, landscape scale disturbances derive primarily from weather (droughts, winds, ice storms, and floods), geophysical activities (…volcanic eruptions, even asteroid strikes), fires, insects, and diseases" (pp. 35–36).

The impacts of natural disasters (extreme disturbances) on tourism and on the wildlife have been the focus of a lot of research, articles, and books to better understand their immediate and long term effects. More often, the theme draws attention to a disaster context, damaged environment, and biodiversity loss, and to help the local tourism industry and related communities to rebuild themselves and to restore the lost natural assets and resources, as well as to develop management plans for reduction of natural disaster impacts (WTO 1998; White and Frew 2015; Richardson et al. 2015). As one destination is affected by a cyclone, hurricane, volcano, quake—a natural disaster—, this is not only an issue related to a site that needs to be rebuilt and to strategies that need to be put into practice to re-establish the place and business of tourism industry, but also has possible severe and lasting effects on the local ecosystems. The extension of damage is above all irreparable if one looks at the loss of local and regional biodiversity; the loss of wild life can be huge in a calamitous natural event (WTO 1998; White and Frew 2015; Richardson et al. 2015). Holden (2016) reinforces that natural disasters have economic and social impacts, affecting the tourism sector of some destinations. For example, on 26 December 2004, the Boxing Day *tsunami,* killed more than 230,000 people across 14 countries in Asia, such as India, Indonesia, Thailand, and Sri Lanka (UNESCO 2006; Buultjens et al. 2015). Holden (2016) explains that "the geographical location of many of the popular environments for tourism, notably coastal areas, small islands and mountains, make them especially vulnerable to extreme weather events and natural disasters; he also mentions the devastating effects of Hurricane Katrina in 2005 by flooding New Orleans in the USA" (p. 241). But lightning strikes are also a potential risk as they can ignite fires on a wildland and destroy the biodiversity, particularly in the rainforests (Mackey et al. 2002), but rainforests don't so often burn as fires usually travel much further and more fiercely through other habitats. Frequent fires have effects on the ecosystem with loss of wildlife habitat, reduction of biodiversity, invasions by non-native species; it can alter the watershed functioning, as well as other fire-associated hazards, including the loss of tourist appeal (Brooks 2008, p. 45) and can threaten fauna and flora populations resulting in habitat loss (Turton 2014), affecting tourism in these areas.

Conversely, non-natural disturbances are anthropogenic-related, caused by humans, mostly in reason of man's overuses of the natural resources; they are human-induced environmental changes that differ from most natural changes, often happen at a faster rate than the natural disturbances which make the living environment to all species an unpleasant and unsustainable place (Candolin 2009). As Walker and Willig (1999) posit it, "Human interferences with natural disturbances (e.g. fire suppression) may actually make them more destructive […]. Some anthropogenic disturbances are well publicised […] such as urbanization, excavation of minerals, soil erosion as a result of agriculture, or logging of forests, may have far greater consequences" (p. 1). Within the mid-shades of tourism impacts, it counts negatively the undesirable corollaries caused by cumulative effects and permanent degradation in a way it compromises tourism sites, particularly, those reliant on natural assets including the wildlife (Green and Higginbottom 2001; Higginbottom 2004) due to facilities and infrastructures, destruction of habitats, aesthetics impacts, and neglected contact with wildlife and unsustainable consumption of fauna and flora which includes plants picking, souvenirs made from wildlife, fishing, and shooting (Intosai 2013; Sunlu 2003), and unregulated recreational hunting (Bauer and Giles 2002; Knight and Cole 1991). Figure 8.8 shows the main natural and anthropogenic disturbances, as well as some effects of disturbances, including ones caused on wildlife by visitors in nature-based tourism activities. Some of the ecological remedial responses (Hughes and Carlsen 2008) to human-induced disturbances on wildlife include conservation, protection, ecological restoration, pests control, zoning and carrying capacity. The last two are widely in tourism as a way to mitigate negative impacts.

Both natural and anthropogenic-related disturbances and the way to overcome them can be issues to be addressed in environmental interpretation and education as these factors draw attention to the relevance of conservation and of ecological restoration of impacted wildlands and wildlife. Interpretation as a process of communicating facts of the natural world, and as Interpretation Canada poses it, "interpretation can play an important role in natural resource management and conservation as well as meeting the goals of sustainable tourism" (1976, cited in Carter et al. 2015, p. 296). On the other hand, unsustainable tourism practices can become sources for impacts on wildlife and on ecosystems. For example, in the literature there is a plethora of publications with criticism on visitors' closeness, touch, holding, and on artificial feeding of wildlife

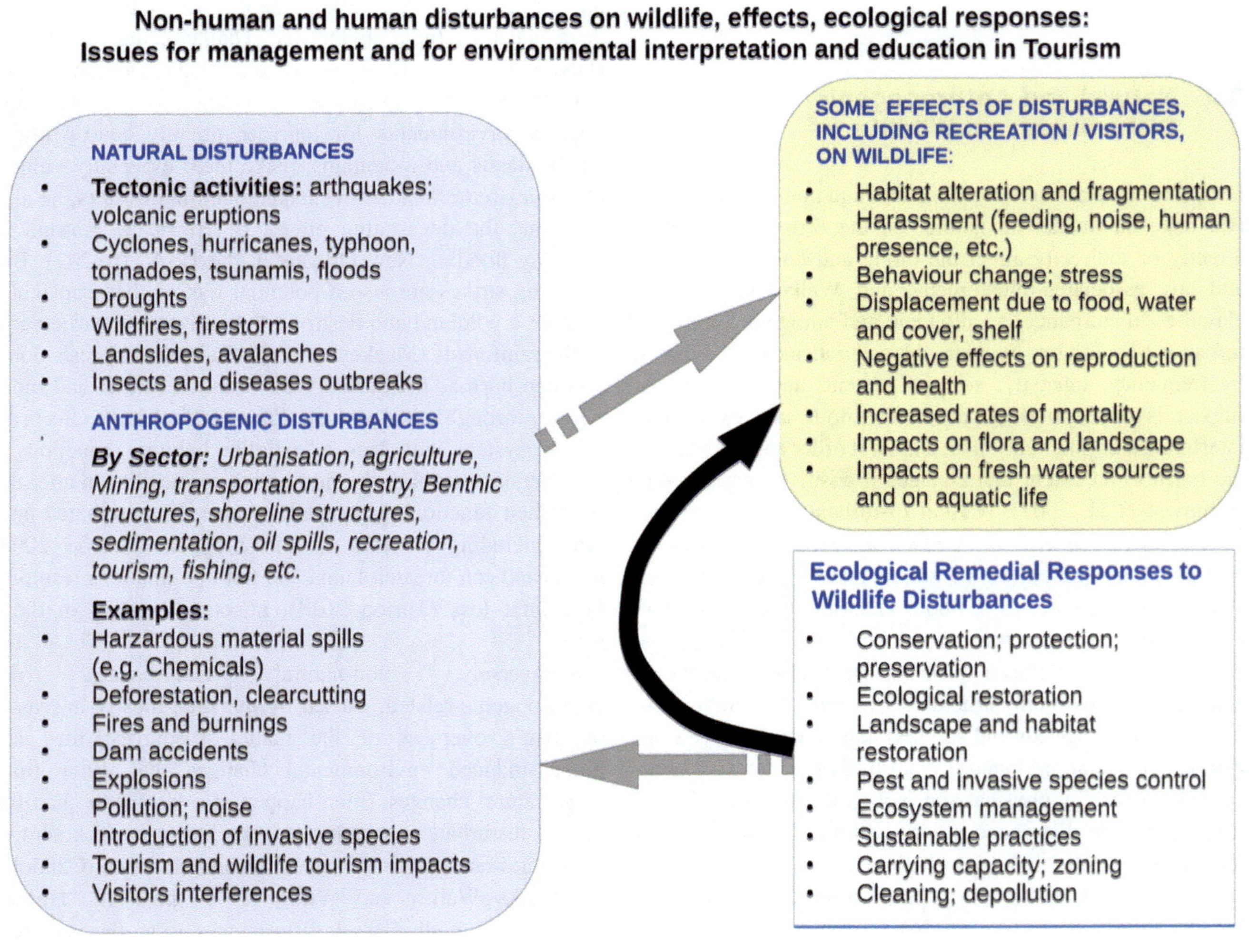

Fig. 8.8 Non-human and human disturbances on wildlife, effects, ecological responses: issues for management and environmental interpretation and education in tourism. *Source* The author, based on multiple sources

(Orams 2000; Newsome et al. 2005; CRC 2009; Fennell 2015; Burns 2015). Orams (2002) found that,

> Deliberate and long-term provision of food to wildlife has been shown to alter natural behaviour patterns and population levels. It has also resulted in the dependency of animals on the human provided food and their habituation to human contact. Intra and inter-species aggression has also occurred where wildlife, in their efforts to obtain food, have harmed one another and harmed tourists (p. 281, abstract).

Wild animals in the wild react differently to human presence, and a series of factors may influence the contact with humans; it depends on the sensitivity of the animal itself, the animal's past experience and characteristics of the habitat in which it occurs, as well as the "frequency, magnitude, timing and location of the disturbance" (Newsome et al. 2002, p. 182), and this topic is thoroughly examined in this Volume by Ronda Green, Chap. 14, entitled 'Disturbing Skippy on Tour: does it really matter? Ecological and ethical implications of disturbing wildlife'. However, wild animals in semi-captivity settings in some wildlife sanctuaries and ecolodges are fed by visitors as observed at Currumbin Wildlife Sanctuary on Gold Coast with kangaroos (see Fig. 8.5) and lorikeets, but as highlighted by Newsome et al. (2005), "there may also be highly structured feeding situations that are directly controlled by management. This may involve the development of a special area or feeding station where controlled amounts of appropriate foods are dispensed to the public for feeding animals at specific times" (p. 76). The three key responses a wild animal may have in reason of deliberate feeding by visitors are 'avoidance', 'attraction', and 'habituation' (Whittaker and Knight 1998), and reactions are regarded as behavioural changes that many animals use as a way to survive in the wild (Newsome et al. 2005). Figure 8.9 presents details of these three responses and the mitigatory actions to manage wildlife feeding as rules for visitors as visiting the wild.

8.10 Harnessing Wildlife, Ecosystems, Conservation and Experiential Learning: A Holistic Approach for Visitors' Environmental Interpretation and Education

In this sphere of wildlife and ecosystems presentation, conservation and experiential learning, ecology, applied ecology and biology as scientific subjects can largely be used by guides and rangers to educate visitors and also to explain the importance, characteristics, services, and particularities of an ecosystem, flora and fauna. These subjects and approaches are also pivotal in environmental interpretation and education to address and present ways of managing human disturbances to

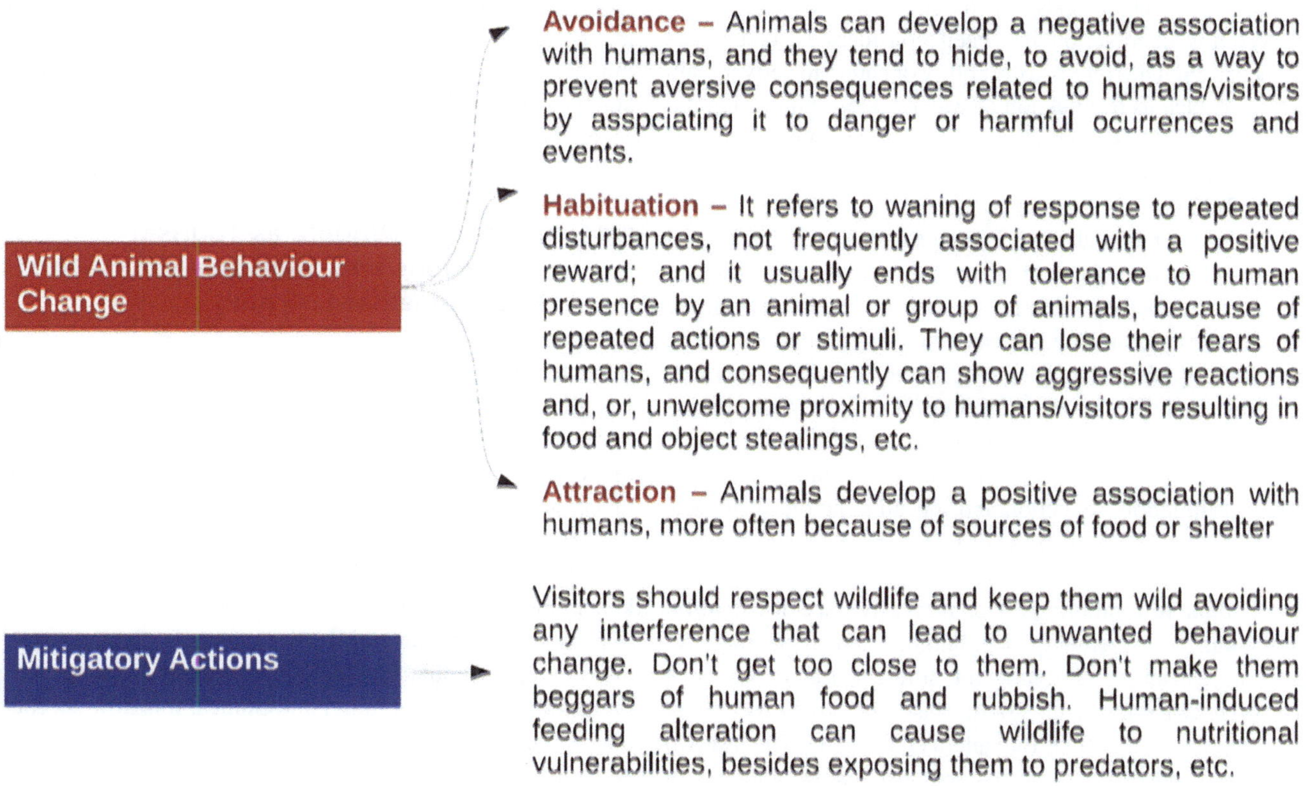

Fig. 8.9 Tourism and human-related interferences and wild animal behaviour change: avoidance, habituation and attraction. *Source* The author, based on multiple publications, e.g., Whittaker and Knight (1998), Newsome et al. (2005)

wildlife either reducing/eliminating impact risks or helping nature to restore its state. Ecology is a term that derives from the Greek work '*Oikos*', which means 'house' or 'place to live', 'household'; and literally, ecology is the study of organisms 'at home', at their environment (Smith 1996; Odum 2006). Ecology is defined by Margalef (1968) as, "the study of systems at the level in which individuals or whole organisms can be considered as elements of interaction, either among themselves, or with a loosely organised environmental matrix. Systems at this level are called ecosystems, and ecology is the biology of ecosystems" (pp. 51–53), and (Cunningham et al. 2005) adds as a definition that 'ecology' is "concerned with the life histories, distribution, and behaviour of individual species as well as the structure and function of natural systems at the level of populations, communities, and ecosystems" (p. 569). Also relevant in this context it is 'applied ecology' as a subset of 'ecology', and is widely used by guides and rangers to address issues related to world of life, the ecosystems and their dwellers, to visitors so they can better understand the natural spaces, interactions, and networks to which the wild species belong to (Cunningham et al. 2005; Odum 2006; Hastings and Gross 2012). The Journal of Applied Ecology, for example, publishes research and academic studies concerning applied ecological problems which include all major themes in this field, such as conservation biology, global change, environmental pollution, wildlife and habitat management, land use and management, aquatic resources, restoration ecology, and the management of pests, weeds and disease. The scale of ecosystems degradation has been considerably in the last decades, and it has affected the ability of nature to deliver main services, such as purification of water cycles, climate regulation, photosynthesis and clean air, waste decomposition, etc. (Balvanera et al. 2001; Elmqvist et al. 2003; Hastings and Gross 2012).

The environmental services are provided by a diversity of organisms within an ecosystem and a holistic guiding is dedicated to mediate an array of visitors experiences in the natural world to an extent it contributes to an understanding and learning about the relevance of the ecosystem services, and sustainability as an equilibrium continuum demanded for human/nature relations and the wonders of wildlife (Balvanera et al. 2001; Odum 2006; Cunningham et al. 2005). Mitigation, reduction and, or, elimination of impacts on the ecosystem and its wildlife can be achieved through conservation, protection, ecological restoration; landscape and habitat restoration; pest and invasive species control; ecosystem management; depolluting, and managing for carrying capacity and zoning for different uses or levels of use (van Driesche et al. 2016; Hastings and Gross 2012; Cunningham et al. 2005; Balvanera et al. 2001). The last two are more noticeable and used for sustainable practices in wildlife tourism. In fact, "wildlife tourism offers unique opportunities for participants to reconnect with nature in a potentially life-changing way and has become increasingly popular in recent years (Ballantyne et al. 2011, p. 2). Reynolds and Braithwaite (2001) identified some categories of wildlife tourism products which can enable a learning experience: nature-based tourism with a wildlife component; locations with good wildlife opportunities; artificial attractions based on wildlife; specialist animal watching; habitat specific tours; thrill-offering tours, e.g., safaris for wildlife viewing. In this sense, environmental interpretation and education is fundamental to help visitors to understand the processes and, more importantly, the sustainable practices necessary for a balance in nature, and even tourism activities should be of attention and inclusion in regards to this aspect, "wildlife watching can only be sustainable if it contributes to the conservation and survival of the watched species and their habitats…to attain long-term sustainability of wildlife watching includes interaction, long-term survival of population and habitats…put in place for sustainably managing wildlife watching tourism, conservation." (Intosai 2013, p. 17). As Ballantyne et al. (2011) posit it, the goals of wildlife tourism should be "to educate visitors about the threats facing wildlife in general, and the actions needed to protect the environment and maintain biodiversity" (p. 770), and their view corroborate the key argument of this chapter that, "the educational aspects of wildlife tourism experiences not only impact on visitor learning and subsequent behaviour, but are also an important contributor to visitor satisfaction with the experience" (p. 772). It is worth noting that interpretation should not be entirely about the problems; it is important to also interpret in such a way as to elicit a fascination with and empathy for the animals so that the messages about the problems mean more to the audience.

8.11 Australian Wildlife as Tourism Attractions and Showcase for Visitors' Environmental Learning

It seems that of all the continents and countries, Australia has the most unusual assortment of animals and plants. Almost half of the world's 314 kinds of marsupials (pouched mammals) are found only in Australia, and most of the others are in neighbouring islands such as New Guinea and the extreme eastern parts of Indonesia, only about 70 kinds occurring in the Americas. Marsupials include kangaroos, koalas, possums, gliders, and bandicoots. The monotremes —platypus and echidna, the only mammals that lay eggs— also live in Australia; the platypus being found nowhere else, but echidnas also inhabiting New Guinea. It is now accepted that the ancestors of the passerines (songbirds) arose in the Australian section of Gondwana, and the most Australian songbirds, despite misleading common names belong to families not found in other continents (Low 2014).

The Australian landscape is very varied. Some of the wetter areas in the south also harbour temperate forests, and there are regions of mountain or coastal heaths, including alpine areas that are often snow-covered in winter (Green 2014). However, Australia is the driest continent on the planet, and the rainfall is the most unpredictable in the world: many lakes and watercourses in the outback (the vast, dry interior of Australia) can remain dry for years and suddenly fill again, attracting many thousands of waterbirds, so it cannot be said it is all dry and desolate (Green 2014). "Australia has many different habitats, from rugged coastlines and sandy beaches to snow-capped mountains, tropical rainforests, huge wetlands, winding rivers and wide open grasslands". The outback is the vast dry interior of Australia, but with significant portion of it covered by hummock or tussock grasses, and in more southern parts by chenopod shrubs (plants of the family *Chenopodiaceae*) or low acacia woodland; moreover, some plants have deep roots that can find water meters down underground (Parish 2006, p. 8).

Australian biodiversity has been acknowledged as very rich, diversified, and a high proportion is endemic to Australia, conveying a great appeal on visitors and have become catchy tourism attractions in zoos, sanctuaries, and parks in the country. The key mammals usually sought-after by visitors are the echidnas, a monotreme; kangaroos; wallabies; koalas; greater bilbies (the bilby); Tasmanian devils; platypuses; dingoes; wombat; flying-foxes (large fruit- and nectar-eating bats) (Egerton and Lochman 2009; Green et al. 2001; Green 2014). The animals of the outback adapt themselves to the harsh conditions of the region, and the majority are endemic species, that is, found only in Australia, which, by the way, it includes the birds, lizards, snakes and frogs of the outback as well a habitat is a place where an animal makes its home. The plants of the outback can provide a colourful contrast to the rusty brown or yellow soil, such as sturt pea with its scarlet and black flowers, or bluebush and saltbush with their pale blue-green or silvery foliage (an adaptation to reflect the heat of the sun) (Green et al. 2001; Green 2014). For an animal to be found naturally in a habitat, it obviously must be able to survive there. It must be able to find enough food, water, oxygen and shelter to help it live, grow and produce young. Some animals can only survive in one type of habitat, while others can live in many different ones across Australia. Australia has a varied landscape, it is formed by rocky cliffs, deserts, coasts, rainforests, woodlands, heathlands, salt and freshwater lakes, rivers and streams (Green et al. 2001; Green 2014; Egerton and Lochman 2009).

Figure 8.10 shows eight ecoregions of Australia with its states and territories, and the landscapes vary from deserts and xeric shrublands, savannah, to temperate, Mediterranean, tropical and subtropical forests. Figure 8.10 also presents the twelve major iconic Australian species with relevant tourism interest and appeal. Most of them are endemic species with significant tourism interest and appeal, and are object of ecological and biological interpretation and education. Table 8.5 has complementary information on the species presented in Fig. 8.10. It brings the popular and scientific names of the species, their conservation e vulnerability status; their geographic location by state and territory, as well as the ecoregion to which they belong to. Of the twelve taxa mentioned in Table 8.5, two are classified as threatened by the IUCN Red List. The focussed species are 'mammals', except for the saltwater crocodile (reptiles) and the cassowary, one of the world's largest flightless birds.

The list just brings some of the most iconic and appealing wild animals, most of them endemic species; the list does not include sharks, whales, dolphins, and other marine species, and it also does not cite the great variety of birds of Australia that attract considerable numbers of visitors every year for bird watching, and some of them are professional 'birders'. A birder is defined as a bird watcher; people who identify and study birds in their natural habitats; it also refers to breeders of birds, In the literature, there are books and manuals specialised, including the CRC Report on birdwatching (Jones and Buckley 2001), at addressing issues related to bird watching and viewing in Australia. In 2015, Leseberg and Campbell published a book on 'top end' [main] birds and animals of Australia that can found and observed in Darwin, Kakadu, Katherine, and Kununurra regions. Australian birds have been extensively discussed in an array of niche publications covering hundreds of species that reside and migrant ones found in the country; the books, manuals guides, etc. usually have detailed birds' key features, distribution, classification of sounds, and behaviour, besides to provide comprehensive habitat explanations (Low 2014; Campbell et al. 2014; Thomas et al. 2011; Clarke and Dolby 2014).

Table 8.5 shows some wild species that usually have tourism appeal in Australia (Fredline 2007; Green et al. 2001) among them kangaroos (red, eastern grey and western grey kangaroo), echidna, flying fox, cassowary, Tasmanian devil, dingo, platypus, wallabies (there are about 30 species of this species), wombat, koala, greater bilby, and crocodile, but the list is not exhaustive as other small and large wild animals also draw attention from visitors. Most of them are endemic, only found in Australian lands. Table 8.5 provides a concise outline of twelve taxa with its scientific names, conservation status (vulnerability), geographic location, and ecoregion where they are usually found. Except for the cassowary and the crocodile, all others are mammals. For a very comprehensive tourism classification of Australian wildlife, the report produced by Green et al. (2001) presents the relevant information and data on major categories of Australian wildlife (terrestrial mammals, birds, reptiles, amphibians, terrestrial invertebrates, freshwater fauna and

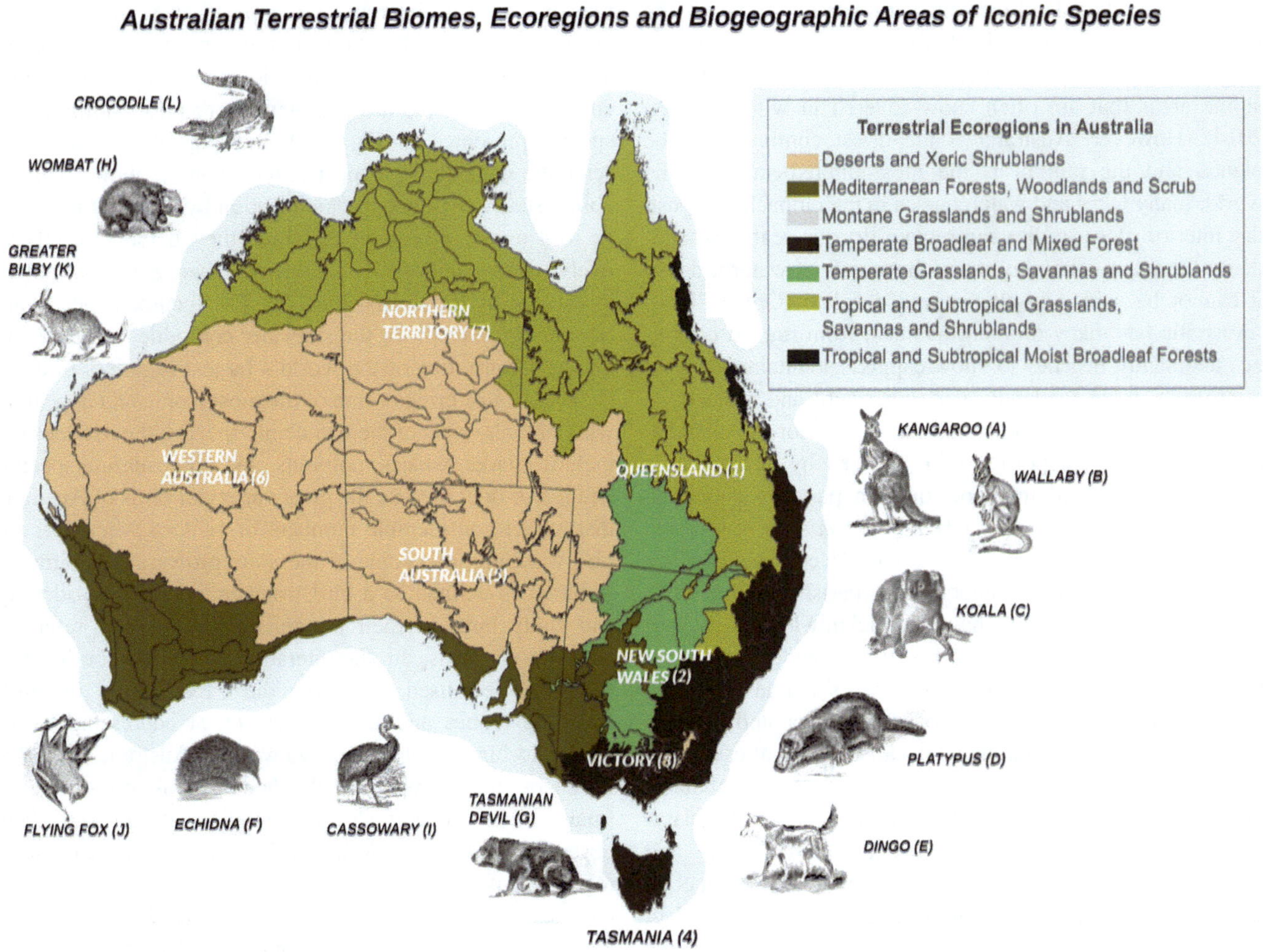

Fig. 8.10 Australian map with its eight ecoregions and the major wild vertebrates. *Source* The author, based on multiple sources. The original map was produced by ERIN (Environmental Resources Information Network), April 2012. Australian Government Department of Sustainability, Environment, Water, Population and Communities. © Commonwealth of Australia, 2012. Available under a Creative Commons Attribution 3.0

marine fauna); the report highlights the kinds of opportunities and constraints on wildlife tourism development within each of these category, as well as the findings can serve to encourage sustainable practices and appropriate development for the wildlife tourism niche in Australia.

8.12 Koalas, Red Kangaroos, and Tasmanian Devils: An Ecological and Biological Detailed Outline of Three Iconic Wild Animals of Australia

This section aims to provide an ecological and biological overview of three iconic species of Australia that are very popular among the visitors, particularly foreign ones. These three species were selected for a detailed outline by reason of their very peculiar characteristics and for being iconic and endemic for Australia. This outline is presented as part of a previous analysis that seeks to show the main characteristics and aspects of them that are pertinent for an environmental interpretation and education within the experiential learning perspective of Kolb. After this extended outline, an analytical diagram will be produced as part of the final considerations for this chapter.

Some iconic native wild animals of Australia such as kangaroos, koalas, platypuses and Tasmanian devils have been for decades a tourism attraction for foreign and domestic visitors alike. The uniquenesses of these wild animals combined with an opportunity of a closer encounter, and even to touch or feed them in some privately or trusted-owned sanctuaries where they live in semi-captivity in Australia are such an experience that has a great appeal; Kangaroos, for example, play a key role for building an Australian tourism marketing imagery (Higginbottom and Northrope 2004), "kangaroos, koalas and crocodiles are the kinds of animals that feature most

Table 8.5 Twelve Australian wild animals with tourism appeal, their biofacts and habitats

Australian species with tourism appeal, and their scientific name	Conservation status/vulnerability	Geographic location (State/Territory) **Refer to* Fig. 8.10 *to check the related area*	Ecoregion/Biome **Refer to* Fig. 8.10 *to check the region*
Kangaroos (A)—Endemic **Red Kangaroo** *Macropus (osphranter) rufus* **Western Grey Kangaroo** *Macropus fuliginosus* **Eastern Grey Kangaroo** *Macropus giganteus*	Least concern, Pop: stable Least concern, Pop: increasing Least concern, Pop: stable	The red kangaroo is only in outback, and not near any of our capitals. The eastern and western greys are in more regions, and seen by many more tourists	Xeric scrubland, grassland, heathlands, and deserts
Red Necked Wallaby or Bennett's Wallaby (B) *(Macropus rufogriseus)* **Whiptail Wallaby (B)** *Macropus parryi* **Agile Wallaby (B)** *Macropus agilis*	Least concern, Pop: stable Least concern, Pop: stable Least concern, Pop: decreasing	1, 2, 3, 4	– Coastal scrub and sclerophyll forest along coastal and highland areas – Temperate broadleaf, mixed forests
Koala ***Endemic*** *(Phascolarctos cinereus).* ***Ps.**: it is not a bear; it is* an arboreal herbivorous marsupial	Vulnerable	Coastal areas of the mainland's eastern and southern and central areas: 1, 2, 3, and 5	Eucalyptus forests and woodlands
Platypus (D) ***Endemic*** *(Ornithorhynchus anatinus)*	Near threatened, Pop: decreasing	1,2, 3, and 4	Watercourses in temperate broadleaf, mixed, and tropical forests
Dingo (E) *(Canis lupus dingo)* *Dingo, also known as warrigal, is **not endemic** to Australia* (Corbett 2008a)	Vulnerable	Throughout Australia, except Tasmania	Dingo's habitat includes alpine, woodland, grassland, desert and coastal areas
Short-beaked Echidna (F) *(Tachyglossus aculeatus)*	Least concern, Pop: stable	Throughout Australia (this and other species in New Guinea)	All ecoregions and habitats
Tasmanian Devil (G) ***Endemic*** *(Sarcophilus harrisii)*	**Endangered**, Pop: decreasing	Tasmania	Temperate broadleaf and mixed forests. Usually, it is found in dry sclerophyll forests and coastal woodlands
Wombat (H) ***Endemic*** **Common wombat** (Vombatus ursinus) **Northern hairy-nosed wombat or yaminon** *(Lasiorhinus krefftii)* **Southern hairy-nosed wombat** *(Lasiorhinus latifrons)*	 Least concern, Pop: stable **Critically endangered**, Pop: stable Least concern, Pop: stable	South-eastern areas, including Tasmania, and an isolated patch of about 300 ha in Central Queensland (Epping Forest National Park): 1, 2, 3, 4, and 5 Ps.: The southern hairy-nosed only found in western WA to western NSW, and they are considered endangered in NSW. Common western Tasmania, southern SA, VIC and NSW, with a tiny population in southern Queensland	Forested, mountainous, and heathland areas. Temperate broadleaf and mixed forests (for common wombat) But, hairy nosed wombats require semi-arid inland regions, which include grassland, open plains, shrubland, savanna and open woodland
Southern Cassowary (I) *(Casuarius casuarius)*	Vulnerable. Pop: decreasing	Northern Queensland: 1	Tropical forests
Grey-Headed Flying Fox (J) ***Endemic*** *(Pteropus poliocephalus)*	Vulnerable. Pop: decreasing	1,2, and 3, but it may be found in different regions; e.g. South Australia	A variety of habitats: woodlands, rainforests and swamps. Temperate broadleaf and mixed forests
Greater Bilby (K) ***Endemic*** *(Macrotis lagotis)* **Long-nosed Bandicoot (K)**	Vulnerable. But, endangered in Queensland, Pop: decreasing	1, 6, 7	Arid, semi-arid **Bandicoot is in forests, woodlands and grasslands*

(continued)

Table 8.5 (continued)

Australian species with tourism appeal, and their scientific name	Conservation status/vulnerability	Geographic location (State/Territory) **Refer to* Fig. 8.10 *to check the related area*	Ecoregion/Biome **Refer to* Fig. 8.10 *to check the region*
Endemic *(Perameles nasuta)*	Least concern, Pop: unknown		
Saltwater Crocodile (L) *Also known as "saltie' (Crocodylusporosus) *Ps.: Crocodiles are found not only in Australia as they populate many other regions of the planet*	Lower risk Least concern	Northernmost parts of the Northern Territory, including the multiple river systems near Darwin; Western Australia; and Queensland. 1, 6, 7	Tropical and subtropical savannah, grasslands and shrublands

For a full understanding of this table, it needs to be viewed in association with Fig. 8.10

Categories of the vulnerability and conservation status of species: extinct (EX); extinct in the wild (EW); critically endangered (CR); endangered (EN); vulnerable to extinction (VU); near threatened; least concern (LC); ***Ps.: Threatened***: *critically endangered (CR) and endangered (EN)*

Ps.: This Table must be used for basic reference only. There are studies that may reveal different figures in regards to wild animals' vulnerability, conservation, geographic locations, and ecoregions Woinarski and Burbidge (2016)

Source The author, and his table was built based on the IUCN Red List of Threatened Species (http://www.iucnredlist.org/multiple) and on various credible sources. GC Grigg, LA 2799 Beard, G Caughley, (1985); BG Norton, M Hutchins, EF Stevens, TL Maple, (1995); Woinarski J, Burbidge AA 2800 (2016) Phascolarctos cinereus. The IUCN Red List of Threatened Species 2016: e.T16892A21960344. Available online at, http://dx. 2801 doi.org/10.2305/IUCN.UK.2016-1. RLTS.T16892A21960344

often in terrestrial wildlife viewing in Australia, as indicated by advertising materials" (Higginbottom and Buckley 2003, p. ii). Moreover, a study carried out by Croft and Leiper (2001) reveals that "mobs of kangaroos, centre stage on a vast outback landscape, are a strong and integral attraction to Outback New South Wales" (p. iii), an opportunity for international tourism.

A couple of years later, another report also prepared by CRC's researchers team, Higginbottom et al. (2003), thoroughly evaluated the existing organised opportunities for viewing free-ranging kangaroos in Australia being concerned with description and classification of kangaroo-related tourism in the country; to describe featured aspects of this niche tourism at a business management level with regards to visitors, interpretation, kangaroo and environmental management, and to propose recommendations for best practices for future development of the sector. For their study, not only those few species correctly named as kangaroos, were considered, but also any existing species of kangaroos and wallabies (family, *Macropodidae*) and rat-kangaroos (families *Potoroidae* and *Hypsiprymnodontidae*), all referred to by biologists as *macropods* (p. 8).

Kangaroos, emus and koalas are national animals; the first two are officially on the coat of arms (Banting 2003; Minahan 2010), and koala is "an unofficial symbol" (Minahan 2010, p. 12). And, "there is a wealth of anecdotal evidence that koalas are an important aspect of a set of unique natural attractions that shape the image of Australia as a tourist destination for both domestic and overseas visitors" (Australian Koala Foundation), and it should be considered the Koalas economic contributions to tourism sector as one of the key attractions in the country (Hundloe and Hamilton 1997). Moreover, Smith et al. (2006) also discuss in their studies issues of production and consumption of wildlife icons. The name "koala" has its origin in coolah or koolah, which means "no drink" (no water), in *Dharug*, an Aboriginal language (Banting 2003, p. 26), as they do not drink much water.

In Australia, wildlife encounters are an experience that interacts with a wide range of other nature-based activities such as bush trails, bush tucker, nature contemplation, etc. Some wild animals are easy to track and observe in their habitats, however time limitations and lack of public transport in many regions may constrain the chances for opportunistic glimpses. Many visitors prefer to go to zoos, aquariums and sanctuaries for wildlife encounters, rather than go to open natural settings, e.g. a National Park. Visitors choose to go to theme parks and zoos because of the convenience of having a ready-to-use structure, multiple enjoyable recreational options, and more importantly, to be sure that they will have a collection of wild species available and on display with entertaining, informative and educational sessions and learning opportunities. It is a guaranteed fun for the whole family, as most wild animals are quite evasive and very hard to spot in the wild (Knight 2013). For example, platypuses and koalas are very elusive; on the other hand, kangaroos are easily found hopping in the outback, and for the two grey species in many coastal woodlands and forests. But, according to the World Animal Foundation alerts that not all zoos and aquariums, for example, are concerned with the needs of the animals; even though, they hold an image as education and conservation-oriented places, most are planned and designed to serve the needs of the visitors with wild animals being captive on display and used in entertaining shows; rather than, educative ones. Many animals in zoos and aquariums reveal some abnormal behaviour as the result of being deprived of their natural habitats and social-group structures (Khan 2013). This Volume brings a valuable contribution to the literature with a chapter on captive wildlife, visitors and the human relations to nature in which Dirk Reiser (refer to Chap. 17) makes critical insights on demystifying zoos.

> Some zoos and aquariums do rescue some animals and work to save endangered species, but most animals in zoos were either captured from the wild or bred in captivity for the purpose of public display, not species protection. The vast majority of captive-bred animals will never be returned to the wild. When the facility breeds too many animals they become "surplus" and often are sold to laboratories, travelling shows, shooting ranches, or to private individuals who may be unqualified to care for them (World Animal Foundation)

Ethics on the Ark, edited by Norton et al., published in 1995, is a book that gathers various contentious viewpoints on the debate about the current situation and the future of zoos and aquariums worldwide; it has multivocal discussions on what should be the priority in these sites where wild animals are semi-captive or captive; wildlife conservation and animal rights are aspects debated by the contributors who collectively take all sides of the issues. However, it is not a mission of this chapter to make a literature review and analysis on whether Zoos and Aquarius are 'evil' or 'providential' for wild animals. The goal of the chapter is to provide a comprehensive literature review, conceptual, theoretical and practical issues vis-à-vis the wildlife resources in Australia, and the discussion on experiential learning for visitors. As part of it, the next section will outline some of the key ecological, biological, physical, behavioural aspects of three Australian species: koalas, Tasmanian devils, and kangaroos taxa. After outlining the main biofacts and information on these three marsupial species, a model of integrated interpretation and experiential learning for visitors will be produced. The main aspects and details to be outlined are: wild animal characteristics; behaviour; food, offspring; threats; and, protection/conservation. Fact sheets and biofacts of the IUCN Red List of Threatened Species of 2016, the World Animal Foundation, the Australian Koala Foundation, the National Geographic and San Diego Zoo Online Library were the main sources for information used to outline and describe the Red Kangaroo, Koala, and Tasmanian Devil. The Mammals of Australia, edited by Ronald Strahan and Steve van Dyck brings comprehensive data on these species.

8.12.1 Koala

8.12.1.1 Koalas, Not a Bear!

The koala looks like a teddy bear, but is a marsupial with a compact round body, soft woolly fur, grey above and white under parts; it has black nose, short limbs, and rounded ears, all of which add to its perceived 'cuteness'. In reason of its *Eucalyptus* leaf-based diet, the koala has very strong chewing muscles, sharp molars, and large jaws; it usually weighs between 4 and 15 kilos, on average 11 kg; its size depends on the latitude the koala lives: smaller in the north; the females are smaller than males. The fingerprints of koalas and of humans are strikingly similar (Henneberg et al. 1997); it has highly sensitive ears; its body length: 60–85 cm (Fig. 8.11).

8.12.1.2 Behaviour

Koalas are not migratory animals, and they are actually solitary except for brief interaction in breeding season and mothers with dependent young Martin and Handasyde (1999) emphasise that the koalas spend a lot of time alone and devote limited time to social interactions. Also the available literature states that the koalas are not territorial, but that there is a dominance hierarchy (Strahan and Van Dicky) and in stable breeding groups, but studies show that females are also fertilised by males that are 'just passing through'. Individual members remain in their own "home range" areas, usually a selection of eucalyptus trees. The animals are mammals with nocturnal habits, and usually they sleep up to 16 h every day. Their life is on the trees; they are arboreal, and do not live in big groups; rather, they prefer to be alone, particularly the females; they are solitary. All koalas sleep on tree fork or on a branch; they do not make nests, and use their massive claws for climbing the trees; the pace is determined by an existing threat or not, but normally they move slowly. The koala is a skilled swimmer, particularly escaping from a threat. Koalas eat eucalyptus leaves from only a few species; each animal eats an incredible amount of one kilogram of leaves per day, even storing them in their cheeks. The trees are home for koala; a place for social relations and mating, food source and shelter. They usually respect each other trees; after a Koala dies it takes up to 12 months to take ownership of it; this is a period time enough for scratches and scents of the former dweller to disappear. Koalas use varied sounds to communicate with each other; a male koala usually defends its territory (tree range) by bellowing against the intruder; this avoids physical confrontation. Males save fighting energy by bellowing their dominance. Female koalas do not bellow as much as males; they do to demonstrate aggressive and sexual behaviour.

,About this picture
Source: Wikipedia Commons
Author: Erik Veland

BABY KOALA

Image captured at Currumbin Wildlife Sanctuary
Gold Coast,Queensland
Australia

Taxonomy

Kingdom	Phylum	Class	Order	Family
Animalia	Chordata	Mammalia	Diprotodontia	Phascolarctidae

Scientific Name: Phascolarctos cinereus
Synonym: Lipurus cinereus Goldfuss, 1817

Habitat and Ecology:

The Koala is an arboreal folivorous marsupial. It occurs in forests and woodlands, typically dominated by eucalyptus species. In inland (semi-arid) portions of its range, it occurs mainly in riparian woodlands (Ellis et al. 2002, Seabrook et al. 2011). Elsewhere distribution may be associated particularly with soil fertility (and hence foliage nutrient content) (Moore and Foley 2000). The Koala has a specialist diet, mostly limited to foliage of Eucalyptus species, with occasional intake of leaves of other plant (mostly Myrtaceous) genera (Martin and Handasyde 1999; Moore and Foley 2000, 2005). At high population densities, Koalas can defoliate preferred tree species, causing tree death and subsequent Koala population crash (Menkhorst 2004, 2008). The Koala is mostly solitary, but individuals have extensive overlap in home ranges. Home range size varies substantially with forest structure and productivity, and males typically have larger home ranges than females. In a coastal forest in New South Wales, average home range size was 10 ha (for females) to 20 ha (for males) (Lassau et al. 2008); in inland Queensland home ranges were 100 ha (for females) to 135 ha (for males) (Ellis et al. 2002). Breeding is seasonal, with births (typically of single young) in October-May. Females can produce young at annual intervals, but births per adult female per year average 0.3-0.8 (McLean 2003). Sexual maturity is reached at 18 months (Jackson 2007). In the wild, longevity of 12 (for males) to 15 years (for females) has been reported (Martin and Handasyde 1999).

* **Generation length**: 6-8 years (Phillips 2000). * **Systems:** Terrestrial * **Mobility:** not migrant

Major Threats: Current threats to this species include continued habitat destruction, fragmentation, and modification (which makes them vulnerable to predation by dogs, vehicle strikes, and other factors), bushfires, and disease, as well as drought associated mortality in habitat fragments. Public concern for the species is high. There are management problems with many populations; remnant populations living at high densities in isolated patches of habitat are at greatest risk (Martin *et al.*2008). Effective management of some of the threats on the mainland could lead to excessive abundance and result in pest problems similar to those occurring on Kangaroo Island and in parts of Victoria. The overall distribution of Koalas has been reduced since European settlement.

Biodiversity Source: Woinarski, J. & Burbidge, A.A. 2016.*Phascolarctos cinereus*. The IUCN Red List of Threatened Species 2016: e.T16892A21960344. http://dx.doi.org/10.2305/IUCN.UK.2016-1.RLTS.T16892A21960344.en. Downloaded on **27 November 2016**.

Fig. 8.11 Koala biofacts

8.12.1.3 Offspring

Koalas breed just once a year, and gestation lasts 35 days; a baby koala is a hairless, blind, and earless joey. At birth, it only has the incredible size of a jelly bean and grows in the pouch on the mother's belly. The joey will stay in the company of the mother for about six months until weaning is complete or so, riding on her back, and feeding on both milk and gum leaves until weaning is complete at about 12 months of age. At the moment a koala reaches its sexual maturation, it leaves its mother's home and finds its own tree range.

8.12.1.4 Threats and Conservation

According to World Animal Foundation, Koalas once in number of millions have faced an extreme decrease of its population, particularly in the 1920s because of hunters looking for their fur; nowadays, the threats are in reason of their habitat destruction, road deaths, and dogs; these combined kill about 4000 koalas every year. Koalas demand large forest areas and corridors in search for territory and for mating. Human population boom of the coastal regions of Australia, consequently resulting in higher demands for urban area development, logging, road construction and agriculture contribute to decrease areas of bush. Also some diseases have been a cause of death in some koala colonies, particularly due to chlamydia and an outbreak of sarcoptic mange (Jackson 2003). On the other hand, overpopulation of koalas also threatens the species. On Kangaroo Island, South Australia, koalas were introduced about 90 years ago and have thrived steadily since then in the absence of predators, "the koala's main enemies are the dingo and the fox" (Hunter 1987, p. 52); this item combined with the fact that koalas are not migrant animals have made Kangaroo Island unsustainable for themselves and for other species; the super-population of koalas represents a threat to an unique ecology of the island. The most viable and likely solution may be a complex sterilisation method; a cull is improbable due to koala's popularity and the negative public opinion it can raise against the government and tourism sector. The translocation resulted in a negligible success, and a relocation to the mainland may not be also successful because there are evidences that koalas may hardly establish in the new area, and Masters et al. (2004) develop a very detailed study on the koalas on Kangaroo Island by examining it

Taxonomy

Kingdom	Phylum	Class	Order	Family
Animalia	Chordata	Mammalia	Diprotodontia	Macropodidae

Scientific Name: *Macropus rufus*

RED KANGAROO

Australia

Habitat and Ecology:

It is found in arid and semi arid habitats. Animals are dependant on green herbage, and populations subsequently decline during drought periods (Croft and Clancy 2008). It is capable of travelling long distances (over 200 km) in response to localized rainfall.

Population: It is an abundant species (Croft and Clancy 2008) that is subject to commercial take under nationally approved management plans. Its abundance is limited by dingo predation, hence it is generally greater in abundance south of the dog fence or within dingo controlled zones. The species has benefited from pastoral infrastructure (i.e., artificial water sources).

Major threat: There appear to be no major threats to this species. There is regulated harvesting of this species for meat and hides.

Conservation actions: This species is present in many protected areas. Harvesting of the species is regulated.

* **Systems:** Terrestrial

Biodiversity source: Ellis, M., van Weenen, J., Copley, P., Dickman, C., Mawson, P. & Woinarski, J. 2016. *Macropus rufus*. The IUCN Red List of Threatened Species 2016: e.T40567A21953534. Downloaded on **27 November 2016**

Fig. 8.12 Red Kangaroo biofacts

from the introduction to pest status in less than a century. The koala on Kangaroo Island became a 'conundrum' because its population management illustrates an example of "conflict between conservation and animal welfare" (Lindenmayer and Burgman 2005, p. 21). However, the investigation of Masters et al. (2004) can shed light on the case as they examine it from the introduction to pest status.

The koala's slow reproduction rate (one young per year) makes it especially vulnerable to population decline. Its specialised diet makes it vulnerable to habitat destruction, and its sense of home range and favourite trees makes it vulnerable to roadkill and dog attack when human residential and road development occurs in its district, as it still tries to visit the areas it used to know (as explained by Ronda Green, this chapter: IUCN Red List data) (Fig. 8.12).

8.12.2 Red Kangaroo

8.12.2.1 Kangaroo Fact Sheet and History

Marsupials probably arrived in Australia between 71.2 and 65.2 million years ago late in the Cretaceous age (Beck 2008). The possum-like marsupial mammals are regarded as the ancestors of the kangaroos (Prideaux and Warburton 2010), and between 50 and 34 million years ago, during the Eocene age, it is believed that their ancestors lived in trees in forests. The fossils of macropod family, kangaroos are dated to about 23 million years ago during early Miocene (Archer and Bartholomai 1978). Some studies have suggested that the hopping has evolved as early as 30 million years ago in forested ecoregions (Dawson and Webster 2010). During the late Pleistocene age, there were two giant kangaroos (Dawson 1995): *Macropus titan* (Marshall and Corruccini 1978) and a large grey kangaroo. "The largest (*Procoptodon goliah*) had an estimated body mass of 240 kg, almost three times the size of the largest living kangaroos, and there is speculation whether a kangaroo of this size would be biomechanically capable of hopping locomotion" (Janis et al. 2014). The Department of the Energy and Environment, Australian Government, released some figures in 2011 with estimated number of kangaroos in four regions of the country and includes four existing species: Red (*Macropus rufus*); Western Grey (*Macropus fuliginosus*); Eastern Grey (*Macropus giganteus*); and Wallaroo/Euro (*Macropus robustus*). Total estimated number of kangaroos for that year was 34,303,677 animals (see Table 8.6). The estimation was done by aerial and ground surveys in areas where commercial harvesting takes place, but the actual national

Table 8.6 Kangaroo population in four Australian regions as estimated in 2011

2011 population estimates for kangaroos within the commercial harvest areas					
State	Red (*Macropus rufus*)	Western grey (*Macropus fuliginosus*)	Eastern grey (*Macropus giganteus*)	Wallaroo/Euro (*Macropus robustus*)	Total
South Australia	1,158,000	674,800	–	494,800	2,327,600
Western Australia	638,185	1,177,534	–	–	1,815,719
New South Wales	3,972,522	496,059	5,258,104	88,430	9,815,115
Queensland	5,745,591	–	10,799,679	3,799,973	20,345,243
Total	11,514,298	2,348,393	16,057,783	4,383,203	34,303,677

Source Department of the Energy and Environment, Australian Government, 2011, online

populations of this species of *macropods* can be significantly higher nationwide. In 1984, a survey carried out a team of researchers (Grigg et al. 1985), published by *Search*, estimated a total of kangaroos' population (not including the wallaroos): 13,283,000 animals. It is believed that there are nearly three times more kangaroos in Australia than cows (Australianwildlife.net).

Currently, it is estimated a total population of up to 60 million Kangaroos living in the country for all 48 kangaroo species (Reference.com). In the *Blogs.Reuters.com*, a short essay entitled 'A necessary evil: the cull of kangaroos', advocates favourably for it by sustaining that "mobs of kangaroos can quickly damage the environment and compete with livestock for scarce food, impacting the livelihood of farmers" (Gray 2013, online). Also as a form of kangaroos' population control with economic and market ends, Spiegel and Wynn (2014) presented a research on the possibilities and implications for promoting kangaroos as a sustainable option for meat production on the rangelands of Australia; the main domestic market problem is that "the consumption of kangaroo [...] by the general population [in Australia] is still uncommon, even though the animal has long been utilised as a bush food by the Aboriginal people" (p. 38), and for a real market and demands this is an issue to be considered.

8.12.2.2 Kangaroo Physical Characteristics

The red kangaroo is the largest marsupial and the largest of its family. It has small head, big ears and dark eyes, and a very long and thick tail used for balance while it hopes (Nowak 1999). The red kangaroo can move its ears through 180° independent of each other (Tyndale-Biscoe 2005). It can be differentiate from other kangaroos by a white underbelly and white patch that extends from its mouth to its ears (Newsome 1995). A male red kangaroo usually weighs between 25 and 85 kg (48–187 lb), and a female, 17–35 kg (37–77 lb). Male kangaroos are generally taller than females of the same species. For males, a body length is 93.5–140 cm (3.1–4.6 ft), and for females: 74.5–110 cm (2.5–3.6 ft). A standing height, it is usually 1.5–2.0 m (4.9–6.6 ft) for males; a male red kangaroo can reach up to 2 m standing higher on its toes when getting aggressive. A male tail can be as long as 1 m (3.3 ft). All *macropods* share the same following characteristics: pouch opens forward and it has four teats; forelimbs are shorter and weaker than hind limbs; long narrow feet; and, it has five digits in its forelimbs. The kangaroo family (macropodidae; macropodids) has "large arytenoid cartilages and very-small vocal or non-existent vocal cords" (Symington 1898, cited in the Australian Journal of Zoology 41, 1993, p. 258).

8.12.2.3 Behaviour and Offspring

Kangaroos can bound at speeds up to 30 miles per hour and can leap some 30 ft. Kangaroos use their long tails for balancing. They can tolerate high temperatures being adapted to dry, infertile areas and to a highly variable climate. On average, kangaroos live in the wild for six to eight years. Kangaroos are grazing herbivores, which means their diet consists mainly of grasses, "just like cattle and sheep, so the grassland consumption of kangaroo populations has been monitored closely" (Reference.com), and can survive long periods without drinking water. Female kangaroos carry newborns, called "joeys" in a pouch on the front of their bellies. As with all marsupials, the joeys are born at a very early stage of development after a gestation that lasts between 31 and 36 days. Newborn joeys have only the forelimbs and the mouth fully developed to allow them to climb to the mother's pouch and to attach to one of her teats, which then swells inside the mouth to keep the joey firmly in place in its early weeks. As for comparison, a human embryo at a similar stage of development would be about 7 weeks old, not mature enough to survive. Kangaroos live and travel in organised groups or "mobs" of ten or more animals, and the mob is dominated by the largest male, called a boomer or buck, it has a certain exclusivity to females for mating; and courtship behaviour in a mob includes the male "checking" the female's cloaca; a female after being checked, it usually urinates, and the boomer then sniffs the urine several times —if he is satisfied and the female shows she is receptive by raising her tail—the male kangaroo starts the mating act.

The arched tail is also an evidence that kangaroos are ready to mate. In reason of their larger size, female kangaroos often reject males by simply moving away from them. Usually, female kangaroos give birth to one joey at a time. A red kangaroo joey does not leave the pouch until it is about eight months old. Amazingly, a female kangaroo can freeze the development of an embryo until the previous joey is big and strong enough to leave the pouch. There are usually three stages at once: a large joey that can come on and off the teat and in and out of the pouch, a younger one that is attached to the teat and receiving a more nutrient-rich milk, and the embryo which is dormant until the younger one detaches from the teat (as explained by Ronda Green, by reviewing this Chapter). The mother's milk varies in its composition according to the needs of the joey; "she is also able to simultaneously produce two different kinds of milk for the newborn and the older joey who still lives in the pouch" (World Animal Foundation 2016, online). According to World Animal Foundation, kangaroos are shy and retiring by nature, and in normal circumstances present no threat to humans. Male kangaroos often "box" amongst each other, playfully, for dominance, or in competition for mates. But, Ronda Green explains, in her review for this chapter that, "they actually don't box; they use their arms to hold the opponent, either in play or a real fight, and kick with their hind legs, using their tails for balance—I've often witnessed it and raised young male kangaroos that tried to play with me in this way when adolescent". Their sharp toenails and long, powerful feet can disembowel an adversary: it has happened to dogs that dare to attack them.

8.12.2.4 Threats

Humans hunt kangaroos for their meat and hides, but this is highly regulated. Also, the presence of domestic herbivores, such as rabbits, cattle, and sheeps can increase disputes for plants and this population of domestic animals can lead to food shortages particularly in times of drought. Many are hit by cars.

8.12.3 Tasmanian Devil

8.12.3.1 Tasmanian Devil Biofacts

See Fig. 8.13.

8.12.3.2 Physical Characteristics, Habits, Feeding and Behaviour

The Tasmanian devil is the largest carnivorous marsupial in the world (National Geographic online) after the extinction of the thylacine reportedly in 1936 and its dietary source is insects, snakes, birds, fish, and other small animals, or carrion of any size. It is notoriously known as a voracious

TASMANIAN DEVIL

Image captured at Currumbin Wildlife Sanctuary in 2015 by **Ismar Lima**

Gold Coast, Queensland Australia

Taxonomy

Kingdom	Phylum	Class	Order	Family
Animalia	Chordata	Mammalia	Dasyuromorphia	Dasyuridae

Scientific Name: *Sarcophilus harrisii*

Habitat and Ecology:
Tasmanian Devils are found throughout Tasmania, in all native terrestrial habitats, as well as in forestry plantations and pasture, from sea level to all but the highest peaks of Tasmania (Jones and Rose 1996, Jones and Barmuta 2000). Densities are lowest in the buttongrass plains of the south-west and, prior to Devil Facial Tumour Disease (DFTD) emergence, highest in the dry and mixed sclerophyll forests and coastal heath of Tasmania's eastern half and north-west coast (Jones and Rose 1996). Open forests and woodlands are preferred, while tall or dense wet forests are avoided (Jones and Rose 1996; Jones and Barmuta 2000).Tasmanian Devils are able to reach very high densities, even in suboptimal habitat, if sufficient food and den sites are available. The 14 km² Badger Island at one time supported 120 Tasmanian Devils. Seabird colonies, such as Short-tailed Shearwaters (or muttonbirds, *Puffinus tenuirostris*), are thought to have traditionally been a preferred habitat for Tasmanian Devils, providing an important food source. These are now much reduced along the east coast, but some sites remain along the west coast (D. Pemberton pers. Comm.). Devils are usullay of nocturnal habits.

Major threat: The major threat to this species at present is Devil Facial Tumour Disease (DFTD), compounded by roadkills, dog kills and persecution.

Conservation actions: As of May 2008, the Tasmanian Devil is listed as Endangered under the Tasmanian Government's *Threatened Species Protection Act 1995*. It is also listed as Vulnerable under the Australian Government *Environment Protection and Biodiversity Conservation Act 1999*. *Insurance strategy:* The highest priority is to establish insurance populations of healthy devils in places isolated from the disease, firstly to avoid total extinction and, secondly, as a source for reintroduction to the wild if devils, and therefore also the disease, become extinct.

*** Systems:** Terrestrial.

Biodiversity source: Hawkins, C.E., McCallum, H., Mooney, N., Jones, M. & Holdsworth, M. 2008. *Sarcophilus harrisii*. The IUCN Red List of Threatened Species 2008: e.T40540A10331066. http://dx.doi.org/10.2305/IUCN.UK.2008.RLTS.T40540A10331066.en. Downloaded on **27 November 2016**.

Fig. 8.13 Tasmanian devil biofacts

animal that consumes nearly everything of a prey, which includes bones, organs, and even hair; it is an eating machine that crushes all for feeding itself! "Feeding devils exhibit twenty known physical postures, including their characteristic vicious yawn, and eleven different vocal sounds. They usually establish dominance by sound and physical posturing" (World Animal Foundation). On average, this marsupial eats roughly 15% of its body weight every day, and amazingly it can devour up to 40% of its body weight in just 30 min if it has enough food for it. In fact, eating is a social event for them. The animal has a threatening appearance particularly as a fighting style; thus, "diminutive as it may be, don't be fooled".

Tasmanian devils have "a notoriously cantankerous disposition and will fly into a maniacal rage when threatened by a predator, fighting for a mate, or defending a meal" (National Geographic, online). Early Europeans during colonial period have named it 'devil' due to its scary displays, "teeth-baring, lunging, and an array of spine-chilling guttural growls" (Livescience online). Tasmanian devil has a muscular body shape, usually of dark colour, with black fur, and it has a sharply strong odour. It is usually quite noisy, blaring, and roaring, particularly when feeding; and it has the strongest bite compared to any other marsupial, and "it hunts prey and scavenges carrion as well as eating household products if humans are living nearby, and unlike most other dasyurids, the devil thermoregulates effectively and its active during the middle of the day without overheating," and is surprisingly able to speed and endure, climbing trees and is very good swimmer in the rivers (World Animal Foundation). Tasmanian devil prefers open forest to tall forest, and dry rather than wet forests. Areas near creeks and thick grass tussocks are chosen as dens, and they use the same dens the whole life. It has nocturnal habits leaving its shelter for hunting, feeding and mating; during the day he solitarily spends the time in burrows, caves and hollow logs. Due to its night-vision capability, the devil's white patches can be easily noticed by its "mates". The "Devils" use "their long whiskers and excellent sense of smell and sight to avoid predators and locate prey and carrion" (National Geographic online). It is worth knowing that the Devil's tail is "important to its physiology, social behavior and locomotion. It acts as a counterbalance to aid stability when the devil is moving quickly. A scent gland at the base of its tail is used to mark the ground behind the animal with its strong, pungent scent" (World Animal Foundation).

8.12.3.3 Offspring and Threat

According to fact sheets of the World Animal Foundation, the Tasmanian devils are not monogamous, and their offspring are usually very competitive with males fighting one another for mating. "Females can ovulate three times in as many weeks during the mating season, and 80% of two-year-old females are seen to be pregnant during the annual mating season. A female devil has on average four breeding seasons per year, and gives birth to 20–30 babies after a Females average four breeding seasons in their life and give birth to 20–30 live young after a gestation of three weeks. "The newborn are pink, lack fur, have indistinct facial features and weigh around 0.0071 oz at birth. As there are only four nipples in the pouch, competition is fierce and few newborns survive. The young grow rapidly and are ejected from the pouch after around 100 days" (World Animal Foundation). Tasmanian devils are regarded as adult at two years of age, and they usually live more than five years in the wild. The main threats currently affecting the animal have been the Devil Facial Tumour Disease (DFTD) (Breed et al. 2009) and on Gold Coast, Australia, Currumbin Wildlife Sanctuary's Insurance Population Program has tried to keep healthy animals on the mainland away from sick ones to prevent the spread infection for the survival of the species (see Fig. 8.14). For a comprehensive inventory on Tasmanian wildlife tourism, refer to the CRC Report prepared by Kriwoken et al. (2002).

8.13 Final Considerations: Interpretation and Education on Wildlife for Visitors Experiential Learning

The issues approached and discussed in this chapter reinstate and corroborate many of the well-known facts about the wildlife and tourism such as the human-caused disturbances and impacts, and the importance of rangers and guides for enriching one's visit in the wild, zoos, and parks. It is noted that more pro-active wildlife tourism with interactive and educational experiences should be part of tourism attractions, sites and destinations as a way of allowing human-nature reconnection. These encounters should occur in a way that could increase visitors' learning, feelings and awareness towards wildlife and the biodiversity. As presented in the former sections through biofacts, behaviour and characteristics, most Australian wild animals are fascinating, intriguing, endemic, and unique in many senses with a great tourism appeal. In Australia, it can be observed some initiatives by government and its bodies, organisations and private sector that have tried to make visitors' experiences more meaningful while protecting the species and their habitats; attractions have underpinned 'environmental learning', 'recreation', and 'conservation' into the same basket, even some commercially-oriented places and tours have tried to balance wildlife displays and shows with significant actions and projects in benefit of the wild animals.

The challenge has been to congregate all information and facts on wildlife and to address them to heterogeneous

Fig. 8.14 Currumbin Wildlife Sanctuary's Program for controlling the spread of 'Facial Tumour Disease' (DFTD) on Tasmanian devils on the mainlands. *Source* Author own work, 2015

groups of visitors. For achieving it, techniques, strategies and interactive activities should be part of a well-elaborated interpretative planning. Aspects such as taxonomy, species distribution, habitats, physical characteristics, population, lifespan, behaviour, diet and feeding, offspring and reproduction, ecology and wildlife web of relations, managed care, threats, and conservation are interrelated ecological, biological and ecosystemic facts and data that can be object of a content to be delivered to visitors according to their age range and visit goals. The challenge is exactly to deliver this type of content in a way that is interesting, entertaining and meaningfully mediated to touch the visitors' sensibility towards nature, without making a visit a dry, exhaustive or boring seminar. More in-depth content can be delivered to curriculum-based visitors and to segmented groups with specific knowledge demands for wildlife and ecosystem. Independently from the type of audience, there are six basic rules of interpretation that must be taken into account by interpreters (guides, rangers, instructors, teachers, etc.): (1) people learn better when they are actively involved; (2) people learn better when they are using as many senses as appropriate; (3) new learning is built on a foundation of knowledge; (4) people prefer to learn that which is of most value to them at the present moment and knowing the usefulness of the knowledge being acquired makes learning more effective; (5) that which people discover for themselves generates a special and vital excitement and satisfaction; (6) learning requires activity on the part of the learner (Fa et al. 2011, p. 227). For the last one, interpreters need to find ways of getting attention and of engaging them in activities that are conducive to learning. The question of motivation in outdoor activities for experiential learning can be also approached through the use of the phenomenology; a philosophical discipline that studies the structures of experience and it draws attention to the ways in which an individual creates a meaningful world (Brown 2003).

As underlined by Green (2014), if tourists are introduced to the beauty, fascination, quirky behaviours and ecological roles (e.g. seed dispersal by cassowaries and flying foxes) of wildlife and then some of the conservation problems are explained to them in a way they understand and relate to, they may be more likely to minimise their own impacts when wildlife-viewing, support conservation initiatives and even become ambassadors for wildlife conservation generally.

Environmental interpretation and education is a tool that rangers, guides, instructors, teachers, etc. should wisely use to deliver biological and ecological content to visitors by serving as mediators—knowledge building bridges—between visitors and the wonders of nature while caring for nature and promoting conservationist messages. Experiential learning theory of Kolb can be adapted to the context of nature-based tourism and guiding, and it is observed that there is a paucity of research in this field. As for the question, what is necessary for a meaningful educative wildlife tourism through experiential learning? A deterministic answer lies in the ultimate fact that the wildlife and ecosystems need conservation and protection for ensuring wildlife resources.

Educative tourism needs to be developed based on visitors' needs, interest, motivation and engagement through meaningful strategies; creative and innovative ways are mandatory to present the natural world to reconnect humans to nature and make them more sensitive to current environmental challenges towards a better world to all living beings, and this understanding has been shared by Green et al. (2001), Green and Higginbottom (2001), Weiler and Black (2015), and by many organisations such as Wildlife Tourism Australia and Interpretation Australia. The Sustainable Tourism CRC Reports gather an outstanding collection of research and studies that were developed for many years and thoroughly assessed the wildlife tourism in Australia since 2001 (Higginbottom et al. Part I & 2, 2001a, b; Davis et al. 2001).

Acknowledgements This paper is part of a broad postdoctoral research project related to 'environmental and cultural learning in tourism development in parks' in Australia, and it has received financial support from the **Brazilian Higher Education and Research Agency, CAPES**, process number 99999.006809/201406, Call: *Portaria 36/2013 Estágio Pós-doutoral—Chamada I 2015* also with Institutional support from the Universidade Estadual de Roraima, **UERR, MultiAmazon Lab**, in Brazil, & **Recinatur Foundation for Science, Nature and Tourism** (Chile, Mexico & Ecuador), and the School of Business and Tourism, at Southern Cross University, **SCU. IUCN** Red List of Threatened Species also contributed to this chapter by permitting the use of biofacts available on its website. My thanks go to all staff—rangers and managers—at David Fleay Wildlife Park for the interviews, clarifications and explanations contributing to data collection, and this is extended to Currumbin Wildlife Sanctuary's rangers in charge of educative sessions. Thanks to prof. Dr. **Betty Weiler**, Dr. **Arianne Reis**, prof. Dr. **Elizabeth Roberts**, and Dr. **Silvia Nelson** for supporting my research, as well as scholar Jeremy Novak, Dr. Simon Wilde and Dr. Flavio Valente for having a share on their academic experiences and friendship during my stay in Australia. Thanks Dr. **Ronda Green** for reviewing this chapter and for her editorial contributions.

References

Anderson DH, Lime DW, Wang TL (1998) Maintaining the quality of park resources and visitor experiences: a handbook for managers. Cooperative Park Studies Unit, Department of Forest Resources, University of Minnesota, St. Paul, Minnesota, USA

Archer M, Bartholomai A (1978) Tertiary mammals of Australia: a synoptic review. Alcheringa 2:1–19

Asthana DK, Asthanba M (2006) A textbook of environmental studies. S. Chand & Company Ltd, New Delhi

Baird IA (1986) The objectives of management planning: a commentary and discussion. Aust Parks Recreat 3(1):11–16

Ballantyne R, Packer J, Sutherland LA (2011) Visitors' memories of wildlife tourism: Implications for the design of powerful interpretative experiences. Tour Manag 32(4):770–779

Balvanera P, Daily GC, Ehrlich PR, Ricketts TH, Bailey S, Kark S, Kremen C, Pereira H (2001) Conserving biodiversity and ecosystem services. Science 291:2047

Banting E (2003) Australia, the land. Crabtree Publishing Company, New York

Bauchop D, Parkin D (2000) Minimal impact education: a new approach to walking softly. Int J Environ Educ Inf 19(2):139–150

Bauer J, Giles J (2002) Recreational hunting: an international perspective. Wildlife Tourism Research Report Series no. 13. Cooperative Research Centre for Sustainable Tourism, CRC, Queensland, Australia

Beaumont NK (1999) Ecotourism: the contribution of educational nature experiences to environmental knowledge, attitudes, and behaviors. PhD thesis. Australian School of Environmental Studies, Griffith University, Nathan, Queensland, Australia

Beck L, Cable T (2002) Interpretation for the 21st century: fifteen guiding principles for interpreting nature and culture. Sagamore Pub, Champaign, IL, USA

Beck MDR (2008) A dated phylogeny of marsupials using a molecular supermatrix and multiple fossil constraints. J Mammal 89:175–189. doi:10.1644/06-MAMM-A-437.1

Beckmann E (1988) Interpretation in Australia: current status and future prospects. Aust Parks Recreat 23(6):6–14

Beckmann E (1991) Environmental interpretation for education and management in Australian National Parks and other Protected Areas. Unpublished PhD thesis, University of New England, NSW, Australia

Bell S (2005) Design for outdoor recreation. Taylor and Francis, London

Breed AC, Plowright RK, Hayman DTS et al (2009) Disease management in endangered mammals. In: Delahay R, Smity GC, Hutchings MR (eds) Management of disease in wild mammals. Springer, Tokyo, pp 215–240

Brochu L (2003) Interpretative planning: the 5-M model for successful planning projects. InterpPress, Fort Collins

Brooks ML (2008) Plant invasions and fire regimes. In: Wildland fire in ecosystems: fire and nonnative invasive plants. General Technical Report vol. 6, Forest Service, United States Department of Agriculture, USDA, pp 33–46

Brown C (2003) The real and the good: phenomenology and the possibility of an axiological rationality. In: Brown C, Toadvine T (eds) Eco-phenomenology: back to the earth itself. State University of New York Press, New York, pp 3–18

Brown TJ, Ham SH, Hughes M (2010) Picking up litter: an application of theory-based communication to influence behaviour in protected areas. J Sustain Tourism 18(7):879–900

Buckley R, Witting N, Guest M (2003) Visitor fees, tour permits and asset and risk management by parks agencies. In: Buckley R, Pickering C, Weaver DB (eds) Nature-based tourism, environment and land management. Ecotourism series 1. CABI Publishing, Wallingford, pp 51–59

Burns GL (2015) Animals as tourism objects ethically refocusing relationships between tourists and wildlife. In: Markwell K (ed) Animals and tourism: understanding diverse relationships. Aspects of tourism. Channel View Publications, Ontario, pp 44–59

Burns JB, McInerney J (2010) Career opportunities in travel and hospitality. Ferguson, New York

Buultjens J, Ratnayake I, Gnanapala AC (2015) From tsunami to recovery: the resilience of the Sri Lankan tourism industry. In: Ritchie BW, Campiranon K (eds) Tourism crisis and disaster management in the Asia-Pacific. CABI, Oxfordshire

Campbell I, Woods S, Leseberg N (2014) Birds of Australia. Princeton University Press. New Jersey, USA

Candolin U (2009) Population responses to anthropogenic disturbance: lessons from three-spined sticklebacks Gasterosteus aculeatus in eutrophic habitats. J Fish Biol 75(8):2108–2121. doi:10.1111/j.1095-8649.2009.02405.x, available online at www.interscience.wiley.com

Carter J (2001) A sense of place: an interpretative planning handbook. Tourism and environment initiatives, inverness, QR, UK. Available online, https://portal.uni-freiburg.de/interpreteurope/service/publications/recommended-publications/carter_sense-of-place.pdf

Cater C, Low T, Bricker K (2015) The encyclopedia of sustainable tourism. CABI, Oxford

Chapin FS III, Mooney HA, Chapin MC, Matson P (2011) Principles of terrestrial ecosystem ecology. Springer, New York

Clarke R, Dolby T (2014) Finding Australian birds: a field guide to birding locations. Csiro Publishing, Collingwood

Cocks S, Simpson S (2015) Anthropocentric and ecocentric: an application of environmental philosophy to outdoor recreation and environmental education. J Exp Educ 38(3):216–227

Coghland A, Kim AK (2012) Interpretative layering in nature-based tourism: a simple approach for complex attractions. J Ecotourism 11 (3):173–187

Cohen E (1985) The tourist guide: the origins, structure and dynamics of a role. Ann Touris Res 12(1):5–29

Colquhoun F (2005) Interpretation handbook and standard: distilling the essence, Department of Conservation, Wellington, New Zealand

Corbett L (2008a) Dingo (*Canis lupus dingo*). Canid Specialist Group, IUCN/SSC. Available at http://www.canids.org/species/Canis_lupus_dingo.htm Accessed on 10 September 2016

CRC (2009). Wildlife tourism: challenges, opportunities and managing the future. In: Research informing planning and practice. Sustainable Tourism Cooperative Research Centre. Australia. Available online, http://www.sustainabletourismonline.com/awms/Upload/Resource/Wildlife%20Tourism%20Snapshot_LoRes.pdf

Croft DB, Leiper N (2001) Assessment of opportunities for international tourism based on wild kangaroos. Cooperative Research Centre for Sustainable Tourism—CRC, Queensland, Australia

Cunningham WP, Cunningham MA, Saigo B (2005) Environmental science: a global concern. McGrawHill, New York

Davidson P, Simon A, Gottschalk L, Hunt J, Wood G, Griffin RW (2006) Management: core concepts and skills. John Wiley & Sons, Milton, Queensland, Australia

Davis D, Tisdell C, Hardy M (2001) The role of economics in managing wildlife tourism. Wildlife Tourism Research Report Series n. 3. Cooperative Research Centre for Sustainable Tourism, CRC, Queensland, Australia

Dawson TJ (1995) Kangaroos: biology of the largest marsupials. Cornell University Press, Ithaca, New York, 162 p

Dawson TJ, Webster KN (2010) Energetic characteristics of macropodoid locomotion. In: Coulson C, Eldridege M (eds) Macropods: the biology of kangaroos, wallabies and rat-kangaroos. CSIRO Publishing, Collingwood, VIC, Australia, pp 99–109

Day J (1995) Skills and training for conservation staff. Ranger 4:12–13

de Lima IB (2011) The micro geopolitics of ecotourism: the intricacies of discursive constructions, power relations and partnership models in planning, promotion and management of ecotourism. Alpha Graphics, Goiania, Brazil & Hamilton, New Zealand, Creative Amazon.com, 400 p

de Lima IB (2016a) Queensland parks as environmental learning settings: visitors, guiding, and environmental interpretation in Australia. Atelie Geografico 10(1):28–51

de Lima IB (2016b) The pivotal role of tour guides for visitors' connection with nature: conceptual and practical issues. Int J Humanit Appl Sci IJHAS 5:18–22

Egerton L, Lochman J (2009) Wildlife of Australia. Jacana Books, Crows Nest

Elmqvist T, Folke C, Nyström M, Peterson G, Bengtsso J, Walker B, Norberg J (2003) Response diversity, ecosystem change and resilience. Front Ecol Environ 1:488–494

EPA (1999) Environmental Protection Agency and Queensland Parks and Wildlife Service 1999–2001 Corporate Plan. EPA, Brisbane, Queensland, Australia

EPA (2001) Environmental Protection Agency. Master Plan for Queensland's Parks System. Queensland Government, Brisbane, Australia

Fa JE, Funk SM, O'Connell D (2011) Zoo conservation biology. Cambridge University Press, Cambridge

Farsani NT, Coelho COA, Costa CMM (2012). Tourism crisis management throuh geotourism development. Revista Turismo e Desenvolvimento (17/18): 1627–1638

Fennell DA (2015) The status of animal ethics in tourism: a review of theory. In: Markwell K (ed) Animals and tourism: understanding diverse relationships. Aspects of Tourism. Channel View Publications, Ontario, pp 27–43

Frawley KJ (1989) Cultural landscapes and national parks: philosophical and planning issues. Aust Parks Recreation 25(3):16–23

Fredline L (2007) An analysis of the domestic wildlife tourism market in Australia. Wildlife Tourism Research Report Series n. 16. Cooperative Research Centre for Sustainable Tourism, CRC, Queensland, Australia

Gray G (2013) A necessary evil—the kangaroo cull. Blogs.Reuters.com. Available at, http://blogs.reuters.com/photographers-blog/2013/04/03/a-necessary-evil-the-kangaroo-cull/. Accessed 29 Nov 2016

Green RJ (2014) Wildlife tourism: a handbook for tour guides, eco-lodges, students and start-up businesses. Creative & Amazon.com, Running Creek, Queensland, Australia

Green R, Higginbottom K (2001) Negative effects of wildlife tourism on wildlife. Wildlife Tourism Research Report Series n. 5. Cooperative Research Centre for Sustainable Tourism, CRC, Queensland, Australia

Green R, Higginbottom K, Northrope C (2001) A tourism classification of Australian Wildlife. Wildlife Tourism Research Report Series n. 7. Cooperative Research Centre for Sustainable Tourism, CRC, Queensland, Australia

Grigg GC, Beard LA, Caughley G et al (1985) The Australian kangaroo populations, 1984. Search 16:277–279

Ham SH (1992) Environmental interpretation: a practical guide for people with big ideas and small budgets. University of Idaho, USA

Hammit WE, Cole DN (1998) Wildland recreation: ecology and management, 2nd ed., John Wiley & Sons Inc, USA
Harpers Ferry Center (1998) Planning for interpretation and visitor experience. Division of interpretative Planning, West Virginia, USA. Available online at, https://www.nps.gov/hfc/acquisition/P13PC00042/Attachment%20B.pdf
Hastings A, Gross L (2012) Encyclopedia of theoretical ecology. University of California Press, Berkeley
Henneberg M, Lambert KM, Leigh CM (1997) Fingerprint homoplasy: koalas and humans. Nat Sci 1(4)
Higginbottom K (2004) Wildlife tourism: impacts, management and planning. CRC Sustainable Tourism & Common Ground, UniversityPress.com, Altona, Victoria, Australia
Higginbottom K, Buckley R (2003) Terrestrial wildlife viewing in Australia. Wildlife tourism research report series n. 9. Cooperative Research Centre for Sustainable Tourism, CRC, Queensland, Australia
Higginbottom K, Northrope C, Green R (2001) Positive effects of wildlife tourism on wildlife. Wildlife tourism research report series n. 6. Cooperative Research Centre for Sustainable Tourism, CRC, Queensland, Australia
Higginbottom K, Rann K, Moscardo G, Davis D, Muloin S (2001) Status assessment of wildlife tourism in Australia: an overview (Part I). Wildlife tourism research report series n. 1. Cooperative Research Centre for Sustainable Tourism, CRC, Queensland, Australia
Higginbottom K, Green R, Leiper N, Moscardo G, Tribe A, Buckley R (2003) Evaluation of organised tourism involving wild kangaroos. Wildlife Tourism in Australia Report Series No. 18. CRC for Sustainable Tourism Pty Ltd, Australia
Hockings M (2000) Evaluating protected area management: a review of systems for assessing management effectiveness of protected areas, University of Queensland, School of Natural and Rural Systems Management, Lawes, Qld, Australia
Hockings M (2003) Systems for assessing the effectiveness of management in protected areas. BioScience 53:823–832
Hogan CB (2010) Abiotic factor. Encyclopedia of Earth. National Council for Science and the Environment, Washington, D.C
Holden A (2016) Environment and tourism. Routledge, London
Howard JL (2013) Managing the natural environment: the role of park rangers and the skills they need. Rural Soc 22(3):242–250
Hu Y (2002) Understanding anthropocentric/biocentric orientations towards natural parks: a survey of student. A Master's degree dissertation. Graduate College of the Oklahoma State University, 115 p
Hughes M, Carlsen J (2008) Human-wildlife interaction guidelines in Western Australia. J Ecotourism 7(2–3):147–159
Hughes M, Ham SH, Brown T (2009). Influencing park visitor behavior: a belief-based approach. J Parks Recreat Adm 27(4):38–53
Hundloe T, Hamilton C (1997). Koalas and tourism: as economic evaluation. Discussion paper number 13. The Australia Institute
Hunter S (1987) The official koala handbook. Chatto and Windus, London
Intosai (2013) Impact of tourism on wildlife conservation. Working report. Intosai—Working Group on Environmental Auditing. Available online at, http://www.environmental-auditing.org/LinkClick.aspx?fileticket=ecrH%2BPLw7Cc%3D&tabid=128&mid=568
IUCN (1994) Guidelines for protected area management categories. IUCN, Gland
Jackson S (2003) Australian mammals: biology and captive management. Csiro Publishing, Collingwood
Janis CM, Buttrill K, Figueirido B (2014) Locomotion in extinct giant kangaroos: were sthenurines hop-less monsters? PLoS ONE 9(10): e109888. doi:10.1371/journal.pone.0109888
Jennings GR, Weiler B (2006) Mediating meaning: perspectives on brokering quality tourism experiences. In: Jennings GR, Gayle R, Nickerson NP (eds) Quality tourism experiences. Elsevier Butterworth-Heinemann, Burlington, MA, USA, pp 57–78
Jones DN, Buckley R (2001) Birdwatching tourism in Australia. Wildlife tourism research report series n. 10. Cooperative Research Centre for Sustainable Tourism, CRC, Queensland, Australia
Katz R (1974) Skills of an effective administrator. An HBR classic. Harv Bus Rev 52(5):90–102
Khan BN (2013) Impact of captivity on social behaviour of chimpanzee (pan troglodytes). J Animal Plant Sci 23(3):779–785
Knight D (2013) The making of modern science: science, technology, medicine and modernity: 1789–1914. Wiley Publishers, Cambridge
Knight RL, Cole DN (1991) Effects of recreational activity on wildlife in wildlands. Trans North Am Wildl Nat Resour Conf 56:238–247
Kolb AD (1984) Experiential learning: experience as the source of learning and development. Prentice-Hall, Englewood Cliffs
Krebs CJ (2007) The message of ecology. University of British Columbia. Indo American Books, Delhi, India
Kriwoken L, Lennox S, Ellis C (2002) Tasmanian wildlife tourism inventory: developing an inventory of wildlife viewing opportunities. Wildlife tourism research report series n. 26. Cooperative Research Centre for Sustainable Tourism, CRC, Queensland, Australia
Lankford SV, Knowles-Lankford J, Wheeler DA (2011) An introduction to park management. Sagamore Publishing LLC, Urbana
Lawrence D (1996) Managing parks/managing 'country': joint management of aboriginal owned protected areas in Australia. Research paper 2 (1996-97). Social Policy Group. Parliament of Australia. Available online at, http://www.aph.gov.au/About_Parliament/Parliamentary_Departments/Parliamentary_Library/pubs/rp/RP9697/97rp2
Lindenmayer D, Burgman M (2005) Practical conservation biology. CSIRO Publishing, Collingwood, VIC, Australia
Littlefair C, Buckley R (2008) Interpretation reduces ecological impacts of visitors to world heritage site. AMBIO 37(5):338–341
Long M (2015) My favorite wildlife experiences in Queensland, Australia. Travel blog opinion essay. Landlopers, available at, http://landlopers.com/2015/07/14/wildlife-experiences-in-queensland. Accessed on 29 Nov 2016
Low T (2014) Where song began: Australia's birds and how they changed the world. Viking & Penguin Books, Australia
Mackey B, Lindenmayer D, Gill M, McCarthy M, Lindesay J (2002) Wildlife, fire and future climate: a forest ecosystem analysis. Csiro Publishing, Collingwook
Manning RE (2002) How much is too much? Carrying capacity of national parks and protected areas. In: Arnberger A, Brandenburg C, Muhar A (eds) Monitoring and management of visitor flows in recreational and protected areas. Conference proceedings, Bodenkultur, University of Vienna, Austria, pp. 306–313
Margalef R (1968) The study of pelagic ecosystems. In: Margalef R (ed) Perspectives in ecological theory. University of Chicago Press, Chicago, pp 51–79
Marion JL, Reid SE (2007) Minimising visitor impacts to protected areas: the efficacy of low impact education programmes. J Sustain Tourism 15(1):5–27
Marshall LG, Corruccini RS (1978) Variability, evolutionary rates, and allometry in dwarfing lineages. Paleobiology 4:101–118
Martin R, Handasyde K (1999) The Koala: natural history, conservation, and management. 2nd ed. Krieger Publishing Company, Malabar, Florida, USA
Masters P, Duka T, Berris S, Moss G (2004) Koalas on Kangaroo Island: from introduction to pest status in less than a century. Wildlife Res 31(3):267–272

McGahan PG, Bassett JA (1999) Training needs analysis of site managers on the participation of local communities in natural world heritage site management in Southeast Asia, West Pacific, Australia and New Zealand. UNESCO, Conservation Tourism Consultants New Zealand, 157 p

McGrath G (2007) Towards developing tour guides as interpreters of cultural heritage: the case of Cusco, Peru. In: Black R, Crabtree A (eds) Quality assurance and certification in ecotourism. CABI, Wallingford, UK, pp 364–394

McIntyre A, Mulligan R, Robinett M, Conley J, Stevens D (2014) Interpretation guide. John Ball Zoo Education Department. Grand Rapids, Michigan, USA. Available online at http://www.jbzoo.org/uploads/files/zoodoers/interp-guide-zoo.pdf

McNeely JA, Miller KR (1984) Protected areas are adapting to new realities. National parks, conservation, and development: the role of protected areas in sustaining society. Proceedings of the World Congress on National Parks, Bali, Indonesia, 11–22 October 1982. Smithsonian Institution Press, Washington, D.C., USA

Medio D, Ormond R, Pearson M (1997) Effect of briefings on rates of damage to corals by scuba divers. Biol Conserv 79:91–95

Minahan J (2010) The complete guide to national symbols and emblems. Greenwood Press, Oxford

Morgan M, Soucy J (2006) Usage and evaluation of nonformal environmental education services at a state park: are anglers catching more than fish? Environ Educ Res 12(5):595–608

Moscardo G (1998) Interpretation and sustainable tourism: function, examples and principles. J Tour Stud 9(1):2–13

Nebel BJ, Wright RT (1993) Environmental science: the way the world works. Prentice Hall, Englewood Cliffs

Newsome A (1995) Red kangaroo *Macropus rufus*. In: Strahan R (ed) Mammals of Australia. Smithsonian Institution Press, Washington D.C., USA, pp 353-355 [Note: Originally published by Reed Books, Chatswood, NSW, Australia]

Newsome D, Dowling R (2010) Setting an agenda for geotourism. In: Newsome D, Dowling R (eds) Geotourism: the tourism of geology and landscape. Goodfellow Publishers Limited, Woodeaton, Oxford

Newsome D, Moore SA, Dowling RK (2002) Natural area tourism: ecology, impacts and management. Channel View Publications, Ontario

Newsome D, Dowling R, Moore S (2005) Wildlife tourism. Channel View Publications, Clevedon

Norton BG, Hutchins M, Stevens EF, Maple TL (1995) Ethics on the ark: zoos, animal welfare and wildlife. Smithsonian Books, Atlanta

Nowak RM (1999) Walker's Mammals of the world, 6th edn., vol 1. The Johns Hopkins University Press, Baltimore and London, pp 84–128

Odum EP (2006) Ecosystem theory. In: Pfafflin JR, Ziegler EN (eds) Encyclopedia of environmental sciences and engineering. A-L. Taylor and Francis, Boca Raton, pp 260–270

O'Neill F, Barnard S, Lee D (2004) Best practice and interpretation in tourist/wildlife encounters: a wild dolphin swim tour example. CRC Wildlife Tourism Research Report Series No. 25. Gold Coast: CRC for Sustainable Tourism, Queensland, Australia

Orams MB (1996) Using interpretation to manage nature-base tourism. J Sustain Tour 4(2):81–94

Orams MB (2000) Tourists getting close to whales, is it what whale-watching is all about? Tourism Manag 21(6):561–569

Orams MB (2002) Feeding wildlife as a tourism attraction: a review of issues and impacts. Tourism Manag 23:281–293

Packer J, Ballantyne R (2013) Encouraging reflective visitors experience in ecotourism. In: Packer J, Ballantyne R (eds) International handbook on ecotourism. Edward Elgar Publishing Limited, Cheltenham

Page SJ (2011) Tourism management: an introduction. Routledge, London

Parish S (2006) Junior encyclopedia of australian wildlife. Pascal Press, Leichhardt, NSW, Australia

Parkin DR (2006) Policy, culture and the achievement of visitor education outcomes: a case study of the Queensland Parks and Wildlife Service. A doctoral thesis. School of Education and Professional Studies, Griffith University, 2006, 243 p

Prestemon JP, Mercer DE, Pye JM (2008) Natural disturbance production functions. In: Holmes TP et al (eds) The economics of forest disturbances: wildfires, storms, and invasive species. Springer, Berlin, pp 35–58

Prideaux GJ, Warburton NM (2010) An osteology-based appraisal of the phylogeny and evolution of kangaroos and wallabies (Macropodidae: Marsupialia). Zool J Linn Soc 159:954–987

Richardson S, March R, Lewis J, Radel K (2015) Analysing the impact of the 2011 natural disasters on the Central Queensland Tourism Industry. In: Ritchie BW, Campiranon K (eds) Tourism crisis and disaster management in the Asia-Pacific. CABI, Oxfordshire, pp 149–160

Reynolds PC, Braithwaite D (2001) Towards a conceptual framework for wildlife tourism. Tour Manage 22:31–42

Sadava DE, Heller HC, Orians GH, Purves WK, Hillis DM (2014) Life: the science of biology. Sinauer Associates Inc, Gordonsville

Scotland Education (2011) Outdoor learning: practical guidance, ideas and support for teachers and practitioners in Scotland. Education Scotland, Glasgow

Sharpe G (ed) (1982) Interpreting the environment (2nd edn). Wiley, New York, NY, USA

Smith RL (1996) Ecology and field biology. Harper Collins College Publishers, New York

Smith AJ, Lee D, Newsome D, Stoeckl N (2006) Production and consumption of wildlife icons: dolphin tourism at Monkey Mia, Western Australia. In: Meethan K, Anderson A, Miles S (eds) Tourism, consumption and representation: narratives of place and self. CAB International, London, pp 113–139

Spiegel NB, Wynn PC (2014) Promoting kangaroo as a sustainable option for meat production on the rangelands of Australia. Anim Front 4(4):38–45

Stone M (2001) Managing visitors in the national parks of Victoria. In: Buckley R (ed) Fenner Conference 2001—Nature Tourism and the Environment. Australian Academy of Science, Canberra

Sunlu U (2003) Environmental impacts of tourism. In: Camarda D, Grassini L (eds) Local resources and global trades: environments and agriculture in the Mediterranean region. Bari CIHEAM, pp 263–270

Thomas R, Thomas S, Andrew D, McBride A (2011) The complete guide to finding the birds of Australia. Csiro Publishing, Collingwood

Tilden F (1957) Interpreting our heritage. University of North Carolina Press, Chapel Hill

TTF (2013) Conceptualising the value of protected areas: a literature review of the value, financing and tourism potential of Australia's protected areas–Report. Tourism and Transport Forum (TTF). Available online http://www.ttf.org.au/wp-content/uploads/2016/06/TTF-Conceptualising-the-Value-of-National-Parks-2013.pdf Accessed on November 2016

Turton SM (2014) Climate chage and rainforest tourism in Australia. In: Prideaux B (2014) Rainforest tourism, conservation and management challenges for sustainable development. Routledge, London and New York

Tyndale-Biscoe H (2005) Life of marsupials. CSIRO Publishing, Collingwood, VIC, Australia, 442 p

UNESCO (2006) Impact of the Tsunami on the Tourism Industry and Ecosystem of the Andaman and Nicobar Islands, India. Working Papers 36, UNESCO (South-South Co-operation Programme), Paris

(France). Available online at, http://www.unesco.org.uy/mab/fileadmin/ciencias%20naturales/mab/wp36.pdf
van Driesche R, Simberloff D et al (eds) (2016) Integrating biological control into conservation practice. Wiley Blackwell, Hoboken
Veverka JA (1998) Planning truly "interpretative" panels. In: Interpretation—the Journal for the Society of the Interpretation of Britain's Heritage, Feb 1998
Veverka JA (2001) Interpretative master planning. Acorn Naturalists, Tustin
Veverka JA (2014) Interpretation guide. John Ball Zoo Report. Available online at, http://www.jbzoo.org/uploads/files/zoodoers/interp-guide-zoo.pdf
Walker L (2012) The biology of disturbed habitats. Oxford University Press, Oxford
Walker LR, Willig MR (1999) An introduction to terrestrial disturbances. In: Walker LR (ed) Ecocystems of disturbed ground. Elsevier, Amsterdam
Ward CW, Wilkinson AE (2006) Conducting meaningful interpretation—a field guide for success. Fulcrum Publishing, Golden
Weiler B (2013) From memorable to meaningful: tour guiding as a tool for sustainable tourism. Third International Research Forum on Guided Tourism. Conference proceedings, 4–6 April 2013, The Netherlands, pp 14–26
Weiler B, Black R (2015) Tour guiding research: insights, issues and implications. Aspects of tourism. Channel View Publications, Ontario
Weiler B, Davis D (1993) An exploratory investigation into the roles of the nature-based tour leader. Tourism Manag 14(2):91–98
White L, Frew E (2015) The devastation of Darwin: representing the recovery and reconstruction of Australia after cyclone Tracy. In: Ritchie BW, Campiranon K (eds) Tourism crisis and disaster management in the Asia-Pacific. CABI, Oxfordshire
Whittaker D, Knight R (1998) Understanding wildlife responses to humans. Wildl Soc Bull 26(2):312–317
Woinarski J, Burbidge AA (2016) Phascolarctos cinereus. The IUCN Red List of Threatened Species 2016: e.T16892A21960344. Available online at, http://dx.doi.org/10.2305/IUCN.UK.2016-1.RLTS.T16892A21960344.en. Accessed on 27 Nov 2016
Worboys G, Lockwood M, De Lacy T (2001) Protected area management: principles and practice. Oxford University Press, south Melbourne, VIC, Australia
World Animal Foundation (2016) Kangaroo fact sheet. Available online at http://www.worldanimalfoundation.net/f/kangaroo.pdf Accessed on October 2016
WTO (1998) Handbook on natural disaster reduction in tourist areas. WTO report. doi:10.18111/9789284402397
Zahran MA (2010) Climate—vegetation: Afro-Asian Mediterranean and Red Sea Coastal Lands. In: Gilbert F (ed). Springer, New York
Zeppel H, Muloin S (2008) Conservation benefits of interpretation on marine wildlife tours. Hum Dimens Wildl 13:280–294

9 Wildlife Tourism for Visitors' Learning Experiences: Some Evidences on the Royal Bengal Tiger in Bangladesh and India

Azizul Hassan and Anukrati Sharma

Abstract

Much wildlife tourism involves seeing animal species in pristine natural settings. However, it does include non-domestic animals on farms, in zoos etc., and in areas which are fairly natural but not really pristine. Resources for promoting wildlife tourism are scattered in different locales of the world. Wildlife tourism needs to be viewed from both theoretical and practical aspects. The chapter presents two classical examples to view wildlife tourism as a learning experience. In Bangladesh context, this chapter explores wildlife tourism aspect of the Sundarbans. The Royal Bengal tiger (*Panthera tigris tigris*) is an important asset of the Sundarbans in Bangladesh. Particular references have been made to learning experiences generated through a new wildlife tourism product to showcase the Royal Bengal tiger. On the other side, in the context of India the chapter outlines wildlife resources in the Hadoti regions and makes some specific suggestions to employ these resources for wildlife tourism. Both empirical and literature-based data are gathered to outline wildlife tourism in these two separate perspectives. Findings outline the necessity of safeguarding wildlife tourism resources in the Sundarbans while promoting wildlife tourism products in the Hadoti region. This research emphasises the importance of involving both academics and tourists with wildlife tourism that can also benefit both ecological and environmental well-being of wildlife resources. On the common ground, the study shows the significance of wildlife tourism education. Then, it suggested that tourism planning in Sundarbans and Hadoti regions should consider the development of educative tour services and products for wildlife visitors aiming to inform them on ecological and biological data of wild animals, as well as to include conservation issues as part of learning experience.

9.1 Introduction

An essential aspect of wildlife tourism involves showing wild animals in their natural habitat. This type of tourism is also seen as having correlation with ecotourism in its simplest way, wild animal watching in their original settings (Newsome et al. 2008; Shackley 1996). Wildlife tourism is classified as an essential part of present day tourism and has experienced rapid growth across the world. This type of tourism is increasing day by day in different parts of the world. According to Moorhouse, Dahlsjo et al. (2015), "tourism accounts for 9% of global GDP and comprises 1.1 billion tourist arrivals per year. Visits to wildlife tourist attractions (WTAs) may account for 20–40% of global tourism, but no studies have audited the diversity of WTAs and their impacts on the conservation status and welfare of subject animals" (p. 1). The rise of Indian wildlife tours mainly resides on the fact that many endangered and "on the brink of extinction" wildlife species dwell in the country's

A. Hassan (✉)
Cardiff Metropolitan University, Cardiff, UK
e-mail: azizulhassancmet@gmail.com

A. Sharma
Department of Commerce and Management, School of Heritage Tourism and Museology, University of Kota, Rajasthan, India
e-mail: dr.anukratisharma@gmail.com

I. Borges de Lima and R.J. Green (eds.), *Wildlife Tourism, Environmental Learning and Ethical Encounters*, Geoheritage, Geoparks and Geotourism, DOI 10.1007/978-3-319-55574-4_9

national parks, reserves and sanctuaries, and as rare species they are on the spotlight for wildlife watching and photograph safaris and expeditions.

India has 103 National Parks, and in 1936 the Hailey National Park (currently called, Corbett National Park) was the first National Park to be created in the country (IISC 2016, online). India has 28 areas specifically established as Tiger Reserves created by Project Tiger launched by the government in 1973 to protect endangered species of tiger; the reserves correspond to 1.09% of the total area of the country (IISC 2016, online). Sundarbans, [1] in West Bengal, was established in 1973–1974 and is the second largest tiger reserve of India with a total area of 2585.10 km^2 (IISC 2016). As commented in the Greener Ideal Forum, India has been a place for "various exotic fauna and flora species… Earlier wildlife tourism brought new challenges for the governmental bodies, but in the present time this rise is indeed supporting the conservation initiatives taken for the welfare of flora and fauna (Frei 2013, online)."

As part of the wildlife tourism assets and resources, the Royal Bengal tiger is an important species in the Sundarbans (Bangladesh) and Hadoti (India). This chapter addresses the importance of this animal as a relevant iconic species of great tourism appeal and its importance for building up educative experiences for the visitors. The research was carried out based on the literature review and on empirical primary data collection in situ. This chapter outlines the importance of Bengal tigers for the Sundarbans and Hadoti tourism identity; two destinations in India and Bangladesh that largely rely on popularising its natural, historical and archaeological sites filled with scenic attractions, water bodies, rare species of birds and so forth.

Thus, relevant marketing and promotional activities are essential to capitalise on its rich legacy of natural and scenic beauty as a wildlife tourist destination (Hassan and Rahman 2015b; Hassan 2012a, b, c; 2014). Wildlife tourism needs to be promoted through social media to attract tourists having interests in this tourism type. Also, conservation and restoration of forests and wildlife resources by using advanced technology is essential because these can possibly support learning experiences (Hassan and Rahman 2015a, b). Research studies indicate the importance and possibility of special interest tours for wildlife tourism learning by tourists (Cobb 2011; Hassan 2012a). This chapter shows two relevant examples covering Bangladesh and India by outlining wildlife tourism as routes for visitors' educative experiences. In the Bangladesh perspective, the study explores essential roles of the Sundarbans as part of wildlife tourism. On the other hand, the Indian context outlines some specific issues and suggestions on effective use of wildlife resources to promote learning experiences through wildlife tourism. The next section will present a short history of the species, biofacts and characteristics of the Royal Bengal tiger as the main wild animal object of this research, before discussing the learning and educational issues and presenting the collected data. As for planning a more educative wildlife tourism, attention should be given to seasonality as sanctuaries and zoo receive a significant demand during the winter and the rainy season. Segmented educative tour packages should be part of wildlife planning and development for Hadoti and Sundarbans regions as a way of enhancing the overall wildlife learning and experiences of visitors, and this includes conservation and animal welfare.

9.2 About the Royal Bengal Tiger Species (*Panthera Tigris Tigris*)

9.2.1 A Short Scientific History on Tigers Migration

In his book entitled, Bones of the Tiger, published in 2010, Hermanta Mishra, provides a thorough historical record of tiger migration in the world. According to Mishra, it is believed that the modern day tiger evolved seven hundred thousand years ago in Siberia, and then they migrated south from Siberia to form the eight subspecies of the 21th Century tigers.

The first tiger group did not go too far and stayed near Siberia to form the Siberian subspecies, also known as Amur tiger (*Panthera tigris altaica*), the largest of all species. A second group spread to southern China to compose the South China tiger (*Panthera tigris amoyensis*). A third group moved to southwest towards the Caspian Sea to form the so-called Caspian tiger (*Panthera tigris virgata*): they mostly settled in Turkey, northern Afghanistan, and Iran. The fourth group dispersed into the Indian subcontinent to compose the Indian subspecies, or Royal Bengal tiger (*Panthera tigris tigris*) of Bangladesh, [India], Nepal, Bhutan and Northwest Burma, and the main wild animal object of this research. The fifth group migrated into Southeast Asia in split into two major streams. The first subgroup settled in Vietnam, Cambodia, eastern Burma, Laos, Thailand, the Malaya Peninsula and southern China to form the Indochinese tiger (*Panthera tigris corbetti*). The second subgroup flourished over Indonesian Islands and became part of three distinct subspecies: the Bali tiger (*Panthera tigris balica*), the Sumatran tiger (*Panthera tigris sumatrae*), and the Javan tiger (*Panthera tigris sondaica*).

Although Siberia was linked to North America in ancient time, tigers never migrated to the Americas nor as far as

[1]It was noted that in some reports and news Sundarbans and Sunderbans are used interchangeably to refer to the same region, forest, mangrove, etc. For this Chapter and book, it is used only Sundarbans as officially referred and used by UNESCO.

African or Australian lands. It is worth noting that the white tiger is indeed a morphological variation of the Royal Bengal tiger. Thus, conversely to a common belief this subspecies—highly prized on displays at zoos and sanctuaries—is not an albino. Mishra (2010) also underlines that "Though rare, black tigers have also been reported in the wild. As with white tigers, the black colour is a result of a "false melanism"—a process that in the case of black tigers, increases the amount of black pigmentation in the skin." (p. 70).

9.2.2 Royal Bengal Tiger: Current Population

It is estimated that the global tiger population represents roughly 5% of its original number a century ago, "in the early 1900s, there were around 100,000 tigers throughout their range" (TigerDay.Org). Since then, tigers' habitats have been vanishing at a rapid escalating speed of 40% in just 10 years of time, rendering nowadays a total of only 2500 animals; this shows how precarious has been the state of tigers in the world, particularly in the wild (Chalise 2012). Once being found in almost all forested areas, the *Panthera tigris tigris* has been now confined to the Sundarbans what raises concerns on its plight in the wild (Nature Environment and Wildlife Society 1999). According to the Begal travel guide edited by Miltra (2011)

> Populating the nebulous terrain of Sundarbans in higher numbers than anywhere else, the undisputed and most often, unseen, monarch of this region is the Royal Bengal tiger". According to conservative estimates around 300, if not more, tigers inhabit these forests across India and Bangladesh. Indeed, it is the terrain… that protects the animal from paochers who find it hard to trace and chase them in the dense forests and clavey mud. The tiger's presence in turn protects the forest keeping intruders at bay (Miltra 2011, p. 20).

9.2.3 Knowing Better the Royal Bengal Tiger: Biofacts and Habitats

Over the last 100 years, hunting and forest destruction have drastically reduced tiger populations in the wild. Tigers have been hunted for different purposes, for example, as trophies for rich people who pay thousands of dollars for a killing safari, and they are also killed in order to have their body parts for medicinal beliefs, "the heightened demand for tiger parts as ingredients in Traditional Chinese Medicine (TCM), and the pressure this has exerted on… [has] diminished tiger populations, is too well known" (Richards and Tyabji 2008, p. 109). Royal Bengal tiger is already enlisted as an endangered animal according to IUCN Red List of Threatened Species in the wild (Fig. 9.1); indeed all five remaining tiger subspecies in the wild are endangered, and many protection programs have been placed in Asia and other regions as an attempt to avoid the worst.

9.2.4 Bengal Tiger Characteristics and Behaviour

According to National Geographic fact sheet, the tigers are solitary animals in the wild and "aggressively scent-mark large territories to keep their rivals away". They have nocturnal habits being powerful hunters that can travel long distances to chase their preys, such as deers, wild boars, monkeys, buffaloes, wild pigs and any other animals such as porcupines, a rodentian mammal. Royal Bengal tiger has coats that can help for a camouflage, and each tiger has a unique set of stripes not found in another animal of its species (Fig. 9.1). For a successful hunting, a tiger lies patiently and in the right moment it moves slowly and cautiously close enough to get their preys with a fatal spring. A tiger can eat as much as 21 Kg of meat in one single day if really hungry (Multiple sources: Tigerday.Org, Oocities.Org, Animal. Corner, National Geographic).

Surprisingly, most tigers avoid humans, but a few may become "man-eaters". As for the offspring, "females give birth to litters of two to six cubs, which they raise with little or no help from the male. Cubs cannot hunt until they are 18 months old and remain with their mothers for two to three years, when they disperse to find their own territory" (National Geographic online). Some peculiarities about the Bengal tiger: its roar can be heard up to two kilometres away at night; the hindquarters of a hunted prey is the starting point for a tiger feast; a tiger can kill thirty buffaloes a year; it usually eats meat that has started to decompose, a feeding preference that is unlike that of other cats; its claws are as big as 10 cm (4 in) in length and are crucial for hunting; the same as the Sabre-tooth cat—its ancestor—the tiger heavily depends on its teeth for its survival, and if it loses its tearing teeth in a combat, attach or because of its old age, it will starve to death. Tigers are very skilled swimmers that appreciate refreshing in lakes, ponds, etc. (Multiple sources: Tigerday.Org, Oocities.Org, Animal. Corner, National Geographic). An adult male Bengal tiger weighs around 220 Kg, and its body length is nearly 3 m; a female tiger is usually a bit smaller weighing 140 kg, and 2.5 m long (Seaworld.Org, 2016, online). Around the world, Royal Bengal tigers have been used in shows and displays in circuses, sanctuaries and zoos as tourism attractions. In Chiang Mai, Thailand, foreign visitors line up for an opportunity to enter the Bengal tigers' settings for a very close physical contact and pictures of an adventurous and likely "risky" experience of touching "the big cat" to show to family and

Fig. 9.1 Royal Bengal tigress (*Panthera tigris tigris*) and cubs in the wild. *Source* Brian Gratwicke, 2009. Creative Commons

friends, and for recording their own travel memories, but this type of attraction always raises ethical and animal well-fare debates both in the media and in Academia, particularly by environmentalists and animal lovers, etc. (Fig. 9.2).

9.3 Literature Review

9.3.1 Wildlife Tourism Learning Perspectives

Tourist interests need to be diversified and tour operators need to make tourists aware of other interesting experiences. Also, tour operators should be able to create a competitive edge for targeting specific tourism consumers. Thus, a tourism product should obviously attract diverse sets of visitors from different areas of tourism. A learning tour is expected to include tourist experiences aiming to fulfil specific educational or academic requirements. A special interest tour actually targets particular tourists groups, offers better experiences, creates interests and thus replaces conventional travelling to see tourism resources (Gladstone 2013). It is obvious that tourists do tend to visit the Sundarbans as a special wildlife tourism destination (Rahman and Hassan 2016). Any addition of new types of products or services can result in yielding higher numbers of tourists. A special interest tour attaching strategic responses from tourist destinations can make changes in business environments, while adopting a bottom-up approach (Cobb 2011).

Such an approach can possibly identify the demands for special interest tourism products or services, while allowing the creation of more experiences for meeting those selected demands. 'Special interest tourism may be defined as the provision of customised leisure and recreational experiences driven by the specific expressed interests of individuals and groups' (Derret 2001, p. 3). According to Pine and Gilmore (1999), human history has four stages: commodity extraction from the ground, industrial goods production, service delivery and experience delivery. The type of tourism that this tiger is promoting is featured as: formed with a small-scale tourism type and complex product typology (i.e. bespoke, custom-made) and focuses on delivering memorable experiences.

A zoo can be a source to upgrade visitors' earlier wildlife learning experience levels. Sir Stamford Raffles built up the world's first zoological park in London in 1862 (Turley 1999). By mid 1990s, there were more than 10,000 zoos worldwide (Mason 2000). This demonstrates that zoos turned into the venues for entertainment that continue to this today. Hayward (1995) at the Monterey Bay Aquarium uncovered that, most visitors seek to learn more about preservation, specifically what an individual can do to safeguard and preserve an indigenous habitat.

Fig. 9.2 Foreign visitors take an opportunity to enter the Royal Bengal tigers' settings assisted by a tamer and animal carer for a close physical contact and photographs. An adventurous and memorable moment for them at Tiger Kingdom, in Chiang Mai, Thailand; though, it is an experience that may lead to several ethical implications with regards to keeping wild animals in captivity and mass wildlife tourism. *Source* Ismar Lima, 2015, field research in Thailand

Researchers (Ballantyne and Packer 2009; Ballantyne et al. 2005; Lee and Moscardo 2005; Tisdell and Wilson 2005; Zeppel and Muloin 2008) affirm that, wildlife tourism can have positive short and long term impacts on wildlife tourism learning experiences by, (i) building up an admiration and appreciation for wildlife life and nature; (ii) bringing awareness about ecological issues; (iii) promoting ecologically sustainable attitudes and activities; and (iv) building tourists' ability for the longer-term appropriation of sustainable living practices. Many research studies conducted in a zoo propose that, visitors discover reasonable or natural enclosures as more fascinating and therefore tend to spend more time watching and getting information about displayed animals (Berger 2009; Fraser 2009; Fraser et al. 2008; Jamieson 2006; Shettel-Neuber 1988).

The idea of 'learning for the sake of entertainment' is investigated by Packer (2006). This researcher recommended that, in free-choice learning environments, even visitors without having any prior learning motivation can be offered with learning knowledge. Such knowledge is pleasant and can be transdevelopmental. This can have vital ramifications for wildlife tourism as it proposes that, visitors will not just be open for accepting information messages. However, the chance to learn about safeguarding the wildlife resources is also essential in such learning experience generation process. Falk (2005) states the dominant part of environmental learning, and learning in general, happens outside formal education settings. A visitor's choice to visit a forest, a zoo or sanctuaries for learning experience in wildlife tourism is voluntary. The visitor in a wildlife environment can select what to learn, as well as whether, where, when and with whom they will learn (Bright and Pierce 2002; Falk and Dierknig 2000; Screven 1995).

Visitor education is a core aspect discussed in this chapter, and it leads to multiple understandings in benefit of biotic natural resources, including the wild species. Parkin (2006), based on the available literature, explained that a visitor education programme can have different names such as "interpretation and education, conservation education, minimal impact education, community outreach, and public contact extension", but all them share common goals. The main target is to "explain the natural phenomena, inform visitors of management issues, provide advice about natural hazards and the safety precautions one can take, and promote the adoption of a minimal impact ethic" (p. 3); however, the backbone of such programmes is to help local managers of public visited places such as Parks, Reserves, and Wildlife Sanctuaries to maintain the state of the natural and wildlife resources while enhancing visitors' experiences with quality products, "The objective is not to 'control' visitor behaviour, but rather to seek to provide a cognitive basis to raise awareness and encourage appropriate visitor behaviour towards protected areas and the environment" (Parkin 2006, p. 3). Even though, Parkin's studies were on National Parks in Queensland, her findings can be very useful for developing wildlife tourism educative products and services in Hadoti and Sundarbans areas.

9.3.2 The Sundarbans and Wildlife Tourism

National tourism policy sets the onset of tourism promotion in Bangladesh (Hassan and Burns 2014). The Sundarbans is the world's largest mangrove forest. This forest lies with an area of 140,000 ha on the Bay of Bengal on the delta of the Ganges, Brahmaputra and Meghna. This forest is in both Bangladesh and India and was inscribed as a UNESCO World Heritage site in 1987. This forest has a complex network with mudflats, tidal waterways and small islands of salt-tolerant mangrove forests. This is at present an excellent example of balanced ecological processes.

This forest area is well-known for its massive range of fauna that includes 260 bird species, the Royal Bengal tiger and many other threatened species (i.e. the Indian python and the estuarine crocodile). The Royal Bengal tigers (*Panthera tigris tigris*) in both Bangladesh and India are one of the beautiful animals with their exquisiteness and magnificence. The tourism type promoted by the Royal Bengal tiger offers experiences that are labour intensive requiring interpreters, service providers' expertise and high yielding.

Bangladesh has been blessed by nature (Hassan et al. 2015). This Delta area is one of the most beautiful natural reservoirs in the world. Regarding the natural beauties along with cultural, the United Nations World Heritage Site (UNESCO) has declared three of her tourism attraction sites as World Heritage Sites. The Sundarbans is the world largest mangrove forest and has been declared as a WHS since 1997. This forest area is known for its wide range of fauna, including 260 bird species, the Royal Bengal tiger and other threatened species such as the estuarine crocodile and the Indian python. The Royal Bengal tigers are widely known to be solitary in their territory barring anyone's entry into their location. Any type of human interference is generally unwelcome by them, regardless of their motive of either hunting or nursing. The incidences of human interference separating cubs from the mother make them more vulnerable.

9.3.3 Hadoti as a Hub of Wildlife Tourism Resources

According to the Economics Times (2012), the Hadoti region of Rajasthan wears an extremely green look, especially in the monsoon. Some places having importance for wildlife tourism as Bhimlat Falls, Rameshwar Falls, Geparnath, Kota Barrage, Guda, Parajhar Mahadev, Gararia Mahadev and Cholia Falls are the basic resources to promote wildlife tourism in this region. Most tourists in these regions derive from Central Asia and Europe especially in winter at Bundi, Baran, Kota and Jhalawar. Some endangered animal species such as thick bar-headed goose, spoonbills, painted stork and several other species of migratory birds can be found here.

Again according to Deccan Herald (2016), Mukundara Hills National Park, situated in this district, was at one time the hunting ground of past rulers. The rulers explored a massive breeding area of wolf, sloth bear, sambar, caracal and puma. The elegant Chambal stream coasts through a thickly lush woods zone, neighbouring the slopes. Sorsan was once celebrated for dark buck, crane and other wild creatures. The Jawahar Sagar Sanctuary has pumas, wild hogs and a few birds of prey. Wild species visit the zone regularly to breed. The Talwas Lake and Kanak Sagar are home to numerous transitory and occupant fowls. The cool climate makes the backwoods more welcoming. So also, the Ramgarh Sanctuary in Bundi is a minimal and huge biological system in this zone. It is a passage to Ranthambore. These less frequented havens have, as it were, developed well in view of low human intercession. To view untamed life in its common cover, these asylums offer a nice scope of touring alternatives.

9.4 Methodology

The study is based on both primary and secondary data. For the Sundarbans, a thorough and detailed literature search was carried out. Thus, data for the Bangladesh context was generated from secondary resources including articles, journals, both published/unpublished resources, newspapers and the sources on the Internet. The analysis and findings presented in this section were generated from the literature review, and it has rendered a summary with a pertinent thematic bibliography with valuable annotations. In the Indian context, results were mainly drawn based on a questionnaire survey and literature review. Also, informal discussions were carried out and presented within quotes. Likert scale was used, and the idea of measurement is principal to almost all experimental research studies (i.e. social science, data systems, marketing and psychology). Likert (1932) contended that the distance of scores, for example, 1, 2, 3, 4, 5 is equivalent and yields information that can be ordinarily dispersed. Even though it appears like that, application of this measurement scale is the most troublesome aspect of behavioural research; it is thus worth noting that application of the Likert scale is evidenced in many studies (Allport and Kerler 2003; Chachamovich et al. 2009). In this research. Likert scale is seen as an arrangement of statements offered for a genuine or theoretical/hypothetical circumstance under study while respondents are solicited to represent their level of agreement (from strongly disagree to strongly agree) with the given statements on a metric scale. Regarding the data collection in the Indian context, every statement is aimed to uncover required data (Singh 2006).

The questionnaire was designed to get the first-hand information about the satisfaction level and experience of the tourists who have visited parks and sanctuaries of the Hadoti region. A brief description was given to all the respondents according to their preference of language in English or Hindi. Brief discussions outline the importance of this study that ends with recommendations. Questionnaires were distributed to 150 respondents at different places covering parks, sanctuaries, roadside restaurants (Dhabahs) especially on highways, Zoo (Kota), different tourist places of the region, Malls and railway stations. Of 150 distributed questionnaires, the researchers received back 120 properly filled. Respondents for this study were divided into three categories: local people; tour operators and travel agents; and, domestic tourists. In the Hadoti region, Kota city is an education hub for the students. This was the primary reason to choose respondents from selected age groups of 15–20 and above 40. This research included two important spots from Kota as the Chambal Sanctuary and the Kota Zoo. Second, this survey was done in the months of May–June, 2016. This is the general school vacation time when most of the schools, colleges and universities have summer break time.

Thus, most students and their families move out for holidays during this period in the Hadoti region, what makes it a great opportunity to conduct surveys in terms of sampling, with different age groups of respondents, and to collect the data in more convenient manner. The researchers drafted the questionnaire for three different categories. For primary descriptive data generation, open ended discussions were also conducted. Tables 9.1, 9.2 and 9.3 provide details on the sampling, number and categories of respondents, and on the sites used in this research.

9.5 Analysis and Discussion: Learning Opportunities and Challenges in Wildlife Tourism in Hadoti and Sundarbans Regions

9.5.1 The Sundarbans, Wildlife Tourism and Visitors' Enhanced Experiences

As a popular wildlife tourism destination, the Sundarbans is well-known for its natural beauties, regarded "the largest single block of tidal halophytic mangrove forest in the world" (Das 2015, p.1), and for this reason it has been acknowledged as one of the UNESCO Natural World Heritage Sites. Evidently, wildlife tourism in this area has as a focus the endangered animal species such as the Royal Bengal tigers, one of the major tourism attractions. The forest is believed to be their sole breeding place, and thus offers a competitive advantage for wildlife tourism (Parveen and Sharma 2015). This nearly extinct animal variety offers a valuable learning experience for tourists and marketing opportunities in the Sundarbans as a wildlife tourism destination in both Bangladesh and India. It is very unfortunate that it is nearly extinct in this region. Several reports provide evidence of the declining numbers. Among many reasons, forestland destruction, climate change, habitat loss, hunting remains the main factor, at least for the last 100 years.

Due to its declining number and the growing threats, the Royal Bengal tiger has been classified as the most endangered wild animal in the country, and is battling constantly to survive (Gooch 2011; Kenney et al. 1994). An oil spillage incident on 9th December 2014 and several bush fires have posed challenges to authorities to reduce the threats to Royal Bengal tigers in the Sundarbans. In some cases, local peoples have also negative attitudes towards this animal species making the general scenario even worse. According to Karanth (2001), the Royal Bengal tiger is one of the world's most beautiful, largest and powerful predators striding on the planet. The tiger is regarded by some people as a 'charismatic species' that holds a rare and magnificent beauty. The animal is well celebrated in mythologies and folk legends in the world, but—paradoxically—its existence has been hugely threatened. The number of individuals of this species is rapidly decreasing and this can be a matter of time for its complete extinction.

9.5.2 The Hadoti Context-Issues of Wildlife Tourism Revealed

In Hadoti area, in India, the authors selected the Zoo as part of the research by considering it as the most appropriate setting for educative wildlife experiences the community has outside the formal educational institutions. The zoo has been

Table 9.1 Primary research survey in Hadoti

• Chambal Sanctuary • Kota Zoo	• Shergarh Sanctuary
Coverage Area of Primary Survey	
• Ramgarh Sanctuary	• Bhensrodgarh Sanctuary

Source this research

Table 9.2 Respondent classification

Local people	Tour operators and travel agents	Visitors
Survey and open ended discussions were conducted at tourist places, railway stations, restaurants, malls and shops	Survey and open ended discussions were conducted at the places (shops) of the tour operators and travel agents as well as at convenient places	Survey and open ended discussions were conducted at the sanctuaries, zoo and other tourist spots of the region

Source this research

Table 9.3 Number of respondents interviewed at different places

Coverage area	Number of local people	Number of tour operators and travel agents	Number of tourists (domestic)
• Chambal Sanctuary	11	19	14
• Kota Zoo	15	5	8
• Shergarh Sanctuary	7	9	2
• Ramgarh Sanctuary	3	2	7
• Bhensrodgarh Sanctuary	4	5	9
Total respondents	40	40	40

Source this research

contributing over the years in more dynamic ways in wildlife protection and learning. A zoo offers an open space to get wildlife experiences, to watch wildlife resources and somehow getting more acquainted with the visitors both psychologically and emotionally (Ballantyne et al. 2005; Lee and Moscardo 2005; Tisdell and Wilson 2005). The price for the entry ticket was revised on the 16th February of 2015 to generate more revenue. According to the responsible board, international visitors need to pay a certain amount of entrance fee (i.e. 150 Indian Rs/app. 2.27 US$) so the zoo can have a source of revenue to maintain basic operational services.

Figures 9.3 and 9.4 show pictures taken by the researchers while conducting field survey. As noted, the boards are written only in Hindi what obviously make them hard to understand and interpret by a foreign visitor not capable with the local language. Moreover, even the information presented in Hindi on the animals, such as age, weight, etc., may be considered as insufficient from a learning point of view. The boards are of little help if a visitor wishes to know more about the number and types of species; alternatively, the visitors may choose to look for a Zoo's staff and luckily to get a detailed animal profile. Additionally, illiterate visitors can hardly understand the warning messages on the boards as international known symbols of danger, etc. are missing.

Hindi is the mother tongue of the locals but, for information sharing it is very important to write it in both languages: the Hindi and the English. In order to show that English language should be widely used on boards, signs and in any other informative means in tourism sites, attractions and destinations, figures on domestic and foreign arrivals from 2012 to 2016 were collected in the Tourist Information Centre of Kota; it is noted that a considerable number of foreigners visit the region. In 2016, the arrivals are lower compared to other years, but this is a partial record with data until June (Table 9.4). The authors had informal discussions with some local visitors, and it was found that more than 20% were students who really want information displayed in English. Renu Sharma, a visitor at the Zoo from Rajasthan state informed that, at the Zoo, it is difficult to realise the importance of a particular animal and to clearly understand the ways for its protection, habits, behaviour, etc.

In the Hadoti region, setting up new goals for wildlife tourism promotion at Hadoti is important for the development of wildlife tourism. Wildlife tourism needs to improve quality and focused tourism structure to attract both international and national wildlife tourists. Thus, financial benefits need to be passed to those relying on wildlife tourism for their livelihood generation. Diverse media communications are required to be incorporated into wildlife tourism

Fig. 9.3 Message boards representing the name of species available inside the zoo (*Source* the authors)

education. Media can play effective roles for exploring insights (McIntyre and Roggenbuck 1998). Conducting training sessions to motivate wildlife tourism among target groups such as researchers or academic learners is also beneficial in this regard. Priority needs to be offered on engaging qualified and skilled tourist guides to benefit the wildlife tourism industry.

Furthermore, arrangements need to be made by different organisations enabling them to confer informal education on wildlife tourism. This should help to create awareness and fascination towards diverse areas of wildlife tourism, expanding tourist attraction and education to other species, while using tigers as an important drawcard. In order to address the demand-supply network necessities, the government may additionally formulate plans to create more workforce having skills in the emerging trends of the tourism industry. There is a clear importance to prepare effective wildlife tourism syllabi. Indeed, one of the primary difficulties wildlife attractions face is to persuade individuals that individual activities can possibly rationalise the world's assets for future generations (Yalerotz 2004). Table 9.5 shows top five attributes that reveal Hadoti is prominent with regards to natural and heritage assets, and five lags in wildlife learning experience identified in that province that deserve attention.

Suggestions made by the three groups of respondents emphasises for the betterment of wildlife interpretation are, the first and the foremost requirement are to provide the accessible, user friendly information about wildlife sanctuaries and park especially the availability of different species. Guides themselves needed to be more informed and trained for giving the tourists a learning experience rather than only a visiting experience. More sign boards are needed around and near the sanctuaries and parks. Audiovisual presentation and virtual tours/videos must be shown prior to the visit of the sanctuaries and parks. This boosts up the morale of the tourists and connects them emotionally with the wildlife. According to Ham and Weiler (2002), the capacity to "connect with" visitors both mentally and emotionally is the foundation of an effective wildlife elucidation. Table 9.6 draws attention to key aspects that should be kept in mind while designing the sign boards/information boards for resourceful interpretation of wildlife.

Table 9.7 shows the level of importance of major attributes as indicated by the survey participants related to wildlife tours and learning. Likert five point scales was used to measure the attributes as 'extremely important', 'highly important', 'moderately important', 'slightly important', and 'not important'.

Fig. 9.4 Sign boards/message boards inside the zoo (*Source* the authors)

Table 9.4 Domestic and foreign tourist arrivals from 2012 to June 2016

Year	Domestic tourist arrivals	Foreign tourist arrivals
2012	62,029	1881
2013	63,015	2889
2014	51,467	3516
2015	90,598	2574
2016 (June)	42,876	1010

Source this research, figures collected by the researcher at the Tourist Information Center in Kota, 2016

Table 9.7 was built based on arguments of three respondent categories. This Table reveals that animal and bird species with 66 responses is the most attributed factor of learning tours for wildlife tourists. Freedom of watching wildlife sanctuary with 49 responses, natural beauty of the sanctuary with 65 responses and undamaged roads with 53 responses were the important attributes. On the other side, safety and security with 34 responses, entry fee and restrictions with 61 responses, basic amenities with 42 responses and transport accessibility to see the sanctuary with 59 responses were the other important attributes. The single moderately important attribute is hygiene and cleanliness with 39 responses. Maximum respondents were local and domestic tourists.

Thus, accommodation with 41 responses at the sanctuaries or at the nearby places was found as the less important attribute. In addition to these, it was important to deliver high levels of wildlife tourism learning to the tourists. Though it was not included on the Table, safety and security issue of the females and child visitors were found as

Table 9.5 Top five attributes where Hadoti is ahead and lags in wildlife learning experience in the state

Top five strengths of Hadoti	Number of respondents	Top five lags of Hadoti	Number of respondents
Natural beauty	38	Poor infrastructure	40
Rare Birds species	35	Lack of information	38
Good range of animals	22	Safety and security issues (especially of women and children)	19
Rich water bodies	15	Lack of basic amenities	16
Unique castles or forts near the sanctuaries	10	Promotion and marketing	7
Total number of respondents	120		120

• Based on the questionnaire's replies collected in the field work
Source this research

Table 9.6 Resources to be considered for full interpretive boards

Demanded Aspects[a]						
Colorful and visible texts	English, Hindi and regional language use for the description of animal and bird species	Motivational quotes on the sign boards for the protection and to safeguard the animals	Origin, importance and lifecycle of animals and birds available in sanctuary and parks	Warning signs on the boards including the statements and symbols	Basic details of exit-entry and other information such as restaurants, local transport	Information regarding the guides and the virtual tour rooms etc.

[a]Major common issues as appointed in the survey responded by 120 participants
Source this research

Table 9.7 Major attributes related to wildlife tours in terms of relevance as classified by the survey respondents

Important attributes	Not important	Slightly important	Moderately important	Highly important	Extremely important	Total
Animal species	11	6	12	25	66	120
Freedom of watching animals	09	11	16	35	49	120
Natural beauty of sanctuary (forest area)	02	09	15	29	65	120
Safety and security at sanctuary (especially of females and children)	15	18	32	34	21	120
Entry fee and restrictions	08	17	25	61	09	120
Hygiene and cleanliness	11	15	39	29	26	120
Basic amenities such as drinking water and toilets	04	10	25	42	39	120
Undamaged roads	03	05	28	31	53	120
Transport accessibility to see the sanctuary	04	07	24	59	26	120
Accommodation at the nearby places of sanctuaries	41	39	28	07	05	120

Likert five point scales was used for measuring it (*Source* this research)

important due to lack of visitor guides and security guards. It was revealed that female tourists preferred to learn more about wildlife but their safety and security assurance came first. As relevant in this regard, Moscardo and Saltzer (2004) found that seeing substantial, uncommon or new species, having the capacity to draw near to wildlife in the natural setting, and having the capacity to find out wildlife species and natural setting are all added to visitor satisfaction.

9.6 The Future of Wildlife Tourism as a Tool for Visitors Education in Hadoti and Sundarbans Regions: Implications, Findings and Recommendations

This research offers some specific recommendations for policy makers and planners; it is emphasized the importance of linking wildlife tourism attractions with other types of tourism products, and by doing so both domestic and foreign visitors could be better contemplated with a web of multiple rewarding experiences during their visit. It is taken for granted that the wildlife tourism sector in Hadoti faces challenges in terms of significantly increasing the number of visitors in the short term, mainly due to a lack of structure and planning; appropriate environment and products should be developed in order to provide greater access to wild animals in their natural habitat as well as on captive and semi-captive settings without causing disturbances to them. It is really necessary to provide information through different media, including virtual tours, so visitors can get informed on the available wild animals and related visit sites, and by circulating such pertinent information, the wildlife attractions can be fairly set as part of tourism circuits and packages. Effective use of the Internet could include techniques such as banner or video advertisements, pop ups etc. to showcase the pictures of animals, birds and sanctuaries. This helps visitors to plan a more encompassing and meaningful visits ahead of time while reducing possible disappointments and frustrations. By using proper marketing tools and means, planners, organisations and companies in the tourism sector can also propitiate the means for visitors to make their mind to visit the sanctuaries and zoos.

Marketing and promotion can play a vital role in enhancing a wildlife watching experience. The promotion of ecologically friendly environment can possibly attract the international tourists to the region. A holistic approach and planning rather than a merely commercial one can bring more investment opportunities for the villages nearby wildlife sanctuaries. Academic institutions, NGOs, other industries and local people must harness efforts for the protection and promotion of the natural habitat of animals. There is an ultimate need of proper destination management and strategies along with better transport and access facilities working through a web of road and air connectivity, as well as a network of different types of accommodation. Basic amenities such as toilets, restrooms, first aid spaces, medical facilities and potable water facilities are also required.

All these mentioned factors should be taken into account in order to strengthening the formulation and implementation of policies and planning for fully developing the wildlife tourism sector. For a fast sector consolidation, public and private partnerships should be promoted by identifying key players with partnered actions and investments being a boon with a plethora of structural improvements for the sanctuaries and surrounding areas. As part of the recommendations, more localised actions should be taken by pertinent agencies and bodies which could include beautification programs at the sanctuaries, the construction of ropeways without harming the natural habitats of wild animals, as well as to develop training and capacity building programs for guides, tour operators, travel agents and hotel staff. Awareness campaigns by academicians, animal lovers, researchers etc. may be organised focusing upon the importance of wildlife watching. Hoarding displays at public places, road shows and skits etc. may be helpful at promoting the wildlife resources and attractions at a local level. Academic Institutions may introduce degree courses in Wildlife Management/Wildlife Tourism. Anecdotal findings reveal the importance of wildlife tourism largely centred around the Sundarbans Royal Bengal tigers. Sharma (2015) has a common understanding with Ritchie on 'educational tourism' by corroborating his views,

> Educational tourism is a tourist activity undertaken by those who are undertaking an overnight vacation and those who are undertaking an excursion for whom education and learning is a primary or secondary part of their trip. It is comprised of several sub-types including ecotourism, heritage tourism, rural/farm tourism, and student exchanges between educational institutions (Sharma 2015, p. 2).

The proposed learning tour by the authors is outlined as a three day long tour to explore the diversified wild lives of the Sundarbans. This proposed tours can be developed by the government that should involve the zoo/forest authorities and private tour operators. Due to restriction of time schedule of both the tourists or guides, the tour is required to be confined within exploring facts related to the zoo and the forest. The tour can also showcase ecological relationships, local ecosystem and aspects related to the forest, competition with other predators, favored prey species and maybe some of the plants its prey species depend on. From the perspective of tourism typologies, this tour can be categorised as both adventure and wildlife tourism. However, this tour is expected to create a link between the adventure tourism, wildlife tourism, ecotourism and river cruise tourism.

The reasons to develop the tour can be diverse as this can create a multi-dimensional image of the Royal Bengal tiger as well as the Sundarbans. The tour could also contribute to raise visitors' awareness aiming to positively impact on their perceptions and views with regards to wild animal species and forested areas. Segmented educative tour packages should be part of wildlife planning and development for Hadoti and Sundarbans regions as a way of enhancing the overall wildlife learning and experiences of visitors, and this includes conservation and animal welfare.

Finally, an educational and educative tour is expected to offer economic benefits to the parties involved by diversifying both image and reputation and the local economic well-being. The tour is also expected to create and attract new market segments, while allowing individual entrepreneurs. Also, the tour anticipates individual entrepreneurs to follow up and appreciates personal interests and operational activities.

9.7 Conclusion

This research outlines two case studies *vis-à-vis* the challenges of exploring the possibilities for visitors learning and experiences in wildlife tourism sites, in Hadoti and Sundarbans regions, adjacent regions respectively located in India and Bangladesh. Brief discussions compared results from both regions by paying attention to replies and feedback provided by visitors, guides and residents who participated in the surveys and informal talks. It was noted that Sundarbans and Hadoti regions hold interesting similarities and dissimilarities as destinations for wild animal tourism. Findings reveal that sanctuaries and zoo have faced difficulties to develop mechanisms for both implementing and delivering satisfactory learning experiences to wildlife visitors. Findings from both case studies show that majority of the visitors acknowledge the importance of wildlife tourism by considering critical to safeguard wild animal assets and natural resources.

As a drawback, wildlife tourism learning resources are quite limited in both regions. It was observed that fully operational promotion and marketing of wildlife resources and attractions, particularly in regards to Royal Bengal tiger in the Sundarbans of Bangladesh, need attention from pertinent government bodies, organisations and tourism sector. This way, wildlife tourism can be strengthened as a subset of nature-based tourism. It was also noted that attractions in Hadoti region are poorly resourced with interpretive sign boards that badly compromise a learning opportunity. In Hadoti, lack of proper infrastructure, basic amenities, transport and other relevant facilities in the zoo and sanctuaries lead visitors to dissatisfaction and disappointments as recorded in the survey and in the informal talks. Mediocre basic amenities combined with the shortage of accessible information in the sanctuary compromises a full understanding about the wild animals on display. The overall available quality for that site can be regarded as below an acceptable standard if it is taken into account the variety of attraction, planning, structure, sign boards available in English, and hygiene. Infrastructural deficiencies reside on damaged and badly maintained roads, and on lack of regular transport facilities to the sanctuaries. Visits are mostly limited to day-tours without an overnight choice as no accommodation exists near the sanctuaries.

In the face of so many drawbacks that badly jeopardise visitors' wildlife encounters and learning experiences, this research recommends the implementation of key wildlife tourism facilities to benefit participants in educational and learning tours. The rise of specialist tour operators is essential. Special tour operators normally tend to satisfactorily tailor wildlife encounters and experiential learning. These operators are used to offer services to visitors of distinct niches and segments in the wildlife tourism market according to their interests. Wildlife is clearly an important part of tourism in both the Sundarbans and the Hadoti region. In terms of strategic planning for wildlife tourism, it is relevant firstly "to connect the tourist desires and exceptions with the wildlife sanctuaries. Secondly it is important to write down their complaints regarding the wildlife visits experience for making the improvements" (Sharma 2014, p. 696).

This study concludes that there is a strong potential for wildlife learning possibilities in Hadoti and Sundarbans, particularly if it is regarded the tigers and other wild animals, and their habitats. A major limitation of this study lies on need of more encompassing primary data on the Sundarbans of Bangladesh. Further studies can fill gaps left by this research at covering further with quantitative and qualitative data the diverse aspects of wildlife tourism and learning opportunities. On the common ground, the study shows the significance of wildlife tourism education. Then, it suggested that tourism planning in Sundarbans and Hadoti regions should consider the development of educative tour services and products for wildlife visitors aiming to inform them on ecological and biological data of wild animals, as well as to include conservation issues as part of learning experience.

References

Allport CD, Kerler IW (2003) A research note regarding the development of the consensus on appropriation scale. Inf Sys Res 14(4):356–359

Ballantyne R, Packer J (2009) Introducing a fifth pedagogy: experience-based strategies for facilitating learning in natural environments. Environ Educ Res 15(2):243–262

Ballantyne R, Packer J, Everett M (2005) Measuring environmental education programme impacts and learning in the field: using an action research cycle to develop a tool for use with young students. Aust J Environ Educ 21:89–100

Berger J (2009) Why look at animals?. Penguin, London

Bright A, Pierce CL (2002) Information and education for managing wildlife viewing. In: Manfredo MJ (ed) Wildlife viewing: a management handbook. Oregon State University Press, Oregon

Chachamovich E, Fleck M, Power M (2009) Literacy affected ability to adequately discriminate among categories in multipoint Likert Scales. J Clini Epi 62(1):37–46

Chalise L (2012) Royal Bengal tiger: *Panthera tigris tigris* in Nepal. Lambert Academic Publishing, p 60

Cobb V (2011) The package tour industry. M-Y Books, New York

Das GK (2015) Estuarine Morphodynamics of the Sundarbans. Springer, Switzerland

Deccan Herald (2016) Home. Retrieved from http://www.deccanherald.com/content/269201/wild-tracks-hadoti.html. Accessed 06 Aug 2016

Derret R (2001) Special interest tourism. In: Douglas N, Douglas N, Derrett R (eds) Special interest tourism. Wiley, New York, pp 1–22

Falk JH (2005) Free-choice environmental learning: framing the discussion. Env Edu Res 11(3):265–280

Falk JH, Dierking LD (2000) Learning from museums: visitor experiences and the making of meaning. AltaMira Press, Walnut creek

Fraser J (2009) The anticipated utility of zoos for developing moral concern in children. Curator 52(4):349–361

Fraser J, Taylor A, Johnson E, Sickler J (2008) The relative credibility of zoo-affiliated spokespeople for delivering conservation messages. Curator: Mus J 51(4): 407–418

Frei J (2013) Increase in wildlife tourism: a positive step towards conservation. Greener Ideal. Retrieved from https://greenerideal.com/science/0530-increase-in-wildlife-tourism-a-positive-step-towards-conservation/. Accessed 8 Dec 2016

Gladstone DL (2013) From pilgrimage to package tour: travel and tourism in the third world. Routledge, New York

Gooch F (2011) Shoot on sight. Xlibris, New York

Hassan A (2012a) Package eco-tour' as special interest tourism product-Bangladesh perspective. Developing Country Stud 2(1):1–8

Hassan A (2012b) Key components for an effective marketing planning: A conceptual analysis. Int J Man Dev Stu 2(1):68–70

Hassan A (2012c) Rationalization of business planning through the current dynamics of tourism. Int J Manag Dev Stud 2(1):61–63

Hassan A (2014) Tour on an imagined heritage trail set in the Mosque City of Bagerhat, Bangladesh: cogitation for market potentials. In: Bansal SP, Walia S, Rizwan SA (eds) Tourism: present and future perspectives. Kanishka Publishers, New Delhi, pp 30–44

Hassan A, Ahmed MU, Shoeb-Ur-Rahman M (2015) The development, nature, and impact of medical tourism in Bangladesh. In: Cooper M, Vafadari K, Hieda M (eds) Current issues and emerging trends in medical tourism. IGI Global, Pennsylvania, pp 296–311

Hassan A, Burns P (2014) Tourism policies of Bangladesh—a contextual analysis. Tour Plan Dev 11(4):463–466

Hassan A, Rahman M (2015a) World Heritage Site as a label in branding a place. J Cultural Heritage Manag Sustain Development 5 (3):210–223

Hassan A, Rahman M (2015b) Macromarketing perspective in promoting tourism: the case of the Buddhist Vihara at Paharpur. Tourism Spectrum 1(2):13–19

Hayward J (1995) Conservation baseline: visitor research for the Monterey Bay Aquarium. People, Places and Design Research, Northampton

IISC (2016) Indian Institute of Science. Centre for Ecological Science. National Parks, Wildlife Sanctuaries and Biosphere Reserves in India. Retrieved from, http://ces.iisc.ernet.in/envis/sdev/parks.htm. Accessed 08 Dec 2016

Jamieson D (2006) Against zoos. In: Singer P (ed) In defence of animals. Blackwell, Oxford, pp 132–143

Karanth KU (2001) Tigers. Colin Baxter Photography, Glasgow

Kenney JS, Smith JLD, Starfield AM, McDougal C (1994) Saving the tiger in the wild. Nature 369:352

Lee WH, Moscardo G (2005) Understanding the impact of ecotourism resort experiences on tourists' environmental attitudes and behavioural intentions. J Sus Tou 13(6):546–565

Likert RA (1932) A technique for the measurement of attitudes. Archives Psychology 22(140):3–55

Mason P (2000) Zoo tourism: the need for more research. J Sus Tou 8 (4):333–339

Mcıntyre N, Roggenbuck JW (1998) Nature/person transactions during an outdoor adventure exeperience: a multiphasic analysis. J Lei Res 30(4):401–422

Miltra S (ed) (2011) Wild train in Bengal: travel guide. Eicher Goodearth Publication, New Delhi

Mishra H (2010) Bones of the tiger: protecting the man-Eaters of Nepal. Lyons Press, USA

Moorhouse TP, Dahlsjö CAL, Baker SE, D'Cruze NC, Macdonald DW (2015) The customer isn't always right—conservation and animal welfare implications of the increasing demand for wildlife tourism. PLoS ONE 10(10):e0138939. doi:10.1371/journal.pone.0138939

Nature Environment and Wildlife Society (1999) Royal Bengal tiger in the 21st Century. Wildlife Wing, Forest Department, Calcutta, India, p 176

Newsome D, Dowling RK, Moore SA (2005) Wildlife tourism. Channel View publications, Clevendon

Packer J (2006) Learning for fun: the unique contribution of educational leisure experiences. Curator 49(3):329–344

Parkin DR (2006) Policy, culture and the achievement of visitor education outcomes: a case study of the queensland parks and wildlife service. A doctoral thesis. School of Education and Professional Studies, Griffith University

Parveen W, Sharma A (2015) Wildlife tourism: prominent panorama at Hadoti region of Rajasthan. Int J 3(9):1135–1149

Pine BJ, Gilmore JH (1999) The experiences economy: work is theatre and every business is a stage. HBS Press, Boston

Rahman M, Hassan A (2016) Sustainable tourism practices and development: the case of the Sundarbans. Tourism Dimensions 3 (1):17–22

Richards MW, Tyabji H (2008) Tigers. New Holland Publishers, London, UK

Screven C (1995) Visitor based exhibit planning: a question of survival. Powerhouse Museum, Sydney

Shackley M (1996) Wildlife tourism. International Thomson Press, London

Sharma A (2014) A study on Rajasthan Wildlife Tourism: conservation and reformation. Elixir Mark Manage 68(2014):91–97

Sharma A (2015) Educational tourism: strategy for sustainable tourism development with reference of Hadauti and Shekhawati Regions of Rajasthan, India (Romanina). J Knowl Manage Econ Info Technol 5 (4):691–697

Shettel-Neuber J (1988) Second and third generation zoo exhibits: a comparison of visitors, staff and animals responses. Env Beh 20 (4):452–473

Singh YK (2006) Fundamental of research methodology and statistics. Newage International (P) Ltd, New Delhi

Tisdell C, Wilson C (2005) Perceived impacts of ecotourism on environmental learning and conservation: Turtle watching as a case study. Env Dev Sust 7(3):291–302

Turley SK (1999) Conservation and tourism in the traditional UK zoo. J Tour Stu 10(2):2–13

Weiler B, Ham SH (2001) Tour guides and interpretation. In: Weaver D (ed) Encyclopedia of ecotourism. CABI Publishing, Wallingford, pp 549–563

Zeppel H, Muloin S (2008) Conservation benefits of interpretation on marine wildlife tours. Hum Dim Wildlife 13(4):280–294

Tamar Project: Conservation and Education in Ecotourism Activities Related to Turtles in Fernando de Noronha Archipelago, Brazil

10

Jasmine Cardozo Moreira and Rafael Azevedo Robles

Abstract

The TAMAR Project (Marine Turtle Project) has focused on the research and conservation of sea turtles along the Brazilian coast since 1980, with 23 research bases and nine visitor centers throughout the country. On the archipelago of Fernando de Noronha (Northeast of Brazil), the TAMAR works since 1984, because it is a strategic location for food and nesting of these kind of turtles. Fernando de Noronha is one of the country's leading ecotourism destinations. It is a National Park, a Biosphere Reserve, and a proposed UNESCO Geopark project is being developed. The turtles were important in the recognition of the area as a World Heritage Site, also designated by UNESCO. In Noronha, TAMAR's efforts are in scientific research activities, conservation and management, environmental education, awareness, and community action. Understanding the importance of environmental awareness, the TAMAR project develops various activities that allow participation by the public. These include turtle nest openings, "tartarugada" (activity during the night, when the turtle makes nest on the beaches) and intentional capture of sea turtles. There are nightly exhibitions and environmental education talks at the Visitor Center. During the daylight hours, the team seeks out and captures turtles by diving and snorkeling for the sole purpose of research to support the sustainability of the turtles. Tourists have the opportunity to follow this exciting field research, and they are encouraged to take pictures and interact with the turtles. This high degree of interaction develops awareness about the importance of field research and the importance of conservation and research in Noronha.

10.1 Introduction

Fernando de Noronha Archipelago (Northeast of Brazil) is one of the country's leading ecotourism destinations. Besides the beaches, the island also has historical monuments, natural beauty and geological monuments that attract Brazilian and foreign tourists. The Archipelago is a popular honeymoon destination and is considered the best place in Brazil to go scuba diving, mainly because of the geological features that can be observed under water and which are forming an impressive backdrop (Moreira and Silva-Júnior 2015).

It is a National Park, an Environmental Protection Area, part of a Biosphere Reserve, and a UNESCO Geopark project is being developed (Moreira et al. 2013, 2014). The sea turtles were important in the recognition of the area as a World Heritage Site, designated by UNESCO (IUCN 2000).

The TAMAR Project (Marine Turtle Project), now named as TAMAR-ICMBio, has focused on the research and conservation of marine turtles along the Brazilian coast since 1980, and has now 23 research bases and nine visitor centers throughout the country. The challenge for Tamar, when the

J.C. Moreira (✉)
Tourism Department, Land Management Grad Program, Ponta Grossa State University, Ponta Grossa, Brazil
e-mail: jasminecardozo@gmail.com

R.A. Robles
Fundação Pró-TAMAR, Natal, Brazil
e-mail: rafaelazevedorobles@gmail.com

I. Borges de Lima and R.J. Green (eds.), *Wildlife Tourism, Environmental Learning and Ethical Encounters*,
Geoheritage, Geoparks and Geotourism, DOI 10.1007/978-3-319-55574-4_10

first research bases were established in Brazil, was to devise an approach to motivate coastal residents with low incomes and few alternatives, who were used to digging up sea turtle eggs for consumption and sale as part of their livelihood strategies (Marcovaldi et al. 2005).

In these bases all over Brazil, Tamar helps as an additional sources of income for local people in production of handicrafts, ecotourism guides, and participation in cottage industries, for example. Some efforts involve a variety of initiatives and topics, including educational and health-related projects, sports activities, kindergartens, community vegetable gardens, and technical assistance to various fisheries sectors, as well as other local activities. Resources are additionally invested in supporting cultural aspects of the communities. It is important to say that a majority of these projects entail informal education, and involve schools and kindergartens (Tamar 2004).

In 1984 the TAMAR Project started to work on the archipelago of Fernando de Noronha, a strategic location for food and nesting of marine turtles in Brazil. TAMAR's efforts are in scientific research activities, conservation and management, environmental education, awareness, and community action.

Understanding the importance of environmental awareness, the TAMAR project develops various activities that allow participation by the public. This includes turtle nest openings, "tartarugada" (activity during the night, when the turtle makes nest on the beaches) and the intentional capture of marine turtles. There are a museum, exhibitions and environmental education talks, nightly, at the Visitor Center.

During the daylight hours, the team captures turtles by diving and snorkeling for the purpose of research to support the sustainability of the turtles. Tourists have the opportunity to follow this field research, and they are encouraged to take pictures and interact with the turtles. This high degree of interaction develops awareness about the importance of field research and the importance of conservation and research in Noronha.

To assess the awareness and motivations of tourists to interact with the turtles, TAMAR conducts research to understand the profile of visitors. This chapter will show and discuss the activities that provide visitors with direct contact with the turtles. Also, the chapter presents conclusions about how TAMAR can provide new perspectives to visitors, with the goal of initiating an environmental awareness process.

The methodology used was bibliographic and documental references and papers, the Management Plan of Protected Areas, as well as participatory observation in the activities. Like in, for the exploratory purpose of this study, we have adopted a qualitative, interpretive approach and used a case-study methodology.

Thus, this chapter was structured in three parts, the first being related to the Fernando de Noronha archipelago, the second on the TAMAR Project, and the third presents the results and the activities that have been conducted by the TAMAR Project in Fernando de Noronha.

10.2 Fernando de Noronha Archipelago

Fernando de Noronha is located in northeastern Brazil, and the archipelago has only 26 km^2 (nine kilometers of marine platform and seventeen kilometers of land area), with a main island, which is surrounded by 18 smaller islands and dozens of rocks (Fig. 10.1). The area is divided into two protected areas: a Marine National Park and an Environmental Protection Area.

The National Park was established in 1988 and covers approximately 70% of the main island and all other islets and rocky outcrops. The Environmental Protection Area was established in 1986 and covers the rest of the main island, including the populated area. With the establishment of these Protected Areas, more strict regulations for the use of natural resources around the archipelago were put into place and rules regarding visitation were implemented.

According to Estima et al. (2014),

> Because there was the need to implement infrastructure, management and supervision of public use and the ICMBio lacked human and financial resources, in 2010 a bidding process was started for a concession contract to provide support services for the public visitation and collection of admission fees in the Park.

This way, since 2012 the area of the Park is a concession to Econoronha, a company of the Cataratas S.A. group, the same one that manages Iguassu Falls National Park (in Brazil). An entrance fee is charged now and there are more rules to be followed.

Noronha is one of the most important regions for the breeding of marine birds in the Atlantic and a nursery for species such as spinner dolphins and sea turtles. The northwestern coast of the island is the 'Inner Sea' because it is protected from the winds and ocean currents by the topography of the island. The 'Outer Sea' is the southeastern coast, which receives wind and waves all year round and where there are three sandy beaches protected by islands. The vegetation is mainly shrub and herbaceous.

10.2.1 Aspects of Tourism, Culture and History in Fernando de Noronha

The islands of the archipelago were officially discovered in 1503 and suffered several invasions, due to its strategic location next to one of the navigation routes from Africa and

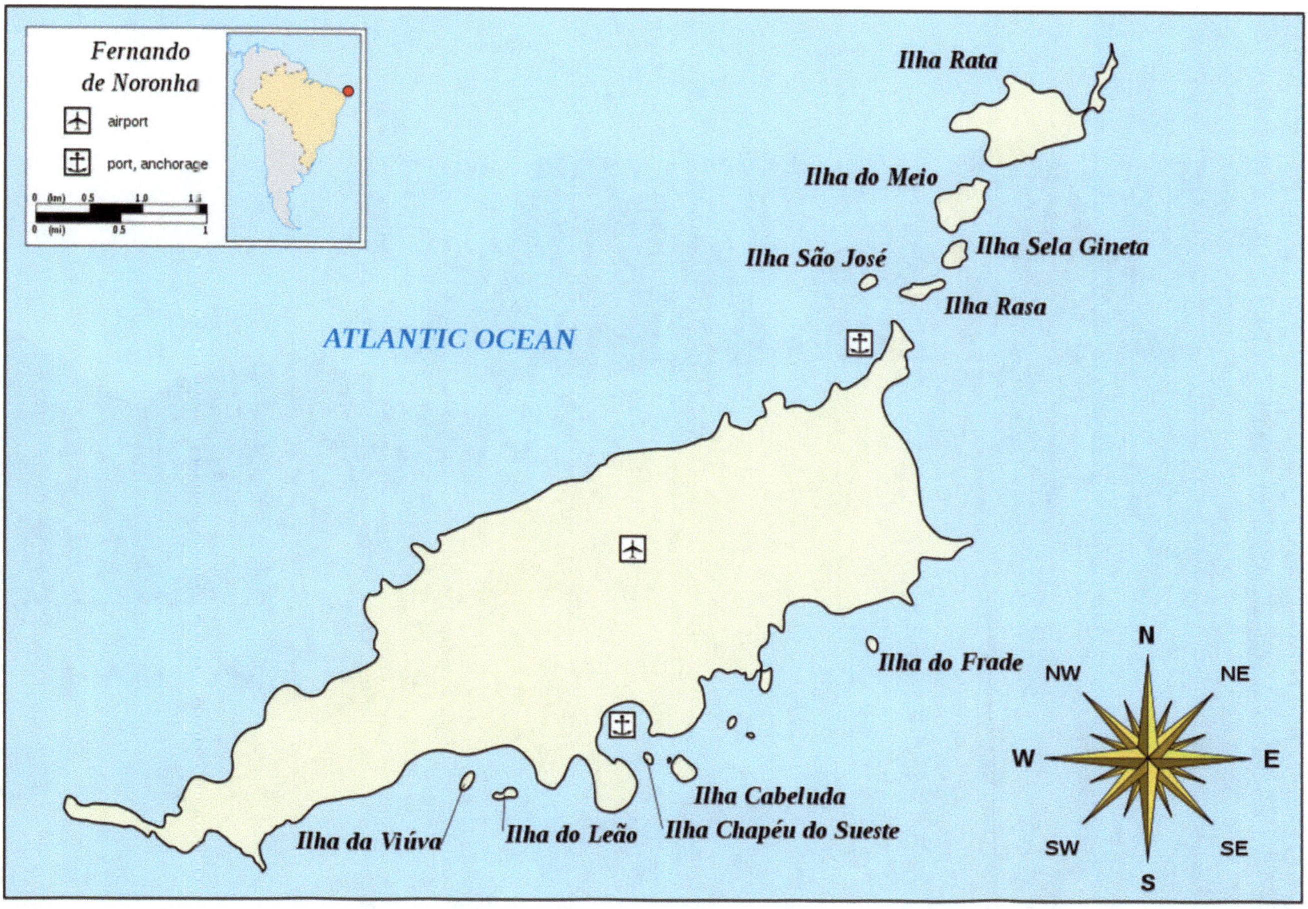

Fig. 10.1 Location map of the Fernando de Noronha Archipelago—Brazil. *Source* Wikimedia Commons 2011

Europe. These occupations were Dutch, French and Portuguese and let rich archaeological and cultural heritage buildings. For over two hundred years the main island housed a convict colony, later transformed into a political prison. During World War II, it was created a Brazilian Military Federal Territory and the USA installed a Support Military Base. Between 1957 and 1965 there was a new North American presence at an observation post for guided missiles (Ibama et al. 2005).

Before the establishment of the protected areas in the 1980s, community living was based around the ocean and the consumptive use of natural resources. Fishing, hunting and gathering were not only common practices but the main source of food and income for most islanders. Currently, tourism is the main economic activity and, although the community is still highly dependent on the ocean and its resources, traditional activities such as fishing and hunting are either strictly regulated or completely forbidden (Mazaro 2009; Moreira et al. 2011).

Nowadays, the archipelago is a major ecotourism destination and very popular for honeymooners. The hostels have different identification according to their category, one to three dolphins. There are museums, visitor centers, variety of entertainment and food including local cuisine, seafood and international cuisine options. There are receptive agencies, rental cars, dive centers and an association of local artisans. It is the only State District in Brazil.

The area also has the most beautiful beaches and bays in Brazil according to the ranking of the main Brazilian tourist guide, the Guia Quatro Rodas (2015), like Porco's Bay (Fig. 10.2). And, according to Trip Advisor (2017), at the Travelers Choice Awards 2014, 2015 and 2017. Sancho's beach was elected the best beach in the world.

At the Management Plan of the protected areas, there is guideline for public use activities that includes, among other restrictions, some 'no access' areas (including some geological monuments, natural swimming pools and beaches), some trails requiring the presence of authorized guides (like Capim-Açu, São José and Atalaia trails), and some areas having strict visitation hours (from 8 am to 6 pm), like Leão and Sancho's beach. Atalaia Beach, for example, has a daily limited carrying capacity. Less than one hundred people per day in only two hours, (16 people per group every thirty minutes), can enter at the natural swimming pools when the

Fig. 10.2 Porco's Bay, one of the most beautiful and famous landscapes in Brazil. *Source* The authors

tide is low. This is due to the fact that the natural pools are important environments for life and growth of reef fish.

The observation of animals on the island is abundant. Numerous species of fish as well as lobsters, octopuses, stingrays and moray eels are seen in dives and snorkeling activities. During the boat trips it can be spotted spinner dolphins, which can also be seen throughout the year at the Dolphin's Bay, where are the researchers of the NGO Golfinho Rotador Center (Spinner Dolphin Center—CGR). Sharks can be seen in the natural pool of Atalaia, the Sueste Beach and especially in Enseada dos Tubarões (Sharks Cove), near the local Shark's Museum.

The turtles are very easy to be seen and can be seen in free dives in almost all beaches. But the best place to observe turtles in free diving is the Sueste Beach. The chance of seeing these animals is certain, and the local guides say to the visitors that they do not have to pay the tour if they do not see a turtle. To protect this environment and combine conservation with the visitation, the beach is divided into three distinct areas separated by buoys. On the right side the free diving activity must necessarily be performed with float vest; in the center of beach bathing is free; in the left side, access is allowed only to authorized researchers. Because it is in the park area, the visit can be done only between 09 h and 16:30, there are restrooms, snack bar, rest area, and equipment rental and gift shop. The facilities follow the principles of sustainability such as wastewater treatment, water reuse, rainwater harvesting and use of construction technologies of low impact and recycling.

The local guides receive free ongoing training that encompasses many topics. It is really necessary to prepare professionals for the contact with the public, knowing and understanding the environment in which he lives, since the use of interpretative trails with guides can be considered an important resource of environmental education (Moreira and Bigarella 2008).

Environmental education takes place in the archipelago through various projects that are developed by different institutions: ICMBio (that manages the two protected areas of the island), the NGOs CGR and the Tamar Project. One of these projects, "Ecological Holidays", enables the contact of students with various topics related to the environment. The Tamar Project develops Tamarzinho and Tamar in School Program, in which the focus is on sea turtles.

The CGR is also responsible for research, conservation and management of cetaceans in Fernando de Noronha region. The CGR develops the "Cidadão Golfinho" (Dolphin Citizen) Project, one of the largest contributors in relation to the training of islanders, offering free courses such as computer science, English, first aid techniques for guides, ecotourism and lodging management, handicraft production, scuba dive, snorkeling guide, birdwatching guide, dolphin watching guide, among others.

10.2.2 Sustainability and Noronha + 20 Project

Population growth in Fernando de Noronha is controlled. The entry and stay of visitors is limited and everyone must pay an Environmental Preservation Tax. This fee is to ensure the maintenance of the environmental conditions of the archipelago, created to preserve the safety and comfort suitable for the wellbeing of the community and visitors.

The tax was established with the primary purpose of implementing mechanisms and procedures for controlling access and establishment of people within the district (Tisdell 1998). Thus, those who do not have the 'permanent resident' card are either recognized as 'exempt' (e.g., researchers, trainees from NGOs, islanders relatives) or are visitors, and therefore pay the tax. As a visitor, the daily rate in 2017 was approximately US$ 20, reaching almost US$ 2000 for 30 days. The tax also provides a framework for tourism planning and development, as is an attempt to restrict tourist visitation to between five and 10 days. Less than five days visitation is neither encouraged nor penalized (the daily rate is the same), but tourists who want to stay more than five days receive a kind of progressive penalty so that for each additional day the amount is increased exponentially. The fee must be paid in advance (online) or on landing.

Due to the serious situation of risk of social and environmental degradation where the island is, described in "Fernando de Noronha Carrying Capacity Study" (ELABORE 2008), an inter-institutional and community agreement was proposed, aiming behavioral changes in institutions and islanders. Therefore, two questions should be answered: Which Fernando de Noronha we will leave to our children and grandchildren? And, what children and grandchildren we will leave to Fernando de Noronha?

To do this, there was a participatory planning process on the island, Noronha + 20 (meaning plus 20 years), which was structured based on participatory workshops, to be an instrument that offers local, national and international visibility to the planning of actions for environmental sustainability and social justice in Fernando de Noronha. The workshops were conducted between 2009 and 2010, with the participation of 230 people and 40 institutions. The result of this work is a long-term planning for the development of the archipelago in a sustainable manner, a total of 84 actions distributed in eight thematic areas (Table 10.1).

To monitor the implementation of these actions and disseminate the results, it was created a "Sustainability Observatory". Compose the Observatory: five representatives of the community, a representative of UNESCO, one from the Ministry of the Environment, one from the state environmental agency and one from the Council District. The whole process was based on the model 'Hawaii 2050' (www.hawaii2050.org). This way, the Noronha + 20 website was created with the intention to ensure the monitoring of actions in the long term, and provide transparency to the program.

10.2.3 Main Attractions of the Archipelago and Geological Aspects: Towards a Geopark Proposal

The archipelago is the submerged part of a large volcanic edifice currently inactive, about four thousand meters deep and about seventy kilometers in diameter. It is part of a chain of underwater volcanic mountains that stretch out from the Atlantic Ridge to the Brazilian continental platform, near the coast of Ceará State (Teixeira 2003).

According to Almeida (2002), it consists of a substrate of pyroclastic rocks penetrated by igneous intrusions, which after a long period of erosion was covered by lava flows. These lavas were very fluid, sufficient to form layers less than three inches thick. The eruptions would begin with violent explosions that cleared the throat of the volcano, ejecting huge quantities of pyroclastic rocks into the air (bombs, lapilli and ash), and then once the pressure of the gases had been released the lava would begin to pour out.

For the Brazilian Geological Survey (CPRM 2011), the area presents important geological, geomorphological and geotouristical aspects, highlighting the exceptional beauty of the landscape. By now, 26 geosites were classified and these aspects when coupled with other attributes observed in the area justify the creation of a Geopark, that can be part of the Global Geoparks Network, by UNESCO.

This way, some actions are being carried out: a Guide training course about Geotourism was held (Moreira and Bigarella 2008), lectures on Geoparks at TAMAR's Visitor Center, and a Working Group is preparing the application dossier and a brochure was distributed to the community. The TAMAR Project is also part of this Working Group.

Due to the small size of the archipelago, it was realized the need to include also some marine geosites.

Table 10.1 Thematic areas included at Noronha + 20 Program. *Source* The authors

Urban and housing issues
Infrastructure—Water, Sewage, Waste, Energy
Public use, Model of tourism, Lodging facilities
Recovery of degraded areas and conservation of terrestrial and marine areas protection, Management and research of terrestrial wildlife, Marine environment and sociocultural aspects
Maritime activity, Port and boats, Fishing activity
Agricultural activity
Quality of life and welfare, Health; Education; Environmental education; Culture; Handicraft development; Inclusion of youth; Professional training
Urbanization control, Migratory monitoring, Vehicles and boats monitoring

The snorkeling areas and scuba diving sites that are listed at the Management Plans were defined as marine geosites, since only these areas may be used for underwater activities (Moreira and Silva 2015).

The next steps involve fundraising in order to make the community aware to the theme, the availability of information on a website, and create an event on the theme. After this event, the community will be consulted so they can manifest themselves if they want the dossier to be sent or not to UNESCO (Moreira et al. 2013).

10.3 TAMAR/ICMBio Project and the Ecotourism Experience

10.3.1 Turtles and TAMAR in Brazil

Environmental policy in Brazil during the 1970s and 1980s was exclusionary. Federal legislation was generic and restricted to the prohibition of wildlife products and derivatives. Governmental efforts were exclusively directed at terrestrial protected areas, and policies involving the conservation of coastal and marine natural resources were nonexistent (Fundação Pro Tamar 2000). According to Marcovaldi et al. (2005), one of the greatest and most complex challenges to the long-term conservation of sea turtles in Brazil, is changing the habits of coastal communities in which intense rates of natural resource use is a vital source of subsistence and income, essential to survival.

Five of the seven extant sea turtle species occur in Brazilian waters: *Caretta caretta* (Loggerhead Turtle), *Chelonia mydas* (Green Turtle), *Dermochelys coriacea* (Leatherback Turtle), *Eretmochelys imbricata* (Hawksbill Turtle), and *Lepidochelys olivacea* (Olive Ridley Turtle). These animals use the beaches along the continental coast and oceanic islands for nesting, feeding and development. Sea turtles with occurrence in Brazil are listed under the International Union for the Conservation of Nature (IUCN) as (ICMBio 2016): *Endangered (EN)—Chelonia mydas* and *Caretta caretta*; *Vulnerable (VU)—Lepidochelys olivacea*; *Critically Endangered (CR)—Dermochelys coriácea* and *Eretmochelys imbricata.*

According to ICMBio Executive Summary (2016), besides having a fundamental importance for marine ecosystems' health, sea turtles are also important in cultural aspects of several Brazilian coastal communities. These animals are associated with mysticism and symbolism, being part of people's life in commemorative and folkloric expressions. In some areas, sea turtles play a strategic social and economic role, since conservation activities towards them generate employment, income, development and tourism. In several areas where sea turtles occur along the Brazilian coast, the expansion of urban and industrial activities have resulted in extensive coastline occupation and increased sources of pollution. Also, the intensive fishing activities rise as one of the major threats to these species due to incidental capture and noncompliance of current laws.

The TAMAR Project is a pioneering initiative in Brazil, focused on research, conservation and management of sea turtles. In the late 1970s and early 1980s, there were no studies on these animals and some oceanography students decided to explore the Brazilian coast searching for information. Thus, in 1980 the Brazilian government created the National Program for the Conservation of Sea Turtles in Brazil, the TAMAR Project. Currently, the project is called TAMAR-ICMBio, and is co-managed by Pro-TAMAR Foundation, a private, nonprofit organization created in 1988.

The name comes from the words in Portuguese, *Tartaruga* (Turtle) and *Marinha* (marine). According to Marcovaldi and Dei Marcovaldi (1999), the initial objectives of TAMAR were to quantify the number of species, distribution and abundance of sea turtles, the seasonality and geographic range of egglaying, and the primary threats to turtle survival. The mission of the program is to develop conservation actions and research in order to ensure the recovery and survival of the five species of sea turtle in Brazil, at healthy levels able to fulfill their ecological roles.

The project is internationally recognized as one of the best experiences in marine conservation in Brazil, this in a country that has reached the end of the twentieth century with serious environmental problems, despite having strict legislation, a ministry of the Government for the environment and large number of non-governmental organizations for the conservation of nature. Through its foundation, the project research and develop special techniques for the conservation and management of sea turtles, as well as genetic studies to increase the level of knowledge on the populations of five species that occur in the Brazilian coast (Fundação Pró-Tamar 2000).

In most beaches where TAMAR is present, the program is a primary source of income, both directly and indirectly, to the local village. Most of the funds raised by Fundação Pró-TAMAR are invested in the communities where TAMAR develops its activities, helping to solidify the bond between TAMAR and the coastal communities (Marcovaldi and Dei Marcovaldi 1999). The same authors also explain that one of the most important aspects of the Project is community outreach and education within the coastal villages. The goal is to increase local awareness of the importance of a healthy marine ecosystem, which include turtles. This way, as a result from TAMAR work, the killing of females and collection of eggs has been drastically reduced in Brazil and there has been an increasing trend in the total number of turtle nests protected on all the mainland beaches patrolled by TAMAR.

Besides working towards the conservation of marine turtles, TAMAR also values social inclusion and supports local coastal communities. It also promotes environmental programs as "Tamarzinhos", which stimulates the participation of teenagers in educational and interactive activities with sea turtles.

The success of the Tamar Project is due to the inclusion of communities in the conservation process, generating jobs and income for productive groups that produce souvenirs that value their culture. As mentioned by ICMBio (2016), after 35 years of its creation, TAMAR Project has shown significant results (see Table 10.2).

Through the projects developed, TAMAR is achieving many of its goals, i.e. involving community members to participate in economically attractive activities. This not only helps them to make an income and improve their education and living conditions, but also facilitates the process of social inclusion, a major issue for developing countries (Marcovaldi and Thomé 1999).

10.3.2 TAMAR Visitor Centers

In places where there is a tourist flow, the TAMAR maintains Visitor Centers (CV), which aims to carry out educational activities and environmental awareness, dissemination of conservation actions and especially inform visitors about sea turtles and the activities of the Project. They have a structure that can include tanks with live marine animals, interpretative panels, marine animal replicas and temporary exhibitions of photos. Also, TAMAR CVs generate income for coastal communities, create job opportunity for locals and gather resources for research and conservation of sea turtles.

The CVs show to the public what the project is, generate economic benefits and increase the model of self-sustainability of the TAMAR. All funds generated by visitor centers and the Project stores are reversed for research and conservation, which are distributed among all the bases. They are indispensable sources of revenue, as more than 40% of the TAMAR annual budget is from the sale of services and products that are result of the work of the communities (Marcovaldi et al. 2007).

The shops are in the Visitor Centers and offer products with unique design. There are many products that have the TAMAR brand, including menswear, womenswear, children

Table 10.2 Some TAMAR results in the last 35 years (the authors, edited from SITAMAR database)

Some TAMAR outcomes
✓ 23 stations along the Brazilian Coast
✓ 1100 km of priority conservation areas monitored
✓ More than 20,000 nests protected annually of five sea turtle species
✓ Around 1,200,000 hatchlings released every year of five sea turtle species
✓ 1610 tagged females of five species (2010/2011 nesting season)
✓ SITAMAR—A database with 286,000 records collected in a standardized way, over 30 years
✓ 1300 direct jobs generated by sea turtle conservation activities, more than 80% are natives of local communities
✓ 900 local people benefits from income-generating programs
✓ 149 scientific publications
✓ More than 1,500,000 people attended the Visitors and Environmental Education Centers, educational campaigns and exhibitions

(shirts, shorts, dresses, shorts, etc.), caps, hats, keychains, calendars, towels, school supplies, postcards, books, bags, backpacks, among others. The products have the turtle as its theme and all income is reverted to research and conservation of sea turtles.

Currently, the TAMAR has nine visitor centers at different beaches, popular Brazilian tourist destinations as Praia do Forte, Arembepe, Guriri, Regencia, Vitória, Fernando de Noronha, Ubatuba, Florianópolis and Aracaju Oceanarium.

10.3.3 TAMAR Project on Noronha

On Fernando de Noronha, the TAMAR Project has been operating since the year 1984 and has helped in the creation of the Fernando de Noronha Marine National Park. Noronha is reproductive and feeding área of the green turtle area (*Chelonia mydas*) and feeding área for the hawksbill turtle (*Eretmochelys imbricata*). According to Marcovaldi and Dei Marcovaldi (1999, p. 38),

> Conservation on the islands is more expensive and logistically challenging than on the mainland. TAMAR has established one permanent island station, located on Noronha. There are several beaches on the main island, which are patrolled nightly, from December through May, to tag and measure nesting females and to mark and monitor nests. All nests are protected in situ.

In this area, TAMAR does research and biometry with the young turtles of the two species by capturing and recapturing in scuba diving or free diving. This base is one of the most important base in the country, because they have excellent conditions for research on the behavior of those species of sea turtles. The nesting beaches have favorable characteristics to a daily and night monitoring. The "Leão" beach concentrates 80% of the cases. The other spawns happen along the called "inside sea", between the beaches of Sancho and Conceição. Each breeding season records an average of 100 spawns, generating approximately 8900 young green turtles (TAMAR 2016).

Neto et al. (2015) did a study monitoring green turtles during the breeding season of 2013/2014 to analyze the potential impact of induced species. Camera traps were used in four selected nests and the authors presented the interim results of this study in which there was an attempt to behavior, predation or detritivore by teju açu lizard and mouse.

According to data from the Capture, Marking and Recapture Program (CMR) maintained by TAMAR-ICMBio in the archipelago since 1987, the annual abundance of sea turtles at Fernando de Noronha was estimated to be approximately 420–1148 individuals. Turtles at Fernando de Noronha would need about 22 years to grow 30–87 cm, to reach the minimum size observed in reproductively active adults of this species in Brazil. Between 1988 and 2013, 1279 individuals of *Chelonia mydas* were marked, in a total of 2979 catches (Colman et al. 2015).

10.3.4 The Tamar Ecotourism Program

Because Noronha is a strategic location in relation to ecotourism in Brazil, in 1996 it was inaugurated the Visitor Center (CV) of the Projeto TAMAR—ICMBio to provide environmental education through lectures and other interpretive activities.

In 2010 it was created the TAMAR Ecotourism Program, which has the objective to approximate the local community and visitors to the activities of research and conservation of sea turtles. On Noronha, the program consists of five activities, and they are all free.

The CV TAMAR consists of a museum (the Marine Turtles Open Museum) and a visitor center, and receives about 40,000 visitors per year (Fundação Pró-Tamar 2011). The architectural project uses certified wood of reforestation, recycling of maritime containers and removable pallets to avoid soil waterproofing.

There is no entrance fee and the visitor center also encompasses the Project gift shop, an auditorium, *Chelonia* cafeteria (food service), an outdoor amphitheater (for concerts and events) and a reception area. The Project aims to transmit knowledge with a clear and accessible language to all sorts of public and to all ages (Vale et al. 2016).

The Marine Turtles Open Museum possesses a "sea turtles area", where replicas of the species found in Brazil can be observed as well as models of nesting and panels containing information regarding the Project and the turtles (Fig. 10.3). The structure of the Museum is available at any time of day or night and there are daily-guided tours by the project researchers.

- *Environmental talks cycle*

  One of the most important activities of TAMAR Project is the Environmental Talks Cycle at the Visitor Center. It has been happening for almost 20 years, every night, and provides visitors with topics of scientific and environmental relevance.

  There is a fixed weekly schedule and before there are videos on environmental topics. Currently, talks are on the following topics: Spinner Dolphins, Sea Turtles, Fernando de Noronha protected areas, Sharks and two nights with special programming.

  On nights with special programming, the TAMAR gives space for researchers who develop their research on the island, to disseminate their results and disseminate further knowledge, primary function of a Visitor Center. Gerhardt et al. (2015) concerning the Environmental Talks Cycle, found that in addition to bring the message

Fig. 10.3 Some of the turtles replicas at the Marine Turtle Open Museum. The entrance is free and the visit is allowed any time, during day and night. *Source* The authors

of environmental preservation for tourists, is a reference point for environmental discussions regarding the region.

The Cycle also shows that the CV TAMAR has success in this field, mainly because it is the main interpretive activity that aims to raise awareness of visitors and the community on Fernando de Noronha (Moreira and Robles 2009). Also, the CV became a national reference in relation to environmental interpretation and use of lectures as a component of the tourism product in Brazil.

- *Intentional capture of sea turtles*

  This activity is performed throughout the year and in it, visitors follow the tagging program, and capture (or recapture) turtles, conducted by the Project biologists. The technical team captures the turtles through free diving and takes the animals to the sand to collect data. At that time the participants have the opportunity to follow the fieldwork and interact with the staff of the TAMAR, who use the opportunity to raise awareness of conservation and the importance of research (Fig. 10.4).

  Biologists measure and check the health of the turtle. If this is the first time it is captured, it will be marked with a ring and a seal with numbers that serve to identify them. This number is registered in TAMAR database and when the animal is found in other locations, information such as its approximate age, location where it was marked, as well as other important information can be accessed online on the system. The seals have different color according to the season, and serve to know which turtles have been taken from the sea to collect information that year.

  This intentional captures of sea turtles has taken place since 1988 in the main beaches on the island, being observed by tourists due to the movement, that saw the activity and asked questions to biologists. But it was only in 2009 that the TAMAR began to define time, place, and started to promote the activity, so the visitors could

Fig. 10.4 Biologists at Porto's Beach, doing the Tamar usual work, that can be followed by the visitors. *Source* The authors

follow the biologist's fieldwork. It was the union of the research carried out for years, with an ecotourism activity, in order to raise awareness to the visitor.

Currently the activity takes place every Monday and Thursday, at high tide on the beaches Sueste or Porto. Every year, around 6000 people follow this activity, according to SIGRE, the Reporting System from TAMAR's Project.

A research by Gomes et al. (2011) showed that there is the environmental awareness by the public who participated. Of the respondents, 89% affirmed understand why Tamar Project exists and the importance of preserving the environment and turtles, and 94% consider very important that this activity is open to the public. This activity takes place in this manner only in Fernando de Noronha, at the other Tamar bases there is no interaction like this.

The same authors also found that when a visitor participates, he gains more knowledge about the sea turtles, because the information that he receives are encouraged by the senses, he can be near the turtle, interact with the biologists and take pictures. This way, the TAMAR Project is providing environmental awareness through ecotourism.

- *Nest openings*

Researchers at TAMAR monitor all turtle nests in Noronha and when they realize that have some where the cubs are close to the surface (Fig. 10.5), they invite the community and visitors to follow the nest opening. When the activity occurs, a presentation is made, clarifying the objectives and the rules needed to participate in the activity. This activity is not programmed, i.e., the activity is reported only on the same day, in the morning and

Fig. 10.5 A biologist from Tamar check all the nests every day. *Source* The authors

happens at the sunset. Usually a lot of people participate, and get to the nest is prohibited, as well as touch the cubs. Only TAMAR biologists can handle them and is placed a cordon in the sand, so that the cubs can go freely to the sea.

The public release of cubs happens when they are transferred from one beach to another, to be released safely. In this case, they are transported minutes before in Styrofoam boxes, and are released on the beach, in small groups, away from artificial light. Cubs are accompanied by biologists and by visitors to entering the sea. One of the biggest challenges is to scare the birds that are waiting to eat the cubs as they arrive to the sea.
This activity takes place only between the months of May and July and in each of them, around 200 people follow the activity.

- *Tartarugada*

  In this activity, which takes place only from December to July, the participants can follow the monitoring work carried out by the biologists. Researchers spend the night in the Lion's Beach, waiting until one or more turtles come up onto the sand to lay their eggs. The activity starts around 8 pm and goes until dawn. Every hour the Tamar biologist walks across the beach to see if any turtle went to the beach to lay eggs. It is forbidden to use flashlights to not disturb the animals. If any turtle is observed, the biologists measures the hull, looking and note the mark on the fin and after laying eggs in the nest, he puts a pole with a number.

  The main objective of "Tartarugada" is to sensitize participants on the need for protection of turtles and the marine ecosystem (Pinheiro et al. 2013).

  Despite spawning happening in other beaches on the island, it is the Lion's Beach which is the largest number of spawning turtles. Almost 80% of the nests in the archipelago occur in this beach (Bellini et al. 2000), which may be due to the fact that this is the most isolated beach on the island.

10.4 Conclusions

Fernando de Noronha is a major ecotourism destination in Brazil and has the potential to be recognized as a Geopark. On its protected areas there are several rules that must be accomplished, which help protect the region and the conservation of biodiversity and geodiversity.

The TAMAR Project has focused on the research and conservation of sea turtles along the Brazilian coast since 1980. Sea turtles conservation is a complex challenge and requires long-term knowledge about essential aspects of its basic biology, such as reproduction, migration and feeding habits (ICMBio 2016). TAMAR's strategy is based on the principle that without the participation of communities, conservation programs may be condemned to failure (Marcovaldi et al. 2005).

On Fernando de Noronha archipelago, the project develops various activities, mainly for scientific research. However, the need for environmental awareness was realized and public participation became to be allowed in a controlled manner. Thus, it was created in 2010, The Tamar Ecotourism Program, composed of five different activities free of charge. All of them seek to raise awareness and educate environmentally participants.

The Visitor Center (CV) offering daily lectures is distinguished for providing reflection of the importance of conservation of the archipelago. The lectures are interpretive activities that stimulate a better knowledge of the environment being visited. The experience of the CV on the valuation of environmental knowledge and the adequacy of language to different audiences assist in the conclusion that the principles of sustainable development in tourism have been achieved. This can be noted in relation to the improvement of part of quality of life of the resident population, in providing interpretive activities that help improve the quality of stay of visitors on the Archipelago, and assistance in environmental conservation.

Moreover, as interpretive, educational environment and local development provider, the Visitor Center and the Museum turned out to be a part of the tourist product of Fernando de Noronha.

The activities that take place on the beach (turtle capture, nest openings and the "tartarugada") are very popular and attract thousands of tourists every year. The community also participates and recognizes the importance of conservation of sea turtles. The turtle is a symbol animal of the Archipelago.

In the case of social benefits observed in the local community, there is the constant training, awareness and environmental education of the community, the increase in leisure activities, and awareness of conservation. The economic benefits are income generation, the creation of jobs for the local community and the increased tax revenue. And the environmental benefits are the provision of information to visitors, which can encourage attitudes related to conservation of the environment they are visiting.

The current importance of ecotourism to the society is not only based on economic variables, but mostly in their educational potential and nature conservation. This way, the TAMAR Project provides to visitors experiences and sensations in nature. By its diversification, it is one of the main activities focused on environmental education of Fernando de Noronha visitors.

References

Almeida FFM (2002) Arquipélago de Fernando de Noronh. In: Schobbenhaus, C, Campos, DA, Queiroz ET, Winge M, Berbert-Born MLC (eds) Sítios Geológicos e Paleontológicos do Brasil. DNPM/CPRM. Brasília, p 554. Available at: http://sigep.cprm.gov.br/SIGEP_Vol_I.pdf

Bellini C, Sanches TM, Lima RP (2000) Temporada reprodutiva da tartaruga marinha Chelonia mydas em Fernando de Noronha, PE, Brasil—Campanha: 98/99. In: Congresso Brasileiro de Zoologia, Cuiabá

Colman LP, Patrício ARC, McGowan A, Santos AJB, Marcovaldi MA, Bellini C, Godley BJ (2015) Estudo de Crescimento de longo Prazo e Sobrevivência de Tartarugas Verdes (Chelonia mydas) no Arquipélago de Fernando de Noronha, Brazil. In: I Encontro de Pesquisa de Fernando de Noronha, São Pedro e São Paulo e Atol das Rocas. Fernando de Noronha 31 Aug–03 Sept 2015

CPRM (2011) Geoparque Fernando de Noronha, Geoparque Fernando de Noronha. Available at: http://www.cprm.gov.br/geoecoturismo/geoparques/noronha/index.php?GEOPARQUE=2 Accessed 5 Dec 2011

ELABORE (2008) Estudo da Capacidade de Suporte da APA de Fernando de Noronha, Elabore Consultoria e Planejamento

Estima DC, Ventura MAM, Rabinovici A, Martins FMCPF (2014) Concession in tourism services and partnerships in the Marine National Park of Fernando de Noronha, Brazil. Rev Gestão Costeira Integrada 14(2):215–232. doi:10.5894/rgci469

Fundação Pró-Tamar (2000) Assim nasceu o Projeto TAMAR, 1st edn. A Fundação, Salvador

Fundação Pró-Tamar (2011) Fernando de Noronha. Available at: http://www.tamar.org.br/base.php?cod=20. Acessed in 21 Dec 2011

Gerhardt C, Robles RA, Rodrigues CGO, Santos AJB, Vale TFV, Moreira, JC. O Ciclo de Palestras Ambientais do Projeto TAMAR de Fernando de Noronha e a Percepção dos Visitantes. In: I Encontro de Pesquisa de Fernando de Noronha, São Pedro e São Paulo e Atol das Rocas. Fernando de Noronha 31 Aug–03 Sept 2015. Available at: http://www.tamar.org.br/publicacoes_html/pdf/2015/2015_O_ciclo_de_palestras_ambientais_do_Projeto_Tamar_de_Fernando_de_Noronha.pdf

Gomes TRC, Moreira JC, Robles AR (2011) A sensibilização ambiental por meio do ecoturismo: captura intencional de tartarugas marinhas em Fernando de Noronha (PE). Rev Bras Ecotur 5(4):856

Guia Quatro Rodas (2015) 'Guia Quatro Rodas'

IBAMA, PNUD, TETRAPLAN (2005) Plano de Manejo da APA de Fernando de Noronha—Rocas—São Pedro e São Paulo. Fernando de Noronha. Available at: http://www.icmbio.gov.br/portal/unidadesdeconservacao/biomas-brasileiros/marinho/unidades-de-conservacao-marinho/2242-apa-de-fernando-de-noronha. Accessed 01 Sept 2016

ICMBio (2011) Programa de Sustentabilidade para o Arquipélago de Fernando de Noronha: Uma Construção Participativa. Available at: http://marineworldheritage.unesco.org/wp-content/uploads/2012/01/Management-Plan_Noronha-2011-portuguese.pdf. Accessed 10 Sept 2016

ICMBio (2016) Executive summary of the National Action Plan for the conservation of the Sea Turtles in Brazil. Available at: http://www.widecast.org/Resources/Docs/ICMBio_Executive_Summary_National_Action_Plan_Turtles_Brazil.pdf. Accessed 07 Sept 2016

IUCN (2000) World Heritage Nomination—IUCN technical evaluation Fernando de Noronha Archipelago/Rocas Atoll tropical insular complex (Brazil). Available at: http://whc.unesco.org/archive/advisory_body_evaluation/1000rev.pdf. Accessed 16 June 2016

Marcovaldi MA, Dei Marcovaldi GG (1999) Marine turtles of Brazil: the history and structure of Projeto TAMAR-IBAMA. Biol Conserv 91(1):35–41. doi:10.1016/S0006-3207(99)00043-9

Marcovaldi MA, Thomé JC (1999) Reducing treats to turtles. In: Eckert KL, Bjorndal KA, Abreu-Grobois FA, Donnelly M (eds) Research and management techniques for the conservation of Sea Turtles, vol 4. Iucn/Ssc Marine Turtle Specialist Group Publication, pp 165–168

Marcovaldi MA, Patiri V, Thomé JC (2005) Projeto Tamar-Ibama: twenty-five years protecting Brazilian sea a community-based conservation programme. Mast 3(2) and 4(1):39–62

Marcovaldi MA, Lopez GG, Soares LS, Santos AJB, Bellini C, Barata PCR (2007) Fifteen years of Hawksbill Sea Turtle (Eretmochelys imbricata) nesting in Northern Brazil. Chelonian Conserv Biol 6(2):223. doi:10.2744/1071-8443(2007)6[223:FYOHST]2.0.CO;2

Mazaro RM (2009) Desde la Quimera a la Realidad: haciendo operativa la sostenibilidad estratégica en destinos turísticos, Rev Turismo Anál 20(3):485–503. Available at: http://repositorio.ufrn.br:8080/jspui/bitstream/123456789/19168/1/DesdeQuimeraRealidad_2009.pdf

Moreira JC, Bigarella JJ (2008) The geotourism guide training at the Fernando de Noronha Archipelago—Brazil. In: Inaugural global Geotourism conference, Fremantle

Moreira JC, Robles RA (2009) As palestras como meio interpretativo: Estudo de caso com palestrantes em Fernando de Noronha —Pernambuco. Rev Bras Ecotur 2(4):322

Moreira JC, Silva-Júnior JM (2015) Proposta de Geosítios Marinhos do Projeto Geopark Aspirante de Fernando de Noronha e a importância da interpretação ambiental relacionada á geodiversidade. In: I Encontro de Pesquisa de Fernando de Noronha, São Pedro e São Paulo e Atol das Rocas. Fernando de Noronha 31 Aug–03 Sept 2015

Moreira JC, Reis AC, Silva-Júnior JM (2011) 'Fernando de Noronha Archipelago: community perspectives on tourism', In: Refereed papers from the 7th international small islands conference. Queensland, pp 34–41. Available at: http://sicri-network.org/

Moreira JC, Robles RA, Araujo R (2013) Fernando de Noronha: proposta de Geopark a ser decidida pela comunidade. In: Actas do III Encuentro Latino-Americano de Geoparks, San Martin de los Andes, pp 1–3

Moreira JC, Robles RA, Rodrigues G (2014) Fernando de Noronha Aspiring Geopark project, Brazil: history, culture, and tourism. In: International conference UNESCO on global Geoparks. Saint John New Brunswick, p 60. doi:10.4138/atlgeol.2014.015

Neto LD, Santos AJB, Bortolon LF (2015) Impactos de Espécies Exóticas em Ninhos de Tartarugas-Verdes no Arquipélago de Fernando de Noronha, PE, como Uso de Armadilhas Fotográficas. In: I Encontro de Pesquisa de Fernando de Noronha, São Pedro e São Paulo e Atol das Rocas. Fernando de Noronha

Pinheiro MÂP, Rodrigues, CGO, Robles RA (2013) '"Tartarugada": uma iniciativa de sensibilização ambiental no Parque Nacional Marinho de Fernando de Noronha (PE)', Rev Bras Ecotur 6 (5):1028–1051. Available at: http://www.sbecotur.org.br/rbecotur/seer/index.php/ecoturismo/article/view/739/607

TAMAR (2004) Relatório Nacional de Utilidade Pública- 2003. Fundação Pró-Tamar, Salvador, Bahia, Brazil

TAMAR (2016) Tamar Available at: http://www.tamar.org.br/base.php?cod=20. Accessed 9 Sept 2016

Teixeira W (2003) Arquipélago Fernando de Noronha: o paraíso do vulcão. Terra Virgem Editora, São Paulo

Tisdell CA (1998) Ecotourism: aspects of its sustainability and its compatibility with conservation, social and other objectives, working papers on economics, ecology and the environment, *Working Paper No. 26* Department of Economics, The University of Queensland, Brisbane

Tripadvisor.com (2017) About trip advisor. Available at: http://www.tripadvisor.com/TravelersChoice-Beaches-cTop-g1. Accessed 3 Apr 2017

Vale TF, Robles RA, Moreira JC (2016) The use of technologies in museums and visitor centers: case study at the TAMAR visitor center in Fernando de Noronha—PE (Brazil). Appl Tourism 1(1):97–112. doi:10.14210/at.v1n1.p97-112

11 Whale and Dolphin Watching, and Visitors' Experiential Responses: A Qualitative Study on Comments in a Travel Forum

Serhat Harman and S. Emre Dilek

Abstract

This chapter will consider the history of whale and dolphin watching as a tourism activity, based on visitors' experiences. The analysis was based on 468 visitors' comments on a familiar travel forum under the four categories developed in Ballantyne, Packer and Sutherland's research in (Tourism Management 32(4):770–779, 2011b). These categories are sensory impressions, emotional affinity, reflective response, and behavioural response. The comments were analysed using a content analysis method. It was found that the behavioural response dimension reported by visitors was lower than the other experience dimensions. The study concluded with some suggestions for both business owners and wildlife tourism researchers.

11.1 Introduction

The watching of animals for recreational purposes has a long history. The Romans, for instance, enjoyed watching animals in a number of different vivaria, including aviaries, fishponds and parks (Kalof 2007; Fennell 2012). The watching of whales and dolphins as a tourism attraction is, however, new. Indeed, Orams (2001) explains the process in which whale hunting has transformed into whale watching. Although whales and dolphins featured in many coastal First Nations' cultures, commercial whaling only began off the west coast of North America in the early 1800s (Ministry of Fisheries and Oceans Canada 1999). In the early 1990s, whale and dolphin watching became popular and was soon a significant part of the tourism industry (Orams and Forestell 1995; Hoyt 2001; Higham et al. 2016). Since then, whale and dolphin watching as a global tourism activity has continued to grow rapidly (Filby et al. 2015; Buultjens et al. 2016). While Hoyt (1995) estimated that the global economic impact of whale and dolphin watching activities in the early 1990s was about US$550 million, by the early 2000s this figure had grown to about US$1 billion (Hoyt 2000). It was reported that while 9 million people from 87 countries were participating in whale watching in 1998, these numbers had risen to over 13 million from 119 countries by 2008, generating US$2.1 billion (IFAW 2009; Chen 2011). Furthermore, 3300 tour operators organise whale and dolphin watching tours and over 13,000 people work in the sector (US Fish and Wildlife Service 2011).

Thus, wildlife watching as a tourism activity has increased year on year, and so have academic studies of the phenomenon around the world. Classifications by academics vary. For example, Cohen's classifications regard tourist contact with whales and dolphins as a non-interactive wildlife watching experience (Cohen 2009). In the description of such experiences, the human–animal relationship setting is "*natural*", the mode of engagement is "*non-interactive*", and the status of the animal is "*wild*". In addition, if it is organised by a tour operator, the mediators are the "*guides*". According to Bullbeck's classification, whale and dolphin watching is "*authentic*", whilst Ballantyne et al. (2007) classify it as "*wildlife tourism*". Moreover, in his book *Tourism and Animal Ethics*, Fennell regards whale and dolphin watching as "*wildlife viewing*". All of these

S. Harman (✉) · S. Emre Dilek
School of Tourism and Hotel Management, Batman University, Campus West Raman, Batman, Turkey
e-mail: harmanserhat@gmail.com

S. Emre Dilek
e-mail: s.emre.d@hotmail.com

I. Borges de Lima and R.J. Green (eds.), *Wildlife Tourism, Environmental Learning and Ethical Encounters*, Geoheritage, Geoparks and Geotourism, DOI 10.1007/978-3-319-55574-4_11

descriptions and classifications are important in order to understand and evaluate animal-based attractions and wildlife watching experiences, such as whale and dolphin viewing. (Higginbottom 2004; US Fish and Wildlife Service 2011), including the 'big five' animal watching experiences (buffalo, elephant, leopard, lion, rhino), as well as gorilla watching, lemur watching, bird watching, whale watching, and dolphin watching. Although these products are important attraction factors for visitors, over time, they have merged with other products (accommodation, transportation, catering, entertainment etc.) and have become one of the most common touristic activities.

Whale watching (in this chapter, this refers to watching cetaceans such as whales, dolphins, and sea porpoises in their natural habitats) is one of the wildlife tourism activities classified under animal-based tourism or wildlife tourism (Fennell 2012; Markwell 2015), and can be defined as any commercial enterprise that allows the public to see cetaceans[1] in their natural habitat (IWC 1994; Hoyt 1995; Parsons et al. 2003a, b). Observing cetaceans in their natural habitat is also about watchers' experiences or visitors' memories (Higham and Carr 2003; Christensen et al. 2009; Ballantyne et al. 2011a, b). For instance, Cloke and Perkins (2005) state that 1.5 million visitors visit Kaikoura Village in New Zealand every year for whale and dolphin watching. In their study, it is stated that visitors to Kaikoura often describe their whale and dolphin watching experiences as a "*special, magical and unforgettable experience*". Swimming with dolphins is also described as being "*felicific, exciting and breathtaking*".

In this chapter, 468 visitors' comments on a popular travel forum are examined using a content analysis method and the four categories developed in Ballantyne, Packer and Sutherland's research in (2011b). These are sensory impressions, emotional affinity, reflective response, and behavioural response.

11.2 Animal-Based Tourism: Literature Review

The global tourism industry continues to grow and spread around the world with every passing year. Both supply and demand in the tourism industry are evolving. As tourists continue to demand more than simply lying on the beach (the same experiences), tourism businesses increasingly supply different tourism products or a variety of tourism attractions to their customers. One of these attractions is animal-based tourism attractions, an important leisure activity in contemporary society (Tribe and Booth 2003; Shani and Pizam 2009).

Animal-based tourism attractions are tourism activities in which animals are used as performers, competitors, indicators or hunters in either captive or non-captive settings (Fennell 2012). Animal-based tourism attractions in captive settings are typically referred to as zoos, aquariums, theme parks, wildlife or wild parks, safari parks, and dolphinariums (Shani and Pizam 2009). Similar attractions in non-captive settings are generally referred to as wildlife watching. These attractions are classified as human–animal engagement by authors in the tourism literature (Shani 2013).

Some authors in the literature differentiate between activities based on the use of the animals. These classifications vary according to the shape of the human–animal interaction in the animal-based tourism activity. For instance, Bullbeck (1999) argues that animal-based tourism experiences can occur in three main types of settings, namely authentic, semi-authentic and staged. On the other hand, Ballantyne et al. (2007) classify the types of settings as first, second, and third generation exhibits, animal shows, feeding programmes, interactive activities, and wildlife tourism. Cohen categorises the settings according to four major themes; the setting of the relationship, the mode of engagement, the status of the animal, and the mediators.

Wildlife is incorporated into the tourism industry in various settings, including both natural and man-made environments (Shani 2013). The classifications mentioned above are in Table 11.1. On the other hand, the different attitudes of humans (visitors) towards animals are also important in order to understand human–animal engagement in tourism. Kellert and Berry's extensive study (1987) provided initial evidence that various animal-based attractions inherently emphasize different wildlife values and thus attract people who espouse diverse attitudes towards wildlife (Fennell 2012; Shani 2013). For instance, while wildlife watchers have moralistic and naturalistic attitudes, camel wrestling enthusiasts or elephant riders have strong domineering and utilitarian attitudes.

As shown in Table 11.1, wildlife watching is one of the classifications of the human–animal relationship in tourism. In addition, wildlife watching can refer to either non-captive (Ballantyne et al. 2007) or fully natural (Cohen 2009) settings with regard to human–animal engagement. Based on the definition of the United Nations Environment Programme (UNEP) and the Convention on the Conservation of Migratory Species of Wild Animals (CMS), wildlife watching tourism is a type of tourism that is organized and undertaken in order to watch or encounter wildlife. It exclusively relates to non-consumptive forms of wildlife-based activities, such as the observing and sometimes touching or feeding of animals, in contrast to

[1]"Cetacean" comes from the Latin "cetus" (whale) and Greek "ketos" (huge-fish) and, particularly in the literature, refers to whales and dolphins (O'Neill et al. 2004; Filby et al. 2015; Higham et al. 2016).

Table 11.1 Classifications of human–animal engagement settings in tourism

Authors	Point of origin	Settings
Shackley (1996)	Animal-based tourism attractions	• Complete confinement • Complete freedom
Orams (1996, 2002)	Tourist-wildlife interactions	• Captive • Semi-captive • Wild
Bullbeck (1999)	Human–animal interaction	• Authentic • Semi-authentic • Staged
Reynolds and Braithwaite (2001)	Animal-based tourism attractions	• General access • Limited access • Restricted access • Contrived access
Beardsworth and Bryman (2001)	Tourist-animal engagements	• Encounter • Representation • Presentation • Qualification
Hall et al. (2002), derived from Hall and Brown (2006)	Human–animal interaction	• Wild creatures for discovery • Tame creatures for interaction • Objects for exhibition • Targets for shooting and fishing • A source of education, training or research • Mythical or symbolic representation • Ancillary roles
Ballantyne et al. (2007)	Human–animal interaction	• Generation 1–2 exhibits • Generation 3 exhibits • Animal shows and feeding programmes • Interactive activities • Non-captive wildlife tourism
Cohen (2009, 2012)	Tourist-animal engagements	• Fully-natural settings • Semi-natural settings • Semi-contrived settings • Fully-contrived settings
Fennell (2012)	Commoditized animal in tourism	• Captives (worker) • Performer • Competitor • Killed for sport • Wildlife watching
Cohen (2014)	Human–animal engagement (in recreational settings)	• Open natural areas • Wildernesses • Game parks • Game farms • Virtual hunting
Webber et al. (2016)	Interactive technology usage and human–animal interaction in the zoo	• Interactive technology appearance and usage in zoos: – affects human–animal encounters in zoos – risks distracting from visitors' encounters with animals – runs counter to expectations of naturalistic zoo landscapes – offers opportunities to enhance important aspects of visitors' experience • Encounters are affected by complex social and organisational forces
Notzke (2014)	Human–animal engagement (especially wild horses)	• Captive • Free/wild

Source The authors

consumptive forms such as hunting and fishing (UNWTO Wildlife Study Report 2014). In this context, wildlife watching is a tourism or leisure activity that includes observing and taking photos of animals in their natural habitat or protected wildlife areas, and even feeding animals.

According to the National Survey of Fishing, Hunting and Wildlife-Associated Recreation by the U.S. Fish and Wildlife Service (2011), there are two types of wildlife watching activities: *away from home* and *around the home.* Away from home activities, include trips or outings of at least 1 mile from the home for the primary purpose of observing, feeding, or photographing fish and wildlife. Fishing, hunting, or scouting trips and trips to zoos, circuses, aquariums, and museums are not considered wildlife-watching activities. Around the home activities include those that are participated within a mile of the home and involves one or more of the following: (1) closely observing or trying to identify birds or other wildlife; (2) photographing wildlife; (3) feeding birds or other wildlife; (4) maintaining natural areas of at least 1/4 acre where a benefit to wildlife is the primary concern; (5) maintaining plants (shrubs, agricultural crops, etc.) where a benefit to wildlife is the primary concern; or (6) visiting parks and natural areas within a mile of the home for the primary purpose of observing, feeding, or photographing wildlife (US Fish and Wildlife Service 2011).

When it comes to categorising wildlife watching as a tourism product, various types of wildlife watching stand out in the tourism industry (Higginbottom 2004; UNWTO Wildlife Study Report 2014) These are the Big Five watching (buffalo, elephant, leopard, lion, rhino), gorilla watching, lemur watching, bird watching, tiger watching, whale watching, dolphin watching, butterfly watching and so on. The popularity of these wildlife watching activities varies across destinations. For example, if a destination has a coastline, it may provide opportunities for marine wildlife watching facilities. Another destination, on the other hand, may have potential for the Big Five watching activities. Briefly, the geographical location, climate, and fauna and flora of a destination have a direct impact on the development of wildlife watching tourism products.

The watching of whales, dolphins and porpoise (cetaceans) in their natural habitats is one of the most popular forms of wildlife tourism (Sun 2014; Buultjens et al. 2016; Guerra and Dawson 2016). There are different definitions for whale watching in the wildlife tourism literature. For example, Hoyt (2001) defines whale watching as a tourism activity with at least some commercial aspect that involves observing by boat, aircraft or from land, or swimming with and/or listening to any of the 83 species of whales, dolphins or porpoises. According to Sun (2014) on the other hand, whale watching refers to any commercial enterprise that allows the public to watch cetaceans in their natural habitats. These definitions focus on whale watching that have a commercial aspect. However, as a recreational activity, whale watching does not always have a commercial aspect. Whale watching should therefore be defined, as it is in Hoyt and Hvenegaard's (2002) study, as watching cetaceans (whales, dolphins and porpoises) in the wild by aircraft, boat or from land and may include swimming with cetaceans.

Historically, humans' interactions and relationships with cetaceans have been mixed. Before the Industrial Revolution, coastal communities because of their oil, which was used for eating and heating (WDC 2016), targeted whales. By the 1980s, after the International Whaling Commission had declared a moratorium on whaling, interactions between humans and whales became less consumptive (Orams 2000; Finkler and Higham 2004). Human-dolphin interactions also have historical roots. In many cultures in geographically diverse locations, dolphins were considered mythological, god-like symbols (Orams 1997). In Icelandic culture, beached whales or dolphins were believed to bring extreme good luck (Einorsson 2009). With increasing concern about environmental problems and animal rights—particularly following the "Save the Whales" movement in the early 1970s (Finkler and Higham 2004)—whale watching began to increase in popularity; the first commercial whale watching boats set sail in 1955 along the Southern Californian coast (Parsons et al. 2003a, b; Constantine and Bejder 2008).

Existing reports on whale watching demonstrate the increasing popularity of this activity. In Hoyt's (2001) study, it was reported that in 1991, only 31 countries were involved in whale watching while 4 million people watched whales in their natural environment. At this point, whale watching generated US$77 million. The number of whale watchers had increased to 9 million by 1998, with 87 countries involved, generating at least US$1 billion. IFAW's report shows that in 2008, 13 million people participated in whale watching, 119 countries were involved in the industry, generating a total expenditure of US$2 billion (O'Connor et al. 2009).

It is argued that the figures in the IFAW report (O'Connor et al. 2009) indicate not only the size of this wildlife tourism sector, but also the diversity of those engaged in whale watching. The diversity of whale watchers can be examined using different variables. For example, in Hoyt and Hvenegaard's (2002) study, it is noted that there are at least three forms of whale watching. These are commercial whale watching (tourists paying whale watching operators for a guided tour in order to see cetaceans), opportunistic whale watching (watching whales in a non-commercial setting), and whale watching for research (non-lethal watching conducted by independent professionals and researchers). The platforms from which watchers observe cetaceans can also be used to differentiate between whale watchers. Finkler and

Higham's (2004) study found that there are significant differences in the experiences of land-based and boat-based whale watchers. Using Finkler and Higham (2004) findings and Hoyt's (2001) definition of whale watching, whale watchers can be grouped as land-based (shore-based), boat-based, and aircraft-based whale watchers. However, it is important to remember that an individual may indulge in many of these activities at different times or even on the same whale watching trip.

Each individual will perceive his or her interactions with whales differently. Thus, their interpretation of the whale watching experience will also be different. It is very important, therefore, to understand what motivates people to watch whales in their natural environment before studying the experience of whale watchers. Motivation is accepted as a driving force that gives value and direction to the travel choice, behaviour, and experience of tourists (Pearce 1988; O'Neill et al. 2004). Whale watching motivations are very diverse. Several studies in the wildlife tourism literature have focused on whale watching motivations (Orams 2000; O'Neill et al. 2004; Filby et al. 2015; Woods-Ballard et al. 2003). Orams (2000) tried to understand whale watching motivations and answer the question "*is whale watching all about getting close to the whales*?" His findings show that whale watching is not simply about getting close to the whales, it is more than this, but it is hard to understand the complex nature of whale watching behaviour.

Parsons et al. (2003a, b) investigated the motivations of whale watchers in Western Scotland. According to their findings, the main whale watching motivational factors that they found were that the whale watchers had always wanted to see whales and dolphins in the wild and they enjoyed wildlife watching trips. O'Neil et al. (2004) argue that dolphins evoke feelings of inspiration and awe as well as a sense of enjoyment and connection with nature. Filby et al. (2015) examined the motivations of tourists who engage in dolphin swimming tours. They found that the main motivations for embarking on dolphin swimming tours are the possibility of seeing a large number of dolphins, getting close to the dolphins, and the opportunity to see dolphins in their natural habitats. Here, the actual interactions with the animals include feeling, touching, and developing an emotional connection with the animals.

The nature of tourist behaviour is complex (Orams 2000), but it is clear that experience is a key concept that must be considered. Curtin (2005) states that lived experience is to the soul what breath is to the body. Experience can be defined in different ways. For example, DeMares (2000) defined experience as an intense and highly valued moment. In the context of tourism, Tung and Ritchie (2011) describe experience as an individual's subjective evaluation (affective, cognitive, and behavioural) and the undergoing of events related to his/her touristic activity, both before (planning and preparation), during (at the destination/during the activity), and after the trip (recollection and memory). Tourism, much like other recreational consumptions, is about purchasing experiences rather than objects (Curtin 2005). In wildlife tourism in particular, activities provide real connections between the tourist and nature or animals. By providing such connections, wildlife tourism can deliver strong and positive educational messages to visitors (Ballantyne et al. 2011b) Understanding tourist experiences is very important for marketers, as a causal relationship has been found between tourist perceptions of the quality of experience, and satisfaction and behavioural loyalty (Kim et al. 2012).

In modern society, most people live in urban areas that are isolated from nature and wildlife (Curtin 2005). There is a growing demand for opportunities to interact, photograph and watch animals in the wild (Curtin 2010). Watching and encountering animals in the wild triggers peak experiences (DeMares 2000), which are undoubtedly the most memorable of experiences.

Whale watching is a sub-section of the wildlife tourism market, and tourist interactions with native wildlife and the natural environment is an integral part of the modern tourist experience (Curtin 2010). In the existing literature, a number of studies can be found that focus on tourist experiences of whale, dolphin and porpoise watching. DeMares (2000) studied the experiences of visitors that had encountered cetaceans. He conducted qualitative research with six visitors that had encountered a whale or dolphin in the past. His findings indicate that participants' peak experiences are triggered by encounters with cetaceans, and are shaped by a sense feeling originated by connection with cetacean, a sense of personal connection, aliveness (a high level of excitement), connectedness (a sense of destiny), and harmony (with the environment).

Ballantyne et al. (2011a) work also provides us with evidence of whale watchers' experiences. Out of the 1286 participants in their research, 304 of them were whale watchers. The findings reveal that the experiences of wildlife tourists, including whale watchers, can be viewed under two headings; experiential excitement (excitement at seeing wildlife, having a good view of the animals etc.) and reflective engagement (emotional connection with the animal, reflecting upon new ideas with companions, and the cognitive/affective processing of the experience).

Higham and Carr (2003) conducted research that analysed wildlife tourists' experiences. A qualitative method was employed in their study in order to understand visitor experiences. The sample consisted of wildlife tourism operators, and found that there are two main dimensions to the wildlife tourism experience; the social dimension and the ecological dimension. Higham and Carr (2003) argue that visitor experiences of wildlife tourism play a critical role in the sustainability of wildlife tourism. Valentine et al. (2004)

examined visitor interactions with whales along the Great Barrier Reef. They found that proximity to the whales during the tour influences the intensity of the visitors' experience. However, we should keep in mind Orams' (2000) claim that "it is not all about getting close to the whales".

Curtin (2010) conducted research on wildlife tourists to answer the question "*What makes for memorable wildlife encounters?*" According to Curtin, animals are anthropomorphic attractions and interactions between humans and animals have always been a part of human existence. Some of the participants in Curtin's (2010) study were whale watchers. A number of factors shaped the participants' memories of their wildlife tourism experiences. These include memories in the making (wildlife moments, witnessing something that makes the visitor excited), the charisma and appeal of the animals, the kill drama of nature, seeing large numbers of animals, close proximity/eye-to-eye contact, and the embodiment and immensity of nature.

Ballantyne et al. (2011b) also examined wildlife tourists' memories of their experiences. They conducted qualitative research of 240 visitors, asking open-ended questions. Some of their participants were whale watchers. Their findings indicate that wildlife tourism experiences can be grouped under the following headings: sensory impressions, emotional affinity, reflective responses, and behavioural responses. Sensory impression refers to the visitor's perception of vivid visual and auditory factors that shape the memory, while emotional affinity refers to their perception of an emotional connection with the animal being observed. The reflective response is the cognitive process of the experience, what they have seen or heard, and their reflections upon the experience. Behavioural response refers to the specific actions that are taken after the wildlife tourism experience, or longer term behavioural changes that the visitor makes as a consequence of their wildlife tourism experience. Examples of this may include participating in whale and dolphin conservation campaigns, joining wildlife related non-governmental organisations, or choosing to no longer eat or use whale or dolphin-related products.

The studies discussed above demonstrate that the experience of whale watching is multi-faceted. It has cognitive and affective facets that can have behavioural impacts. It is also clear that employing only quantitative research methods to an examination of the experience is not enough to understand the complex nature of the experience.

11.3 Analysis of Travel Forums: Content Analysis as Methodology

The research for this chapter was designed based on qualitative methods. Content analysis was employed in order to collect data on the experiences of whale watchers. Content analysis is defined as, "describing, with optimum objectivity, precision, and generality, what is said on a given subject in a given place at a given time" (Laswell et al. 1952; derived from Stepchenkova 2012). It is used to systematically evaluate the actual and symbolic content of all forms of recorded communication (Hall and Valentin 2005). Data for content analysis can be gained from different sources, including textbooks, book chapters, journal articles, commercial publications, publications about companies and destinations, press releases, and company documents (Altınay and Paraskevas 2008). Content analysis can also be usefully undertaken on the World Wide Web, which, alongside being a tool for undertaking content analysis, can also be the subject of the research (Hall and Valentine 2005). Textual content generated by tourists on the World Wide Web have been the focus of some tourism studies (for example, see Yagi 2001; Chen et al. 2001).

Travel forums such as Trip Advisor and Lonely Planet give tourists the opportunity to share their travel experiences. A travel forum was selected as the data source for this study. To limit the data being collected, we selected the whale watching company that had received the most comments, in the most popular region (USA) according to the IFAW report (2009). Furthermore, only comments made between 2012 and 2016 were included in the sample frame. Consequently, 467 visitors' comments on this travel forum were examined under the four categories of experience presented in Ballantyne et al. (2011b), namely sensory impressions, emotional affinity, reflective response, and behavioural response. To ensure the reliability of the qualitative data, inter-coder consensus was reached. Three independent coders, who are leading experts on sociology and tourism, were employed, and there were only three comments for which the three independent coders did not arrive at a consensus. This indicates that the reliability of the data is high.

11.4 Findings

11.4.1 Travel Companions of Visitors

The data collected amounts to 127 pages and 41,770 words. The content analysis began with an examination of the whale watchers' comments, to determine whether they were related to the whale watching experience. It was found that five of the comments did not include any words or sentences related to their whale watching experience; these comments were excluded from the content analysis. Content analysis was implemented on the remaining 462 comments on whale watching experiences. Table 11.2 shows the distribution of the whale watching travel companions of the commenters. According to Table 11.2, most of the whale watchers participated in this activity with their partner or family. It was

Table 11.2 Distribution of travel companions of whale watchers (N: 462)

Travel companion	Frequency	Percentage (%)
Wife/Boy/Girl Friend/Partner	195	42.2
Family with kids	135	29.2
Single	68	14.7
Friends	64	13.8

Source The authors

also found that nearly 15% of the commenters went whale watching alone. The high percentage (around 85%) of those that participated in the activity with someone else indicates that whale watching has a social facilitation dimension, as argued by Ballantyne et al. (2011a, b).

11.4.2 Categorizing the Whale Watching Experiences of Visitors

As mentioned above, whale watchers' comments were analysed according to Ballantyne et al.'s (2011a, b) wildlife tourism experience framework. Table 11.3 presents the distribution of the comments under the headings "sensory impressions", "emotional affinity", "reflective response" and "behavioural response". It was found that the most observed experience dimension is sensory impression (99%), followed by reflective response (60%), emotional affinity (58.4%) and behavioural response (5%). The findings therefore indicate that whale watchers perceive sensory stimuli whilst whale watching, which move them to think (cognitive processing). They then attach themselves emotionally to the cetaceans and, depending on the intensity of their emotional connection, exhibit some behavioural responses. These four experience dimensions will be illustrated below using some of the original comments of the whale watchers (Table 11.3).

11.4.3 Sensory Impressions

Sensory impression refers to whale watchers' vivid visual, auditory, olfactory and tactile memories of their whale watching experiences. The whale watchers often reported making eye contact with the whales, and sometimes reported seeing other sea animals such as sea birds:

Comment 4: "We had a great experience, saw finback minke, and humpback whales."
Comment 5: "We also saw some dolphins and birds that were on the endangered list."
Comment 17: "We ended up with multiple sightings of fin and humpback whales that put on quite the show for us."
Comment 34: "Saw so many whales that I quit counting!"
Comment 137: "Great whale watching and don't forget about birding... saw some great birds on our way out to Whale watch."

Sea animals are not the only aspects that contribute to the shaping of the whale watching experience; the marine environment also plays a role in the sensory impression dimension of the whale watching experience:

Comment 329: "I booked this trip as my 5-year-old is beyond obsessed with whales and marine life."
Comment 347: "We were blessed with excellent weather conditions clear blue sky and virtually millpond water."
Comment 418: "The weather was beautiful (80F/25C) and the ocean very calm."

Being physically close to the cetaceans is also a part of the sensory impression dimension of the experience. Some whale watchers pointed this out in their comments:

Comment 444: "We saw them blowing bubble rings and eating, close to the boat, which was awesome to see."
Comment 451: "In fact, sometimes the whales seemed to make an effort to get closer to provide us with some REAL close-ups!!!"
Comment 453: "We saw everything from feeding to flipper communication, jumping and mother baby sleeping and nursing, in total about 7 individual humpbacks and real close up."

Table 11.3 Distribution of whale watchers comments under Ballantyne et al.'s categories

Experience category	Frequency	Percentage (%)[a]
Sensory impression	459	99.3
Reflective response	279	60.4
Emotional affinity	270	58.4
Behavioural response	24	5.2

Source The authors
[a]Due to the multiple categorizing of one comment, the total percentage exceeds 100%

11.4.4 Reflective Response

Reflective response refers to the cognitive processing of the whale watching experience; their reflections upon what they have seen and heard. Reflective responses do not always depend on the interpretive content of communication or signage, but are sometimes aroused as natural extensions of sensory and emotive experiences, or are facilitated by social interactions (Ballantyne et al. 2011a, b).

Reflective responses can include visitors' ideas based on interpretive commentary, concern and respect not only for the individual animal but for the species as a whole, awareness of their endangered status, feelings of empathy and protectiveness, and a reinforcement of environmental awareness. These can all lead to changes in behaviour, and discussions with companions and staff interpreters or volunteer guides:

Comment 317: "Protect the whales… do no harm."
Comment 316: "Safety on board the vessel was attended to very well and the safety of the whales and other creatures of the sea were also a priority."
Comment 304: "I just don't agree with that. I am a vegan for animal rights reason, so this issue is very important to me. Yes, they are making money, to pay the scientists and crew, to maintain and power the boat and to fund research. I did NOT see a lack of concern on the part of the staff. The captain kept the boat still when whales were near. And the whales were never THAT near."
Comment 305: "So not only do they get too close to them, they go out multiple times a day which disturbs them and on top of that they pollute the area where the whales are."
Comment 133: "The naturalist aboard was very knowledgeable and accommodating for questions and the whale spotter and captain got us in close to several pods of whales over the Stillwagen Bank. We have been on whale watching excursions in Alaska and San Diego and this was BY FAR the best experience for seeing these magnificent creatures."
Comment 144: "The information provided about whales and other marine life has more significance when you are seeing the whales' close up in their environment."
Comment 258: "We also learned tons of stuff about whales and feel a lot closer to the animal!"
Comment 343: "We were thrilled to give our whale watching money to a business that does not exploit the animals, but rather uses the opportunity to further Conservation Science while making a living doing what they love. I was especially pleased by their connection to the whale sense program origination."
Comment 368: "There is a naturalist on board the vessel who talks about the different whales: what they eat, where they live and travel, how they reproduce, et cetera. Even when there aren't whales to see immediately, they fill in the lulls with interesting information."

11.4.5 Emotional Affinity

Emotional affinity refers to the whale watchers' emotional connectedness with the cetaceans. These emotions are sometimes the strongest aspect of the whale watching experience. It must also be said that emotional affinity is highly related to visitors' sensory impressions:

Comment 455: "We must have seen 30–40 whales, two of which swam directly under our boat. These animals are more majestic than you can ever imagine."
Comment 423: "Our experience was phenomenal, but it does depend heavily on feeding patterns and water temps."
Comment 93: "We saw seals and lots of whales… breath-taking."
Comment 252: "Whale watching is an absolutely unforgettable, breath-taking, magic experience!"
Comment 51: "This was my first whale watching excursion. I thought we'd be lucky to see one whale. OMG!!! We saw so many I lost track. What a fabulous adventure."
Comment 110: "It really is a worthwhile experience to see these amazing creatures in the wild."

Whale watchers may also attribute human characteristics to the whales, which can make their emotional connection even stronger:

Comment 214: "Seeing a Finback is absolutely torture for the viewer and the whale. It feels more like a Whale chase than a Whale watch… and I always feel a bit bad for the poor whale…"
Comment 149: "We saw at least 8 whales including a mother and baby who played alongside the ship."
Comment 247: "This included a mother/calf combination sleeping on the surface. Never realized how cool it is to hear a whale breath!"
Comment 305: "According to the announcer there was a mother and calf who was nursing and we also got too close next to them."

Sensory impression and emotional affinity are highly related categories. Emotionally connecting with the animals firstly requires the visitor to experience sensory impressions. For this reason, some comments may be examined under both sensory impression and emotional affinity categories, for example, comments 444 and 305.

11.4.6 Behavioural Response

After having been on a whale watching trip and experience lived during whale watching trip, whale watchers may make changes in their lives or take certain actions. Behavioural response refers to the taking of specific actions or heightened awareness of the need for such actions (Ballantyne et al. 2011a, b). These can include changes in household practises or purchasing practices, taking responsibility for the environment in their everyday lives, seeking further information, discussing environmental issues, volunteering on environmental issues, or returning to watch whales etc.:

Comment 4: "The biologist gave a great educational lesson during the event. We would definitely do that again!"
Comment 253: "By joining them on a trip you are supporting the conservation and data collection of the whales around the Cape Cod area."
Comment 282: "I kept wishing we were back already so I could explore more."
Comment 304: "I am a vegan for animal rights reason, so this issue is very important to me. Yes, they are making money, to pay the scientists and crew, to maintain and power the boat and to fund research. I did NOT see a lack of concern on the part of the staff."
Comment 305: "So not only do they get too close to them, they go out multiple times a day which disturbs them and on top of that they pollute the area where the whales are. I am reporting them as a scam to the NOAA, and the WDC. They need to be held accountable for their actions!!"
Comment 418: "It is a trip I would definitely do again; hoping to see more whales next time."

Although behavioural responses may include changes in participants' daily lives, such as supporting a conservation project or only buying dolphin-friendly fish products, such comments were not found in this sample. Our content analysis included only 467 visitors' comments between 2012 and 2016; if the sample frame was larger, it is likely that other interesting comments related to the behavioural response facet would be found.

11.5 Discussion and Conclusion

Whale watching is one of the most popular forms of animal-based marine wildlife tourism. A report by IFAW (2009) on whale watching tourism indicates that 13 million people participated in whale watching in 2009. The report also shows that North America is the most preferred location for whale watching; 6,256,277 people watched whales in their natural habitat in North America that year (O'Connor et al. 2009).

Whale watching becomes a commercial activity when visitors pay a whale watching company for a guided tour in order to see the cetaceans (Hoyt and Hvenegaard 2002). Ensuring the sustainability of the environment and the business is vital for all commercial activities. Commercial activities should therefore be managed and marketed professionally. Broadly speaking, marketing refers to the meeting of customers' needs and wants with marketing mix. In order to do this, business managers must first understand the complex nature of consumer behaviour. Consumer motivation and experience are key concepts that are part of consumer behaviour. It has been shown that there is a causal relationship between consumer experience and satisfaction and loyalty. Thus, if a business wants to succeed in the competitive market environment, it must understand consumer experiences and monitor the factors that shape those experiences.

Whale watching generated US$2.1 billion in 2008 (O'Connor et al. 2009), and when it is remembered that the average annual economic growth is US$0.4 billion (Cisneros-Montemayor et al. 2010; New et al. 2015), it can be shown that the average expenditure of the whale watching industry in 2015 was US$5 billion. Examining the experiences of whale watchers is therefore crucial in terms of the marketing efforts of whale watching companies and the wildlife tourism literature. In this study, we examined the whale watching experiences of whale watchers that participated in whale watching tours in North America. We selected North America, and more specifically, the region where whale watchers had shared more of their whale watching experiences; Princeton, MA. We examined whale watchers' experiences of whale watching by conducting content analysis of their comments in a travel forum, under four headings. The headings were derived from Ballantyne et al.'s (2011a, b) study. They are sensory impressions, emotional affinity, reflective response, and behavioural response.

The content analysis revealed that the most observed experience heading is sensory impression, which deals with visitors' vivid visual, auditory, olfactory or tactile memories. The second most observed experience dimension is reflective response, which refers to visitors' cognitive processes that are stimulated by his/her sensory impressions. The third experience dimension is emotional affinity, which means visitors' emotional connection with the animal being observed. The last experience dimension, behavioural response, was observed the least. This is the most interesting finding of the study, and is in line with Ballantyne et al.'s (2011a, b) findings. In their study, only 7% of respondents reported behavioural changes following their participation in wildlife tourism. In our study, similarly, only 5% of the 462 whale watchers reported a behavioural change. Thus, we can say that the whale watching experience sensually impresses visitors, moves them to think about whales and their natural environment, and allows them to connect emotionally with the animals in question, but that their whale watching is not so intense and impactful that it forces them to change their behaviour. Behavioural changes in participants' daily lives after having a whale watching experience is, although interesting, difficult to observe. Participants in such research would need to be engaged for longer, for example through a longitudinal study.

The findings of this study may be of use to whale watching companies. In particular, they may wish to use sensory and emotive messages about whales in their promotional campaigns. In addition, the reflective findings of the study suggest that companies should concentrate on the quality of the educational programs that they provide. Indeed, whale watching can act as a platform from which whale watching companies (commercial tour operators) can educate whale watchers about the long-term sustainability of whale watching (Wearing et al. 2014; Argüelles et al. 2016). In this way, visitors' attitudes towards cetaceans and their natural habitats may be changed by the educational aspects of whale watching. This study emphasises the importance of the educational dimension of the whale watching experience with reference to whale watchers' comments. Whale watching tours are therefore not merely "entertainment" but are in fact "edutainment" (Pratt and Suntikul 2016).

Our study raises other research questions that may be answered by further research. For instance, why is the behavioural response dimension of whale watching lower than other experience dimensions? What factors contribute to this situation? This study focused on visitors' comments in the North American setting; there may be differences among visitor experience dimensions in other whale watching locations, which could be explored by further research.

References

Altinay L, Paraskevas A (2008) Planning research in hospitality and tourism. Elsevier, Burlington, MA

Argüelles MB, Coscarella M, Fazia A, Bertellotti M (2016) Impact of whale watching on the short-term behaviour of Southern right whales (*Eubalaena australis*) in Patagonia, Argentina. Tourism Manage Perspect 18:118–124

Ballantyne R, Packer J, Hughes K, Dierking L (2007) Conservation learning in wildlife tourism settings: lessons from research in zoos and aquariums. Environ Educ Res 13(3):367–383

Ballantyne R, Packer J, Falk J (2011a) Visitors' learning for environmental sustainability: testing short-and long-term impacts of wildlife tourism experiences using structural equation modelling. Tourism Manage 32(6):1243–1252

Ballantyne R, Packer J, Sutherland LA (2011b) Visitors' memories of wildlife tourism: implications for design of powerful interpretive experiences. Tourism Manage 32(4):770–779

Beardsworth A, Bryman A (2001) The wild animal in late modernity: the case of the disneyization of zoos. Tourism Stud 1(1):83–104

Bullbeck C (1999) The 'nature dispositions' of visitors to animal encounter sites in Australia and New Zealand. J Soc 35(2):129–148

Buultjens J, Ratnayke I, Gnanapala A (2016) Whale watching in Sri Lanka: perceptions of sustainability. Tourism Manage Perspect 18:125–133

Chen CL (2011) From catching to watching: moving towards quality assurance of whale/dolphin watching tourism in Taiwan. Mar Policy 35(1):10–17

Chen JS, Kerstetter DL, Graefe AR (2001) Tourists' reasons for visiting industrial heritage sites. J Hosp Leisure Mark 8(1/2):19–32

Christensen A, Needham MD, Rowe S (2009) Whale watchers' past experience, value orientations, and awareness of consequences of actions on the marine environment. Tourism Mar Environ 5(4): 271–285

Cisneros-Montemayor AM, Sumaila UR, Kaschner K, Pauly D (2010) The global potential for whale watching. Mar Policy 34:1273–1278

Cloke P, Perkins HC (2005) Cetacean performance and tourism in Kaikoura, New Zealand. Environ Plann D: Soc Space 23(6): 903–924

Cohen E (2009) The wild and the humanized: animals in Thai tourism. Anatolia 20(1):100–118

Cohen E (2012) Tiger tourism: from shooting to petting. Tourism Recreat Res 37(3):193–204

Cohen E (2014) Recreational hunting: ethics, experiences and commoditization. Tourism Recreat Res 39(1):3–17

Constantine R, Bejder L (2008) Managing the whale- and dolphin-watching industry: time for a paradigm shift. In: Higham JE, Lück M (eds) Marine wildlife and tourism management: insights from the natural and social sciences. CABI, London, pp 321–334

Curtin S (2005) Nature, wild animals and tourism: an experiential view. J Ecotourism 4(1):1–15

Curtin S (2010) What makes for memorable wildlife encounters? Revelations from 'serious' wildlife tourists. J Ecotour 9(2):149–168

DeMares R (2000) Human peak experience triggered by encounters with cetaceans. Anthrozoos 13(2):89–103

Einarsson N (2009) From good to eat yo good to watch: whale watching, adaptation and change in Icelandic fishing communities. Polar Res 28(1):129–138

Fennell DA (2012) Tourism and animal ethics. Routledge, New York

Filby NE, Stockin KA, Scarpaci C (2015) Social science as a vehicle to improve dolphin-swim tour operation compliance? Mar Policy 51:40–47

Finkler W, Higham J (2004) The human dimensions of whale watching: an analysis based on viewing platforms. Hum Dimensions Wildl 9(2):103–117

Guerra M, Dawson SM (2016) Boat-based tourism and bottlenose dolphins in Doubtful Sound, New Zealand: the role of management in decreasing dolphin-boat interactions. Tourism Manage 57:3–9

Hall CM, Valentin A (2005) Content analysis. In: Ritchie B, Burns P, Palmer C (eds) Tourism research methods. CABI, Oxfordshire, pp 191–209

Hall DR, Brown F (2006) Tourism and welfare: ethics, responsibility and sustained well-being. CABI, Oxfordshire

Higginbottom K (2004) Wildlife tourism: an introduction. In: Higginbottom K (ed) Wildlife tourism: impacts, management and planning. Common Ground Publishing in Assocation with the CRC for Sustainable Tourism, Gold Coast Australia, pp 1–14

Higham JE, Carr AM (2003) Sustainable wildlife tourism in New Zealand: an analysis of visitor experiences. Hum Dimensions Wildl 8(1):25–36

Hoyt E (1995) The worldwide value and extent of whale watching 1995. Whale and Dolphin Conservation Society, Bath, UK

Hoyt E (2000) Whale watching 2000: worldwide tourism numbers, expenditures and expanding socioeconomic benefits. International Fund for Animal Welfare, Crowborough, UK

Hoyt E (2001) Whale watching 2001: world tourism numbers, expenditures, and expanding socioeconomic benefits. International Fund for Animal Welfare, Yarmouth, MA

Hoyt E, Hvenegaard GT (2002) A rewiev of whale-watching and whaling with applications for the Caribbean. Coast Manage 30(4):381–399

International Fund for Animal Welfare (2009) Whale watching worldwide: tourism numbers, expenditures and expanding economic benefits. http://www.ifaw.org/sites/default/files/whale_watching_worldwide.pdf. Accessed 19 June 2016

International Whaling Commission (1994) Forty-fifth report of the International Whaling Commission. IWC, Cambridge

Higham JES, Bejder L, Allen SJ, Corkeron PJ, Lusseau D (2016) Managing whale-watching as a non-lethal consumptive activity. J Sustain Tourism 24(1):73–90

Kalof L (2007) Looking at animals in human history. Reaktion Books, London

Kellert SR, Berry JK (1987) Attitudes, knowledge, and behaviors toward wildlife as affected by gender. Wildl Soc Bull 15(3): 363–371

Kim JH, Ritchie JRB, McCormick B (2012) Development of a scale to measure memorable tourism experience. J Travel Res 51(1):12–25

Markwell K (ed) (2015) Animals and tourism: understanding diverse relationships. Aspects of tourism, vol 67. Channel View Publications, Bristol

Minister of Fisheries and Oceans Canada (1999) Whales, dolphins & porpoises of British Columbia, Canada. www.dfo-mpo.gc.ca/Library/241969.pdf. Accessed 18 June 2016

New LF, Hall AJ, Harcourt R, Kaufman G, Parsons ECM, Pearson HC, Cosentino AM, Schick RS (2015) The modelling and assessment of whale-watching impacts. Ocean Coast Manage 115:10–16

Notzke C (2014) Wild horse-based tourism as wildlife tourism: the wild horse as the other. Curr Issues Tourism 19(12):1235–1259

O'Connor S, Campell R, Cortez H, Knowles T (2009) Whale watching worldwide: tourism numbers, expenditures and expanding economic benefits. International Fund for Animal Welfare, Yarmouth, MA

O'Neill F, Barnard S, Lee D (2004) Best practice and interpretation in tourist/wildlife encounters: a wild dolphin swim tour example. Wildlife tourism research report, No 25. Sustainable Tourism Cooperative Research Centre, Gold Coast, Queensland

Orams MB (1996) Using interpretation to manage nature-based tourism. J Sustain Tourism 4(2):81–94

Orams MB (1997) The effectiveness of environmental education: can we turn tourists into 'greenies'? Prog Tourism Hosp Res 3(4):295–306

Orams MB (2000) Tourist getting close to whales, is it what whale-watching is all about? Tourism Manage 21(6):561–569

Orams MB (2001) From whale hunting to whale watching in Tonga: a sustainable future? J Sustain Tourism 9(2):128–146

Orams MB (2002) Feeding wildlife as a tourism attraction: a review of the issues and impacts. Tourism Manage 23(3):281–293

Orams MB, Forestell PH (1995) From whale harvesting to whale watching: Tangalooma 30 years on. Recent Adv Mar Sci Technol 94:667–673

Parsons ECM, Warburton CA, Woods-Ballard A, Hughes A, Johnston P (2003a) The value of conserving whales: the impacts of cetacean-related tourism on the economy of rural West Scotland. Aquat Conserv 13(5):397–415

Parsons ECM, Warburton CA, Woods-Ballard A, Hughes A, Johnston P, Bates H, Lück M (2003b) Whale-watching tourists in West Scotland. J Ecotourism 2(2):93–113

Pearce PL (1988) The ulysses factor: evaluating visitors in tourist settings. Springer-Verlag, New York

Pratt S, Suntikul W (2016) Can marine wildlife tourism provide an 'edutaining' experience? J Travel Tourism Mark 33(6):867–884

Reynolds PC, Braithwaite D (2001) Towards a conceptual framework for wildlife tourism. Tourism Manage 22(1):31–42

Shackley M (1996) Wildlife tourism. International Thomson Business Press, London

Shani A (2013) Differentiating settings of tourist-animal interactions: an anthrozoological perspective. Tourism Recreat Res 38(1): 104–107

Shani A, Pizam A (2009) Tourists' attitudes toward the use of animals in tourist attractions. Tourism Anal 14(1):85–101

Stepchenkova S (2012) Content analysis. In: Dwyer L, Gill A, Seetaram N (eds) Handbook of research methods in tourism: quantitative and qualitative approaches. Edward Elgar Publishing, UK, pp 443–457

Sun X (2014) A content analysis of ecotourism attributes of New Zealand whale and dolphin watching operators' presence on the Internet. Unpublished master thesis of International Tourism Management, School of Hospitality and Tourism, Auckland University, New Zealand

Tribe A, Booth R (2003) Assessing the role of zoos in wildlife conservation. Hum Dimensions Wildl 8(1):65–74

Tung VWS, Ritchie JRB (2011) Exploring the essence of memorable tourism experiences. Ann Tourism Res 38(4):1367–1386

United Nations World Tourism Organization (2014) Towards measuring the economic value of wildlife watching tourism in Africa. Briefing paper. http://sdt.unwto.org/content/unwto-briefing-paper-towards-measuring-economic-value-wildlife-watching-tourism-africa. Accessed 14 June 2016

US Fish and Wild Service (2011) National survey of fishing, hunting and wildlife-associated recreation. https://www.census.gov/prod/2012pubs/fhw11-nat.pdf. Accessed 18 June 2016

Valentine PS, Birtles A, Curnock M, Arnold P, Dunstan A (2004) Getting closer to whales-passenger expectations and experiences and the management of swim with dwarf minke whale interactions in the Great Barrier Reef. Tourism Manage 25(6):647–655

Wearing SL, Cunningham PA, Schweinsberg S, Jobberns C (2014) Whale watching as ecotourism: how sustainable is it? Cosmopolitan Civil Soc J 6(1):38–55

Webber S, Carter M, Smith W, Vetere F (2016) Interactive technology and human–animal encounters at the zoo. Int J Hum-Comput Stud (in press)

Whale and Dolphin Conservation (2016) Whaling. http://us.whales.org/wdc-in-action/whaling. Accessed 13 June 2016

Woods-Ballard AJ, Parsons ECM, Hughes AJ, Velander KA, Ladle RJ, Warburton CA (2003) The sustainability of whale-watching in Scotland. J Sustain Tourism 11(1):40–55

Yagi C (2001) How tourists see other tourists: analysis of online travelogues. J Tourism Stud 12(2):22–31

Major Wildlife Attractions and Choices in Turkey and the Tourists' Reports about Their Experiences and Learning in this Tourism Niche

12

Reyhan Arslan Ayazlar

Abstract

Contemporary life has caused human to become estranged from nature and in the present day we observe people's efforts to return to nature. These efforts have generated changes in the tourism industry and alternative holidays are planned for tourists who seek unique, authentic experiences which require a more active participation. At this point wildlife tourism is capable of responding to the requirements and needs of tourists as well as the industry. Wildlife interactions occur in a wide range of settings worldwide. Tourists can observe, feed, touch and/or photograph the wild animals and their experiences have been getting more important to develop this area. Although there are various investigations in countries such as Nepal, Indonesia, Uganda, Kenya, Costa Rica, there is little research about wildlife tourism in Turkey. Therefore, this chapter presents information about wildlife tourism, its attractions and visitors' motivations and experiences in Turkey. 'Wildlife tourism' in Turkey is not named segmentedly like that, and it is instead mentioned as being part of hunting tourism and ecotourism in the country. Thus, there aren't special strategies or plans for developing this tourism type in the country. It hasn't also been considered and attempt to enhance visitors' learning experiences. However, zoo visitors' learning experiences at zoo have been reported in this study. Therefore academic research about wildlife tourism and visitors' experiences should be increased in the country. Clarification of wildlife tourism, determination of main attractions and marketing as a tourism product must be extended in Turkey.

12.1 Introduction

The demands of current tourists have been changed and the natural experiences have been popular in our day. In this context the demand for being close or interacting with wild and/or semi-captive wild animals in particular has increased (Reynolds and Braitwaite 2001). Observing animals in their habitats, feeding or touching them is defined as an inspirational experience (Karis et al. 2013). On the other side, many countries rich in biodiversity but poor economically have been promoting tourism as a conservation tool in their protected areas (He et al. 2008). Wildlife tourism plays an important role meeting both tourists' and countries' needs.

The demands for wildlife tourism activities and its revenue has increased substantially. In 2008, Europe generated a revenue of 97 million US$ with whale watching events. During the same year the Azores, an autonomous region of Portugal which is a significant whale watching center contributed 23% to tourism revenue (O'Connor et al. 2009). In 2011 approximately 7000 divers visited the Azores to dive with blue sharks and in 2012 approximately 364,000 tourists came to watch whales (Bentz et al. 2016). Tourism contributed 3.3% to the GDP of Nigeria in 2011 (Adefalu et al. 2015). It can be said that wildlife tourism has been developing in Turkey. However, there is no research specially giving information about wildlife tourism in Turkey. On the

R.A. Ayazlar (✉)
Tourism Faculty, Muğla Sıtkı Koçman University, Campus Kötekli, 48000 Muğla, Turkey
e-mail: reyhanayazlar@gmail.com

I. Borges de Lima and R.J. Green (eds.), *Wildlife Tourism, Environmental Learning and Ethical Encounters*,
Geoheritage, Geoparks and Geotourism, DOI 10.1007/978-3-319-55574-4_12

other side, the understanding of this tourism type has become more complex with the larger spectrum of information sources. Because there is little study investigating Turkey's wildlife tourism context, this chapter focuses on three specific research objectives as following:

1. to examine the wildlife tourism situation in Turkey.
2. to identify the wildlife tourism attractions in the country.
3. to identify visitor motivations for interacting with the wildlife in Turkey and their learning experiences.

12.2 Wildlife in Turkey

Wildlife tourism is comprised of a wide range of setups which are in either natural or manmade environments. Animals may be free or captive in these setups and they are also separated into different setups for species such as endangered, dangerous and rare ones (Higginbottom 2004; Ballantyne et al. 2011; Cong et al. 2014). The activities carried out within the context of wildlife tourism are consumptive and non-consumptive. Activities such as watching wild animals, photographing and feeding them are non-consumptive activities whereas hunting and fishing are described as consumptive activities (Higginbottom 2004; Adefalu et al. 2015). Participation of the visitors in the relevant activities may be passive or active. Visitors can engage with wild animals or watch them from a certain distance. The variety of these experiences ensure an increase in the number of world destinations (Cong et al. 2014).

Wildlife tourism has been categorized in several studies. Table 12.1 shows one of these categorizations created by Reynolds and Braithwaite (2001). According to the table, some wildlife tourism activities require more experienced participants whereas others don't need to be experienced.

The area of Turkey within the crossroads of the continents of Asia, Europe and Africa is 785,345 km^2 and it is surrounded by seas on three sides. Three percent of the mentioned area are located on the European continent and is known as Thrace while 97% is in Asia and is named Anatolia. Twenty-seven percent of Turkey is comprised of forest areas. There are three different bio-geographical zones hosting their own endemic species and natural ecosystems: Eastern Black Sea Mountain Forests, Central and Eastern Steppe Prairies and the Mediterranean Region. Turkey has a rich biodiversity due to its geological and morphological structure as well as its geographical position. This diversity is also valid for hunting and wildlife. Over 80% of the plants and animal species of Europe are available in Turkey (Kantarlı 2013).

The first thing that comes to the mind about wildlife tourism is the watching, feeding or photographing of large predator species, although wildlife tourism is a type of tourism which covers many endangered species regardless if they are large or small. Countries which organize wildlife tourism activities based on large predator species are known for such events whereas countries with a different kind of rich potential remain in the background. In this context Turkey hosts various and special species which are specific to this geographical area. There are bears, wolves, wild boar, red deer, fallow deer, roe deer, antelope, mountain gazelle,

Table 12.1 Wildlife tourism types

Wildlife tourism activity	Content
Nature-based tourism with wildlife component	Accordingly, many nature based tourism products contain incidental wildlife components
Locations with good wildlife opportunities	Some accommodation facilities may be established close to areas with a rich wildlife. Such facilities have the potential to establish feeding wildlife and other attractions
Artificial attractions based on wildlife	Some species are dependent on manmade protection areas. Some attractions near such areas may have harmful impacts on animals
Specialist animal watching	Tours can be organized for some species or groups of species. Bird watching is a good example of this
Habitat specific tours	Some tours involve entering wildlife habitats and such tours need special vehicles to continue
Thrill-offering tours	They are comprised of tours into habitats accommodating dangerous species in nature
Hunting/Fishing tours	Such tours are organized in natural habitats, semi-consumptive or farm conditions. During such tours animals may be killed or released into nature

Source Reynolds and Braithwaite (2001: 33)

chamois, wild goats, wild sheep and special species such as Anatolian Wild Sheep, Caracal and Caretta Caretta turtles. Some of these wild animals in the country can be hunted that is permitted by the Ministry of Forestry and Water Affairs. Furthermore the country has a significant potential for water activities such as fishing, diving, observing aquatic animals as it is surrounded by seas on three sides. At the same time the observation of wildlife in terms of tourism as well as photographing events are carried out.

12.2.1 Wildlife Attractions in Turkey

Due to its climate and topography Turkey has a rich biodiversity. The rich flora and fauna is comprised of more than 11,000 species of plants, 162 varieties of mammals, 460 types of birds, 716 different fish and 141 reptile species. The area of Turkey corresponds to 0.1% of the World's area and 2.9% of the fish and mammals in the world are available in the country. Furthermore two of the main migration routes (Fig. 12.1) used by millions of migrating birds which are situated between the West Palearctic-Africa transit through Turkey (Kantarlı 2013).

There are numerous sweet and cold water resources in the forests in Turkey hosting local trout species and presenting opportunities for amateur angling activities (Kantarlı 2013). The wetlands in Turkey provide a home for many local and migrant bird species. Together with the dam lakes the total wetlands in the country covers an area of 1,851,000 ha including natural lake surface areas. These areas provide food, habitats and shelter to many aquatic birds and other aquatic species. There are around 3000 wetlands in the country out of which 135 have international significance.

It is a nesting and breeding area for many endangered species on a global scale. For example, the Dalmatian Pelican (*Pelecanus crispus*) nests on Lake Manyas. The majority of the white-headed ducks (*Oxyura leucocephala*) (70%) of the worldwide population winter on Burdur Lake. Tuz Gölü (Salt Lake) with more than 10,000 nests is a significant incubation area for Flamingos (*Phoenicopterus ruber*). The coasts of the Mediterranean and the Aegean in Turkey provide shelter for the endangered species of Mediterranean Seal (*Monachus monachus*) and Sea Turtles (*Caretta caretta* and *Chelonia mydas*) (Kantarlı 2013) (Table 12.2).

Hunting is a very popular and one of the major wildlife tourism activities in Turkey. It started in 1977 with wild boar hunting and the hunting of wild goats was also included within the scope of hunting tourism in 1981. The authority to control, organize and regulate hunting in Turkey has been given to the Ministry of Forestry and Water Affairs. All applications to be carried out within the scope of hunting tourism in the country are determined to take place during the 1st of April and the 31st of March. These decisions involve the species for which hunting licenses shall be given, the areas where hunting shall be allowed, dates for hunting, the fees, principles of hunting as well as forbidden procedures and forms of hunting (Ukav 2012). Twelve species of mammals and 134 species of birds have been determined as hunting by the Ministry. It was allowed to hunt seven mammals and 31 bird species in 2013–2014 hunting season. Hunting is restricted to certain days and hours. The permission to hunt large mammals within the scope of hunting tourism is also granted by the Ministry of Forestry and Water Affairs (Kantarlı 2013).

Foreign hunters can travel to the country with group A licensed travel agencies. The tourists travelling with the travel agencies can benefit from the hunting grounds in Ankara, Afyon, Bolu, Eskişehir, Kütahya, Adana, Adıyaman, Artvin, Antalya, Mersin, Nigde, Tunceli and Konya. Foreign tourists who come to the country with travel agencies or independently must obtain a 'Temporary Hunting License' from the General Directorate of National Parks. Within the various big mammals in the country (Table 12.3), local and foreign tourists are allowed to hunt chamois, wild goat, wild boar, lynx, wolves, jackals and foxes. Foreign hunters are not allowed to hunt other animals outside dedicated hunting grounds (Ukav 2012).

Hunting tourism generates good income as well as opportunities for employment in the country. The revenue generated in Turkey during the 2014–2015 seasons from hunting tourism was 3,257,212 US$. There are eleven A group travel agencies licensed for hunting tourism and focused only hunting organizations. They employ 90 permanent personnels and temporarily engage 2–10 guides/assistants for each hunting organizations. Every foreign hunter spends around 3000 and 200,000 US$ in foreign currency in the country. The amount of tourism revenue per capita was 640 US$ in 2012. According to this figure the income generated by wild goats is at least 15–35 fold in hunting tourism, 5–10 fold for wild boar, 25–30 fold for red deer and at least 120 fold in revenue for wild sheep. When all these figures are considered together the amount of foreign currency entering the country is approximately 12 million US$ and this is generated only by the 1500 hunting animals (Ulusoy 2015).

Fig. 12.1 African-Eurasian Migratory Waterbirds (AEWA). *Source* WMBD—World Migratory Bird Day (2014)

Table 12.2 Major waterbirds of Turkey

No.	Name	Latin	Winter migrant	Summer migrant	Local	Passing
1	Armenian Gull	*Larus armenicus*			x	
2	Crane	*Grus grus*		x	x	
3	Gull-billed Tern	*Gelochelidon nilotica*		x		
4	Ruddy Shelduck	*Tadoma ferruginea*			x	
5	Dalmatian Pelican	*Pelecanus crispus*			x	
6	Pygmy Cormorant	*Phalacrocorax pygmeus*		x		
7	Flamingo	*Phoenicopterus ruber*	x	x		
8	Pruple Heron	*ArdeapPurpurea*		x		
9	Black Stork	*Ciconia nigra*		x		
10	Squacco Heron	*Ardeola ralloides*		x		
11	Purple Gallinelo	*Porphiro porphiro*			x	
12	Marbled Teal	*Marmaronetta Angustirostris*		x		
13	Audouin's Gull	*Larus audonii*		x		
14	Avocet	*Recurvirostra avosetta*			x	
15	Spur-winged Plover	*Hoplopterus spinosus*		x		
16	White-headed Duck	*Oxyura leucocephala*	x	x		
17	Caspian Tern	*Sterna caspia*	x	x		
18	Ferruginous Duck	*Aythya nyroca*			x	
19	Little Tern	*Sterna albifrons*		x		
20	Little Stint	*Calidris minuta*	x			x
21	Sandwich Tern	*Sterna sandvicensis*	x	x		x
22	Little Egret	*Egretta garzetta*	x	x		
23	Night Heron	*Nycticorax nycticorax*		x		
24	Spoon Bill	*Platalea luecoradia*		x		
25	Coot	*Fulica atra*			x	
26	Little Ringed Plover	*Charatrius dubius*		x		
27	Black-winged Stilt	*Himantopus*		x		
28	Gallinago Gallinago	*Gallinago gallinago*	x	x		
29	Snipe	*Gallinago media*				x
30	Lapwing	*Vanellus vanellus*		x		

Source Kantarlı (2013: 11)

Table 12.3 Major big mammals of Turkey

No.	Name	Latin
1	Red Deer	*Cervus elephus*
2	Fallow Deer	*Dama dama*
3	Roe Deer	*Capreolus capreolus*
4	Antelope	*Gazella gazella*
5	Wolf	*Canis lupus*
6	Gazelle	*Gazella marica*
7	Rupicapra Rupicapra Ornata	*Rubicapra rubicapra*
8	Caracal	*Caracal caracal*
9	Capra Aegagrus	*Capra aegagrus*
10	Lynx	*Lynx lynx*
11	Hyena	*Hyaena hyaena*
12	Anatolian Wild Sheep	*Ovis gmelinii anatolica*
13	Wild Sheep	*Ovis gmelinii gmelinii*
14	Bear	*Ursus arctos*
15	Wild Boar	*Sus scrofa*

Source Kantarlı (2013: 26)

Box 12.1. Culture of Sparrow Hawk Hunting in Turkey

Sparrowhawk hunting in Turkey has been a tradition since the era of the Ottoman Empire to date. Sparrowhawk hunting is carried our in the Northeast Region of the country in the province of Rize and Artvin where sparrowhawks are used as birds of prey to hunt mainly quail. Individuals must obtain falconry certificates to exercise this sport. Turkey is on the migration route of sparrowhawks.

According to the protocol made between the Ministry for Forestry and Water Affairs and the Ministry of National Education it is obligatory to attend a hunting course in order to practice sparrowhawk hunting in Turkey. During one year hunters are allowed to capture and train only two sparrowhawks. Those engaging in sparrowhawks can be separated into four groups: the first group only goes into the mountains to catch and release sparrowhawks, the second group capture and train sparrowhawks and subsequently use them for quail hunting, the third groups only promenade the bird in August and November (sparrowhawk season) the fourth group only keeps sparrowhawks.

No touristic activities are available in Turkey regarding the observation, capturing or photographing of sparrowhawks. However, promoting this culture which is specific to the country and establishing the necessary infrastructure has potential for a significant wildlife activity.

Source Çolak (2013), Kantarlı (2013)

One of the primary activities carried out in Turkey within the scope of wildlife tourism is bird watching. Under favour of bird photography, the bird watchers numbers have reached 500,000 in 2016 (Kocabıyık 2016). The number of bird species which can be observed in the country is 461 and the number of birds included in the ÖKA (important bird species) with significance on a national and international level within the 'Important Bird areas in Turkey' project has been determined as 184 (Sevindi 2013). Basic centers of bird watching in Turkey can be listed as Lake Manyas (Balıkesir), Lake Bafa (Aydın-Muğla), Akyatan Lagoon (Adana) and Lake Kuyucuk (Kars).

Lake Manyas which was declared Manyas Bird Sanctuary National Park in 1959 is en route of birds migrating from Anatolia to Europe and hosts thousands of birds in breeding colonies comprised of pelicans, herons, spoonbills, cormorants, wild geese, wild ducks and songbirds (MFWA—Ministry of Forestry and Water Affairs 2016a). Lake Bafa was declared a national park in 1989 and hosts 20 species of orchids and endemic plant species in addition to 224 species of birds. Therefore in addition to bird watching it is possible to take part in activities such as photographing endemic plants around the lake as well as participate in hunting and boat trips on the lake. 250 bird species have been determined in Akyatan Lagoon while 182 bird species have been detected in Kuyucuk Bird Sanctuary (MCT—Ministy of Culture and Tourism 2016a, b). One of important endangered bird species in Turkey are the Bald Ibis (*Geronticus eremita*). According to the legend the Bald Ibis taken on board Noah's Ark by Prophet Noah as a symbol of plentitude have chosen Şanlıurfa, Birecik as their breeding grounds (Birecik İbis Breeding Farm) (MFWA 2016b).

Another wildlife observation activity carried out in Turkey is watching turtles. Although the whole Mediterranean coast offers significant potential for wildlife activities in terms of turtles there is very limited actual activity. Within this context day trips are organized to the İztuzu Beach in Dalyan district of Muğla province which is a protected area to observe nesting Caretta Caretta turtles. Within the scope of the tour participants can observe the turtles from boats operated with solar and electric energy, take photographs and observe them feeding on blue crabs.

Zoos in Turkey also attract the attention of local as well as foreign tourists. The top ranking largest five zoos in the country which are increasing daily are the Gaziantep Zoo, İzmir Wildlife Park, Soğanlı Zoo (Bursa), Atatürk Orman Çiftliği (Ankara) ve Boğaziçi Zoo (Kocaeli). Gaziantep Safari Park which was opened in 2015 is the first safari park in Turkey where approximately 250 animals of 25 species roam freely in a natural habitat. The year that the park was opened it had 375,000 visitors. The safari which is a first of its kind in Turkey ranks as the third largest zoo in Europe and the fourth largest zoo in the world (Özsöyler 2015).

There are various wildlife activities and attractions based on water not surprisingly since Turkey is surrounded by seas on three sides. For example, diving has a significant potential and offers participants the opportunity to observe various species including turtles, stingrays, schools of tuna, dolphins and seals. Major diving regions in Turkey shown in Fig. 12.2 are Saros, Gökçeada and Bozcaada in Çanakkale province, Ayvalık in Balıkesir province, Datça, Marmaris, Bodrum in Muğla province and Kemer, Kalkan, Sıçan Island, Gök Cave, Suluin Cave in Antalya province (MCT 2016b). Diving activities in Turkey are carried out under the "Blue Cruise" brand and with traditional rumca boats. However, no statistical information is available regarding the number of participants engaged in diving activities in Turkey.

According to O'Connor et al. (2009) whale watching is not available in Turkey however opportunities are available for dolphin watching in the Aegean Sea and the Black Sea. But, they indicated that no further developments regarding marine mammals have taken place from 1998 to 2008. The researchers underline the significant potential presented by groups of dolphins observed around the Bay of Kalkan in particular and recommend the wildlife tourism potential of Turkey is wide-ranging.

There are various dolphin watching tours as well as swimming with them for therapy in Turkey. Although foreign and domestic tourists prefer these tours, marketing activities are scarce to introduce this potential of the country. In addition, three species of dolphins are frequently observed in Turkish seas. They are known as the *Delphinus delphis, Tursiops runcates* and *Phocoena phocoena. Delphinus delphis* which is called common dolphin species have been put on the list of endangered species in the Mediterranean by the International Union for Conservation of Nature (IUCN) in 2003 (WWF—World Wide Fund of Nature 2016). There are dolphin parks in Turkey where it is possible to interact with dolphins as well as receive rehabilitation services. In addition dolphin watching tours are organized in various areas (such as Istanbul, Antalya). For example the Turkish Marine Research Foundation (TÜDAV) organizes dolphin watching tours in Istanbul along the Beykoz-Poyraz route (BELA—Buğday Ecologic Life Association 2016).

Fishing activities are available along Turkey's coastline of 8333 km. Regular commercial tours for amateur fishermen are organized along the South Aegean coast in places such as Çeşme, Seferihisar, Karaburun, Didim, Akbük, Bodrum and Antalya. The participants in the tours made with boats of different size and features. The organisers supply their fishing rods and other necessary equipment. Depending on the size of a boat the tour can comprise between 13 and 25 participants (Boz and Tuncer 2016). A comparison of fishing in Turkey and the examples in the world reveal that fishing in Turkey is not carried out with

Fig. 12.2 Major diving points in Turkey

traditional fishing boats. The tours are carried out with daily tour boats due to the lack of permission from the Ministry of Transport. Whereas if the tours were carried out with traditional fishing vessels the activity would gain popularity as a tourism event as well as would support the local fishermen.

The available potential is particularly evident taking into consideration that the Black Sea and the Bosphorus is a transit point for migrating fish. For example the volume of anchovies and marine bonito in the Black Sea, bluefish in the Bosphorus and species such as bream in the Aegean is sufficient to attract domestic and foreign tourists. In addition it is possible to catch fish such as amberjack, yellowtail, grouper, sea bream, bonito, horse mackerel, sea bass, barracuda, garfish, Spanish mackerel and Mediterranean red coral weighing between 5 and 40 kg which are particular to the Mediterranean Region.

12.2.2 Management of Wildlife in Turkey

The objective of hunting and wildlife management plans is to ensure minimum human impact on wildlife habitats to ensure the sustainability of the wildness and benefit from animals exceeding the carrying capacity of the ecosystem. In this context focus must be placed on conservation. The main area of activity is the protection of rare and endangered species. However, it is not enough to grant a conservation status to an area. In order to utilize a conservation status efficiently it is necessary to develop a protection-utilization system. Priority is given to the active participation of people in protection efforts for the sustainability of conservation (Kantarlı 2013).

The legal framework regulating the protection and management of hunting and wildlife in Turkey are governed by laws and regulations of the Constitutional Law of the Republic of Turkey as well as international conventions such as the Paris, Ramsar, Bern, Washington (CITES), Barcelona and Biological Diversity conventions. Execution authority is with the Ministry of Forestry and Water Affairs. The status of conservation zones is determined in accordance with the National Parks Act and necessary legal regulations depict the installation, development and management of these areas. According to this Law four different protection statuses have been determined in the country:

- National Parks (40 in total with an area of 848,119 ha).
- Nature Parks (184 in total with an area of 81,989 ha).
- Nature Conservation Areas (31 in total with an area of 63,694 ha).
- Nature Monuments (107 in total with an area of 5560 ha).

In addition to the conservation areas indicated above, 80 wildlife development areas with a total area of 1.2 million ha to protect the habitats of rare and endangered species have been established. Together with dam lakes 60% of wetlands have been placed under protection under various statutes. Thus the protected area in Turkey amounts to 7.24% of the total area (Kantarlı 2013).

The General Directorate of Nature Conservation and National Parks (DKMPGM) has developed a program to initiate and develop national and international hunting tourism. Within the scope of this program inventory data regarding the size and dynamics of the hunting population in a selected area are obtained, the ecosystem carrying capacity for each animal is calculated and the number, gender, age and permitted hunting seasons of the animals exceeding the carrying capacity are determined. Areas for which hunting and fishing management plans have been completed are opened for hunting tourism to benefit rural development. The villagers in areas selected for hunting tourism participate in conservation works and therefore various shares from hunting tourism revenues are allocated to them depending on the hunting species. The villagers also generate income by being hired as guides and porters by tourists (Kantarlı 2013).

Box 12.2. Anatolian Wild Sheep

The Anatolian wild sheep is one of the 15 subspecies of Asian Moufflon which is one of the 5 species of wild sheep in the world. It is only available in Turkey. In other words it is an endemic species to Turkey. The fact that the Anatolian wild sheep could be the ancestor of domesticated sheep is also a noteworthy consideration. The Anatolian wild sheep is the only species of wild sheep in which the female is not equipped with horns.

Works to rehabilitate wildlife populations in Turkey the numbers of which have been reduced or are threatened with extinction have continued for over 50 years. With the acceptance of the CITES Convention (Convention on International Trade in Endangered Species of Wild Fauna and Flora) the interest and funds allocated for rehabilitation in this area have increased. Within this scope the studies carried out in Konya province for the survival of Anatolian Wild Sheep have been found worthy of the EDMOND BLANC AWARD which is awarded annually by the International Council for Game and Wildlife Conservation (CIC) for the best protected wildlife area.

Source Kantarlı (2013), TRT—The Turkish Radio and Television Corporation (2004)

12.2.3 Experiencing Wildlife in Turkey

Various wildlife tourism models have been developed and are available in the literature outlining wildlife tourism experiences. Within this context the valuable studies of Duffus and Dearden (1990), Orams (1996), Reynolds and Braithwaite (2001), Higginbottom (2004) are prominent. These studies focus on the dynamic structure of wildlife tourism. For example Reynolds and Braithwaite (2001) stated that a satisfactory wildlife tourism experience requires various quality related factors such as authenticity, intensity, uniqueness, duration, species popularity and species status.

Higginbottom (2004) has focused on the motivation of wildlife tourists and determined that the three primary elements which motivate them the most are 'seeing wildlife in their natural environment (67%)', 'seeing wildlife behaving naturally (36%)' and 'rare and unique wildlife (33%)'. Other features which motivate wildlife tourists are listed as follows:

- Being able to get close to wildlife (29%);
- Being in a virgin natural environment (26%);
- A large variety of wildlife to see (25%);
- A knowledgeable guide is available (19%);
- Interesting information available about the wildlife (18%);
- A large number of wildlife to see (13%);
- Being in a pleasant environment (12%);
- The wildlife are easy to see (10%);
- Feeling safe (8%);
- Being able to touch or handle wildlife (7%).

Woods (1998) has determined the factors which determine the good and bad experiences of tourists in terms of captive and free animals. There were similarities between both groups in terms of positive experiences. The fundamental differences between the two groups was that those who had free animal encounters found the wildlife experience more authentic because they could observe animals in their natural habitats while those who interacted with captive animal felt the experience was positive because they had the opportunity to feed the animals (Table 12.4).

In general the participants in wildlife encounters can be separated into two varieties depending on whether they are passive or active. Passive participants prefer to view animals from a distance while active participants display behavior in which they want to establish physical contact with the animals by feeding, touching and even holding them. There are also participants which are between passive and active and can be manipulated in terms of the relevant experience by the behavior of the animal. For example during an activity animals can be fed by the guide or another mechanism. Furthermore, it should not be forgotten that interaction with animals may be challenging from time to time, even dangerous. Ryan (1998) stated that people may consider a saltwater crocodile un-human and dangerous while a gorilla dangerous but human-like. Therefore, some participants tend

Table 12.4 Factors which led to best and worst experiences with wildlife

Captive animal wild encounters		Non-captive wild animal encounters	
Best experiences	Worst experiences	Best experiences	Worst experiences
• See live animals previously only seen in photos or on TV • Educational/learned new things • Get close to animals/touch/feed • Well kept animals/space for animals • Aspects of the natural environment	• Dirty cages/small cages/inappropriate • Boring • Animals appear unhappy • Animals badly treated • Animals was threatening/attacked	• Be in natural environment/beautiful scenery • See animals close up/get close • See variety of animals interesting/educational • In animals world/in the wild/not a staged experience	• Weather/ illness/other reasons beyond operator control • Saw no or few animals • Animal was threatening/attacked • Boring

Source Fredline and Faulkner (2001: 16)

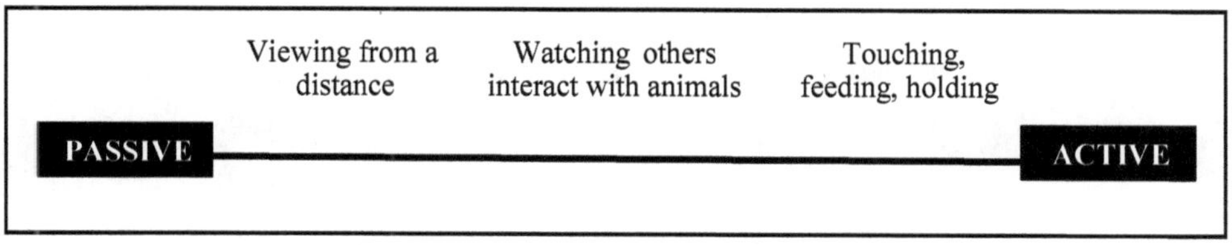

Fig. 12.3 Continuum of wildlife encounters ranging from passive to active. *Source* Fredline and Faulkner (2001: 13)

to be passive. In truth, the number of participants who can tolerate close contact with animals is rather small. Swimming with dolphins or other aquatic animals can be given as examples of this situation (Fredline and Faulkner 2001) (Fig. 12.3).

There are some A group licence zoos in some parts of Turkey. However there is no zoo in Black Sea and Eastern Anatolia Regions. The zoos such as Gaziantep, İzmir, Boğaziçi, Bursa and Antalya are as large as European samples with regard to area. For example, Gaziantep Zoo is the first in Turkey and Middle East, second in Europe and fourth zoo in the world. Gaziantep Zoo is the precursor of establishing Turkish Association of Zoos. On the other side, İzmir Wildlife Park is the second member of European Association of Zoos and Aquaria (EAZA) (Özen 2016).

It has been noted that studies with a view on describing the wildlife experiences of tourists in Turkey focus on tourist profiles and motivations. Studies are carried out in specific areas in this context. For example Iğırcık et al. (2005), Bekiroğlu and Okan (2009) determined the hunters' demographical and socio-cultural characteristics in Marmara and Aegean Regions. According to the results, individuals' main motivation factors in both groups are love of nature, the opportunity for physical exercise and spending time with friends, respectively. Another study which is carried out by Ünal et al. (2010) determined the amateur fishermen's experiences in the Çanakkale (Dardanelles) Strait. There are 51 species of fish in the Çanakkale Strait and its vicinity and amateur fishing mainly involves the catching of 3–4 species such as bluefish (30%) Spanish mackerel, mackerel and common sea bream (Ünal et al. 2010). It was also investigated the bird watchers' profile in Turkey (Çakıcı and Harman 2006). Accordingly, the participants go bird watching on their own as well as in groups. Most of the birdwatchers (90.9%) have spent 27 days on average watching birds during the past year. Most of the birdwatchers exercised their hobby in the country (86.7%) while 13.2% preferred to travel abroad to watch birds.

Interaction with captive or non-captive wildlife animals provide educational experiences to the participants. Zoos are good example to setup these educational experiences. Because they are opening doors to the nature in the city, provide to the visitors to encounter the wildlife animals but also not dangerous, zoos are preferred by many city-dwellers. According to Anderson et al. (2003) modern zoos may combine recreation and education for visitors thus providing a context for learning in the form of entertainment. Rabb (2004) stated that zoos encourage the visitors to protect the environment and natural resources. Because of no research about wildlife tourists' learning experiences in Turkey, this study investigates zoo visitors' motivational factors and learning experiences in İzmir Wildlife Park, Turkey. To gather in-depth knowledge about visitors' motivation factors and learning experiences, content analysis was utilized. Because it is one of the most used websites in social media (Cong et al. 2014) TripAdvisor.com was used. The participants' comments who visited İzmir Wildlife Park were analysed in this study (Fig. 12.4). The reasons for selecting İzmir Wildlife Park were twofold: (1) İzmir Wildlife park has a large number of animals (2) this zoo attracts domestic and international visitors (but it's used only domestic visitors' comments to determine their learning experiences). As a result 328 comments with 15,059 words were analysed within this framework.

NVivo 11, a qualitative analysis tool, was used. After the text was entered into the NVivo software program, three steps which are free codes, interpretive codes and overarching themes were followed. It couldn't get an accurate information about the participants' demographic features. Because some participants give their demographic information while others not. According to the results, the participants learning experience was divided into two features: animal conservation and learning about animals (Table 12.5).

Besides this findings, participants mentioned "kids", "kids learning", "fun with kids" words so many times. When one takes into account the Turkish family structure, its kid-focus structure can be easily recognized. That means, participants focus their kids' learning more than themselves. Other results related to participants motivations and experiences are presented in Table 12.6.

Fig. 12.4 Map of İzmir Wildlife Park. *Source* İzmir Doğal Yaşam Parkı (n.d.)

12.3 The Future of Wildlife in Turkey

The wildlife tourism is squeezed between ecotourism and hunting tourism in Turkey. However, while ecotourism is related to visit a destination to experience nature, wildlife tourism contains visiting a destination to observe wildlife (Reynolds and Braithwaite 2001). Although these concepts are different, this difference is not recognized in the country. Thus, it's difficult to find pure information and statistics about wildlife tourism in Turkey. This unawareness reflects to both strategies and plans about wildlife tourism. The Ministry of Culture and Tourism divided wildlife tourism into two categorizes as hunting tourism and bird watching which are done in Turkey. For example, the protection and development of wildlife in the country is included in the Tourism Strategy of Turkey 2023 however it is handled under ecotourism. It's noted that the Black Sea Region and Antalya locality are prioritized in the Tourism Strategy of Turkey 2023. There are plans to develop these regions in terms of national parks, nature conservation areas, hunting and wildlife conservation areas and archeologic sites as well as focus on ecotourism. However, there is no plan to regulate and market these wildlife areas for national and international tourists' interests in the context of wildlife tourism.

There is a regulation called "The regulation about the conservation of wildlife and development areas of wildlife" which is generated by Ministry of Forestry and Water Affairs. This regulation aims to protect wildlife animals, establish, manage and control the wildlife development areas. It also aims to determine activities which are allowed and inhibited. There are some initiatives to protect wildlife areas in Turkey. A significant conservation program has been developed in the country for the future of wildlife. The project called 'Green Corridor for Wildlife' is the first wildlife corridor project in Turkey. The objective of the project scope is to integrate small forest areas and enable the safe passage of wildlife trapped in these areas. The project target is focused on connecting fragmented forest blocks starting from Kars and going through Erzurum, Ardahan and Artvin provinces all the way to Georgia with forestation and other forestry activities (MFWA 2015).

There isn't any program to encourage wildlife tourists' learning at ministerial level in Turkey. However, it can be mentioned some private initiatives such as dolphin therapy

Table 12.5 The participants' learning experience of zoo in İzmir Wildlife Park

Overarching themes	Interpretive codes	Free codes	Frequency
Learning experience	Animal conservation	Social consciousness	1
		Habitat	43
		Zoolatry	1
		Conservation	3
		Respect for animals	1
	Learning about animals	Recognize animals	6
		Learning	1
		Educational	2
		Instructive	3
		Meeting with animals	2
		Knowledge	8

Source The author

Table 12.6 Zoo visitors' experiences

Overarching themes	Interpretive codes	Free codes
Animal features	Animal behavior	Animal behavior
	Animals' living situation	Habitat, living situation
	Animal diversity	Animal diversity
Setting features	Physical structure of zoo	Modern, clean, design, zoo environment, zoo management, size, fun place, entrance
	Staff behavior	Service staff
	Destination attributes	İzmir, municipality, climate, weather
	Marketing	Marketing
	Comparing	Comparing with old zoo, comparing with other zoos, comparing with other countries
	Facilities	Cafe facilities, resting places, WC, signages, souvenir shop, zoo activity, car park, cooling points, invalid chair
	Lack of zoo	Aquarium, number of cafe, lack of cafe service
	Transportation	Bicycle, bus, car, transportation
	Price	Car park price, cafe price, entrance price, price
	Security	Danger, security
Visitor features	Visitor motivations	Back to nature, explore, escape, family, kids
	Visitor experience	Tired, touching animals, observing animals, taking photo, feeding animals, unforgettable memories, being close to animals
	Visitor behavior	Length of stay, visit frequency, visit time, visiting hours

Source The author

centers and zoos. There are 14 university departments in associate degree which are included daytime and evening education for students. There is also one university department called faculty of wildlife ecology and management which trains wildlife engineers and they graduated their first students in 2016.

While there is a potential for diversity in terms of wildlife tourism in Turkey it is evident that the necessary initiatives are inadequate. The first thing to be done is to determine the limits and difference of wildlife tourism and make planning according to this differentiation at ministerial level. The limited number of studies in national literature in an academic context about wildlife tourism is also evident. Most of these studies are focused on the determination of the tourist profile. Furthermore, studies with a focus on the manifesting of the wildlife tourism potential of certain regions, motivation sources and experiences of tourists in terms of demand should be increased.

Box 12.3. Caretta Caretta's

There are two important species of marine turtles in Turkey. One of them is known as Caretta Caretta (common sea turtle) and the green sea turtle (*Chelonia mydas*). Within this context EKAD (Ecologic Research Association) has carried out an inventory for the sea turtles which nest on the beaches since 1987 as well as research and protection activities. The following works are carried out during these studies which take place in May and September;

- The population is monitored at night with marking studies in the field,
- The location of sea turtle nests are determined during daytime field work and they are controlled continuously,
- The hatching periods are monitored with regular daily checks, the offspring are counted, offspring who are stuck in the nest or are lost on the beach are rescued and offspring success statistics are kept,
- Caging works are carried out to protect sea turtle eggs from the predation of some natural enemies such as foxes, dogs and crabs,
- Cages are also used to protect nests made in front of hotels from human pressure,
- Awareness and information services to the local population and tourists is provided,
- Training is given to local administrations, tourism facility administrators and personnel, tradespeople in the region, small enterprises and schools about the correct and efficient use of the coastline, the impact of light and noise pollution and their effect on marine turtles,
- Regulating works are carried out with relevant organizations and enterprises regarding the use of the beach and beach perimeter,
- Works are carried out to equip sea turtles with satellite monitoring devices (EKAD has also been the first organization in Turkey to carry out this work),
- The works regarding the establishment of a 'Sea Turtle Rehabilitation Center' in Turkey which will undertake a significant role in the protection and rescue of sea turtles are continuing.

In addition to the mentioned works EKAD has signed various projects since the year 2000 involving sea turtles. These works have spread over a comprehensive mainly Belek, Kızılot, Patara, Gazipaşa, Kumluca, Alanya in Antalya province; Gökova, Fethiye, Köyceğiz, Dalyan, Dalaman in Muğla province; Foça in İzmir province; Ayvalık in Balıkesir province; Göksu and Kazanlı in Mersin province and Yumurtalık in Adana province.

German tour operator TUİ organizes a festival to draw attention to the caretta caretta in Dalyan. The festival lasts 3 days with various activities such as water games, concerts.

Source EKAD—Ecological Research Society (2007)

References

Adefalu LL, Omotesho KF, Alao OS (2015) Determinants of visitors' preference for wild animal species (a case study of Unilorin Zoo, Ilorin, Kwara State, Nigeria). J Res For Wildl Environ 7(1): 124–135

Anderson US, Kelling AS, Pressley-Keough R, Bloomsmith MA, Maple TL (2003) Enhancing the zoo visitor's experience by public animal training and oral interpretation at an otter exhibit. Environ Behav 35(6):826–841

Ballantyne R, Packer J, Falk J (2011) Visitors' learning for environmental sustainability: testing short- and long-term impacts of wildlife tourism experience using structural equation modelling. Tourism Manage 32(6):1243–1252

Bekiroğlu S, Okan T (2009) Avcıların demografik ve sosyoekonomik yapısı: Marmara ve Ege Bölgesi örneği. Acta Turcica Çevrimiçi Tema Türko Derg 1(1):297–308

BELA (2016) Yunus gözlem turları. http://www.bugday.org/portal/haber_detay.php?hid=4003. Accessed 21 May 2016

Bentz J, Lopes F, Calado H, Dearden P (2016) Managing marine wildlife tourism activities: analysis of motivations and specialization levels of divers and whale watchers. Tourism Manage Perspect 18:74–83

Boz M, Tunçer S (2016) Turizm talebinde yeni eğilimler: Balıkçılık turizmi. Paper presented at the III. Ulusal Deniz Turizmi Sempozyumu, University of 9 Eylül, 26–27 Feb 2016

Çakıcı C, Harman S. (2006) Kuş gözlemciliğinin önemi: Türkiye'de kuş gözlemciliğinin profili. Anatolia: Turizm Araştırmaları Dergisi 17(2):161–168

Çolak C (2013) Vazgeçilmez tutku atmacacılık. http://www.haberler.com/vazgecilmez-tutku-atmacacilik-5215550-haberi/. Accessed 25 Apr 2016

Cong L, Wu B, Morrison AM, Shu H, Wang M (2014) Analysis of wildlife tourism experiences with endangered species: an exploratory study of encounters with giant pandas in Chengdu, China. Tourism Manage 40:300–310

DKMPMG (2013) Türkiye'de Av ve Yaban Hayatı. Doğa Koruma ve Milli Parklar Genel Müdürlüğü, Ankara. http://www.milliparklar.gov.tr/kitap/77/AV_YABAN_TR.pdf. Accessed 23 Apr 2016

Duffus DA, Dearden P (1990) Non-consumptive wildlife-oriented recreation: a conceptual framework. Biol Conserv 53:213–231

EKAD (2007) Deniz kaplumbağalarını koruma ve izleme projesi. http://www.ekad.org.tr/files/gonullu_programi_kitapcigi_web.pdf. Accessed 18 May 2016

Fredline L, Faulkner B (2001) International market analysis of wildlife tourism, CRC Sustainable Tourism, Wildlife Tourism Research Report Series, No: 22. http://sustain.pata.org/wp-content/uploads/2014/12/WT22_Fredline-IntnlMkts.pdf. Accessed 20 July 2016

He GM, Chen XD, Liu W, Bearer S, Zhou SQ, Cheng YQ (2008) Distribution of economic benefits from ecotourism: a case study of Wolong nature reserve for giant pandas in China. Environ Manage 42(6):1017–1025

Higginbottom K (2004) Wildlife tourism: impacts, management and planning. Common Ground Publishing Pty Ltd, Australia

Iğırcık M, Yadigar S, Bekiroğlu S, Okan T, Aktaş ME (2005) Marmara Bölgesi Avcı Profili. TC Çevre ve Orman Bakanlığı Ege Ormancılık Araştırma Müdürlüğü, Teknik Bülten No: 29, İzmir http://yayin.ogm.gov.tr/yaydepo/t381.pdf. Accessed 20 May 2016

İzmir Doğal Yaşam Parkı (n.d.). İngilizce Broşür. http://www.izmirdogalyasamparki.org.tr/Pages/Content.aspx?id=37&mid=1506 . Accessed 27 July 2016

Kantarlı M (2013) Türkiye'de Av ve Yaban Hayatı. TC Orman ve Su İşleri Bakanlığı Doğa Koruma ve Milli parklar Genel Müdürlüğü, Ankara http://www.milliparklar.gov.tr/kitap/77/AV_YABAN_TR.pdf. Accessed 21 May 2016

Karis A, Utouh L, Liphafa L (2013) Impact of tourism on wildlife conservation. Working Group on Environmental Auditing (WGEA) Publication. http://www.environmental-auditing.org/LinkClick.aspx?fileticket=ecrH%2BPLw7Cc%3D&tabid=128&mid=568. Accessed 28 Apr 2016

Kocabıyık Z (2016) Kuşların peşinde 80 bin km. Türkiye Gazetesi. http://www.turkiyegazetesi.com.tr/saglik/347949.aspx. Accessed 23 Mar 2016

MCT (2016a) Kuyucuk kuş cenneti ve doğal güzellikler. http://www.karskulturturizm.gov.tr/TR,54879/kuyucuk-kus-cenneti-ve-dogal-guzellikler.html. Accessed 11 May 2016

MCT (2016b) Su altı dalış - Bazı dalış bölgeleri. http://yigm.kulturturizm.gov.tr/TR,11518/bazi-dalis-bolgeleri.html. Accessed 11 May 2016

MFWA (2015) Yaban hayvanları için yeşil koridor. http://www2.ormansu.gov.tr/osb/haberduyuru/engelsizhaber/15-08-22/Yaban_Hayvanlar%C4%B1_%C4%B0%C3%A7in_Ye%C5%9Fil_Koridor.aspx?sflang=tr. Accessed 10 May 2016

MFWA (2016a) Kuş cenneti milli parkı. http://www.milliparklar.gov.tr/mp/kuscenneti/index.htm. Accessed 11 May 2016

MFWA (2016b) Kelaynak kuşları - *Geronticus eremita*. http://sanliurfa.ormansu.gov.tr/Sanliurfa/AnaSayfa/BirecikSefligi/kelaynaklar.aspx?sflang=tr. Accessed 11 May 2016

Mintel (2008) Wildlife tourism international. http://reports.mintel.com/display/351669/. Accessed 07 June 2016

O'Connor S, Campbell R, Cortez H, Knowles T (2009) Whale watching worldwide: tourism numbers, expenditures and expanding economic benefits. International Fund of Animal Welfare, Yarmouth, MA

Orams MB (1996) A conceptual model of tourist-wildlife interaction: the case for education as a management strategy. Aust Geogr 27 (1):39–51

Özen AS (2016) Türkiye'deki A grubu hayvanat bahçeleri ve akvaryumların durumu (Araştırma Makalesi). Eurasian J Vet Sci 32(3):149–155

Özsöyler C (2015) Türkiye'nin ilk safari parkı. http://www.hurriyet.com.tr/turkiyenin-ilk-safari-parki-29280094. Accessed 15 May 2016

Rabb GB (2004) The evolution of zoos from menageries to centres of conservation and caring. Curator 47(3):237–246

Reynolds PC, Braitwaite D (2001) Towards a conceptual framework for wildlife tourism. Tourism Manage 22(1):31–41

Ryan C (1998) Saltwater crocodiles as tourist attractions. J Sustain Tourism 6(4):314–327

Sevindi C (2013) Ekoturizm ve kuş gözlemciliği açısından Kuyucuk Gölü Kuş Cenneti (Arpaçay-Kars). Türk Coğrafya Dergisi 61:63–76

TRT (2004) Bozkırın çocukları – Anadolu yaban koyunu. http://www.bozkirincocuklari.com/yaban_koyunlari.htm. Accessed 01 June 2016

Ukav İ (2012) Adıyaman'da av turizmi. II. Disiplinlerarası Turizm Araştırmaları Kongresi, 12–15 Apr 2012, pp 3–18

Ulusoy H (2015) Av turizminin kırsal turizm açısından kırsal kalkınma üzerindeki etkisinin irdelenmesi. Türk Bilimsel Derlemeler Dergisi 8(2):74–80

Ünal V, Acarlı D, Gordoa A (2010) Characteristics of marine recreational fishing in the Çanakkale Strait (Turkey). Mediterr Mar Sci 11(2):315–330

Woods B (1998) Animals on display: principles for interpreting captive wildlife. J Tourism Stud 9(1):28–39

WMBD (2014) Working together for international migratory bird conservation. http://www.worldmigratorybirdday.org/2014/index91d9.html?option=com. Accessed 02 Oct 2016

WWF (2016) Doğa koruma - Türler - Yunus. http://www.wwf.org.tr/ne_yapiyoruz/doga_koruma/turler/yunus/. Accessed 01 June 2016

Part III

Ethical Issues and Wildlife Tourism: Dealing with the Stalemates

Ethics and Responsibility in Wildlife Tourism: Lessons from Compassionate Conservation in the Anthropocene

13

Georgette Leah Burns

Abstract

Whether captive or non-captive, consumptive or non-consumptive, targeted or non-targeted, guided or non-guided, wildlife tourism activities have traditionally been dominated by an anthropocentric worldview that recognizes wildlife only for its extrinsic value. This chapter argues that the advent of the Anthropocene provides an opportunity for humans to accept responsibility for how they engage with animals in tourism settings and ethically reassess this engagement. Reviewing theories of ethics dealing with animals, tourism, the environment and conservation, the conclusion is drawn that in order to effectively manage wildlife tourism for the equitable benefit of both humans and wildlife, and thus create a viable wildlife tourism ethic, valuable lessons can be extracted from an approach that embodies compassionate conservation.

13.1 Introduction

Human and non-human animals have a long and varied history of engagement (Burns and Paterson 2014). Engaging with wildlife as an object of tourism is just one part of this history, yet it is an increasingly important one. As the Anthropocene recasts humans as the dominant species on the planet and the World Wildlife Fund reports that the Earth has lost half its wildlife in the last 40 years (Carrington 2014), a reassessment of the ways humans and wildlife engage seems long overdue. Cautionary tales of negative impacts of wildlife tourism for both people and animals (e.g., Burns and Howard 2003; Green and Giese 2004) parallel those of the positive contributions this form of tourism can make to conservation (e.g., Higginbottom and Tribe 2004), and to sustainability of local communities (e.g., Burns 2004).

The aim of this chapter to not to review these positive and negative aspects of wildlife tourism, but rather to review the ethical frameworks within which they exist and highlight possible pathways to increase satisfaction with wildlife tourism activities for all stakeholders, including the wildlife. This is achieved by exploring the ethical intersection where tourism and animal coexist, and arguing for increased attention to ethics and responsibility in wildlife tourism.

Tourism activities lie at the intersection of an array of deep and complex ethical concerns (Burns 2015a). In the context of wildlife tourism, these include concerns about proper treatment of animals, habitat and ecosystems, and local communities and business, as well as appropriate behaviour of tourists. Focus here is on ethics relevant to interactions between tourists and wildlife, so concern in this context is less on the business of tourism, or the actions of operators and hosts—hence the exclusion of business ethics in this discussion, for example. Thus, the perspective presented draws on four established fields; ***animal ethics*** (e.g., Regan 2001, 2004; Singer 1975, 2005), ***tourism ethics*** (e.g., Burns 2015a; Fennell 2006; Lovelock and Lovelock 2013), ***environmental ethics*** (e.g., Holden 2005; Minteer 2011), and ***conservation ethics*** (e.g., Ramp 2013; Ramp and Bekoff 2015; Wallach et al. 2015) in the search for an appropriate ethical framework for wildlife tourism contexts.

G.L. Burns (✉)
Environmental Futures Research Institute and Faculty of Environmental Science, Griffith University, Nathan, Australia
e-mail: leah.burns@griffith.edu.au

I. Borges de Lima and R.J. Green (eds.), *Wildlife Tourism, Environmental Learning and Ethical Encounters*, Geoheritage, Geoparks and Geotourism, DOI 10.1007/978-3-319-55574-4_13

13.2 Wildlife Tourism

Wildlife tourism involves encounters between undomesticated animals and humans during times when the humans are at leisure, or undertaking recreation. These encounters occur in settings where the wildlife is captive and ex situ, such as zoos and wildlife sanctuaries, and non-captive and in situ, such as in national parks, freehold or public spaces.

One justification for keeping wildlife in captive settings for display is to facilitate human learning about them. Ideally, exposure to animals that most people do not encounter in their daily lives leads to a more conservation oriented ethic and many zoos have actively sought to achieve this (Carr and Cohen 2011; Frost 2011). Once an anthropocentric setting purely to objectify exotic wildlife for tourist entertainment, over time zoos and other wildlife-based attraction sites, such as wildlife parks and sanctuaries, have repositioned themselves as educationally-oriented exhibits and experiences that connect visitors with animals (Mason 2000; Skibins and Powell 2013). By doing so, they seek to foster pro-wildlife attitudes and behaviours.

Zoos and other captive settings are also at least partially responsible for raising public awareness about animal welfare issues in tourism and other settings. Several studies have examined the value of zoo experiences for connecting people with wildlife and increasing their awareness of conservation issues (e.g., Ballantyne et al. 2007; Bruni et al. 2008; Gusset and Dick 2010; Tribe and Booth 2003). Zoos also exist for preservation purposes, to house wildlife that cannot be returned to non-captive settings due to lack of habitat (Frost 2011), and in the hope that safe habitat areas will become available in the future.

Attitudes now increasingly exposed through public media may demonstrate that the captive wildlife tourism goals of education and awareness have had some success. This could be connected with a wider surge in public interest in animal welfare, during which ethical considerations concerning animals have evolved. This has been noted by Bock and Buller (2013), for example, in connection with concern about the welfare of farm animals and is further evidenced by changes in welfare and legislation for companion and livestock animals via animal care and ethic committees across the world (Littin et al. 2004; Littin and Mellor 2005).

The activities of the humans in wildlife tourism settings may be consumptive, involving the deliberate killing of animals, or non-consumptive, for example watching wildlife. Non-lethal interactions have also been argued as a form of consumption (Higham and Hopkins 2015) because the wildlife is framed as an object of the tourist gaze, and thus consumed visually by the tourist in this context (Burns 2015b). For ease of discussion and to comply with common discourse, however, the use of 'consumption' here will be confined to those forms of tourism that deliberately result in death for the wildlife.

This chapter does not focus on consumptive wildlife tourism, although the types of activities that occur under this heading, such as fishing and hunting, are certainly an important part of encounters, particularly in non-captive settings. These forms of wildlife tourism are, at least in part, responsible for fueling public interest and debate around appropriate ways of engaging with wildlife in tourism settings. Many recent examples of this could be cited, but perhaps the most well known is the killing of Cecil the African lion by a trophy hunter in 2015 (Nelson et al. 2016). Another recent example circulated on public media was the August 2016 story of a hunter in Canada who, after setting a bait trap for a black bear, wounded it with a spear before leaving it to die overnight (ABC 2016). Events such as these are not illustrative of the attitudes of all hunters and fishers to animal welfare issues as part of their tourism experience. Rather, they demonstrate wider public attitudes about, and increasing demand for, appropriate forms of wildlife tourism engagement.

Encounters with wildlife can be targeted or non-targeted. In a targeted encounter, the leisured person deliberately journeys to a site to engage with wildlife. A non-targeted encounter occurs when the person travels for another purpose but during that experience they encounter wildlife; for example, when a person on a ski tour in Norway sees an arctic fox in an encounter that was unplanned and not the major reason for the tourist activity. Regardless of the intention to interact, the consequences may be the same for the wildlife and the person.

Wildlife tourism encounters may also be guided or non-guided, where guided encounters contain some form of interpretation of, or education about, the wildlife while non-guided encounters do not. Such interpretation can range in form from information delivered via signs or brochures, to information delivered in person, by a guide or ranger (Moscardo et al. 2004). Non-guided encounters occur where no interpretative messaging is provided. Non-targeted encounters are more likely to be non-guided than targeted encounters, and are more likely to occur in non-captive settings.

Historically, wildlife tourism concerned with non-captive settings and non-lethal activities has received less attention in the literature and academic discourse. Such research is, however, increasing in popularity, in part due to increasing pressure from global tourism for travel to more remote areas, where the wildlife is located. For example, Iceland is now heavily marketed as a nature based tourism destination although its background is one where foreign tourism was almost nonexistent until the middle of the twentieth century (Jóhannesson and Huijbens 2010). The number of arrivals to

Iceland is currently increasing by an average of 22% per year (Óladóttir 2016). On the Vatnsnes peninsula, in a region promoted as the 'land of seals', the town of Hvammstangi with 558 residents (Ragnarsson 2015, p 10) recorded 39,079 visitors in 2016 (Þórisson 2017). Such increases in the numbers of tourists visiting remote and rural areas can have profound social and ecological consequences that require sound ethical guidance to ensure effective management.

13.3 Ethical Wildlife Tourism

Stemming from the Greek word "ethos", the broad field of Ethics is a key component of western philosophical thought. Key ethical questions implore humans to consider what is good, what is right, and how we should act in given situations. Thus, in the context of this discussion the key ethical question is this: how should humans behave in tourism based interactions with wildlife? Ethics differ cross-culturally, and play a key role in individual reasoning, unpinning decision making, though often subconsciously, in all contexts (Burns 2015a). In the search for an appropriate ethical framework for wildlife tourism, it is useful to draw on dominant theories from the fields of ethics formulated in the context of animals, tourism, the environmental, and conservation.

13.3.1 Animal Ethics

The field of animal ethics can be broadly divided between those which consider animal rights (e.g., Regan 2001, 2004; Singer 1975, 2005) and those which consider animal welfare. Ethics based upon the perspective of animal rights advocate affording non-human animals the same rights as humans and considers all, or at least some, animal species equally deserving of moral regard and respect (Fennell 2015a, p 29). As such, it is based on the notion of intrinsic, or internal, value in individual animals. In contrast, ethics based upon the perspective of animal welfare accept prioritization of human interests over those of non-human animals (Garner 1993) provided that the quality of life of the animal is taken into account (Bekoff and Nystrom 2004).

The ethical viewpoint expanded on in this chapter leans more toward the animal welfare perspective than the animal rights one. This is based upon pragmatic necessity that, since an animal rights perspective might argue that we should not hold wildlife captive under any circumstances, or undertake any activities in non-captive settings that disturb the animals, an argument might be advanced that wildlife tourism, as an essentially hedonistic human activity, should not exist in any form. In contrast, an animal welfare approach is more accepting of continued use of animals but under the condition that we consider what is best for their health and well-being in our interactions with them.

13.3.2 Tourism Ethics

> One of the rising challenges in the 21st century will be to find an ethical stance that facilitates tourism scholarship moving beyond the paradigm of objectivity and frontier thinking in order to contribute to a more thoughtful, reflexive, and sustainable platform (Macbeth 2006, p. 963).

In recent years, scholarship around the topic of ethics in tourism has increased significantly and branched into increasingly more areas of tourism (Burns 2015b; Fennell 2006). Searching for a place for ethics, Macbeth (2006) scrutinized Jafari's four platforms of tourism (Jafari 1990, 2001) and suggested the addition of a fifth platform on sustainable development and a sixth on ethics. This ethical, more reflexive and more sustainable, type of platform is particularly pertinent to wildlife tourism.

Although ethical issues in animal tourism have been the subject of increasing academic interest in recent years (Fennell 2012; Lovelock and Lovelock 2013), scholarship in wildlife tourism has yet to incorporate environmental ethics in any substantial manner (Burns 2015b; Dobson 2011) and such ethical consideration has been largely absent from key wildlife tourism texts (e.g., Higginbottom 2004a; Newsome et al. 2005; Shackley 1996). Scientific endeavours focus instead on different aspects of the importance of the wildlife experience for visitors (Ballantyne et al. 2009, 2011), and continue the long-standing scholarly interest in the impacts of tourism on wildlife (Green and Giese 2004; Higginbottom 2004b; Higham and Shelton 2011; Newsome and Rodger 2007). This is reflective of the wider field of tourism in general, where the study of the impacts of tourism is the dominant theme of the research, and consideration and development of ethics and ethical frameworks are largely absent (Burns 2015a).

13.3.3 Environmental Ethics

Summarising the status of animal ethics research in tourism, Fennell (2015a) draws heavily on theories founded in environmental ethics, including utilitarianism and ecocentrism. Environmental ethics emerged in the early 1970s as a new sub-discipline of philosophy in reaction to the domination of traditional anthropocentrism that encompassed several anthropocentric ethics (Minteer 2011); such as, the frontier ethic (Macbeth 2006) and the Judeo-Christian ethic (White 1967). In contrast, environmental ethics question the assumed moral superiority of human begins over other species and investigate the possibility of rational arguments for assigning intrinsic value to the environment and the non-human species within it.

Ethical thought that is essentially anthropocentric places humans at the centre of any decision making. This placement ensures that human interests are prioritized, and ultimate decision making is based on outcomes considered best for the majority of humans. A key notion of sustainability, of preserving resources for future generations, is an intergenerational ethic (MacClellan 2013) that perpetuates the human focus. Similarly, focus on ethics that are extrinsic, or utilitarian, legitimizes human's privileged use, and exploitation, of wildlife. In contrast, recognising ethics based on more intrinsic values, inside a more holistic and ecocentric view of the world, allows us to venture beyond the perceived human right to exploit and, therefore, appreciate the imperative to manage wildlife in a more equitable way. These different approaches can be understood on a continuum, where an extrinsic, or utilitarian value system exists at one end and intrinsic value at the other. Similarly, anthropocentric, or human-centred, ethics exist at one end of a continuum of thought and action and ecocentric, or ecosystem, based ethics at the other.

The domination of extrinsic and anthropocentric values frames a discourse in which the natural environment, and the wildlife within it, exist for the use of humans. Hardin's (1968) seminal text on the tragedy of the commons was meant as a warning cry against human overuse of natural resources, but he offered a solution by suggesting "a moral shift in human behaviour as the simplest way to avoid over-use" (Holden 2005, p 339). Enacting such a moral shift, however, which would need to take place globally across multiple cultures and people with multiple words views, has proven to be anything but simple over the fifty years since this suggestion. In 2005 Holden proposed that recognition of the intrinsic value of nature "may be necessary for the ultimate sustainability of common pool resources" (p 339). To make this moral shift, discourse surrounding wildlife tourism needs to focus more on ethical approaches to management that encompass ecocentric and intrinsic values.

13.3.4 Conservation Ethics

Conservation, or conservation biology, "addresses the biology of species, communities, and ecosystems that are perturbed, either directly or indirectly, by human activities or other agents" with the goal of preserving biological diversity (Soulé 1985, p 727). Therefore, theories used for managing conservation issues, and the ethics they encompass, are also applicable to managing the disturbance of wildlife as a consequence of tourism. The two ethical theories discussed here are consequentialism and compassionate conservation.

Consequentialism posits that the basis for judging whether a policy or action is right or wrong depends solely on its consequences (Nelson et al. 2016). This thinking is common in conservation (Gore et al. 2011) but is not without shortcomings. Three shortcoming of consequentialist arguments of particular relevance to our considerations about wildlife tourism are (1) that the means do not always justify the ends, (2) the theory lacks recognition of the importance of motivation when determining right and wrong, and (3) it assumes that we can accurately predict future consequences (Nelson et al. 2016). In this later aspect in particular, consequentialism does not take into account the precautionary principle.

The precautionary principle, conceived in the 1970s for the purpose of exercising foresight in environmental policy and protection (Fennell 2015b, p 68), calls for caution in the absence of certainty of consequences (Burns et al. 2011). It is a key element in compassionate conservation. As a management tool, compassionate conservation aims to bridge the gap between animal welfare advocates and conservation biologists (Moore et al. 2014, p 93), between an animal focused approach and an environmental approach, by looking for synergies between them. As described above, animal ethics approaches are often concerned with the welfare of animals on an individual level, which contrasts with environmental ethics approaches that are more concerned about the conservation of a species (Ramp 2013). Comparative conservation "seeks to identify, enhance and promote the commonalities between animal welfare and conservation" (Moore et al. 2014, p 94) by recognising the benefits of preserving both the species and the individuals of that species.

13.4 Wildlife Management

In the context of wildlife tourism, the contact between human and non-human animals requires consideration and management. Thus, an essential part of this discussion concerns ethical management in the wildlife tourism context: who has responsibility for this management and how should it be done? "A dominant focus of wildlife management continues to be the treatment of wildlife as 'crops' to be cultivated and harvested for human use" (Nelson et al. 2016, p 302) in a framework that perpetuates anthropocentric and extrinsic values of wildlife.

Research in the field of human dimensions of wildlife management (Kellert 1985, 2003; Manfredo 1989, 2008) has recognized an important shift in the orientation of wildlife values in the Western world away from the management of wildlife for human benefit and toward a philosophy embedded in the rights of wildlife (Fulton et al. 1996; Manfredo et al. 2003). This shift toward 'mutualism obligation' (Manfredo 2008), replacing a domination or utilitarian view of wildlife (Manfredo et al. 2009; Teel and Manfredo 2009), arose from post-modern human relationships which place greater emphasis on egalitarianism, belongingness, and self-expressive values (Manfredo et al. 2003, 2004; Manfredo 2008; Manfredo et al. 2009; Vaske et al. 2006). "Postmodern understandings of relations between humans and animals are characterized by stronger emotional and moral content" (Burns 2014, p 8). This paves the way for alternative management approaches that reflect changing perspectives: approaches that decentralise the position of humans in favour of more ecosystem centered approaches.

Wildlife, particularly in non-captive settings where rights to ownership of them may be ambiguous, can be viewed as a common pool resource (CPR). Common pool resources "possess characteristics that make their exclusion from human use difficult or impractical" (Holden 2005, p 339) and "…exploitation by one user reduces the availability of the resource to others" (Moore and Rodger 2010, p 831). "These two characteristics—difficulty of exclusion and subtractability— create potential common pool resource dilemmas in which people following their own short-term interests produce outcomes that are not in anyone's long-term interest" (Ostrom et al. 1999, p 1). In the context of wildlife tourism, the result is compounded complication about management, and ethical consideration, of activities.

Common pool resources can be held at a national or international level. Plans and policies on how to equitably deal with dilemmas caused by CPRs often focus on international CPRs, also known as global commons (Moore and Rodger 2010), such as fishing grounds. For wildlife tourism in national, non-captive and non-protected settings where users cannot be excluded, threats to sustainability are highly likely. Threats from overcrowding and environmental disturbance may become apparent as a reduction in animal numbers, decline in the health of the animals, and decreased tolerance of human behaviour by both wildlife and other tourists (Granquist and Sigurjonsdottir 2014; Marschall et al. 2017). The reverse, increased tolerance by wildlife to human behaviour, often termed 'habituation' can also occur and cause detrimental changes in animal behaviour (e.g., Burns and Howard 2003).

An additional complication is that non-captive wildlife is a mobile resource. This is exemplified by migratory birds which move across human recognised boundaries where they may be confronted by multiple owners and/or people responsible for their management who hold differing attitudes and behaviours toward them. The mobility is particularly problematic where the wildlife has competing values. For example, it may be of aesthetic interest to tourists but also consumed as food for locals, or hunted as pests (Moore and Rodger 2010, p 832). Amidst these complexities lies the issue of responsibility, responsibility for the right way to manage wildlife as a tourism resource.

13.5 Human Responsibility for Ethical Engagement

The emergence of *Homo sapiens* during the Holocene coincided with an extinction phase incorporating habitat and biodiversity loss on an unprecedented scale (Gössling and Hall 2006; Higham and Hopkins 2015) and we have now entered a phase referred to as the Sixth mass extinction, amidst forecasts of further dramatic species loss (Ceballos et al. 2010, 2015). The pronounced human dominance of global ecosystems has been recognized as a new evolutionary epoch, the Anthropocene, in which humans are the only species responsible for the start of the current mass extinction event (Ceballos et al. 2010, 2015; Gascon et al. 2015). The advent of the Anthropocene formalises recognition of humans as the dominant species on the planet, with demonstrated global and comprehensive ecological, and even geological, influence (Burns 2014; Ceballos et al. 2010).

This recognition can be considered positively in terms of the need for acceptance of human responsibility. Human responsibility stems too from the fact that humans are the only species thought to have the ability to make cognitive ethical decisions (Dorresteijn et al. 2015; Manfredo 2008; Wallach et al. 2015), and these decisions can have profound consequences for other animals (Lewis et al. 2016). As noted by Gamborg et al. (2012), humans have both the capacity and the responsibility to manage wildlife with moral and ethical insight. Connected with this insight is an increased willingness to accept some level of ethical responsibility for our actions (Lewis et al. 2016). Thus, recognition of the Anthropocene epoch provides a timely reminder to reassess human responsibility to engage ethically with wildlife in tourism settings. One way to do this is through the lens of compassionate conservation.

13.6 Compassionate Conservation

From a management perspective, we can look to the theory of compassionate conservation for ethical illumination. In the context of wildlife tourism, most stakeholders are likely to accept the conservation focus of this approach, because conservation has long been seen as a vital component of the justification of all wildlife tourism, non-captive and captive, non-consumptive and even consumptive (Higginbottom and Tribe 2004). Compassionate conservation offers an alternative management approach, encompassing a new paradigm for rethinking ways to effectively and ethically managing wildlife. For these reasons, it can make a valuable contribution toward the search for a suitable wildlife tourism ethic.

Compassionate conservation is an ethic that evolved for the purpose of ensuring coexistence, of humans relinquishing some of their domination and instead, necessarily, sharing space with nature (Ramp and Bekoff 2015). For wildlife tourism it offers a new model of best practice for conservation outcomes, especially in relation to wildlife management actions. Compassionate conservation ideology embraces the precautionary principle (Burns et al. 2011) that advocates "first, do no intentional harm" (Bekoff 2010). It also recognizes the intrinsic value of all animals, independent of their utility to humans (Fox and Bekoff 2011) and considers the welfare of individual animals (Bekoff 2010; Bekoff and Ramp 2014; Draper and Bekoff 2013, 2015; Ramp 2013; Wallach et al. 2015).

Most of the ideologies behind compassionate conservation are not new. Its foundation spans centuries of philosophers and thinkers such as Aristotle, Kant, Gandhi, Bentham, Regan and Singer who debate the concepts of sentience, animal awareness, and the importance of human perceptions of, and interactions with, non-human animals (Bekoff and Jamieson 1996). In challenging humans to make management decisions from an ethical standpoint, compassionate conservation offers an ethical conversation, from a non-anthropocentric perspective, about moral values that can inform thoughts and actions (Hadidian et al. 2006). In doing so, it promotes co-existence and the recognition of a shared planet for which humans are responsible.

13.7 Conclusion

As the number of species of wildlife on the planet continues to decrease, the need for effective management of human interactions with wildlife is becoming increasing apparent. The advent of the Anthropocene, in which humans are the unsurpassed species with global social, ecological and geological impact upon the planet, highlights an opportunity to reassess our responsibility toward other species. Humans have responsibility, as the dominant species on the planet, as the species empowered with cognitive processes, to engage ethically with other species. Wildlife tourism is just one of many ways humans engage with non-human animals, but it is a critical one. As species decline in numbers, tourism interest in them increases and humans look to 'last chance tourism' (e.g., Dawson et al. 2011) as a way to both assuage their environmental guilt and experience a unique tourism product.

The need for a moral shift in humanity was suggested by Hardin in 1968 and such changes in how humans value wildlife has been monitored by Manfredo and others (2003; 2004; 2008; 2009). Recent attention in public and social media to perceived mistreatment of wildlife, such as the case of Cecil the Lion, may suggest that we are now seeing a shift toward increased recognition of intrinsic and ecocentric values. Certainly, evidence is increasing that the public are becoming more empathetic toward wildlife (Nelson et al. 2016, p 304).

Management of, and engagement with, wildlife in tourism settings has been anthropocentric (Burns et al. 2011). Moving away from that ethical framework entirely may be impossible. But we need to strive for, or at least recognise the potential for, more ecocentric perspectives and approaches that offer the possibility for more balance in power relations between humans and other species as well as a more holistic approach to management that takes into consideration the needs of other species. A more sophisticated understanding of ethical perspectives could assist with management of wildlife tourism in such a way that responds to recent debates in the public sphere and respects the lives of wild animals.

What does an ethical approach to wildlife tourism look like? Ideally, it needs to empower humans, as the drivers and moral guardians of this activity, to make the right decisions about how we engage with wildlife in tourism settings. It is unlikely that such decisions will win acceptance globally across all stakeholders. However, as a starting point, recognition of our place as the species with the cognitive capacity to both cause and understand our unprecedented damage to the planet is required. Such recognition highlights the need for us to manage our tourism activities and their consequences in ways that are responsible and equitable to the wildlife we engage with. The right ethical approach for wildlife tourism, that would serve as a guide for the treatment of people and wildlife in tourism settings, should draw on the history of ethics from other schools of thought. In addition, we should add compassionate conservation to this list as wildlife tourism strives to make an ethical stance of its own: Something more holistic, less anthropocentric, and accepting of the intrinsic value of wildlife.

References

ABC (Australian Broadcasting Commission) (2016) Hunter who killed black bear with spear and left it to die overnight could face charges, 17 August 2016. http://www.abc.net.au/news/2016-08-17/outcry-over-spear-hunted-bear/7750326

Ballantyne R, Packer J, Falk J (2011) Visitors' learning for environmental sustainability: testing short-and long-term impacts of wildlife tourism experiences using structural equation modeling. Tour Manage 32(6):1243–1252

Ballantyne R, Packer J, Hughes K (2009) Tourists' support for conservation message and sustainable management practices in wildlife tourism experiences. Tour Manage 30(5):658–664

Ballantyne R, Packer J, Hughes K, Dierking L (2007) Conservation learning in wildlife tourism settings: lessons from research in zoos and aquariums. Environ Educ Res 13(3):367–383

Bekoff M (2010) First do no harm. New Scientist, August 24–25

Bekoff M, Jamieson D (1996) Ethics and the study of carnivores: doing science while respecting animals. In: Gittleman JL (ed) Carnivore behavior, ecology, and evolution. Cambridge University Press, Cambridge, pp 15–45

Bekoff M, Nystrom J (2004) The other side of silence: Rachel Carson's views of animals. Zygon 39(4):861–883

Bekoff M, Ramp D (2014) Cruel to be kind. New Scientist, June 26–27

Bock B, Buller H (2013) Healthy, happy and humane: evidence in farm animal welfare policy. Sociologia Ruralis 53(3):390–411

Bruni CM, Fraser J, Schultz PW (2008) The value of zoo experience for connecting people with nature. Visitor Stud 11(2):139–150

Burns GL (2004) The host community and wildlife tourism. In: Higginbottom K (ed) Wildlife tourism: impacts, management and Planning. Common Ground Publishing, Altona, pp 125–144

Burns GL (2014) Animals and anthropomorphism in the Anthropocene. In: Burns GL, Paterson M (eds) Engaging with animals: interpretations of a shared existence. Sydney University Press, Sydney, pp 3–20

Burns GL (2015a) Ethics in tourism. In: Hall CM, Gössling S, Scott D (eds) The Routledge handbook of tourism and sustainability. Routledge, London, pp 117–126

Burns GL (2015b) Animals as tourism objects: Ethically refocusing relationships between tourists and wildlife. In: Markwell K (ed) Animals and tourism: understanding diverse relationships. Channel View Publications, Bristol, pp 44–59

Burns GL, Howard P (2003) When wildlife tourism goes wrong: a case study of stakeholder and management issues regarding Dingoes on Fraser Island. Aust Tourism Manage 24(6):699–712

Burns GL, Macbeth J, Moore S (2011) Should dingoes die? Principles for engaging ecocentric ethics in wildlife tourism management. J Ecotourism 10(3):179–196

Burns GL, Paterson M (2014) Introduction. In: Burns GL, Paterson M (eds) Engaging with animals: Interpretations of a shared existence. Sydney University Press, Sydney, pp ix–xiv

Carr N, Cohen S (2011) The public face of zoos: images of entertainment, education, and conservation. Anthrozoos 24 (2):175–189

Carrington D (2014) Earth has lost half of its wildlife in the last 40 years, says WWF. Guardian. 30 Sept 2014. http://www.theguardian.com/environment/2014/sep/29/earth-lost-50-wildlife-in-40-years-wwf

Ceballos G, Garcia A, Ehrlich P (2010) The sixth extinction crisis: loss of animal populations and species. J Cosmol 8:1821–1831

Ceballos G, Ehrlich PR, Barnosky AD, Garcia A, Pringle RM, Palmer TM (2015) Accelerated modern human-induced species losses: Entering the sixth mass extinction. Sci Adv 1(5):1–5

Dawson J, Johnston MJ, Stewart EJ, Lemieux CJ, Lemelin RH, Maher PT, Grimwood BSR (2011) Ethical considerations of last chance tourism. J Ecotourism 10(3):250–265

Dobson J (2011) Towards a utilitarian ethic for marine wildlife tourism. Tourism Mar Environ 7(3–4):213–222

Dorresteijn I, Schultner J, Nimmo DG, Fischer J, Hanspach J, Kuemmerle T, Kehoe L, Ritchie EG (2015) Incorporating anthropogenic effects into trophic ecology: predator-prey interactions in a human-dominated landscape. Proc Bio Sci 282(1814):301–306

Draper C, Bekoff (2013) Animal welfare and the importance of compassionate conservation—a comment on McMahon et al. 2012. Biol Conserv 158: 422–423

Fennell DA (2006) Tourism ethics. Channel View Publications, Clevedon

Fennell DA (2012) Tourism and animal ethics. Routledge, London

Fennell DA (2015a) The status of animal ethics research in tourism. In: Markwell K (ed) Animals and tourism: understanding diverse relationships. Channel View Publications, Bristol, pp 27–43

Fennell DA (2015b) Tourism and the precautionary principle in theory and practice. In: Hall CM, Gössling S, Scott D (eds) The Routledge handbook of tourism and sustainability. Routledge, London, pp 67–77

Fox CH, Bekoff M (2011) Integrating values and ethics into wildlife policy and management—Lessons from North America. Animals 1 (4):126–143

Frost W (2011) Zoos and tourism: conservation, education, entertainment? Channel View Publications, Bristol

Fulton D, Manfredo M, Lipscomb J (1996) Wildlife value orientations: a conceptual and measurement framework. Human Dimensions Wildlife 1(2):24–47

Gamborg C, Palmer C, Sandoe P (2012) Ethics of wildlife management and conservation: what should we try to protect? Nature Edu Know 3(10):8

Garner R (1993) Political animals: a survey of the animal protection movement in Britain. Parliamentary Affairs 46(3):333–352

Gascon C, Brooks TM, Contreras-MacBeath T, Heard N, Konstant W, Lamoreux J, Vié JC (2015) The importance and benefits of species. Curr Biol 25(10):431–438

Gore ML, Nelson MP, Vucetich JA, Smith A, Clark M (2011) Exploring the ethical basis for conservation policy: the case of inbred wolves on Isle Royale, USA. Conserv Lett 4:394–401

Gössling S, Hall CM (eds) (2006) Tourism and global environmental change: ecological, social, economic and political interrelationships. Routledge, London

Granquist SM, Sigurjonsdottir H (2014) The effect of land based seal watching tourism on the haul-out behaviour of harbour seals (Phoca vitulina) in Iceland. Appl Animal Behav Sci 156:85–93

Green R, Giese M (2004) Negative effects of wildlife tourism on wildlife. In: Higginbottom K (ed) Wildlife tourism: impacts, management and planning. Common Ground Publishing, Altona, pp 81–98

Gusset M, Dick G (2010) 'Building a Future for Wildlife'? Evaluating the contribution of the world zoo and aquarium community to in situ conservation. Int Zoo Year Book 44:183–191

Hadidian J, Fox CH, Lynn WS (2006) The ethics of wildlife control in humanized landscapes. In: Timm RM, O'Brien JM (eds) Proceedings of the 22nd Vertebrate Pest Conference. University of California, Davis, pp 500–504

Hardin G (1968) Tragedy of the commons. Science 162(3859):1243–1248

Higginbottom K (ed) (2004a) Wildlife tourism: Impacts, management and planning. Common Ground Publishing, Altona

Higginbottom K (2004b) Managing impacts of wildlife tourism on wildlife. In: Higginbottom K (ed) Wildlife tourism: impacts, management and planning. Common Ground Publishing, Altona, pp 211–230

Higginbottom K, Tribe A (2004) Contributions of wildlife tourism to conservation. In: Higginbottom K (ed) Wildlife tourism: impacts, management and Planning. Common Ground Publishing, Altona, pp 99–124

Higham J, Hopkins D (2015) Wildlife tourism: "call it consumption". In: Hall CM, Gössling S, Scott D (eds) The Routledge handbook of tourism and sustainability. Routledge, London, pp 280–293

Higham J, Shelton EJ (2011) Tourism and wildlife habituation: reduced population fitness or cessation of impact? Tour Manag 32(6):1290–1298

Holden A (2005) Achieving a sustainable relationship between common pool resources and tourism: the role of environmental ethics. J Sustain Tourism 13(4):339–352

Jafari J (1990) The Basis of tourism education. J Tourism Stud 1:33–41

Jafari J (2001) The Scientification of tourism. In: Smith V, Brent M (eds) Hosts and guests revisited: Tourism issues of the 21st Century. Cognizant Communications, Elmsford, pp 28–41
Jóhannesson GT, Huijbens EH (2010) Tourism in times of crisis: exploring the discourse of tourism development in Iceland. Curr Issues Tourism 13:419–434
Kellert SR (1985) Public perceptions of predators, particularly the Wolf and Coyote. Biol Conserv 31:167–189
Kellert SR (2003) Kinship to mastery: Biophilia in human evolution and development. Island Press, Washington, DC
Lewis P-M, Burns GL, Jones D (2016) Response and responsibility: Humans as apex predators and ethical actors in a changing societal environment. Food Webs
Littin KE, Mellor DJ, Warburton B, Eason CT (2004) Animal welfare and ethical issues relevant to the humane control of vertebrate pests. New Zealand Vet J 52(1):1–10
Littin KE, Mellor DJ (2005) Strategic animal welfare issues: Ethical and animal welfare issues arising from the killing of wildlife for disease control and environmental reasons. Rev Sci Tech (Int Off Epizootics) 24(2):767–782
Lovelock B, Lovelock KM (2013) The Ethics of tourism: critical and applied perspectives. Routledge, New York
Manfredo MJ (1989) Human dimensions of wildlife. Wildl Soc Bull 17 (4):447–449
Manfredo MJ (2008) Who cares about wildlife? Social science concepts for exploring human-wildlife relationships and conservation issues. Springer, Fort Collins
Manfredo MJ, Teel TL, Bright AD (2003) Why are public values toward wildlife changing? Human Dimensions Wildlife 8:287–306
Manfredo MJ, Teel TL, Bright AD (2004) Application of the concepts of values and attitudes in human dimensions of natural resources research. In: Manfredo MJ, Vaske JJ, Field D, Brown PJ (eds) Society and Natural Resources: a summary of knowledge prepared for the 10th International Symposium on Society and Natural Resources. Jefferson, MO, Modern Litho, pp 271–282
Manfredo MJ, Teel TL, Zinn HC (2009) Understanding global values toward wildlife. In: Manfredo MJ, Vaske JJ, Brown PJ, Decker DJ, Duke EA (eds) Wildlife and society: The Science of human dimensions. Island Press, Washington, DC
Marschall S, Granquist S, Burns GL (2017) Interpretation in wildlife tourism: Assessing the effectiveness of signage to modify visitor behaviour at a seal watching site in Iceland. J Outdoor Recreation Tourism 17:11–19
Mason P (2000) Zoo tourism: the need for more research. J Sust Tourism 8:333–339
Macbeth J (2006) Towards an ethics platform for tourism. Annals Tourism Res 32(4):962–984
MacClellan J (2013) How (not) to defend a Rawlsian approach to intergenerational ethics. Ethics Environ 18(1):67–85
Minteer B (2011) Refounding environmental ethics: Pragmatism, principle, and practice. Temple University Press, Philadelphia
Moore S, Rodger K (2010) Wildlife as a common pool resource issue: enabling conditions for sustainability governance. J Sustain Tourism 18(7):831–844
Moore, RS, Wihermanto, Nekaris, KAI (2014) Compassionate conservation, rehabilitation and translocation of Indonesian slow lorises. Endangered Species Res 26:93–102
Moscardo G, Woods B, Saltzer R (2004) The Role of interpretation in wildlife tourism. In: Higginbottom K (ed) Wildlife tourism: Impacts, management and planning. Common Ground Publishing, Altona, pp 231–251
Nelson MP, Bruskotter JT, Vucetich JA, Chapron G (2016) Emotions and the ethics of consequence in conservation decisions: Lessons from Cecil the Lion. Conserv Lett 9(4):302–306
Newsome D, Dowling R, Moore S (2005) Wildlife tourism. Channel View Publications, Clevedon
Newsome D, Rodger K (2007) Impacts of tourism on pinnipeds and implications for tourism management. In: Higham J, Lück M (eds) Marine wildlife and tourism management: insights from the natural and social sciences. CABI, Oxfordshire, pp 182–205
Óladóttir O (2016) Tourism in Iceland in figures—May 2016. Icelandic Tourist Board, Reykjavik
Ostrom E, Burger J, Field CB, Norgaard RB, Policansky D (1999) Revisiting the commons: local lessons, global challenges. Science 284:278–282
Ragnarsson A (2015) Norðurland Vestra, Stöðugreining 2014 [Northwest Status Analysis 2014]. Byggðastofnun, Iceland
Ramp D (2013) Bringing compassion to the ethical dilemma in killing kangaroos for conservation: comment on "Conservation through sustainable use" by Rob Irvine. J Bioethical Inquiry 10 (2):267–272
Ramp D, Bekoff M (2015) Compassion as a practical and evolved ethic for conservation. Bioscience 65:323–327
Regan T (2001) Defending animal rights. University of Illinois Press, Chicago
Regan T (2004) The Case for animal rights: Updated with a new preface. University of California Press, Berkeley
Shackley M (1996) Wildlife tourism. International Thomson Business Press, London
Singer P (1975) Animal liberation: a new ethics for our treatment of animals. Harper Collins, New York
Singer P (ed) (2005) In defense of animals: the second wave. Blackwell, Oxford
Skibins JC, Powell RB (2013) Conservation caring: measuring the influence of zoo visitors` connection to wildlife on pro-conservation behaviors. Zoo Biology 32:528–540
Soulé M (1985) What is conservation biology? Bioscience 35(11):727–734
Teel TL, Manfredo M (2009) Linking society and environment: a multi-level model of shifting wildlife value orientations in the western U.S. Soc Sci Q 90(2):407–427
Tribe A, Booth R (2003) Assessing the role of zoos in wildlife conservation. Human Dimensions Wildlife 8(1):65–74
Vaske JJ, Shelby LB, Manfredo MJ (2006) Bibliometric reflections on the first decade of human dimensions of wildlife. Human Dimensions Wildlife 11(2):79–87
Wallach AD, Bekoff M, Nelson PM, Ramp D (2015) Promoting predators and compassionate conservation. Conserv Biol 29 (5):1481–1484
White L (1967) The Historical roots of our ecological crisis. Science 155(3767):1203–1207
Þórisson SL (2017) Selasetur Íslands Annual Report 2016. Icelandic Seal Center, Hvammstangi

Author Biography

Georgette Leah Burns is a Senior Lecturer in the Faculty of Environmental Sciences at Griffith University in Brisbane, foundation member of Griffith's Environmental Future's Research Institute, and an Advisor on the Griffith University Human Research Ethics Committee. Her research focuses on understanding and managing the interactions between people and wildlife in nature-based tourism settings. In this quest she has encountered dingoes, penguins, sharks, crocodiles, flying foxes and koalas in Australia, as well as seals, whales, foxes and migratory birds in Iceland. In addition to numerous book chapters and journal articles on this topic, Leah is co-editor of the book *Engaging with Animals: Interpretations of a Shared Existence* (2014).

Bio link: https://experts.griffith.edu.au/academic/leah.burns

Disturbing Skippy on Tour: Does It Really Matter? Ecological and Ethical Implications of Disturbing Wildlife

14

Ronda J. Green

Abstract

"What does it matter if a few Skippies are disturbed?" was the rhetorical retort of one tourism operator to the idea of minimal-impact wildlife-viewing. The most reasonable answer to this would probably be "maybe not at all or maybe quite a lot, depending on the situation". Wildlife tourism, even with some unavoidable impacts, is often better than alternative land uses, but should be conducted responsibly, with a view to minimising impact on animal welfare and conservation as well as with consideration of other tour operations and local residents. This chapter explores some of the ethical considerations of disturbance of wildlife on tour, including driving animals away from feeding, breeding or resting areas, feeding of wildlife and stress related to close approach or other activity, in both wild and captive situations. Most of what we know so far involves the effects on individual animals. Further research is needed on how much the disturbances influence population numbers if we are to understand conservation implications, which can also be important for the satisfaction of human residents and visitors who want to continue watching wildlife. More research is also needed on animal welfare aspects such as activities that may cause serious levels of stress and how to determine same.

14.1 Background: Potential for Positive and Negative Impacts of Wildlife Tourism

I've been asked by a number of people whether it is wise for me to talk about negative effects of wildlife tourism, since this is the category my own business falls into, and since as chair of Wildlife Tourism Australia Inc. I actively promote the expansion and diversification of the wildlife tourism industry. It is certainly not my intention to denigrate the industry as a whole, but to promote the more sustainable and ethical examples of it above others.

There are certainly some examples that need to improve their environmental, animal welfare and human fairness to be considered ethical, and some so poor in this regard I would be very happy to see closed down. But despite the problems to be discussed below, the use of land for wildlife tourism can often be far better for fauna and flora conservation than alternative land uses such as clearing for industrial estates, expansion of suburbia, monoculture farming or large scale mining, and it can also impart various positive impacts (Higginbottom et al. 2001). As a conservationist and as a wildlife ecologist I thus continue to support the concept and practice of wildlife tourism throughout the world. However, just as car manufacturers need to know what might happen to passengers in a collision, and restaurant-owners need to understand the causes of food poisoning, so also do tour operators or managers of eco-lodges and wildlife parks need to be aware of potential problems relating to conservation and animal welfare. Far from damaging such industries, the knowledge not only

R.J. Green (✉)
Griffith University, Brisbane, Australia
e-mail: chair@wildlifetourism.org.au

R.J. Green
Wildlife Tourism Australia Inc., Rathdowney, Australia

I. Borges de Lima and R.J. Green (eds.), *Wildlife Tourism, Environmental Learning and Ethical Encounters*, Geoheritage, Geoparks and Geotourism, DOI 10.1007/978-3-319-55574-4_14

assists in minimising negative effects, but by demonstrating this awareness and showing that steps are taken to mitigate problems, the industry should achieve greater public acceptance amongst conservation managers and members of conservation-oriented NGOs (who are often cynical about the whole idea of environmentally-sustainable tourism), and amongst conservation-minded and animal-loving tourists. Indeed, not only could it assist the public face of the industry as a whole, but a potential competitive advantage may also accrue for those operations that can demonstrate a higher level of responsibility than their competitors.

Wildlife in the wild can be negatively influenced by various forms of tourism, including wildlife tourism itself in numerous ways (Green and Higginbottom 2001), including: (1) accidental or deliberate killing of wildlife (e.g. hunting and fishing, eradication of wildlife regarded as pests near accommodation, roadkill in national parks), (2) destruction or modification of habitat (e.g. clearing for a resort, draining of swamps for mosquito control, over-collection of firewood in campsites, inadvertent food supplements at rubbish dumps or picnic sites) and (3) direct disturbance of animals by tourist activity (the major focus of this chapter).

Disturbance can also be an issue for captive animals (see references below). Well-run zoos and wildlife parks can contribute positively to wildlife conservation, education and research, but they have variously been criticised, sometimes to the point of demanding they all be closed because, for instance, they prevent wild creatures from living lives of freedom in their natural habitats, frequently keep non-breeding animals that don't contribute to conservation programs, do not rehabilitate enough captive-bred animals into the wild, keep animals in inadequate enclosures, and often separate social animals from their conspecifics or mix individuals that do not form a compatible social group (Frost 2011, and Reiser, this volume, who raises doubts about reintroduction of zoo-bred animals to the wild, and considers zoo animals too altered from the wild condition to provide useful subjects for meaningful research and that visitors do not stay long enough at any enclosure to truly learn anything). Disturbance issues in captivity include close encounters with humans such as hand-feeding (whether or not encouraged by management), petting, cuddling or riding, or simply viewing at close quarters where the animal has no opportunity to seek shelter.

This chapter considers the effects of disturbance of wildlife on both the welfare of individual animals and conservation of wildlife populations, as well as effects on human stake-holders (tourists, tourism operators and local residents), and in this context discusses animal welfare rather than animal rights. A comparison of various aspects of ethics, including environmental ethics and animal welfare in wildlife tourism can be found in the literature (see Fennel 2015; Burns et al. 2011, Burns in this volume). Conservation of biodiversity and the welfare of individual animals are separate issues (and occasionally in conflict) but both need consideration if wildlife tourism ventures are to be considered ethical. Despite many attempts over the decades for authors to 'bend over backwards' in a rather excessive attempt to avoid anthropomorphism, there is nowadays ample evidence that many non-human animals experience a variety of emotions, can recall past events and anticipate future ones, and are capable of suffering (e.g. de Waal 2016; Kaplan 2015), no matter how difficult it may be for us to know their precise feelings.

For any wildlife tourism activity to be truly ethical, the needs and desires of human stakeholders (especially tourists, tourism operators and host communities) also need to be considered and respected when animal numbers are reduced or the animals become shy or aggressive in the presence of humans.

14.2 Disturbing Animals in the Wild: Animal Welfare Implications

"What does it matter if a few Skippies are disturbed?" was the somewhat contemptuous response of one tourism operator to the idea of minimal-impact wildlife-viewing, when I mentioned a wildlife tour where the guide had deliberately set several kangaroos bounding along in front of the vehicle for several minutes to thrill his passengers. A wildlife ecologist I mentioned it to later said she felt the same way, as no kangaroo species is currently threatened with extinction, and there was no evidence that behavioural disturbance affected populations rather than individuals.

It is true that there is an enormous knowledge gap on how disturbance of individual animals translates into population effects, and also that none of our three kangaroo species (as opposed to certain wallabies and other relatives) are threatened with global extinction (Pople and Grigg 1999, and all listed as 'of least concern in IUCN's 2016 red list). This however should not be used to argue that we can simply ignore the effects of disturbance. Local extinctions or population declines present their own problems, including effects on other species that share the habitat, and it may displease residents and visitors who derive pleasure from their presence. There is also a growing awareness internationally that wildlife tour operators have a responsibility to respect not only biodiversity issues but also the welfare of the individual animals their businesses depend on.

So, how much does it matter if animals are chased to offer thrills or photographic opportunities for tourists? Is this really any different to natural events such as the sudden appearance of a predator, with the animal soon settling down to normal behaviour when the tourists depart?

Taking the example of fleeing kangaroos, there are several logical possibilities see:

- the animals may simply flee for a short time, then settle down and resume their normal activities with no harm done (this may well be assumed by most tourists and many guides: research by Taylor and Knight 2003, suggests many recreationists underestimate their effects on wildlife)
- a fleeing animal may be struck by another vehicle (e.g. Fox 1982) or, (if panicked and traveling fast, especially if young) become tangled in a fence
- a joey could become separated from its mother, causing high stress levels to both, and with subsequent exposure of the joey to predators such as eagles, dingos or feral dogs
- if the disturbance occurs frequently, the animals may abandon their favoured feeding area for vegetation providing less nutrition (not much research has been conducted on kangaroos in this regard, but see below for some evidence for other species)
- they may experience a slight rise in stress levels for a short time but with no discernible welfare problem (e.g. as for captive free-range kangaroos studied by Sherwen et al. 2015c)
- if the disturbance is very alarming, loud and/or prolonged, or perimeter fences and other barriers prevent escape, severe stress is possible and could be lethal (kangaroos appear to be especially susceptible to myopathy and other stress disorders: Garlick and Austen 2014; Rose 2005).

There is some evidence that in some situations disturbance of mammals, birds and other wildlife by tourists, including close approach with no obviously visible effect, can interfere with feeding, breeding and resting, e.g.:

- Harris and Leiper (1995) cited reports of high stress levels of sea-lions at Seal Bay, Kangaroo Island, indicated by an increase in aggressive behaviour and failure to breed on a tourist beach, before the current systems for regulation of visitor access to the breeding colony. A local national park ranger (pers. comm.) told me visitors would often deliberately wake the animals for photos, sometimes resulting in exhausted animals in need of rest after many hours of fishing heading back into the sea.
- Humans walking through a gull colony can frighten the parents from their nests, resulting in predation of eggs and young by neighbouring gulls (Anderson and Keith 1980).
- Adelie penguin stress levels (as measured by heart rate) rise before they show visible signs such as head movements, and this early stage is not obvious to tourists or tour operators Giese 2000). If the penguins reach the point of standing up, any chicks or eggs underneath them can receive a blast of cold air which, especially if repeated, could be lethal.
- Other effects may be cumulatively important but too subtle to observe. Albatrosses in New Zealand have shifted to less optimal nesting sites, apparently as a result of tourist activity, but, too gradually to be discerned in short term studies (Higham 1998).
- The rule that visitors on gorilla tours should not approach closer than 7 metres is often breached, but projectiles from a human sneeze can travel up to 6 metres, and several gorillas appear to have contracted fatal lung diseases from tourists (Hanes 2012).
- South American monkey species were found by Kauffman (2010) to differ in their responses to tourists, some showing a greater tendency than others to flee or give alarm calls.
- Wolf and Croft (2010) found kangaroos more likely to flee if visitors made unexpected moves such as leaving the designated walking tracks.
- Begder et al. (2006) found a dolphin population declined apparently in response to tourist boats, but this was a local effect, probably indicating the more sensitive of the animals moving elsewhere rather than a general population decline.
- Recreational hunting and fishing causes obvious impact on the animals directly taken (including prolonged pain and fear if death is not swift) but can also cause immediate and longer-lasting stress to members of the social group (e.g. to mothers or offspring, or the loss of a dominant and protective animal) and disturbance to other wildlife close enough to see or hear the events. Non-lethal sport fishing (where the fish is un-hooked and released after capture) can also result in stress and post-release mortality (Schlenker et al. 2016). The owner of a catch-and-release family fishing attraction (using barbless hooks) told me the fish never seemed to gather at the bridge where people preferred to fish (probably with good reason).
- Animals not actively sought by tourists may also be disturbed by their activities. While spotlighting for koalas, possums and frogmouths along a bushland track, how many bandicoots, antechinus and low-nesting birds are fleeing the site but un-noticed by the tour group? Green and Higgibottom (2001) suggested a decade and a half ago that more research is needed on this, but we still know very little about such effects.

Many other examples of wildlife disturbance by tourists can be found in Green and Higginbottom (2001), Higginbottom (2004), Newsome et al. (2002, 2005) and Oberbillig (2000).

Feeding of wildlife is a popular activity in many places, including hand-feeding of wild animals at designated feeding stations (e.g. dolphins, parrots, fish), or against regulations at picnic tables (e.g. kookaburras, lace monitors), feeding captive animals from wallabies to elephants in zoos and wildlife parks (some encouraged by management, some in breach of rules), or the provisioning of food to enable watching wild animals safely but at reasonably close quarters, including bears, Tasmanian devils, Komodo dragons and sharks. Well-managed feeding activities may sometimes help animals through a very lean season or following depletion of resources due to a severe bushfire. It can also provide a way of ensuring a wild animal will be available for viewing at a predictable time, so as not to disappoint tourists with limited time available, bring animals close for photography or for interpretive activities, and provide a human-animal interaction with potential for increasing the visitors' positive feelings towards the animal and its conservation needs. There are also many potential negative effects.

Green and Higginbottom (2001) summarise arguments for and against feeding, including ecological and animal welfare reasons as well as reasons of philosophical attitude (an almost spiritual dimension of communicating with wild beings or, by contrast, objections to any intrusion into their wild-ness) and human desire (wanting to feel or photograph the animal) or safety. Some of the animal welfare issues, of which many examples are presented, and which could also lead to ecological problems, include inappropriate food, spread of disease, disruption of normal behaviour, increase in population numbers of aggressive or predatory species to the detriment of others, overcrowding leading to aggression and high stress levels, nuisance or danger to humans from animals that become demanding of hand-outs leading to subsequent removal of the animals by authorities, and creation of dependencies leading to stress if the source is later removed (Fig. 14.1).

Attitudes towards feeding tend to be polarised between 'feeding wildlife is great for the animal and the human' and 'no one should ever feed wildlife'. Some conservation authorities endorse a total ban, on the assumption that it is generally impossible to train people that it is fine to feed animals in some situations but not in others. Others point to the fact that many will still want to feed them for a variety of reasons, and will secretly continue to do so even if officially banned, and that it is probably best to let them know what is and isn't appropriate rather than impose a total ban (CALM 2000; Jones 2011). Delegates at a Wildlife Tourism Australia workshop some years ago discussed the dilemma, resulting in a set of guidelines (Wildlife Tourism Australia Inc. 2016a).

If a tour guide chases, feeds or otherwise disturbs animals on a tour, this may send the wrong message to tourists, and they may think it is acceptable to do the same in other areas. Many tourists come from big cities, and lack the previous experience with wild animals that could have taught them otherwise. On the other hand, many tourists nowadays, well-versed in nature documentaries and the like, tend to be more sophisticated in their awareness of wildlife conservation and welfare and may object to such behaviour by guides, as witnessed by many comments in social media platforms such as Facebook and Trip Advisor as well as pers. obs., and conversations with keepers. I was once even chastised by a tourist for taking a flashlight photo of a scorpion (from behind) because he had seen a television documentary showing their sensitivity to light. It would be to the guides' advantage to realise the negative attitudes they might cause (and which may be published for all to see on social media) by behaviour visitors consider irresponsible.

Tour guides are not the only offenders in this way. Field zoologists are often shown on social media and other websites interacting with wild animals. Ward-Paige (2016)

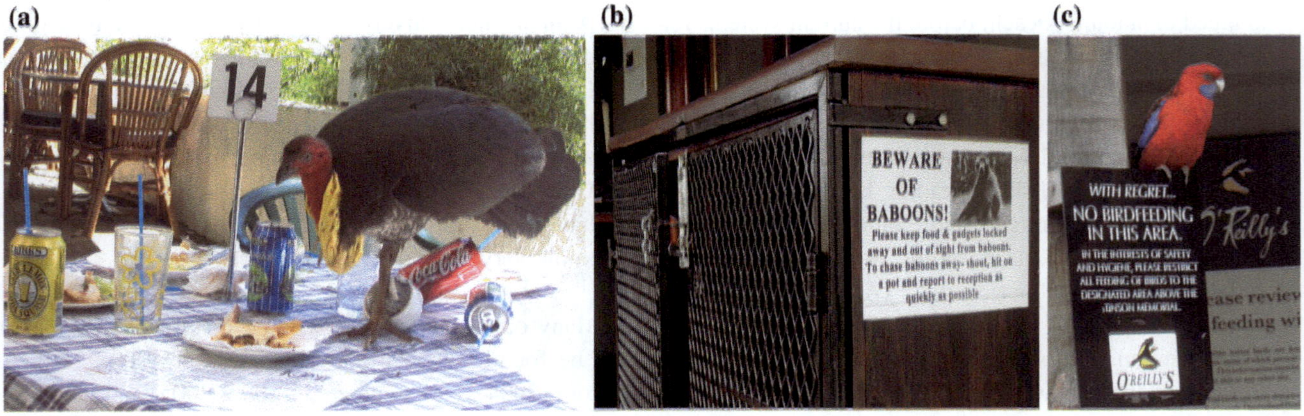

Fig. 14.1 **a** Some animals, such as this Australian brush turkey, can become nuisances if they come to associate humans with food (cafe in southeast Queensland). **b** Some can become very intrusive and even dangerous (Kruger National Park). **c** It can be possible to train visitors that it is acceptable to feed wildlife in one area but not in others (O'Reillys Rainforest Retreat). Photos by author

points out that although this can have a positive effect by encouraging viewers to empathise with and admire species they previously thought of as dangerous, it can also enhance desires for travellers to approach and attempt 'selfie' photos with wild animals, which can be dangerous for the animal as well as for the humans (citing for instance the cases of a young dolphin dying and a woman gored by a bison while taking selfies). Tour operators and travel agencies and scientists alike would do well to consider whether their photos may also encourage unwise approaches to wild animals.

14.3 Disturbing Animals in Captivity: Animal Welfare Implications

Close encounters such as cuddling koalas (a topic of much debate, for example, Hansard 1995), feeding parrots, patting young crocodiles and riding camels or elephants raise many questions. There appears to be much strong emotion and too little research (although some does exist, as cited below) on the degree of harmful stress caused to animals by such activities. Some animal rights groups would ban all such encounters. Right Tourism (2012) for instance advises travellers "Do not patronise places that use any wild animal as a prop for tourist photographs Never pay to feed or pose with a wild animal". The Intrepid Travel website (https://www.intrepidtravel.com) states "Responsible operators like Intrepid won't offer activities that involve animal welfare issues like riding elephants. Intrepid has committed to animal welfare (we were one of the first to put a stop to elephant rides on our trips)". Shcneider (2016) quotes the Director of Wildlife at World Animal Protection as saying "We need to stop the demand for elephant rides and shows, hugs and selfies with tigers and lions by exposing the hidden suffering behind wildlife attractions. If you can ride it, hug it or have a selfie with a wild animal, then you can be sure it is cruel. Vote with your feet and don't go". This is probably very good advice in many regions, and much of the mistreatment they report on does undoubtedly happen, but the word 'never' and the phrase 'you can be sure it is cruel' may be a little extreme.

Certainly there are cases across the world where stress levels and general suffering of animals are unacceptable (both during interaction and the housing of the animal at other times), but it does not immediately follow that the same applies to all similar attractions. If some zoos starve their animals this does not mean they all do. If some African wildlife parks allow interaction with young animals which instead of being rehabilitated into the wild are subsequently sold for canned hunting it does not necessarily imply that all do the same. Some animals used for photographs or riding are treated cruelly, but it does not logically follow that all trainers and managers are cruel, and many that I have met personally appear to have much compassion for their animals (although of course some may still unwittingly cause suffering of various levels, or unconsciously send the message to their audience that all wild animals can be approached or handled in similar fashion). Conversely, the fact that many animal handlers take good care of their charges should not lead us to complacently assume that all do so, and while many responsible hunters take conservation and welfare seriously it does not follow that all hunters pay heed to either.

It is well-known among animal handlers (including veterinarians) that animals often conceal their pain and other stress, and this has been shown in the wild, e.g. Giese's (2000) demonstration that penguins begin to experience physiological signs of stress before it becomes obvious to observers. This concealment is often considered to be an adaptation for not allowing predators or competitors to notice their vulnerability. Animal carers (pers. comm.) have thus claimed that it is impossible for an observer to tell if an animal that appears calm is still feeling considerable stress, and some go further by stating that they are invariably stressed in the presence of humans, or at least humans they do not know very well, and therefore oppose all public interaction at zoos, wildlife parks and traveling wildlife displays, or any approach closer than an arbitrary distance either in captivity or in the wild. Many zoo keepers and others who handle such animals however (pers. comm. with managers and keepers at various operations, including Bush Buddies, Currumbin Wildlife Sanctuary, Melbourne Zoo, Dreamworld, Elephant Whispers, Geckos Wildlife and others) report that the degree of stress varies greatly with the species, the individual (basic temperament, past experience, training) and daily mood or health level, that animals roam freely in spacious enclosures between handling, and that with due care (rotation to allow ample resting between interactions, patient habituation, careful selection of individuals on any particular day), stress can be minimised during such activities. Dreamworld (Australia) has taken the problem seriously enough to donate $10,000 to Griffith University researchers to conduct further research into stress induced by handling captive koalas. Animal handlers (pers. comm. As above) often consider close encounters between humans and wildlife to be important for teaching people to care about the animals and also for raising money to care not only for the animals they own but many injured and orphaned animals that are brought to them by the public and in situ conservation projects they support elsewhere.

There thus arise subjective and highly polarised views as to whether individual animals are indeed suffering high levels of stress. Animals themselves, whether wild or captive, sometimes approach humans quite closely, sometimes to the point of deliberately making contact, whether out of curiosity, aggression or expectation of food, while others spontaneously hide or flee (extensive pers. obs., and many reported examples in the literature and in social media). Others suggest that occasional mild stress may cumulatively be less than the same animal would be likely to experience in the wild, with associated predation attempts, scarcity of food on some days, parasites, inclement weather and other uncomfortable situations.

There has in fact been an increase in research on the effects of visitors on zoo animals in recent years (e.g. Baird et al. 2016; Benesch 2007; Carder and Semple 2008; Davey 2007; Fernandez et al. 2009, Sherwen et al. 2015a, b, c). Some have demonstrated avoidance of human visitors by captive animals, or signs of stress such as increased aggression or stereotypic behaviour, others have failed to show an effect and some suggest that stimulation by visitors can be a form of enrichment. Analysis for substances such as cortisol in serum, saliva or urine (Bayazit 2009) provide a way of comparing stress levels in different situations or between animals, as have behavioural factors such as time spent resting, feeding or watching visitors, increased aggression within the social group, or avoidance of humans. Measurement of stress in animals and the interpretation of results are not always straightforward (Tribe 2008; Sade 2013; Sherwen 2015c), and the degree of stress can be expected to vary greatly between species, between individuals within the species, and between different situations. There remain wide gaps in our knowledge, and further research is needed (while for ethical reasons avoiding causing them extreme stress during such research) to determine likely stress levels for different species, individual variation within a species, and different situations (Sherwen et al. 2015a).

Inspections of conditions is now common in major zoos. Draper and Harris (2012) describe factors looked for in British zoos relating to provision of food and water, suitable environment, animal health care, opportunity to express most normal behaviour and protection from fear and distress, but raise concerns as to how adequately these are assessed, with poor standardising of methods, and tendency to base assessments on welfare inputs rather than on outcomes. Keepers (pers. comm.) who wish to remain anonymous have also told me that far too little genuine inspection occurs in some captive attractions in Australia.

Attractions featuring captive wildlife do face an obvious dilemma. Tourists visiting zoos or wildlife parks want to see the animals they paid entry fees for, but animals do not always want to be permanently on display. There is a welcome trend nowadays towards larger, more natural enclosures, where animals desiring some privacy can curl up in a log or rest behind the vegetation, but of course this can disappoint some visitors. Concealed viewing platforms with "peep-holes", video cameras in the animal's hideaway location displaying its activities to the public, and windows through which the spectator can see the animal but not vice versa have been employed at some venues, to simultaneously satisfy the animal's need of privacy and the desire of the visitor to view it.

14.4 Disrupting Animal Activities in the Wild: Wildlife Conservation Implications

There is considerable conjecture about effects of disturbance of animals on population levels but (as lamented by Higginbottom (2004), Green and Higginbottom (2001) and Rodger and Calver (2005), and a number of other researchers) insufficient evidence in most cases. Calver (2005) suggests that "[an] emphasis on testing predictions derived from hypotheses may be the most important approach wildlife biology can give to applied studies of the impacts of wildlife tourism".

If feeding, breeding or resting behaviour of exhausted animals is frequently disrupted, it is possible that populations could decline, especially where individual deaths have in fact been recorded (e.g. the transition of fatal diseases to gorillas as mentioned above), and a responsible policy would be to err on the side of caution.

A few local extinctions due to visitor impact have indeed been recorded. Rock wallabies on one of the peaks in the Warrumbungle Ranges of Australia appeared to constitute a healthy and stable population, but when a road below the peak and a walking track to the summit were constructed, the wallabies were frequently disturbed by walkers, resulting in frequent roadkill, and by the early sixties 'the last rock wallaby had gone from this mountain, run over on the road below as it tried to evacuate' Fox (1982).

Preventing predators from catching their prey can lead to declines. Roe et al. (1997) reported that vehicles at Masai Mara National Park sometimes disrupt a predator's hunting or subsequent feeding activity, causing obvious immediate loss of sustenance but also an unidentified stress-related disease similar to HIV, and a shift from hunting at night, which was more energy efficient, to hunting in the heat of the day when tourists were absent, and they report a 30% decline in the cheetah population over four years,. Further examples of the effects of tourist disturbance of animals that have or potentially could lead to population decline can be found in Green and Higginbottom (2001).

Any of the disturbances to individual animals mentioned above could have the potential to cause a decline in population, but without sufficient research, or at least well-planned and standardised monitoring of new developments, it is difficult to resolve difference of opinions and confidently plan for wildlife conservation in tourist areas.

14.5 Responsibility Towards Human Stakeholders: Tourists, Tour Operations and Local Residents

A large and noisy party of tourists, shouting at a koala or thumping the tree to wake it up for better photographs could well persuade the koala to stay further away from walking tracks and ecolodges in the future. Even a quiet walk at night spotlighting animals could have an effect if there is much crunching of gravel or twigs (Wilson 1999), or if animals are disturbed by flashlight photos or spotlights too close to them. Shouting and pointing at shy birds or platypus can cause them to immediately hide. The local population decline of dolphins recorded by Begder et al. (2006) in response to tourist boats probably did not indicate a general population decline, but as they point out it could.have a negative effect on the tourist industry that is so important to the local economy. The tourists at the time of the initial disturbance may well be satisfied that they have good views and photographs, and be unaware of their effect on the animals, but even if the animal itself is not seriously affected, the behaviour of such visitors could be spoiling things for the next group, which affects both the satisfaction of future tourists and possibly the livelihood of other tour guides if they repeatedly find their viewing areas are no longer suitable.

Safety of tourists and local residents is another consideration. Feeding large or potentially aggressive animals can cause problems when these animals subsequently see other tourists (or residents) and seek or demand handouts from them. Some animals can be quite dangerous, others more of a nuisance. There are signs in several parts of Kruger National Park warning people not to feed hyenas or baboons, both capable of inflicting severe injury. Feeding of dingos by tourists in Australia has caused some to lose their fear of humans, and people have been attacked and at least two children killed (Burns et al. 2011; OMICS 2014). Kookaburras and ibis are unlikely to inflict serious wounds, but they and many other species can invade picnics, land on tables, steal food and generally cause considerable annoyance (pers. obs).

Local residents vary widely in their attitudes towards wildlife (especially some controversial animals such as bats: e.g. Roberts et al. 2011), but many do appreciate and enjoy their presence (e.g. Fitzgibbon and Jones 2006), and want their children to continue to do so, and are disappointed to see this wildlife disappearing from their area. None of the three species of large kangaroos are threatened globally, but local kangaroo populations are under threat in some areas. Koala populations are flourishing in some regions, and they are breeding well in some zoos, but diminishing quite dramatically in numbers at other sites. Brisbane was once dubbed the "Koala Capital" by the Brisbane City Council and Lonely Planet, but there has been rapid decline in recent years (Rhodes et al. 2015), and reserves in Greater Brisbane where sightings were almost guaranteed less than a decade ago now seem almost or entirely devoid of koalas (pers. obs. based on many visits over the years), which is disappointing for tourists and local animal-lovers alike. Other species are likewise diminishing in numbers in both suburban and rural areas (e.g. FitzGibbon and Jones 2006).

An example of uncontrolled tourism affecting not only the wildlife but also local residents is the situation at a site never intended for tourism: a psychiatric hospital. According to Gordon (2014), the population of kangaroos coming to graze on the lawns of the Morisset Hospital has become very popular with tourists, many of whom have promoted the site on social media, and it has been recommended on Lonely Planet, Trip Advisor and other websites. This is causing problems for the kangaroos in that tourists are disturbing them from their activities, feeding them inappropriate foods such as bread, running into them with vehicles and at least one occasion even attempting to 'box' with them. It is a problem for the locals, including the privacy of the patients and local road conditions, as crowds of visitors walk up the road to the hospital, blocking traffic to the hospital for staff and visitors, and also to local schools at times when children and being delivered to or picked up from those schools. It is to be hoped that most tour guides would consider responsibility towards the local residents as well as the animals, and refrain from causing such disruption.

Tourists attempting to get close to animals can also affect local residents by trespassing on private lands, leaving gates open, feeding animals which then expect local residents to do the same, scaring animals away from homes where locals enjoy seeing, wildlife or into yards with dogs that may attack them, and causing other undesirable impacts. There is also a risk that attracting animals for tourism can affect neighbouring livelihoods (e.g. parrots or elephants near crops, leopards near farm animals, or bears near the herding routes for reindeer described by Iivari (this volume). An increase of large animals such as lions or buffalo near villages, or their habituation to human presence without actually being tame, can be a serious problem for food supplies (whether in crops, paddocks or food storage facilities) but also directly dangerous for the residents themselves. Such problems can become complex and demand holistic solutions which are not always easy to find (but see https://biglife.org/predator-compensation for one apparently successful project) (Fig. 14.2).

14.6 Doing It Right

How then do we best minimise the negative impacts of tourists on wildlife and accentuate the positive ones? The first step is to understand the impacts and collate what we

Fig. 14.2 When wildlife enters neighbouring residential areas it can be a challenge for tour guides to allow guests to view and photograph the wildlife without intruding on the privacy of the residents. Photo by author

know so far about ways of avoiding or at least mitigating the negative, and what might work for the positive impact. Much further research is still needed, but there is already a lot that we do know, and can base some management plans on.

Wolf and Croft (2010) compared the effect of different styles of visitor approaches to kangaroos after observations on the most common visitor behaviour: driving and walking on and off trails, direct or tangential approach, continuous approach *versus* stop-and-go, and quiet approach *versus* talking while walking. Kangaroos were less readily disturbed (as measured by distance of observer to kangaroo when the animal fled) by the approach of vehicles than by walkers. Comparing walking methods, they were least disturbed when walkers stayed on the trails, did not talk, and used the stop-and-go method rather than continuous approach. Taylor and Knight (2003) found ungulates in North America were less likely to flee if approached obliquely rather than directly, and that approaches by mountain bikes and hikers were similar in their effects.

There appears to be growing interest in nocturnal viewing of Australian wildlife by visitors, whether on tour or staying at ecolodges or farm-stays. Wolf and Croft (2012) found kangaroos to be less likely to flee and more likely to continue normal behaviour (social interaction, body maintenance) at night if viewed with infrared rather than either red or white light. Visitors they interviewed also expressed a preference for using night vision equipment over the conventional spotlights. Wilson (1999) studied the effects of various sounds as well as light intensity on possums, finding that quiet adult voices and vehicles traveling past had less effect than the crunching of gravel or snapping of twigs, possibly because these are sounds associated throughout their evolutionary history with the approach of predators. These kinds of findings need to be more widely known by visitors and tour operators.

Giese (1996) compared different ways of approaching penguins, and the approach distance at which early signs of stress (as measured by heart rate) occurred, and subsequently made an educational video (Giese 2000) to be shown to visitors on their way to the Antarctic.

Many guidelines are already offered for tour operators, based on our current knowledge. A common guideline for any tour involving swimming with marine animals or encountering wild apes is to never approach the animal, but to allow it to approach you, and this needs to be combined with education on how to behave if an animal does in fact approach.

There are strict guidelines for marine wildlife tourism operators such as whale watching or swimming with whale sharks (e.g. Ecocean, date not supplied), although this can be difficult to police in the vastness of the ocean.

Higginbottom et al. (2003a) provide a study of operators including kangaroos or other macropods in their tourism activities throughout Australia, and an extensive list of ways in which such operations could be improved in terms of reducing negative impacts as well as education of tourists and remaining financially viable. Green (2013) also suggests a number of ways of contributing positively to conservation and improving interpretation of wildlife to visitors.

Operators can contribute to conservation and welfare by quality interpretation enhancing their guests' understanding of the needs of the animals, and behaviour appropriate to not negatively affecting this. A good message needs to be well-delivered if it is to be understood and memorable. I have seen guides talking in rapid English although most of their group were not well-versed in the language, or talking in a monotone while their audience paid attention to the surrounding scenery, the flies, or anything but what was being said. Others are excellent in grabbing the attention and interest of all present, and expressing things clearly, sometimes humorously, and in ways that will not soon be forgotten. Attractive, clear signage also helps, as do brochures, if the message is clear and the appearance is attention-grabbing (e.g. Ballantyne et al. 2007; Moscardo et al. 2004). Simply writing 'Don't feed the wildlife' can sound too authoritative and even mean to those unfamiliar with the issues, and it is probably more effective to briefly explain the reasons. There are some who will never care, but others who will, and simply do not realise potential impacts of their behaviour. Much has been written in recent years on good and memorable interpretation that encourages environmental responsibility (e.g. Ham and Weiler 2002; Hughes and Ballantyne 2013; Weiler and Black 2014).

Innovative measures such as video cameras in dens, two-way mirrors, peep-holes in fences or hides can provide ways of seeing captive animals or wild animals near eco-lodges up close without disturbing them.

Managers of eco-accommodation can conserve and restore native vegetation on their properties, including areas of restricted human access where animals can shelter in peace when they feel the need, and encourage neighbours (including local councils) to do the same. They can also fence off areas that are hazardous to wildlife, such as routes to busy roads, while affording free access to surrounding grasslands and forest. Provision of water could be considered, especially during drought or after bush-fires. Monetary donations or active involvement in local or national conservation projects is useful in itself and an additional educational point for guests, some of whom might also donate time or money to conservation efforts (Fig. 14.3).

Wildlife Tourism Australia Inc. (2016a, b) has developed guidelines to interactions with wildlife after a workshop discussion around the circumstances under which tourists should be permitted to feed wildlife or indulge in other interactions, and also provide links to guidelines by other organisations.

Monitoring is important, to determine whether management actions are having the desired effect. Newsome et al. (2005) recognise monitoring as an essential part of adaptive management (i.e. management that is progressively refined by new information) but comment that monitoring is a part of management that seems to always "fall between the cracks" when resources are limited. Higgenbottom et al. (2003a, b) offer tips on managing wildlife tourism for minimising effects on wildlife and monitoring such efforts. Much useful advice and information on monitoring is also found in Lindenmeyer and Gibbons (2012).

14.7 Research Needed

There are still many knowledge gaps regarding the impacts of wildlife tourism. Rodger and Calver (2005) point to barriers that impede wildlife biology research relevant to wildlife tourism, including an apparent lack of interest in tourism impacts (as opposed to other aspects of conservation biology) by wildlife scientists and the expenses and complexities involved in conducting relevant research. Another complication is that researchers need to investigate relevant broad topics while conservation managers need specific information on the area under their control, but it is often possible to develop research programs which do both simultaneously.

Some of the issues on which we need further research include:

- the degree of stress experienced by wild and captive wildlife involved in various kinds of interactions with humans, perhaps including the exploration of animal preferences along the lines of Dawkins (2008);
- what kinds of disturbance result in potentially serious effects such as animals abandoning their favoured feeding grounds, and what effect this may have on their health, especially during lean seasons;
- under which situations and for which species repeated disturbance by tourist activity (whether general noise and activity, approach to nests, interference with prey stalking, inappropriate feeding, vehicular traffic or consumptive activities such as hunting and fishing) results in a

Fig. 14.3 Tourists can be trained to approach wildlife with minimal impact. Here, Araucaria Ecotour guests have been advised to walk obliquely past the kangaroos, gradually getting closer, not directly towards them, and to retreat quietly at any sign of nervousness by the animals. Photo by author

decrease in local populations (and if so, whether such decrease is due to reduced birthrates, increased mortality, or simply animals leaving the district). This would generally need to be a long term project, and there may be scope for synergies between researchers and tour operations (see Marcoll and Tribe, this volume, and the Wildlife Reseatch Network referred to below).

- the effects of wildlife tourism activities on cryptic species that are seldom noticed by tourists or even the operators;
- the degree to which close encounters affect subsequent attitudes to conservation and resultant behaviour by tourists (this should guide the extent to which compromises might be made between tourist desires and attitudes on one hand, and wildlife welfare and conservation on the other).
- how disturbance of individual animals of various species relates to population declines.

It would not be ethical to examine the above questions by deliberately disturbing animals to the point where their populations are decreased, but it should be possible to find a variety of localities experiencing different levels and different kinds of disturbance, and comparing these over several years (see Newsome et al. 2005). One could also conduct surveys in sites about to undergo development and compare both condition of individuals and population numbers before and after (preferably on several occasions before the development and regular intervals thereafter), and also observe the kinds of disturbance, if any, to both conspicuous and more cryptic species. A further possibility would be to seek out areas of apparently high disturbance and make improvements, with everything else remaining the same as far as possible, then observe if populations increase.

Waiting for population changes, especially as we would always predict year-to-year variation based on external factors such as weather, would necessitate a very long-term study. Conditions of animals however can also provide clues within a shorter timeframe: are they showing signs of starvation (e.g. visible rib pattern or hollowing of hips) or other health problems (lumpy jaw, matted coats, scouring) and how well are they breeding (what proportion of female mammals have young at foot, are birds taking food to nestlings)?

It may be difficult to observe effects on threatened species, partly because they may be rarely encountered and partly because some kinds of study could increase pressures on them. There is still merit in conducting research on common animals (e.g. Wolf 2010), which could still be subject to local declines, and the results used to extrapolate (with due caution) to less common relatives. And even if not a threatened species, a decline in local population numbers of key species such as pollinators, seed dispersers or top predators could cause disruptions to ecological processes.

Beckmann and Shine (2012) sound a warning for those relying on responses to survey questions. Their respondents reported that they would deliberately run over introduced cane toads far more often than native frogs or snakes, but experimental placings of models of all three in the centre of the road (so drivers could very easily avoid them) showed no difference in the tendency to run over the animal. The results are somewhat depressing in themselves, and also indicate a need to use methods other than questionnaires, at least as a backup, where possible.

Calver (2005) offers some quite detailed advice on: developing a hypothesis-testing framework for wildlife biology relevant to wildlife tourism, developing testable predictions, choosing study techniques (including experimental and other approaches), analysing data and determining how the results can be used into adaptive management of wildlife tourism.

Wildlife Tourism Australia is developing a wildlife research network (Wildlife Tourism Australia 2016c) designed to facilitate communication between: (1) tour operators conducting research and interested in collaborating or exchanging ideas, (2) researchers and tour operators willing to provide low-cost or free regular transport to research sites (marine and terrestrial), accommodation or entry to captive wildlife facilities and (3) tour operators involved in research with tourists wishing to be temporary research assistants (and often willing to pay for same, if some of their financial contribution assists further research or conservation measures).

14.8 Conclusions

Very little wildlife tourism could claim a zero impact on wildlife, but its impact is often less than alternative land uses, and small disturbances may make very little difference to many individual animals or wildlife populations. There is however potential serious impact of some wildlife-related tourist activities, with negative effects on animal welfare and local wildlife populations, and also on local human residents and other tour operators and their guests. Tour operators, developers and governments should be aware of possible problems and plan accordingly, aiming at minimal disturbance and where possible engaging in well-thought out monitoring activities and appropriate interpretation for visitors. We need to disseminate available knowledge more widely. We also need to encourage more ecological research to determine under what circumstances the disturbance of individual animals is likely to lead to local population decline, more behavioural and physiological research into stress factors associated with human-wildlife interaction, and more research into the effects of wildlife encounters on subsequent human attitudes towards wildlife conservation and welfare. Such research will face challenges and may need innovative approaches and collaboration between stakeholders to ask the right questions, interpret results in meaningful ways and achieve solutions that will benefit both humans and wildlife.

References

Anderson DW, Keith JO (1980) The human influence on seabird nesting success: conservation implications. Biol Conserv 18:65–80

Bacon HJ (online content) Assessment of animal welfare in zoos. Retrieved from http://www.ed.ac.uk/files/imports/fileManager/Assessment%20of%20animal%20welfare%20in%20zoos%20(1).pdf. Nov 2016

Baird BA, Kuhar CW, Lukas KE, Amendolagine LA, Fuller GA, Nemet J, Willis MA, Schook MW (2016) Program animal welfare: using behavioral and physiological measures to assess the well-being of animals used for education programs in zoos. Appl Anim Behav Sci 176:150–162

Ballantyne R, Moscardo G, Hughes K (2007) Designing interpretive signs: principles in practice. Fulcrum Publishing, Golden, CO, USA

Bayazit V (2009) Evaluation of cortisol and stress in captive animals. Aust J Basic Appl Sci 53(3):1022–1031

Beckmann C, Shine R (2012) Do drivers intentionally target wildlife on roads? Austral Ecol 37(5):629–632

Bejder L, Samuels A, Whitehead H, Gales N, Connor MJ, Heithaus R, Watson-Capps J, Laherty C, Krutze M (2006) Decline in relative abundance of bottlenose dolphins exposed to long-term disturbance. Conserv Biol 20(6):1791–1798

Benesch AR (2007) Chronoethological assessment of well-being and husbandry in captive koalas, *Phascolarctos cinereus*, Goldfuss 1817. Ph.D. thesis. Goethe Universitätist, Frankfurt. Available online at http://publikationen.ub.uni-frankfurt.de/frontdoor/index/index/year/2007/docId/902

Burns GL (this volume) Ethics and responsibility in wildlife tourism: lessons from compassionate conservation in the anthropocene. In: de Lima IB, Green RJ (eds)

Burns GL, Moore S, Macbeth J (2011) Should dingoes die? Principles for engaging ecocentric ethics in wildlife tourism management. J Ecotour 10(3):179–196

CALM (Department of Conservation and Land Management, Western Australia) (2000) Recreation, tourism and visitor services: draft policy guidelines. Department of Conservation and Land Management, Bentley, Western Australia

Carder G, Semple S (2008) Visitor effects on anxiety in two captive groups of western lowland gorillas. Appl Anim Behav Sci 115(3): 211–220

Calver M (2005) Researching ecological impacts. In: Newsome D, Dowling RK, Moore SA (eds) Aspects of tourism: wildlife tourism. Channel View Publications, Clevedon

Davey G (2007) Visitors' effects on the welfare of animals in the zoo: a review. J Appl Anim Welfare Sci 10(2):169–183

Dawkins MS (2008) The science of animal suffering. Ethology 114(10): 937–945

de Waal F (2016) Are we smart enough to know how smart animals are?. W.W. Norton & Co., New York

Department of the Environment and Heritage (2005) Australian national guidelines for Whale and Dolphin watching. Retrieved from https://www.environment.gov.au/system/files/resources/fed9ff86-0571-43ff-bb18-32205fc6a62c/files/whale-watching-guidelines-2005.pdf. Nov 2016

Draper C, Harris S (2012) The assessment of animal welfare in British zoos by government-appointed inspectors. Animals 2(4):507–528

Ecocean (2016) Best practice whale shark ecotourism manual. Retrieved from http://www.whaleshark.org/ECOCEAN%20Best%20Practice%20Whale%20Shark%20Ecotourism%20UNEP%20MANUAL.pdf. Oct 2016

Fennell D (2015) The status of animal ethics research in tourism: a review of theory. In: Markwell K (ed) Animals and tourism: understanding diverse relationships. Channel View Publications, Clevedon, pp 27–43

Fernandez EJ, Tamborski MA, Pickens SR, Timberlake W (2009) Animal–visitor interactions in the modern zoo: conflicts and interventions. Appl Anim Behav Sci 120(1–2):1–8

Fitzgibbon SI, Jones DN (2006) A community-based wildlife survey: the knowledge and attitudes of residents of suburban Brisbane, with a focus on bandicoots. Wildlife Res 33(3):233–241

Fox A (1982) Conservation vs. recreation: national parks at the crossroads. Australian Science Magazine (Apr/May/June 1982), pp 16–19

Frost W (ed) (2011) Zoos and tourism: conservation, education, entertainment?. Channel View Publications, Clevedon

Garlick S, Austen R (2014) Post-traumatic stress disorder in kangaroos. Paper presented at australian wildlife rehabilitation conference, Hobart, Tasmania, May 2014

Giese M (1996) Effects of human activity on Adelie penguin *Pygoscelis adeliae* breeding success. Biol Conserv 75(2):157–164

Giese M (2000) Giving penguins their space: visiting breeding Adelie penguins without causing disturbance [Video presentation]. Department of environment, Commonwealth of Australia and the Australian Antarctic Division, Tasmania

Gordon J (2014) Tourist influx threatens kangaroos, poll, photos. Newcastle Herald. Retrieved from http://www.theherald.com.au/story/2016694/tourist-influx-threatens-kangaroos-poll-photos/ Aug 2016

Green RJ (2013) Wildlife tourism: a handbook for guides, tour operators, job-seekers and business start-ups [Kindle Edition], Unknown Publisher Amazon.com

Green RJ, Higginbottom K (2001) The negative effects of wildlife tourism on wildlife. Wildlife tourism research report no. 5. CRC for Sustainable Tourism. Gold Coast, Queensland

Ham SH, Weiler B (2002) Interpretation as the centrepiece of sustainable wildlife tourism. In: Harris R, Griffin T, Williams P (eds) Sustainable tourism: a global perspective. Butterworth Heinemann, Oxford, pp 35–44

Hanes AC (2012) Seven years later: can we maintain the 7-metre gorilla tracking distance? Gorilla J 44:9–11

Hansard (1995) Exhibited animals protection act: disallowance of regulation. Egislatve Council, Parliament of New South Wales. 21 Nov 1995

Harris R, Leiper N (1995) Sustainable tourism: an Australian perspective. Butterworth-Heinemann, Chatswood, New South Wales

Higginbottom K (ed) (2004) Wildlife tourism: impacts, management and planning. Common Ground Publishing, CRC for Sustainable Tourism, Gold Coast

Higginbottom K, Green RJ, Leiper N, Moscardo G, Tribe A, Buckley R (2003a) Evaluation of organised tourism involving wild kangaroos. Wildlife tourism research report no. 18. CRC for Sustainable Tourism. Gold Coast, Queensland

Higginbottom K, Green RJ, Northrope CL (2003b) A framework for managing the negative impacts of wildlife tourism on wildlife. Hum Dimens Wildlife 8(1):1–24

Higginbottom K, Northrope CL, Green RJ (2001) Positive effects of wildlife tourism on wildlife. Wildlife Tourism research report no. 6. CRC for Sustainable Tourism. Gold Coast, Queensland

Higham JES (1998) Tourists and albatrosses: the dynamics of tourism at the Northern Royal Albatross colony, Taiaroa Head, New Zealand. Tourism Manag 19(6):521–531

Hough IJ (2006) Jaw disease in macropods. Marsupial Society. Retrieved from http://www.marsupialsociety.org/jaw_disease2.html. Aug 2016

Hughes K, Ballantyne R (2013) Winning hearts and minds through interpreting ecotoursim experiences: walking the talk. In: Ballantyne R, Packer J (eds) International handbook on ecotourism. Edward Elgar, Cheltenham, UK, pp 322–335

Iivari P (this volume) Land-use conflict and perspectives for its resolution—wildlife watching meets reindeer herding in the European North, In: de Lima IB, Green RJ (eds)

Jones DN (2011) An appetite for connection: why we need to understand the effect and value of feeding wild birds. Emu 111:i–vii

Kaplan G (2015) Bird minds: cognition and behaviour of Australian native birds. CSIRO Publishing, Clayton North, Victoria

Kauffman L (2010) Primate response to tourist presence at Raleighvallen, Central Suriname Nature Reserve. Doctoral thesis. University of Florida

Lindenmeyer D, Gibbons P (eds) (2012) Biodiversity monitoring in Australia. CSIRO Publishing, Clayton North, Victoria

Marcoll and Tribe (this volume) Wildlife tourism and conservation: the hidden Vale Project. In: de Lima IB, Green JR (eds)

Moscardo G, Woods B, Saltzer R (2004) The role of interpretation in wildlife tourism. In: Higginbottom K (ed) Wildlife tourism: impacts, management and planning. Common Ground Publishing, Altona, VIC, Australia, pp 231–251

Newsome D, Moore SA, Dowling RK (2002) Natural area tourism: ecology, impacts and management. Channel View Publications, Clevedon

Newsome D, Dowling RK, Moore SA (2005) Aspects of tourism: wildlife tourism. Channel View Publications, Clevedon

Oberbillig DR (2000) Providing positive wildlife viewing experiences, a practical handbook. Colorado Division of Wildlife and Watchable Wildlife, Inc

OMICS International (2014) Dingo attack. Open access article. Available at http://research.omicsgroup.org/index.php/Dingo_attack. Oct 2016

Pople A, Grigg G (1999) Commercial harvesting of Kangaroos in Australia. Department of Zoology, The University of Queensland, for Environment Australia

Reiser (this volume) Will the ark sink? Captive wildlife, tourism and the human relationship to nature: demystifying zoos. In: de Lima IB, Green JR (eds)

Rhodes JR, Beyer HL, Preece HJ, McAlpine CA (2015) South East Queensland Koala population modelling study. UniQuest, Brisbane, Australia

Right Tourism (2012) Photo taking with animals. Available at http://right-tourism.com/issues/animals-attractions/photo-prop-animals/#sthash.WnkazUcb.dpbs. Nov 2016

Roberts B, Eby P, Catterall C, Kanowski JJ, Bennett G (2011) The outcomes and costs of relocating flying-fox camps: insights from the case of Maclean, Australia. In: Law B, Eby P, Lunney D, Lumsden L (eds) The biology and conservation of Australasian bats. Royal Zoological Society of NSW, Mosman, pp 277–287

Rodger K, Calver M (2005) Natural science and wildlife tourism. In: Newsome D, Dowling RK, Moore SA (eds) Aspects of tourism: wildlife tourism. Channel View Publications, Clevedon, pp 217–234

Roe D, Leader-Williams N, Dalal-Clayton B (1997) Take only photographs, leave only footprints: the environmental impacts of wildlife tourism. In: IIED wildlife and development series, vol 10. International Institute for Environment and Development, London, UK

Rose K (2005) Common diseases of Urban wildlife: common diseases—Myopathy and Trauma June. Australian Registry of Wildlife Health

Sade C (2013) Visitor effects on zoo animals. Plymouth Stud Sci 6 (1):423–433

Schlenker LS, Latour RJ, Brill RW, Graves JE (2016) Physiological stress and post-release mortality of white marlin (*Kajikia albida*) caught in the United States recreational fishery. Conserv Physiol 4(1):1–15

Sherwen SL, Harvey TJ, Magrath MJ, Butler KL, Fanson KV, Hemsworth PH (2015a) Effects of visual contact with zoo visitors on Black-capped Capuchin welfare. Appl Anim Behav Sci 167(June 2015):65–73 doi:10.1016/j.applanim.2015.03.004

Sherwen S, Trista JH, Magrath MJL, Butler KL, Fanson KE, Hemsworth PH (2015b) Effects of visual contact with zoo visitors on black-capped capuchin welfare. Appl Anim Behav Sci 167:65–73

Sherwen SL, Hemsworth PH, Butler KL, Fanson KV, Magrath MJL (2015c) Impacts of visitor number on kangaroos housed in free-range exhibits. Zoo Biol 34(4):287–295. doi:10.1002/zoo.21226

Taylor AR, Knight RL (2003) Wildlife responses to recreation and associated visitor perceptions. Ecol Appl 13(4):951–963

Tribe A (2008) Zoos and animal welfare. In: Proceedings of: AAWS 2008: the welfare of animals—it's everyone's business. AAWS the Australian Approach. AAWS 08: international animal welfare conference. Gold Coast, QLD, Australia, 31 Aug–3 Sept 2008, pp 1–10

Ward-Paige CA (2016) Even scientists take selfies with wild animals. Here's why they shouldn't. The Conversation. 28 June. Retrieved from https://theconversation.com/even-scientists-take-selfies-with-wild-animals-heres-why-they-shouldnt-61252. Nov 2016

Weiler B, Black R (2014) Tour guiding research insights, issues and implications. Channel View Publications, Clevedon

Wildlife Tourism Australia Inc. (2016a) Policy: tourist wildlife interactions http://www.wildlifetourism.org.au/conservation/policies/policies/tourist-wildlife-interactions/

Wildlife Tourism Austrlia Inc. (2016b) Australia wildlife viewing guidelines http://www.wildlifetourism.org.au/conservation/policies/best-practices/australian-wildlife-viewing-guidelines/

Wildlife Tourism Austrlia Inc. (2016c) Wildlife research network (http://www.wildlifetourism.org.au/resources/research-by-tour-operators-etc/research-by-tour-operators-etc-background-information/)

Wilson R (1999) Possums in the spotlight. Nat Aust 26(4):35–41

Wolf ID, Croft DB (2010) Minimizing disturbance to wildlife by tourists approaching on foot or in a car: a study of kangaroos in the Australian rangelands. Appl Anim Behav Sci 126(1):75–84

Wolf ID, Croft DB (2012) Observation techniques that minimize impacts on wildlife and maximize visitor satisfaction in night-time tours. Tourism Manag Perspect 4(2012):164–175

15 Land-Use Conflict and Perspectives for Its Resolution—Wildlife Watching Meets Reindeer Herding in the European North

Pekka Iivari

Abstract

The purpose of this study is to analyse the land-use conflict between wildlife watching tourism and reindeer herding, and to present an analytical framework for studying conflict resolution mechanisms. The paper addresses the need for careful management of conflicts in wilderness and remote areas in circumstances where a new industry meets traditional users of lands. Particularly interesting from the viewpoint of this paper are areas and resources that have the potential to become predictive of conflictual intersectoral relations. Conflict resolution mechanisms elaborated by impartial experts and crucial stakeholders consist of legislative proposals and voluntary actions which address the sensitivity of the issue, as well as local traditions in land-use. This paper proposes a conflict resolution model, in which collaborative conflict anatomy identification, impartial expert involvement, tourism industry codes of practices, and legislative proposals aim at bringing the intersectoral conflict to a manageable level. The model adds to the social sustainability of competing land usage by offering a venue for collaboration and local community influence on decision-making.

15.1 Introduction

Tourism is one of the major driving forces behind land use and economies in Northern Finland. Tourism's impact on the environment is significant due to the fact that it operates in areas that are environmentally and socio-culturally sensitive, and where competing livelihoods each have their own demands on land use. Rural communities in remote sub-arctic regions of Northern Finland consider nature-based tourism an asset in the improvement of local livelihoods (Uusitalo 2010).

Development and marketing of wildlife tourism as a form of nature-based tourism has led to rapid growth in wildlife watching tourism in Northern Finland during the past two decades. The process has incited an intersectoral land-use conflict between wildlife watching tourism and reindeer herding, as tourism activities expand into remote areas traditionally used by local reindeer herders. Wildlife tourism has both positive and negative impacts on local economies. An increased number of tourists and expansion of wildlife tourism companies may not only disturb the indigenous culture, but also change the sensitive environment in which the reindeer herders live (Kitti et al. 2006; Tolvanen and Kangas 2016). The emerging conflict has spawned demands to modify current management regimes regarding land use, particularly when it comes to attracting large carnivores for commercial wildlife watching purposes.

Predator impacts on reindeer and local livelihoods are almost never taken into account when wildlife tourism is promoted (Saberwal et al. 1994), although the Finnish Tourism Strategy considers wildlife and nature-based tourism as the spearhead of the tourism industry. Moreover, the effects of wildlife tourism on the reindeer herding environment are still poorly understood, and there is an apparent lack of scientific oversight of their interaction and conflict management mechanism of competing livelihoods.

P. Iivari (✉)
Multidimensional Tourism Institute, Lapland University of Applied Sciences, Rovaniemi, Finland
e-mail: pekka.iivari@lapinamk.fi

I. Borges de Lima and R.J. Green (eds.), *Wildlife Tourism, Environmental Learning and Ethical Encounters*, Geoheritage, Geoparks and Geotourism, DOI 10.1007/978-3-319-55574-4_15

15.2 Literature Review

According to Melling (1994), the term "conflict resolution" has been described as "a process by which two or more conflicting parties improve their situation by co-operative action… [allowing] the parties to expand the pie, or to prevent it from shrinking, giving each party a larger slice". The practical content of co-operative action has many options, and conflict resolution does not simply mean cessation of conflict. Tourism and land-use conflicts have been studied from the vantage point of the tourism industry, conflicts with forestry (McKercher 1992; Chaplin and Brabyn 2013; Raitio 2008; Hilsendager et al. 2016), agriculture and farming (Campbell et al. 2000; Gaughan et al. 2009), the power industry (Sæþórsdóttir 2010), local inhabitants and residents (Al Haija 2011; Stylidis et al. 2014), indigenous communities (Hall 2007; Hoffman and Rohde 2007; Pettersson 2006; Yang et al. 2013; Olsen 2016) and protected areas and landscape (Tzatzanis and Wrbka 2002; Wray et al. 2010; Haukeland et al. 2011). Many authors have described the mechanism by which tourism has created conflicts between multiple interest groups (Saremba and Gill 1991; Teye 1992; Eagles and McCool 2002; Dredge 2010; Lee et al. 2010) and how, as a consequence, shared decision-making in land-use planning has been addressed (Williams et al. 1998; Pearlman 1990; Manning 2004; Maguigad 2013; Tyrväinen et al. 2014a).

Tourism's ecological, social and political sustainability in the context of land-use planning is highlighted in an array of studies (Atik et al. 2010; Cohen 1978; Davenport and Davenport 2006; Fagence 1990; Tyrväinen et al. 2014b). Human-wildlife conflict (see, for example Conover 2002; Dickman and Hazzah 2015) is often scrutinised from the perspectives of farmers versus wildlife (Walpole and Thouless 2005), hunting (Gunnarsdotter 2006); and the impact of wildlife tourism on residents, flora and fauna. Competing land usage has occasionally escalated to human vs human conflicts over wildlife (Hall and Page 2006; Higham and Shelton 2011; Watson and Kajala 1995). Concerns about the sustainability of wildlife tourism and its negative impacts have been voiced by, for example, Higham and Lück (2008), Newsome et al. (2005), and Moore and Rodger (2010). In Finland, large carnivores have been presented as a case in point in the Finnish reindeer herding zone from the conflict management viewpoint by, for example, Bisi et al. (2007), Helle and Jaakkola (2008) and Kojola et al. (2004).

Conflict and co-operation are two aspects of human relations. Coexistence of human beings always has the potential for conflictual relationships brought about by clashes of interests, goals, values, actions, views or directions. People tend to compete for natural resources they need or want to ensure their livelihoods. Such competition for the use of, control over and access to natural resources results in disagreement, disputes and conflicts (Filley 1975; De Bono 1985; Tyler 1999; Sidaway 2005). Land-use conflicts, as a subset of natural resource conflicts, arise when competing claims over the use of land provoke clashes between groups of people. A vast amount of literature deals with management of resource conflicts from land-use, ethnic and human viewpoints (Tosi et al. 1986; Grimble and Wellard 1997; Upreti 2001; Henle et al. 2008; Ångman et al. 2016).

15.3 Wildlife Watching Meets Reindeer Herding

Traditionally, extensive use of land for reindeer herding entails complex land management issues in the northern Scandinavian reindeer herding zone (Sandström et al. 2003). In the Scandinavian countries and Finland—together forming the geographical region of Fennoscandia—reindeer husbandry has been a traditional livelihood since the seventeenth century and it still conveys important cultural and social values. Herding practices were originally adopted from the indigenous Forest Sámi (Heikkinen et al. 2012; Sköld 2015), and currently about 40% of Fennoscandia's land area is used as reindeer pasture (Tyler et al. 2007; Moen 2008), so tourism's impact on reindeer herding cannot be avoided (Forbes et al. 2004; Helle and Särkelä 2008; Hukkinen et al. 2006; Kivinen 2015). Vast nature areas—and with them traditional livelihoods—support the development of nature-based tourism products (Newsome et al. 2005; MINTEL 2008). Not only do traditional livelihoods such as reindeer herding or fishing provide a base for the development of nature-based tourism products, but in return the products help preserve these traditional forms of employment.

Reindeer are half-domesticated animals descending from deer. A fully-grown reindeer weighs 100–150 kg and it can live to the age of 18–20 years. The reindeer requires extensive pastures. They graze all the year around in the coniferous forest region where the snow depth during wintertime is normally quite small, snow structure is light and porous and where the autumn and spring migrations are short or non-existent (non-migratory). Forest areas are important for the reindeer since they are easily accessible pasture under a light snow cover, and they provide lichen on the old threes that can be a matter of survival when the snow cover happens to be hard (Nieminen and Heiskari 1989). In the summer, reindeer eat leaves, grass, sedges and rushes. In late summer and autumn, they like mushrooms whereas in the winter they eat moss and varieties of lichen growing in the clean nature of the North (Skuncke 1969; Kumpula 2001; Bezard et al. 2015).

In Finland, reindeer graze freely in the reindeer herding zone (Fig. 15.1), and the livelihood has been able to adjust to major changes in overall land use during the past decades. Dialogues between reindeer herders and other categories of primary production, such as agriculture and forestry, have gradually found their established forms. However, sporadic disputes and conflicts have occurred in Northern Finland, especially along with the commercialisation and technological development of industrial forestry (Berg 2010; Sandström et al. 2006). The same socio-technological drivers are also changing reindeer herding, which is losing some aspects of its traditional culture; yet farming, agriculture and forestry have still a subsistence role for many reindeer herding families (Müller-Wille and Pelto 1971; Heikkinen 2006; Raitio 2008).

Discourse on challenges (or opportunities) of the wildlife watching industry is irrelevant to Sámi reindeer herders living in a designated Sámi Homeland area in Finnish Lapland. The industry has not yet expanded to the area. It should be stressed here that the land-use conflict between wildlife tourism and reindeer herding does not have a Sámi human rights dimension in Finland, where anyone living within the area of Finnish reindeer husbandry and who is a citizen of the European Union has the right to own reindeer (Bernes et al. 2013). The Sami population does not enjoy a specific legal position in the reindeer ownership and herding in Finland. However, they have special rights in influencing land-use decisions in the Sámi Homeland area.

Tourism is a relatively recent phenomenon, and wildlife watching as its subset is a new type of land use, in reindeer grazing areas. Wildlife watching tourism has gained ground in Finland since the beginning of the 1990s. Finland, belonging almost entirely to the boreal coniferous forest zone, is one of only a few countries where it is possible to see the flagship species of Finnish wildlife tourism—brown bear (*Ursus arctos*), wolf (*Canis lupus*) and wolverine (*Gulo gulo*)—simultaneously at the same watching site. Wildlife watching based on large carnivores is one of the most popular tourism products in Finnish North Karelia and the districts of Kainuu and East Ostrobothnia bordering Russia. The sparsely populated and largely forested municipalities of Kuusamo, Suomussalmi and Kuhmo stand out for having the largest share of the large carnivore watching industry in Finland. According to Järviluoma (2012), key markets for Finnish wildlife watching products can be found in the UK, Germany and the Netherlands. However, the turnover of the industry is still quite modest—less than €10 M—but it is on the rise, and the indirect impact locally can be considerable (Järviluoma 2012, p. 12).

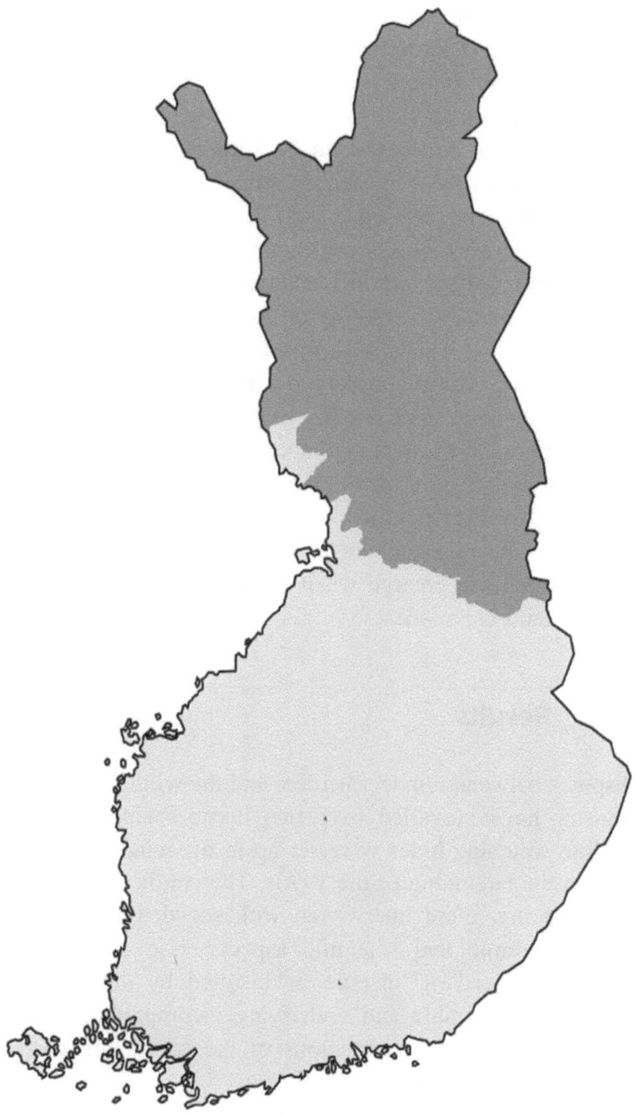

Fig. 15.1 Reindeer herding zone in Finland (marked in dark grey)

15.4 Methodology

The primary data for the research was collected by employing descriptive case study method, including expert interviews and participatory design. The case description is a written account, or merely a historiography of the sequence of steps the land use conflict and its resolution have evolved. Participatory design was notable in that the author of the paper has participated in different stages of the conflict resolution process. The paper draws on collaborative approaches to integrate four inter-related phases of conflict management. Procedures and policies are invented through inter-agency collaboration, expert reviews and scientific social and cultural data.

An appropriate conflict resolution framework is reviewed in this paper in order to better understand the parameters that have an effect on efficient management of intersectoral land-use conflicts incited by the expanding tourism industry. The framework is constructed by focusing on the conflict management process in which strategic solutions to the land-use problems are suggested.

The framework builds on four phases: (1) collaborative identification of the conflict anatomy, (2) involvement of impartial experts, (3) agreeing on tourism industry codes of practices and (4) legislative proposals. The framework seeks to bring intersectoral conflicts to a manageable level and it adds to the social sustainability of competing forms of land use by offering a venue for collaboration and by allowing the local community to have their say in decision-making. Additionally, the framework described in the paper contributes to the management of other possible disputes and conflicts that can emerge when modern industries face traditional land use modes.

15.5 Results

Disputes between reindeer herders and the wildlife watching industry have prevailed over two decades since the first wildlife watching hides were set up in the reindeer herding zone in the beginning of the 1990s. The conflict is complex and enduring, often interwoven with social, political, cultural, economic and scientific aspects. Use of the term 'conflict' instead of 'dispute' is justified by the fact that fundamental, durable and underlying incompatibilities can be distinguished in the relations of the conflicting parties (Putnam and Wondolleck 2003).

The growth of wildlife tourism is dependent on a high concentration of predators, which in turn puts pressure on reindeer herding, particularly in the eastern part of the zone. Reindeer herders started questioning the feasibility of bear, wolf and wolverine watching at the end of the 1990s after the negative effects of carnivore attractants on livestock surfaced. Although it is rarely possible to find only one primary reason for a conflict or dispute, here the dispute boils down to the use of attractants and particularly animal carcasses on state-owned lands and private properties. The use of attractants, such as feeding wild animals with pig and salmon carcasses, is a common practice in the Finnish wildlife watching industry. However, regular use of carcasses has a potentially severe impact on reindeer herding because the carcasses attract large carnivores to sites that are proximate to vital reindeer pasturing areas, including calving sites that are particularly vulnerable to large predators. This explains the opposition by reindeer owners to the idea of letting populations of large predators grow in the reindeer herding zone.

The current Act on Damage Caused by Game Animals (2009) compensates for damage to reindeer by carnivorous animals. The maximum amount of compensation for damage to reindeer is the amount corresponding to one and a half times the current value of a reindeer killed by a large carnivore or put down due to damage caused by a large carnivore. The complainant who seeks compensation must indicate that the reindeer were killed by one of the large predators (Ministry of Agriculture and Forestry 2016). However, finding killed reindeer is difficult because bears usually bury their prey and therefore full compensation is not possible (Pakkanen and Valkonen 2011). It is estimated that half of the killed individuals are found and identified (Suvantola 2013).

In 2000, the Reindeer Herders' Association, a regional state body and an interest group for herders, decided to sharply oppose wildlife tourism that is based on attracting large carnivores in the reindeer herding zone. The conflict has led to court proceedings (The Supreme Administrative Court 2009), and numerous initiatives to solve the problem have been submitted to ministries and policy-makers by the Reindeer Herders' Association. According to the Association, organised watching of large carnivores by using carcasses should be moved outside the zone (Association Board decision 2000). The decision was justified by the strong increase in damages to reindeer herding inflicted by large carnivores, particularly in reindeer herding co-operatives located in the proximity of the eastern border of Finland. Additionally, as the Association noted, wildlife watching concentrates large carnivores in individual co-operative sites, which also bear most of the damages.

On the other hand, a wildlife watching company operating in the Kuusamo area asserts that special bear hunting licenses issued by the Finnish Wildlife Agency for game management or damage prevention in reindeer herding areas in the immediate vicinity of wildlife watching sites damage the tourism business. According to the wildlife watching company, it should be offered an opportunity to influence the issuance of bear hunting licenses (The Supreme Administrative Court 2009). It can be claimed that the case more or less reflects the opinions of an array of wildlife companies in the region. Settlement of the conflict is a priority also for wildlife watching operators, because domestic and international tourists value genuine experiences as well as implementation of various principles of social and ecological sustainability at their travel destination.

Ethical issues are an important aspect of the discussion of the principles involved in wildlife watching and are interwoven in the conflict management framework. Disputes over land-use have always an ethical dimension because they are inherently a failure also in ethical considerations of human activities and management of human-to-human relations (Glover et al. 2008; Gritten et al. 2009). Ethical aspects of wildlife watching are derivable from philosophies of responsible tourism in which initiatives are taken to maximize the positive impacts, and minimize the negative ones (Burgin and Hardiman 2015). Initially, the industry has failed to consider what negative impacts it may have on reindeer herding and how to tackle them in a best possible way. The imagined and real societal problems attached to large predators do not convey only differing views and opinions on predator behaviour, ecology and number of animal individuals, but an array of social, political, economic, cultural and ethical linkages. The opponents of wildlife watching activities in reindeer-herding zone have questioned ethics of wildlife watching by claiming that the industry has not been able to adapt to the social and cultural traditions and practices. Viewed from the angle of predator behaviour, launch of habituation process is also an ethical dilemma (Pohja-Mykrä and Kurki 2009; Kojola and Heikkinen 2013).

The Reindeer Herders' Association stresses an imminent need for more stringent regulation of the expanding wildlife watching industry. In 2001 the Association requested the Ministry of Agriculture and Forestry to prohibit the use of carcasses as attractants of large carnivores in the wildlife watching industry in the reindeer herding zone. The Association and the Finnish Forestry Agency Metsähallitus signed a contract that confines issuance by Metsähallitus of licenses for keeping carcasses in lands owned by it in the reindeer herding zone, with an exception for carcasses used as attractants for research and conservation purposes. The prohibition initiative did not, however, proceed in the Ministry, and the Reindeer Herders' Association renewed its request in 2008 by turning to the Finnish Department for Fisheries and Game under the Ministry of Agriculture and Forestry (Reindeer Herders' Association 2008; Pohja-Mykrä and Kurki 2009, pp. 22–23). The Ministry and its department were not willing to issue a prohibition, but the concern of reindeer herders became known to both political bodies and the media. The continuous increase of carcass sites also on private properties within the reindeer herding zone is apt to prolong the conflict and evoke new disputes.

Discussion around damages to reindeer herding and the need for wildlife watching regulation or alternatively attractant regulation has continued in the 2010s and the dialogue between the two conflicting parties has intensified during the recent years. A more structured procedure for finding a long-term solution started to take shape as the stakeholders realised that the existing regime or merely the lack of a collaborative approach would lead to nowhere. In 2011 Kajaani University of Applied Sciences in Northern Finland launched a development project that focused on safety management in wildlife tourism. The project offered a platform for a deeper dialogue between parties. As an outcome of the project, best practices for organised wildlife watching were published and the project boosted collaboration with other crucial stakeholders.

15.6 Processes that Aim at Solving the Land-Used Conflicts and a Four-Step Model for a Conflict Resolution Framework

15.6.1 Identification of the Conflict Anatomy and Stakeholders

The first phase of the conflict solving procedure consists of identifying the conflict anatomy. Essential questions here are why there is a conflict between two (or more) stakeholders and what is the root cause of the conflict? Additionally, the phase identifies signals of emergence of a conflict. A joint understanding of the very conflict (that there really is a conflict) and its causes has to be reached between the disputants. This tackles the problem where natural resource conflicts always convey opposing ways of interpreting policy issues. Complex problems tend to evolve as narratives, storylines, discourses or frames (Fischer and Forester 1993; Schön and Rein 1994; Hajer 1995; Dryzek 2005). In general, the first phase examines the nature of the conflict and the circumstances in which it arises, as well as the meaning of, and justification for, conflict resolution.

As mentioned earlier, it has been obvious since the 1990s that a conflict prevails between reindeer herders and the wildlife watching industry. Documentation, such as court decisions, proposals to the Ministry of Agriculture and Forestry and public voices confirm the existence of a conflict. Use of animal attractants is identified as the root cause of the conflict in the context of the expanding wildlife industry.

Identification of the conflict anatomy brings with it tools for identifying key stakeholders that have their say in the conflict's resolution. The core stakeholders (actors) are the disputants, but stakeholder analysis also includes other parties with respective interests in the field. According to a survey conducted by Kajaani University of Applied Sciences in Northern Finland, attitudes towards commercial wildlife watching tourism vary between different socio-cultural groups, the most favourable of them being municipal authorities, nature conservation representatives and tourism entrepreneurs. Alarmingly, however, negative attitudes towards the industry prevail among farmers, hunters and reindeer herders, the latter expressing the most dislike. About 60% of reindeer herders have a very negative or relatively negative attitude towards wildlife watching tourism (Järviluoma 2012, p. 22). Table 15.1 identifies the opposing parties of the conflict; Proponents (facilitators) and

Table 15.1 Proponents and opponents of commercial wildlife watching

Proponents	Opponents
Municipalities	Reindeer herders
Nature conservationists	Hunters
Local tourism entrepreneurs	The Finnish Frontier Guard
District authorities	The Finnish Forestry Agency Metsähallitus
Visit Finland	

Source this research

Opponents (confiners) of wildlife watching. The conflict's stakeholders could be divided into a core group formed by the main opposing parties, surrounded by other stakeholders forming concentric circles around the core. The attitude factor offers valuable information on potential intersectoral conflict that, in this case, points towards emerging conflicts between the wildlife watching industry and reindeer herders.

In the case of reindeer herders and tourism entrepreneurs, stakeholder identification was conducted in a workshop attended by the conflicting parties. The first-ever meeting between reindeer herders and wildlife watching entrepreneurs was organised in Kuusamo, one of the hot spots of the intersectoral conflict in eastern Finland in 2012. During the meeting, facilitated by the development project on safety issues in wildlife watching, the opposing parties discussed and agreed upon the anatomy of the conflict. In addition, other stakeholders that have their say in the conflict were identified in the meeting and the participants agreed upon the fact that the use of attractants and animal carcasses forms the focal point of the conflict. The meeting in Kuusamo did not resolve intersectoral problems, but it gave voices and faces to all the parties. A roadmap for proceedings was also an important outcome. In this meeting, the opposing parties ended up with common understanding of the need to have more scrutinised data on the behaviour of wild animals when exposed to feeding.

One has to bear in mind that also others, not only wildlife photographers, hold carcasses. The most common use is in small game hunting, but carcasses for this type of activity have not been questioned by reindeer herders. In addition to that, private users of game cameras keep attractants as a hobby when photographing wildlife on their own lands.

Nonetheless, the wildlife watching industry still does not accept the suggestion voiced by reindeer herders, that wildlife watching should be moved out of the reindeer herding zone. Tourism entrepreneurs and reindeer herders came to the conclusion that more regulation is needed to tackle the cons of relatively free use of attractants in Finland. Currently, only permission from the land owner is required to establish a carcass site. However, legislation stipulates the use and disposal of animal by-products not intended for human consumption, and animal by-products are classified into three categories according to the degree of risk involved (EC, No. 1069/2009).[1] The rules limit the type of carcass and the place of its disposal that can be used for feeding wild animals. Disposal regulation in Finland is still more relaxed than in, for example, the neighbouring countries of Sweden and Norway, or Canada. According to the Swedish Board of Agriculture, feeding of wild animals is allowed mainly for hunting and fishing purposes. A license from the Swedish Board of Agriculture is required if an animal attractant is used for other purposes, for example wildlife photography. Municipalities have a mandate to issue a statement on the deposition of attractants (Statens jordbruksverks föreskrifter 2014). Thus, there are national variations in interpretation of EC Regulation 1069/2009 and its article number 18. The Swedish Board of Agriculture has very rarely issued licenses for wildlife photography (Liljenström 2016).

Wildlife photography as commercial tourism in Norway is not as organised as in Finland or Sweden. Attitudes towards using carcasses in photography are reserved, particularly among nature conservationists. The Ministry of the Environment of Norway (2008) has proposed a tightening of the use of carcasses. Professional use of animal attractants is a new phenomenon in Norwegian wildlife photography and regulations are open to interpretation (Norwegian Environment Agency). There are also regional differences in Norway regarding implementation of regulations (Viltloven 1982).

15.6.2 Codes of Conduct

The second phase of the process touches an urgent need to unanimously agree on the codes of practices as voluntary guidelines for settling the tone of the conflict and enhancing the sustainability of tourism operations. The guidelines reflect the requirements coming from reindeer herders and local inhabitants, and highlight the responsibilities of the tourism industry. The Finnish wildlife watching industry has formulated safety and quality principles into a self-regulating

[1](EC) No 1069/2009 of the European Parliament and of the Council of 21 October 2009 laying down health rules as regards animal by-products and derived products not intended for human consumption and repealing Regulation (EC) No 1774/2002 (Animal by-products Regulation).

document for distribution to companies offering wildlife watching products (Iivari 2013). The practices serve as an incentive for stakeholders to work together to establish a set of rules and codes of conduct in the disputed field. In the preparation of the practices, key stakeholders identified during the first phase of the conflict resolution process were heard. An extensive commenting round stressed the participatory nature of the process and resulted in valuable observations and suggestions from local and national stakeholders concerning the development of joint rules and regulations. National coverage of the best practices has been achieved by publishing them on Visit Finland's internet pages and in the form of a printed booklet.

The principles, referred to as best practices, facilitate the implementation of safe and high-quality wildlife watching. Employees, customers and local inhabitants are the end users of safety. At the same time, the practices are flexible enough to cover the variety of situations and activities of wildlife tourism in Finland. The safety and quality principles complement the rights and obligations set by legislation, acting as a guideline and offering criteria for all professionally organised watching activities. Species-specific differences between practices are mainly described in terms of animal behaviour and the various risks such behaviour may pose.

It was clear from the outset that the practices should take a stance on the use of attractants in the reindeer herding zone. As the principles emphasise, animal carcasses should be set up and used according to legislation and instructions. An existing carcass used for photography purposes must not be used for hunting and the attractant should be placed so that customers and preying animals do not meet each other. For example, carcasses must not be placed along the routes that tourists use to move from one place to another. Attention should be paid to the quality and aesthetic aspects of using carcasses for feeding purposes. As the practices stipulate, the food bait used to attract wildlife is intended to be the natural food of a wild animal.

According to the jointly agreed practices, wildlife watching must not disturb areas critical to reindeer herding, such as calving and marking areas, summer and autumn forest pastures as well as transportation and access routes. The photography and watching hides, as well as bird watching towers, should be located far enough away from residential areas, trails, reindeer husbandry areas and other livelihood-related facilities. Other activities occurring in the vicinity of the hide may disturb bears' normal behaviour and increase safety risks. Such activity should be observed and its impact assessed. This activity could be, e.g., a random berry picker wandering near the hide. In addition, border guards, reindeer herders or other professionals may be present in the environs of the watching site and their safety should also be assessed in case of such situations. The best practices are considered a temporary solution to the very conflict, but they have offered a more practical and systematic approach to the conflict by both parties. Agreeing upon practices gave room for initiating more permanent solutions in terms of legislation amendments and more intensive involvement of impartial stakeholders.

15.6.3 Expert Survey

During the third phase of the process, a survey was conducted among impartial experts in order to find and evaluate the best possible alternatives for the conflict solution. The target group of the survey were rural development officials in municipalities and veterinary services in the reindeer herding zone. The survey was carried out in January 2016 and the results of the survey are published here for the first time. A questionnaire focusing on the use of animal attractants, risks and regulation needs was sent to 88 experts, of which 29 answered. The response rate is thus 33%, and over half of the respondents are veterinarians and the rest consist of rural developers.

Veterinary services organised by municipalities are primarily the responsibility of statutory co-operative areas for environmental health. The Act on Veterinary Service in Finland requires municipalities to ensure the provision of veterinary services and control of food safety, and to allocate resources for supervision of animal health and welfare, for which the State is responsible. Veterinary services are available for production animals and other domestic animals. Municipalities may choose to provide veterinary services beyond the basic veterinary services for domestic animals, including various specialist veterinary services.

Rural development officials in municipalities work to diversify rural enterprising and achieve higher employment. Their area of responsibility also covers development of better services and improvement of possibilities for residents in rural regions to participate. Rural development officials also contribute to improving the competitiveness of agricultural production that is based on producing high-quality food and improving animal welfare.

The impartial experts have profound knowledge of the issue concerned and both parties listen to their opinions carefully. A majority of the respondents are of the opinion that the owner of a carcass bears responsibility for damages caused by keeping the animal attractant. Three out of four respondents (76%) say that carcasses change the behaviour of predators, for example, by domesticating them and increasing their cub production. The rest (one quarter) of respondents were not able to say how carcasses affect predator behaviour. However, according to the respondents, there is a lack of research on this theme in Finland. As they say, more research needs to be done to scrutinise the potential of a carcass site to attract a predator to the area and how big a danger carcasses pose to reindeer in the form of predator concentration. Over half of the experts say that current regulation is insufficient. They suggest that a license should be required for keeping carcasses. Many of the experts say that legislation is sufficient but that carcass keepers do not properly abide by it. Opinions vary also among experts, since some say that keeping carcasses in the reindeer herding zone should be prohibited.

The expert survey encouraged finding solutions to the land-use conflict between wildlife watching and reindeer herding by developing principles of attractant use from voluntary and regulatory approaches. The regulatory approach highlights restrictions and prohibitions, but also amplifies reporting, licensing and hearing procedures. The increasing use of attractants, particularly in the reindeer herding zone, is apt to evoke new clashes between local economies if voluntary and regulatory means are not implemented. The existing situation is not satisfactory for either side, and both photographers and reindeer herders have to live in uncertainty. Photographers who rent private lands for their livelihood cannot be sure of the continuation of their business. Reindeer herders, on the other hand, expect to have more power to influence the establishment of new carcass sites for business purposes. Both sides need more scientific knowledge on possible changes in predator behaviour due to the use of artificial feeding.

15.6.4 Regulatory and Voluntary Measures

Phase four of the conflict solution framework suggests several options that facilitate achieving more enduring collaborative processes. Ultimately, the solution has a political dimension that extends the conflict to become a nation-wide issue. Regulatory measures suggested by experts consist of licensing, requirement of environmental impact assessments and intervention by authorities in the negative effects of carcass sites. Other measures have a directive and instructive nature. Examples include long-term land-use planning and strategies in municipalities, instructions for carcass holders, communication events, wildlife research, establishment of an inter-agency negotiation board and development of good practices.

The regulatory measures could bring amendments and changes to acts and legislation that concern, for example, hunting and environmental protection, extension of hearing rounds in carcass license consideration and requirement of environmental impact assessments in small-scale but sensitive projects. Legislative proposals are always a part of national political processes and sometimes these initiatives proceed slowly. Proposed allocation of resources to wildlife research, especially research on the impact of feeding on predator behaviour, very often depends on national agencies but offers crucial information for decision-making. However, voluntary measures are very often in the hands of conflicting parties of local stakeholders, therefore opening an avenue to more rapid solutions. Yet these solutions may prove to be temporary if national political processes leading to regulatory measures are not implemented.

15.7 Discussion

The conflict between reindeer herders and the wildlife watching industry should not be exaggerated. The disputed regions are merely individual spots that are geographically confined. Conflictual relations do not extend to the whole reindeer herding zone; they are concentrated in its south-eastern part. However, there is a potential for the conflict to expand if preventive measures and proper conflict management solutions are not implemented. Worth mentioning in this case is that the conflict has not yet spilled over to become an international issue in which, for example, appeals to the Court of Justice of the European Union or the UN Human Rights Committee (UNHRC) could be an option. There have been no signs of this kind of escalation of the conflict and such a development is not expected.

The lack of a human rights dimension has kept the conflict on a local and regional level. Sámi reindeer herders living further north from the case area are not yet involved with the conflict as disputants or stakeholders. Internationalisation of the conflict has also been avoided by early involvement of both disputants and stakeholders in the conflict. Reindeer herders and tourism operators started to see the emergence of a conflict in a very early phase, contributing to careful approaches on both sides. The structure and origin of the disputants provide an interesting aspect to the genesis of the conflict. Both parties are local and in many cases the individuals born and living in this sparsely populated area know each other personally. Thus, it is not about some faceless national or international operator penetrating the area, which is often the case in natural resource conflicts (see, for example Muradian et al. 2011; Raitio 2008). Conflict management also in this case is going to be a continuous process, rather than a temporary solution.

15.8 Conclusion

The four phases of the conflict management procedure produce all the necessary information for establishment of a framework that can be applied to management of natural resource disputes and conflicts of a large scale. Conflict management is a continuously evolving process and its principal purpose is to prevent further disputes from erupting.

The conflict resolution framework places the focus on the multi-faceted elements and phases that interact during conflict management efforts. It also helps identify challenges and opportunities related to each of the elements in the development of the conflict management process and it contributes to a collaborative approach to solutions. This framework will ensure that social and cultural sustainability is protected, while at the same time safeguarding local livelihoods by enabling them to proceed by maintaining a dialogue and community involvement.

The framework suggests that both reindeer herders and the wildlife watching industry will profit from a managed co-existence and that they may even benefit each other. It is also in the interest of local communities to develop an inter-sectoral dialogue that would create a platform for solving other possible disputes in the use of natural resources. The interacting elements (phases) shown in the case are necessary to achieve the expected outcome of the framework. It may still take years to reach the desired and optimal outcome, which would be a clearly regulated and participatory scheme. Finally, political processes can take place after facts (anatomy of the conflict) and stakeholders are identified and sufficient research information is available for decision-makers.

References

Act on Damage Caused by Game Animals (2009) Available online 22 June 2016: http://www.finlex.fi/fi/laki/ajantasa/2009/20090105, 27.2.2009/105

Ångman E, Buijs AE, Arts I, Ljunggren Bergeå H, Verschoor G (2016) Communicating emotions in conflict management over natural resource in the Netherlands and Sweden. In: Peterson TR, Ljunggren Bergeå H, Feldpausch-Parker AM, Raitio K (eds) Environmental communication and community. Constructive and destructive dynamics of social transformation. Routledge Studies in Environmental Communication and Media, London and New York, pp 184–203

Al Haija AA (2011) Jordan: tourism and conflict with local communities. Habitat Int 35(1):93–100

Association Board decision (2000) Decision and suggestions on wildlife watching in reindeer herding area. Reindeer Herders' Association Board, 7–9 Sept 2000

Atik M, Altan T, Artar M (2010) Land use changes in relation to coastal tourism developments in turkish mediterranean. Polish J Environ Stud 19(1):21–33. Available online 1 June 2016: http://www.pjoes.com/pdf/19.1/21-33.pdf

Berg A (2010) Reindeer herding and modern forestry—the historical impacts on forests of two main land users in Northern Sweden. Doctoral Thesis, Swedish University of Agricultural Sciences, Umeå 2010

Bernes C, Bråthen KA, Forbes BC, Hofgaard A, Moen J, Speed JDM (2013) What are the impacts of reindeer/caribou (Rangifer tarandus L.) on arctic and alpine vegetation? A systematic review protocol. Official J Collab Environ Evid 2:6

Bezard P, Brilliand S, Kumpula J (2015) Composition of late summer diet by semi-domesticated reindeer in different grazing conditions in northernmost Finland. Rangifer 35(1):29–52

Bisi J, Kurki S, Svensberg M, Liukkonen T (2007) Human dimensions of wolf (Canis lupus) conflicts in Finland. Eur J Wildlife Res 53:304–314

Burgin S, Hardiman N (2015) Effects of non-consumptive wildlife-oriented tourism on marine species and prospects for their sustainable management. J Environ Manag 151:210–220

Campbell DJ, Gichohi H, Mwangi A, Chege L (2000) Land use conflict in Kajiado District, Kenya. Land Use Policy 17(4):337–348

Chaplin J, Brabyn L (2013) Using remote sensing and GIS to investigate the impacts of tourism on forest cover in the Annapurna conservation Area, Nepal. Appl Geogr 43:159–168

Cohen E (1978) The impact of tourism on the physical environment. Ann Tour Res 5(2):215–237

Conover M (2002) Resolving human-wildlife conflicts. The science of wildlife damage management. Lewis Publishers

Davenport J, Davenport JL (2006) The impact of tourism and personal leisure transport on coastal environments: a review. Estuar Coast Shelf Sci 67(1–2):280–292

De Bono E (1985) Conflicts. A better way to resolve them, Harrap, London

Dickman AJ, Hazzah L (2015) Money, myths and man-eaters: complexities of human–wildlife conflict. Problematic Wildlife, pp 339–356

Dredge D (2010) Place change and tourism development conflict: evaluating public interest. Tour Manag 31(1):104–112

Dryzek JS (2005) The politics of the earth: environmental discourses. Oxford University Press, Chicago

Eagles PFJ, McCool SF (2002) Tourism in national parks and protected areas. Planning and management. CABI Publishing, New York

Fagence M (1990) Geographically-referenced planning strategies to resolve potential conflict between environmental values and commercial interests in tourism development in environmentally sensitive areas. J Environ Manag 31(1):1–18

Filley AC (1975) Interpersonal conflict resolution. Foresman, Glenview IL, Scott

Fischer F, Forester J (1993) The argumentative turn in policy analysis and planning. Duke University Press, Durham, N.C

Forbes BC, Monz CA, Tolvanen A (2004) Ecological impacts of tourism in terrestrial polar ecosystems. In: Buckley R (ed) Environmental impacts of ecotourism. CABI Publishing, Wallingford, pp 155–170

Gaughan AE, Binford MW, Southworth J (2009) Tourism, forest conversion, and land transformations in the Angkor basin, Cambodia. Appl Geogr 29(2):212–223

Glover TD, Stewart WP, Gladdys K (2008) Social ethics of landscape change. Toward community-based land-use planning. Qual Inq 14(3)3: 384–401

Grimble R, Wellard K (1997) Stakeholder methodologies in natural resource management: a review of principles, contexts, experiences and opportunities. Agric Syst 55(2):173–193

Gritten D, Saastamoinen O, Sajama S (2009) Ethical analysis: a structured approach to facilitate the resolution of forest conflicts. For Policy Econ 11(8):555–560

Gunnarsdotter Y (2006) Hunting tourism as ecotourism—conflicts and opportunities. In: Gossling S, Hultman J (eds) Ecotourism in scandinavia: lessons in theory and practice. Cabi Publishing, Oxfordshire, pp 178–192

Hajer MA (1995) The politics of environmental discourse: ecological modernization and the policy process. Oxford University Press, New York, NY

Hall CM, Page SJ (2006) The geography of tourism and recreation. Environment, place and space. Routledge, London

Hall CM (2007) Politics, power and indigenous tourism. Tourism and indigenous peoples. Issues and Implications, pp 305–318

Haukeland JV, Daugstad K, Vistad OI (2011) Harmony of conflict? A focus group study on traditional use and tourism development in and around Rondane and Jotunheimen National Parks in Norway. Scand J Hospitality Tour 11(supl 1):13–37

Heikkinen HI (2006) Neo-entrepreneurship as an adaptation model of reindeer herding in Finland. Nomadic Peoples 10(2):187–208

Heikkinen HI, Sarkki S, Nuttall M (2012) Users or producers of ecosystem services? A scenario exercise for integrating conservation and reindeer herding in northeast Finland. Pastoralism: research, policy and practice. Pastoralism 2(1):1–24

Helle TP, Jaakkola LM (2008) Transitions in herd management of semi-domesticated reindeer in northern Finland. Ann Zool Fenn 45 (2):81–101

Helle T, Särkelä M (2008) The effects of outdoor recreation on range use by semi-domesticated reindeer. Scand J For Res 8(1–4): 123–133

Henle K, Alard D, Clitherow J, Cobb P, Firbank L, Kull T, McCracken D, Moritz RFA, Niemelä J, Rebane M, Wascher D, Watt A, Young J (2008) Identifying and managing the conflicts between agriculture and biodiversity conservation in Europe—a review. Agric Ecosyst Environ 124(1–2):60–71

Higham J, Lück M (2008) (eds.) Marine wildlife and tourism management. Insights from the natural and social sciences. Biddles Ltd., King's Lynn

Higham JES, Shelton EJ (2011) Tourism and wildlife habituation: reduced population fitness or cessation of impact? Tour Manag 32 (6):1290–1298

Hilsendager KW, Harshaw HW, Kozak RA (2016) Reducing the impact of forest harvesting on the vancouver island tourism industry. For Chronicle 92(1):101–111

Hoffman MT, Rohde RF (2007) From pastoralism to tourism: the historical impact of changing land use practices in Namaqualand. J Arid Environ 70(4):641–658

Hukkinen J, Heikkinen H, Raitio K, Müller-Wille L (2006) Dismantling the barriers to entrepreneurship in reindeer management in Finland. Int J Entrepreneurship Small Bus 3(6):705–727

Iivari P (2013) 100 best practices for organised wildlife watching. Kajaani University of Applied Sciences, Outdoors Finland

Järviluoma J (2012) Survey on current state of commercial wildlife watching tourism. Kajaani University of Applied Sciences Publication Series B, Reports and surveys 24 (In Finnish)

Kitti H, Gunslay N, Forbes BC (2006) Defining the quality of reindeer pastures: the perspectives of sámi reindeer herders. Reindeer Management in Northernmost Europe 184, Ecological Studies, pp 141–165

Kivinen S (2015) Many a little makes a mickle: cumulative land cover changes and traditional land use in the Kyrö reindeer herding district, northern Finland. Appl Geogr 63:204–211

Kojola I, Huitu O, Toppinen K, Heikura K, Heikkinen S, Ronkainen S (2004) Predation on european wild forest reindeer (Rangifer tarandus fennicus) by wolves (Canis lupus) in Finland. J Zool 263:229–235

Kojola I, Heikkinen S (2013) Problem brown bears Ursus arctos in Finland in relation to bear feeding for tourism purposes and the density of bears and humans. Wildlife Biol 18(3):258–263

Kumpula J (2001) Winter grazing of reindeer in woodland lichen pasture: effect of lichen availability on the condition of reindeer. Small Ruminant Res 39(2):121–130

Lee TJ, Riley M, Hampton MP (2010) Conflict and progress: tourism development in Korea. Ann Tour Res 37(2):355–376

Liljenström S (2016) Use of animal by-products for photographing purposes in Sweden. Expert comment of the Swedish Agricultural Board via e-mail to the writer on the 18 Jan 2016

Maguigad VM (2013) Tourism planning in archipelagic Philippines: a case review. Tour Manag Perspect 7:25–33

Manning RE (2004) Recreation planning frameworks. In: Manfredo MJ, Vaske JJ, Bruyere BL, Field DR, Brown ṔJ (eds) Society and natural resources: a summary of knowledge modern Litho. Jefferson, MO, pp 83–96

McKercher B (1992) Tourism as a conflicting land use. Ann Tour Res 19(3):467–481

Melling T (1994) Dispute resolution within legislative institutions. Stanford Law Rev 46(6):1677–1715

Ministry of Agriculture and Forestry (2016) Damage to reindeer. Available online 2 May 2016: http://mmm.fi/en/wildlife-and-game/game-animal-damages/large-carnivore-damage/damage-to-reindeer

Ministry of the Environment of Norway (2008) Hearing promemoria to proposal for changes to the diversity act and wildlife act. Available online 2 May 2016: https://www.regjeringen.no/no/dokumenter/Prop-75-L-20092010/id593685/?ch=4

MINTEL (2008) Wildlife tourism—international. Travel & Tourism Analyst 10

Moen J (2008) Climate change: effects on the ecological basis for reindeer husbandry in Sweden. Ambio 37:304–311

Moore SA, Rodger K (2010) Wildlife tourism as a common pool resource issue: enabling conditions for sustainability governance. J Sustain Tour 18(7):831–844

Muradian R, Martinez-Alier J, Correa H (2011) International capital versus local population: the environmental conflict of the tambogrande mining project, Peru. Soc Nat Resour Int J 16(9):775–792

Müller-Wille L, Pelto PJ (1971) Technological change and its impact in Arctic regions: Lapps introduce snowmobiles into reindeer herding (Utsjoki and Inari, Northern Finland). Polarforsch Bremerhaven Alfred Wegener Inst Polar Mar Res Ger Soc Polar Res 41(1/2): 142–148

Newsome D, Kingston DR, Moore SA (2005) Wildlife tourism. Channel View Publications, Clevedon

Nieminen M, Heiskari U (1989) Diets of freely grazing and captive reindeer during summer and winter. Rangifer 9(1):17–34

Olsen LS (2016) Sami tourism in destination development: conflict and collaboration. Polar Geography 39(3):179–195

Pakkanen A, Valkonen J (2011) Porotalouden hyvinvointi ja tulevaisuuskuvat eteläisissä paliskunnissa. Wellbeing and Future Perspectives of Reindeer Herding in Southern Reindeer Cooperatives. Available online. Accessed on 6 June 2016, http://www.petohanke.fi/instancedata/prime_product_julkaisu/suomussalmi/embeds/petohanke/14754_Porotalouden_hyvinvointi_ja_tulevaisuuskuvat_etelaisissa_paliskunnissa.pdf

Pearlman MV (1990) Conflicts and constraints in Bulgaria's tourism sector. Ann Tour Res 17(1):103–122

Pettersson R (2006) Ecotourism and indigenous tourism: positive and negative impacts of sami tourism. I ecotourism in scandinavia: lessons in theory and practice. CABI Publishing (Ecotourism series), Wallingford, pp 166–177

Pohja-Mykrä M, Kurki S (2009) Feeding large carnivores by carcasses—challenges on social sustainability. University of Helsinki, Ruralia Institute (In Finnish)

Putnam LL, Wondolleck JM (2003) Intractability: definitions, dimensions, and distinctions. In: Lewicki RJ, Gray RJ, Elliot M (eds) Making sense of intractable environmental conflicts. Concepts and cases. Island Press, Washington, pp 35–59

Raitio K (2008) You can't please everyone—conflict management practices. University of Joensuu, Publications in Social Sciences, Frames and Institutions in Finnish State Forests, p 86

Reindeer Herders Association (2008) Request for prohibition of photographing at animal carcasses in Reindeer Herding Area. Ministry diary number 879/2008

Saberwal VK, Gibbs JP, Chellam R, Johnsingh AJT (1994) Lion–human conflict in the Gir-forest, India. Conserv Biol 8:501–507

Sæþórsdóttir AD (2010) Tourism struggling as the wilderness is developed. Scand J Hospitality Tour 10(3):334–357

Sandström P, Granqvist Pahlén T, Edenius L, Tømmervik H, Hagner O, Hemberg L, Olsson H, Baer K, Stenlund T, Brandt LG, Egberth M (2003) Conflict resolution by participatory management: remote sensing and GIS as tools for communicating land-use needs for reindeer herding in Northern Sweden. J Human Environ 32(8): 557–567

Sandström C, Moen J, Widmark C, Danell Ö (2006) Progressing toward co-management through collaborative learning: forestry and reindeer husbandry in dialogue. Int J Biodivers Sci Manag 2 (4):2006

Saremba J, Gill A (1991) Value conflicts in mountain park settings. Ann Tour Res 18(3):455–472

Sidaway R (2005) Resolving environmental disputes from conflict to consensus. Earthscan Publishers, London

Skuncke F (1969) Reindeer ecology and management in Sweden. Biological Papers of the University of the Alaska 8 Feb 1969

Sköld P (2015) Perpetual adaption? Challenges for the Sami and Reindeer Husbandry in Sweden. In: Evengård B, Nyman Larsen J, Paasche Ø (eds) The new Arctic. Springer International Publishing, Switzerland, pp 39–55

Statens jordbruksverks föreskrifter (2014) Regulations of the Swedish board of agriculture. SJVFS 2014:43 19 a, 21, 21 b §§

Stylidis D, Biran A, Sit J, Szivas EM (2014) Residents' support for tourism development: the role of residents' place image and perceived tourism impacts. Tour Manag 45(Dec 2014):260–274

Suvantola L (2013) The golden eagle compensation scheme in Finland as an example of incentive measures. In: Klenke Reinhard A, Ring I, Andreas K, Niels J, Felix R, Klaus H (eds) Human—wildlife conflicts in Europe: fisheries and fish-eating vertebrates as a model case. Springer, pp 201–220

Teye VB (1992) Land transportation and tourism in Bermuda. Tour Manag 13(4):395–405

The Supreme Administrative Court (2009) Finnish supreme administrative court's decision on bear hunting. On 13 Jan 2009

Tolvanen A, Kangas K (2016) Tourism, biodiversity and protected areas—review from northern Fennoscandia. J Environ Manag 169 (15 Mar 2016):58–66

Tosi HL, Rizzo JR, Carroll SJ (1986) Managing organizational behavior. Pitman, Boston

Tyler SR (1999) Policy implications of natural resource conflict management. Policy implications of natural resource conflict management. In: Buckles D (ed) Cultivating peace. IDRC, Ottawa, pp 263–280

Tyler NJC, Turi JM, Sundset MA, Strom BK, Sara MN, Reinert E, Oskal N, Nellemann C, McCarthyi JJ, Mathiesene SD, Martellok ML, Maggae OH, Hovelsrudl GK, Hanssen-Bauerm I, Eiran NI, Eirae IMG, Corello RW (2007) Saami reindeer pastoralism under climate change: applying a generalized framework for vulnerability studies to a sub-arctic social-ecological system. Glob Environ Change 17(2):191–206

Tzatzanis M, Wbrka T (2002) Sun beds versus sand dunes: a conservation—tourism conflict. In: Brebbia CA (ed) Environmental problems in coastal regions IV, vol 8. Environmental studies. WIT Press, Southampton, Boston, pp 25–34

Tyrväinen L, Uusitalo M, Silvennoinen H, Hasu E (2014a) Towards sustainable growth in nature-based tourism destinations: clients' views of land use options in Finnish Lapland. Landscape Urban Plann 122(Feb 2014):1–15

Tyrväinen L, Mäntymaa E, Ovaskainen V (2014b) Demand for enhanced forest amenities in private lands: the case of the Ruka-Kuusamo tourism area, Finland. For Policy Econ 47(Oct 2014):4–13

Upreti BR (2001) Conflict management in natural resources: a study of land, water and forest conflicts in Nepal. PhD thesis

Uusitalo M (2010) Differences in tourists' and local residents' perceptions of tourism landscapes: a case study from Ylläs, Finnish Lapland. Scand J Hospitality Tour 10(3):310–333

Viltloven (1982) Norwegian law on hunting and capturing wild animals

Walpole MJ, Thouless CR (2005) Increasing the value of wildlife through nonconsumptive use? Deconstructing the myths of eco-tourism and community-based tourism in the tropics. In: Woodroffe R, Thirgood SJ, Rabinowitz A (eds) People and wildlife: conflict or coexistence?. Cambridge University Press, Cambridge

Watson AE, Kajala L (1995) Intergroup conflict in wilderness. Balancing opportunities for experience with preservation responsibility. In: Sippola A-L, Alaraudanjoki P, Forbes B, Hallikainen V (eds) Northern wilderness areas: ecology, sustainability, values. Arctic Centre Publications 7, University of Lapland, Rovaniemi, pp 251–270

Williams PW, Penrose RW, Hawkes S (1998) Shared decision-making in tourism land use planning. Ann Tour Res 25(4):860–889

Wray K, Espiner S, Perkins HC (2010) Cultural clash: interpreting established use and new tourism activities in protected areas. Scand J Hospitality Tour 10(3):272–290

Yang J, Ryan C, Lingyun Z (2013) Social conflict in communities impacted by tourism. Tour Manag 35:82–93. doi:10.1016/j.tourman.2012.06.002

16 Not Quite Wild, But Not Domesticated Either: Contradicting Management Decisions on Free-Ranging Sika Deer (*Cervus nippon*) at Two Tourism Sites in Japan

Rie Usui and Carolin Funck

Abstract

Previous studies on the conventional form of wildlife tourism have limited the discussion on wildlife management to the context in which wild animals are regarded exclusively as ecologically significant beings. With ever-increasing tourism development around the world, wildlife tourism takes place in a wide range of forms and settings beyond their natural environment. In many Asian countries, where certain species of animals have a long history of close associations with humans, the animals are often found in human modified environments. Thus, their cultural significance is expected to play an important role in management decision-making process. In this chapter, we shift our attention to wild animals whose habitats extensively overlap with human modified space using two case studies from Nara Park and Miyajima Island in Japan. Wild sika deer (*Cervus nippon*) found in these sites are well-adapted to the human modified environment and have close interactions with humans. A rapid growth in tourism at these locations in modernity created the challenge of managing the deer that hold ecological and cultural values simultaneously. We review the history of deer at each site, their management practices, and challenges that they face. While both Nara and Miyajima deer are considered wild, they are managed in dissimilar manner. We argue that understanding the context of wildlife tourism and the history of the human and animal relationship is a necessary component of more sustainable wildlife management for tourism.

16.1 Introduction

With ever-increasing tourism development around the world, wildlife tourism takes place in a wide range of settings. Geographical locations are no longer limited to Australian and African continents where their unique and popular wildlife attracts a large number of tourists on a global scale. Nowadays, wildlife tourism studies are also produced from many other countries including, but not limited to, Iceland (Granquist and Nilsson 2016), Argentina (Argüelles et al 2016), and China (Cong et al. 2014a). A wide spread of wildlife tourism activities led to the involvement of more varieties of wildlife species that are subjected to tourism. While wild animals from Australia and Africa (e.g., koalas, kangaroos, elephants, and lions) continue to be popular attractions, some groups of tourists are fascinated with creatures such as bats (Paksuz and Ozkan 2012), crocodiles (Burgin and Hardiman 2016), and insects (Lemelin 2015). Depending on the species of animals involved, forms of tourist-wildlife encounters vary. Wildlife viewing, one of the most common forms of wildlife tourism, is organized using provisioning at some locations. For instance, Monkey Mia in Australia (Bach and Burton 2016; Smith et al. 2008) offers tourists an opportunity to interact with dolphins during food provisioning sessions. At Japanese monkey parks across Japan, provisioning is used as the basis for tourism operations. The park operators herd free-ranging macaques

R. Usui (✉) · C. Funck
Graduate School of Integrated Arts and Sciences, Hiroshima University, Higashihiroshima, Japan
e-mail: leahrieusui@hiroshima-u.ac.jp

C. Funck
e-mail: funckc@hiroshima-u.ac.jp

I. Borges de Lima and R.J. Green (eds.), *Wildlife Tourism, Environmental Learning and Ethical Encounters*, Geoheritage, Geoparks and Geotourism, DOI 10.1007/978-3-319-55574-4_16

with baited food (Knight 2011; Kurita 2012). In the case of dangerous aquatic species (e.g., white sharks), tourists dive in a cage to experience close viewing of such wildlife (Apps et al. 2016).

Because this widespread operation of wildlife tourism involves diverse participants and species of wildlife, management of wildlife is presumably more complex than ever. No matter where tourist and animal interactions take place, management plays an essential role in minimizing potential harm that animals and humans impose on each other. However, previous studies on the conventional form of wildlife tourism (i.e., viewing of wild animals in natural settings) has limited the discussion on wildlife management to the context where wildlife is placed as ecologically significant beings.

In one of the first comprehensive books on wildlife tourism, Newsome et al. (2005) state, "the key strategy for wildlife tourism planning and development is that priority must be given to ecological sustainability (p. 30)." Wildlife tourism industry sets three primary objectives for managing tourist-wildlife interactions: (1) to reduce potential negative impacts on environment, (2) to make contributions to the local people in communities, and (3) to provide satisfactory experiences to tourists. Therefore, a management plan is implemented to achieve the above objectives. For example, at a site level management, Orams (1996) examined four strategies (i.e., physical, regulatory, economic, and educational) for managing wild animals at tourism destinations. Among them, physical and regulatory strategies are the two most commonly used methods. Physical management strategy refers to the method that uses objects such as glass barriers or cages to control tourist and animal contact (e.g., captive or semi-captive environments). In wild or semi-natural environments, park guards or rangers control the behavior of tourists and their proximity to wild animals as part of a physical management strategy. Regulatory management strategies are often practiced in semi-captive environments and control tourists' behaviors through posting of signs, implementation of park rules, and enforcement of laws by authorities.

Unfortunately in many cases, rules and regulations are ignored or violated by tourists (Fuentes et al. 2007; Ruesto et al. 2010) and park rangers tend to exert little effort in regulating tourists' interactions with wild animals (Fuentes et al. 2007; Usui et al. 2014). Such management dysfunction can be explained by cultural differences in the perception of wildlife between western and non-western societies. Qingming et al. (2012) pointed out the importance of investigating cultural aspects for more effective wildlife tourism management. They explained that Chinese people perceived close human-animal interactions as a harmonious human relation with wildlife, thus a management strategy that ignores such cultural attributes will not be effective. The study reported by Cong et al. (2014b) supported this argument. They found tourists' satisfactions with wildlife tourism experience increased as they had closer interactions with giant pandas in Chengdu, China.

The above findings highlighted the importance of taking into account cultural aspects when planning the management of wildlife for tourism and in fact, several scholars began realizing the needs in site-specific wildlife management models. Some raised concerns about the "globalization of wildlife management (Ikeya 2009: 296)" and criticized the idea of separating human and nature in order to protect nature (Ikeya 2009; Saarinen 2016). Still, these discussions are only applicable to the context in which human and wildlife encounters take place in natural or semi-natural settings. In many Asian countries, where certain species of animals have a long history of close associations with humans, the animals are often found in a human modified environment [e.g., cattle in India (Sinha and Sinha 2008) and monkeys in Bali, Indonesia (Wheatly 1999)]. In such cases, cultural significance of the animals is expected to play an important role in the management decision-making process. However, little research has been reported on how animals whose habitats extensively overlap with human modified space are managed for tourism. Thus, this chapter attempts to fill this gap using two case studies from Japan: Nara Park and Miyajima Island (Fig. 16.1).

Fig. 16.1 Geographical locations of Nara Park and Miyajima Island

Wild sika deer (*Cervus nippon*) inhabiting Nara Park and Miyajima Island stroll through historical town sites and have close contact with tourists. They have had a long history of close associations with humans and have been playing an important cultural role in both communities. A rapid growth in tourism at these locations in modernity created the challenge of managing the deer that hold ecological and cultural values simultaneously. The aim of this chapter, therefore, is to explore how the sika deer in Nara Park and Miyajima Island are managed for tourism. Our specific objectives are: (1) to overview the history of the deer; (2) to identify problems and challenges they encounter; and (3) to examine management characteristics (e.g., management policy, how the deer are managed, who manages them). In the following sections, we examine management practices of Nara Park and Miyajima Island based on the primary (e.g., fieldwork) and secondary data (e.g., review of the available documents and pertinent literature).

16.2 Wild Sika Deer at Nara Park and Miyajima Island in Japan

16.2.1 Nara Park

16.2.1.1 About Nara Park

Nara Park (approximately 6.6 km^2 in size) is a public park found in the City of Nara surrounded by the urban area of the city on the west side and Mt. Kasuga forest reserve on the east side (Fig. 16.2). Since Nara was the capital of Japan in the eighth century, the present park preserves old architectures including Tōdaiji Temple and Kasuga Shrine. These

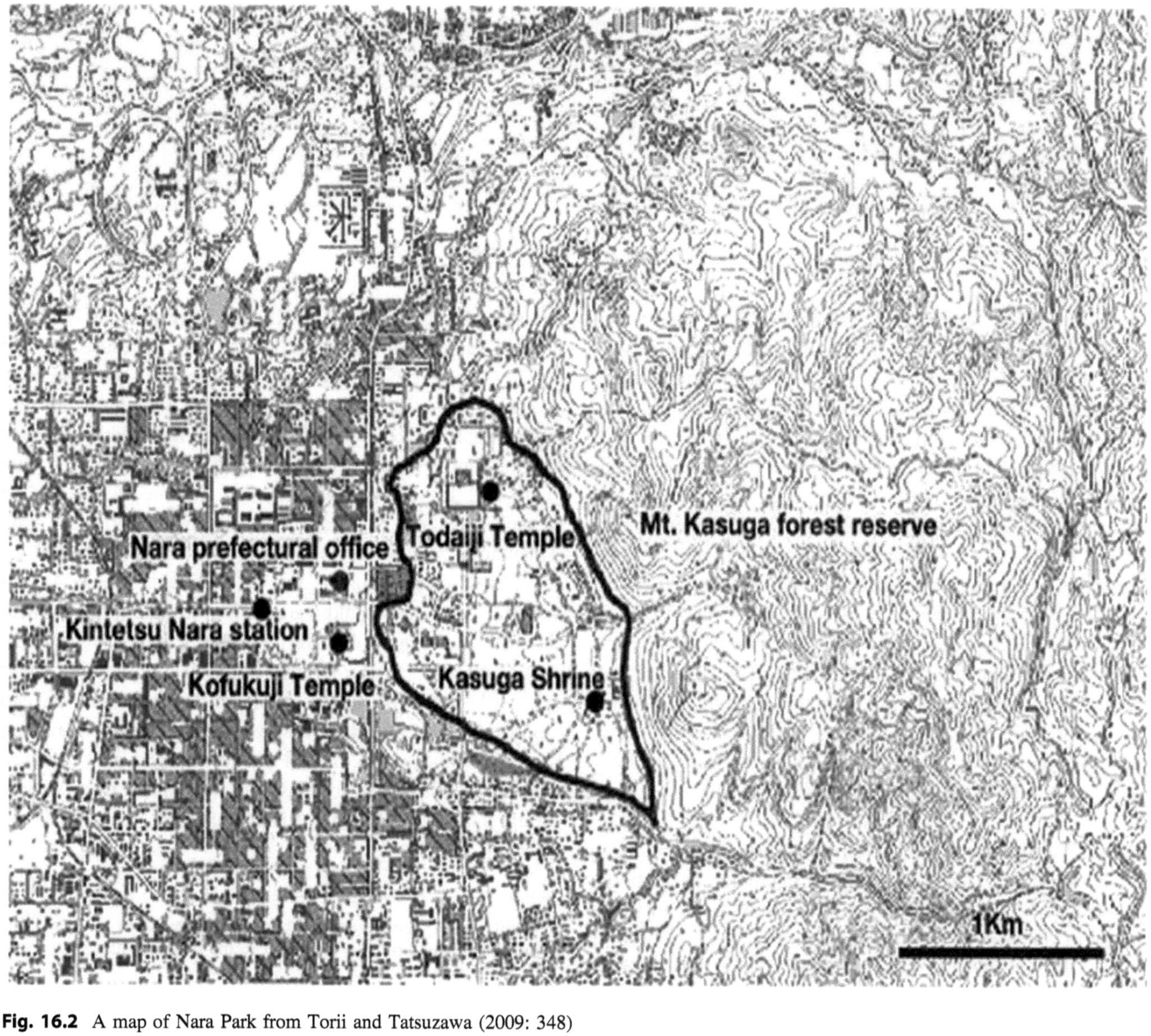

Fig. 16.2 A map of Nara Park from Torii and Tatsuzawa (2009: 348)

Fig. 16.3 Deer and tourists in Nara Park with Mt. Kasuga in the background. Photo by Rie Usui

ancient buildings were added to the UNESCO World Cultural Heritage list as the Historic Monument of Ancient Nara (UNESCO 2016). Because of its historical and cultural significance, Nara Park attracts over 14 million Japanese and international tourists annually (City of Nara 2014).

One unique feature of the park is the presence of habituated wild sika deer (Nara Prefecture 2016a). Throughout the park, tourists find deer grazing on the lawn, resting on the sidewalks, and roaming around temples and shrines. According to the latest population census conducted by the Foundation for the Protection of Deer in Nara (FPDN) in 2015, there are approximately 1500 deer living in the park, which gives a density of 226 deer/km^2. The deer are recognized as a component that provides harmonious image typical for Nara (Fig. 16.3). Hence, Nara deer were designated as a national natural treasure, "Deer of Nara" in 1957 (Hatakama 2010; Torii and Tatsuzawa 2009).

16.2.1.2 History of the Deer in Nara Park

As the history of Nara deer is well-documented in the book, "The Deer of Nara (FPDN 2010)," the following history was mainly taken from there:

Shin-Roku (sacred deer) is a synonym often used to refer to Nara deer. According to the legend documented on some historical records, a god of Kashima Shrine[1] came to Mt. Kasuga[2] along with a white deer in 768. Because the god rode on the deer, Nara deer became associated with the deity and were bestowed a divine status. For a century, the religious beliefs protected the deer from being hunted or harmed and the land where the deer inhabited was considered pure (Torii and Tatsuzawa 2009). As we shall see, hereafter the sacred status of Nara deer was used for political gain (Hatakama 2010).

Twelfth century Japan saw a decline in the power of imperial government and a formation of a new government called B*akufu* (garrison government). Under the new system of governance, *Bushi* (warriors) ruled the nation. Their upheaval shook the social status of Buddhist temples and Shinto shrines. Afraid of losing their authoritative power, Kasuga Shrine and Kofukuji Temple issued an ordinance regarding killing of deer. This was when Nara deer became strongly associated with Kasuga Shrine[3] and Kofukuji

[1]Kashima Shrine is a shrine found in Ibaraki Prefecture of Japan. The shrine is estimated to have been built in the mid-seventh century.
[2]The mountain that is located adjacent to Nara Park.
[3]Kasuga Shrine is a Shinto shrine located in Nara Park.

Temple[4] (Hatakama 2010). Killing of deer was considered one of the three deadly sins and anyone who killed or harmed the deer was administered severe punishment. Even dogs were included as possible subject for punishment and *Inugari* (dog hunt) was implemented to protect the deer from being preyed upon by dogs. This absolute protection of deer led to an increase in the deer population, which triggered conflicts between humans and deer (Torii and Tatsuzawa 2009). Nonetheless, people who suffered from deer-related problems had no choice but to endure the situation (Torii and Tatsuzawa 2009).

By the 17th century, the significance of deer shifted from the focus of protection to the subjects of management. During this time period, Nara underwent its first urbanization. For instance, there was an increase in resident population as a township was constructed; Nara became an attractive destination for travelers; and large areas of the land were cultivated for farming as agricultural technology advanced. As a consequence, human-deer conflicts became more apparent and they could no longer be ignored. To keep humans from being injured and to prevent the damages to night lanterns on the streets by deer, *Tsunokiri* (antler-cutting) began in 1671 (Torii and Tatsuzawa 2009). Furthermore, for protecting agricultural crops from deer, *Shishigaki* (deer proof fence) was built around the town of Nara. The deer that had long been protected as sacred deer were swallowed in the wave of urbanization and their life has been altered drastically ever since.

The 19th century was the period of modernization and the human-deer relationship entered into a new phase. To this date, religion played somewhat of an essential role in society. Modernization mindsets dismissed religious beliefs as merely superstitions (Torii and Tatsuzawa 2009). As people's religious beliefs faded away, the idea of *Shin-Roku* (sacred deer) was given little importance. Eventually, Nara deer became considered just as a wild animal species (Torii and Tatsuzawa 2009). This conversion of thoughts about the deer was reflected on how the deer were treated. Some reports indicate that government officials ate deer meat with the intention of changing people's perceptions about the deer (Watanabe 2010). Crop raiding deer were no longer allowed to roam freely in the park and were confined in a caged space to ensure that they were off nearby farmlands (Torii and Tatsuzawa 2009). Unfortunately, this attempt led to overcrowding of the deer in the limited space provided, resulting in a decline of the deer population to 38 individuals in 1873. Concerned about Nara deer diminishing, the officials released the deer out of the cage in the very same year. Not too long after that, Nara Park with its present-day form was established in 1880.

[4]Kōfukuji Temple is a Buddhist temple located in Nara Park.

Toward the end of World War II (WWII) when people suffered from food shortages, the population of Nara deer dropped due to poaching (Torii and Tatsuzawa 2009). With a great amount of protection to the deer after WWII, the deer population soon revived vigorously (Torii and Tatsuzawa 2009) and has been fairly constant at between 1100 and 1500 individuals (Nara Prefecture 2016b).

16.2.1.3 Deer-Related Problems

Problems concerning deer can largely be categorized into agricultural- and tourism-related issues. Some problems already existed in the 17th century and others were generated as a result of urbanization of the landscape. It is likely that the current problems are more complex than ever because of multi-stakeholders' and multi-national tourists' involvements.

Rokugai (crop raiding) has been one of the serious social problems in Nara for centuries. The challenge lies in the fact that some crop raiding deer are residents of Nara Park while others are regular wild deer that do not belong to the park. Thus, how to differentiate Nara deer from other groups of wild deer has been problematic (Torii and Tatsuzawa 2009). To prevent erroneous capturing of Nara deer, the Kasuga Shin-Roku Preservation Society—which later became FPDN—was founded in 1891 (Watanabe 2010). They received a local government subsidy from 1902 to 1918 to operate their management duties which included monitoring the deer, building deer-proof fences and paying compensation for crop damages to local farmers (Torii and Tatsuzawa 2009). However in 1979, the crop-raiding problem became extremely unbearable such that farmers filed a lawsuit against the national and local government as well as Kasuga Shrine and FPDN (Watanabe 2001). The lawsuit was settled in 1985 with an agreement to implement a new management rule. Still, the total number of *Rokugai* has remained the same since and some farmers exhibit a high level of frustration over the problem related to *Rokugai* (Watanabe 2007).

Tourism-related issues include traffic accidents, physical injuries and feeding. As the park is adjacent to the City of Nara and large roads run across the park, traffic accidents are a critical problem, with some resulting in fatal injuries. Annually, more than 150 deer reportedly die from collisions with cars (FPDN 2016). Close human-deer contact creates a risk of physical injury for both parties. In several reported cases, tourists were harmed by deer antlers or children were knocked down by deer (Hatakama 2010). Conversely, there were incidents where deer were the victims of malevolent tourists. For instance, some tourists reportedly tease deer by hanging their purses on deer antlers. FPDN (2016) once found a male deer struggling to get a tangled purse off his antler. Intake of non-digestible items (e.g., plastic bags) or inappropriate items (e.g., paper and snacks) by deer is another concerning issue at Nara Park (FPDN 2016). They can block the digestive tract, which may lead to malnutrition (Nara Prefecture 2016b). As deer are herbivores, providing them with meat or snacks could cause digestive dysfunction (e.g., diarrhea), which is likely since Takahashi (1999) found that almost one in every two tourists did not have appropriate knowledge about deer's natural feeding habits.

16.2.1.4 Management of the Deer for Tourism

Policy and Protection Laws

Policy for the management of Nara deer has been implemented by Nara Deer Conservation Management Plan Discussion Committee, which was organized in 2013. The committee members include officials of Nara Prefecture and City, Kasuga Shrine, FPDN, Nara Deer Supporters Club (NDSC), Nara Deer Consultation Office (NDCO) and academics. The vision of their management goal states, "hundred years later, the deer in Nara Park continue to inhabit healthy (Nara Prefecture 2016b)." With this management prospectus, the committee carries out a protection-based management plan that aims to balance conservation of the habituated deer in Nara Park and mitigation of human-deer conflicts (Nara Prefecture 2016b).

Nara deer are protected under several laws. Since 1890, Nara Park has been designated as a Sacred Deer Killing Prohibition Area. "Deer in Nara" as natural treasures have also been protected under the Law for the Protection of Cultural Properties since 1957 (Nara Prefecture 2016b). This law also strictly bans harming or hunting deer. However, because the definition of "Deer in Nara" is vague, it is unclear as to which deer the law applies (Watanabe 2012). To solve this problem, Agency for Cultural Affairs[5]proposed an improved management plan that divides the deer habitat into four areas in 1979 and specific management rules were implemented for each area (Nara Prefecture 2016b; Watanabe 2001).

Onsite Management

At a site level, eleven full-time FPDN workers, volunteers of the supporting organization including NDSC, and NDCO (a prefectural office) cooperate in daily management of deer (Ikeda 2010). Perhaps, the most distinguishing management characteristic of the deer in Nara is the availability of *Sika Senbei* (deer crackers) that are made exclusively for the deer in Nara Park using rice bran and flour that are healthy for the deer (Takeda 2010). Several FPDN vendors can be found selling deer crackers in the park (Fig. 16.4). Tourists can feed purchased deer crackers to deer and a small portion of the profit is donated to FPDN in order to support their management activities. One rationale for feeding the deer crackers is that providing them in the park (concentrated food source) can prevent the deer from dispersing to surrounding areas where *Rokugai* (crop raiding) is a severe problem (Torii and Tatsuzawa 2009).

FPDN's work is not limited to selling the deer crackers. On daily basis, the staff at FPDN patrol in and around Nara Park to look for crop raiding or injured Nara deer (Ikeda 2010). During certain seasons of the year, they capture pregnant females or antlered males. The captured individuals are kept in *Rokuen* (deer garden) temporarily. Those deer that damaged agricultural crops are kept permanently in *Rokuen* (Ikeda 2010). To prevent potential injuries by antlered male deer, *Tsunokiri* (antler cutting) is performed as an annual event. The removed antlers are sold and the profit is used to support the FPDN work. Because a limited number of staff at FPDN work on a considerable amount of tasks, NDSC supports the FPDN's work by hosting an educational seminar and eco-tour, conducting fieldwork, helping patrol the park, performing the annual deer census, and outreaching to the general public (NDSC 2016). NDCO is a prefectural office located in Nara Park. Two full-time staff members mainly answer phone calls that are related to deer problems. It was founded in 2010 in response to the growing number of tourists ('We often hear' 2010).

16.2.2 Miyajima Island

16.2.2.1 About Miyajima Island

Miyajima Island (Fig. 16.5) is located in the Seto Inland Sea of Hiroshima, Japan. The island is 30 km^2 in size and is primarily covered with primeval forest. The tourism district and residential area (approx. 4.3 km^2) are in the northern part of the island. The population of the island has been in decline since 1945. Currently, less than 1700 residents live on the island (Hatsukaichi City 2016a). This is partly due to strict land-use regulations, which resulted in young generations migrating off the island (Asano 2002; Asano and Funck 2001).

Tourism is the major industry on Miyajima Island (Asano and Funck 2001). In 1996, a part of the tourism district, including Itsukushima Shrine, was designated as a UNESCO World Cultural Heritage Site (Miyajima Tourism Association 2009). It attracts as many as four million domestic and international tourists annually (Hatsukaichi City 2016b). The majority of tourists visit Miyajima Island to see the Itsukushima Shrine (Funck 2013). Many tourists enjoy the nature on the island as well (Funck and Maruyama 2011).

[5]Agency of Cultural Affairs or *bunkachō* (in Japanese) is a government agency of Japan.

Fig. 16.4 Deer crackers are sold by the Foundation for the Protection of Deer in Nara. Photo by Rie Usui

One feature that characterizes Miyajima Island is the presence of free-ranging wild sika deer. According to Hatsukaichi City (2014), there are estimated to be 300–500 wild sika deer inhabiting Miyajima Island. A vast majority of them lives in or near the tourism district. Like Nara deer, Miyajima deer are tame. Tourists visiting the island will see deer resting on sidewalks or following people around as soon as they arrive at the ferry terminal. They will also observe different groups of deer while making their way to the shrine. While Nara deer are registered as a national natural treasure, Miyajima deer are not given any special identity.

16.2.2.2 History of Deer at Miyajima Island

Compared to Nara, little is known about Miyajima deer history. In the following, we briefly summarized the background story of deer at Miyajima Island based on some published research, official documents of Hatsukaichi City, newspaper articles and personal communications:

Since ancient time, people have considered Miyajima Island itself as a place where gods reside (Miyajima Tourism Association, 2009). Because of religious beliefs rooted in *Shintoism*,[6] deer on the island were protected from harm for many centuries (Hatsukaichi City 2014). As the island was regarded as a holy land, people were prohibited from living on or visiting the island in order to maintain its purity. In the year 593, Itsukushima Shrine in its original form was built by Saeki no Kuramoto[7] (Official Website of Miyajia Tourism). According to "Itsukushima Zue[8] (Fukuda 1973)," the ancestor of the Kuramoto clan was demoted after he

[6]*Shintoism* is a religion native to Japanese culture, which is characterized by the beliefs that regard nature and every object has a spiritual power.

[7]A samurai (warrior) who possessed the power in the region back then.

[8]A document that depicts Miyajima history with drawings and short stories.

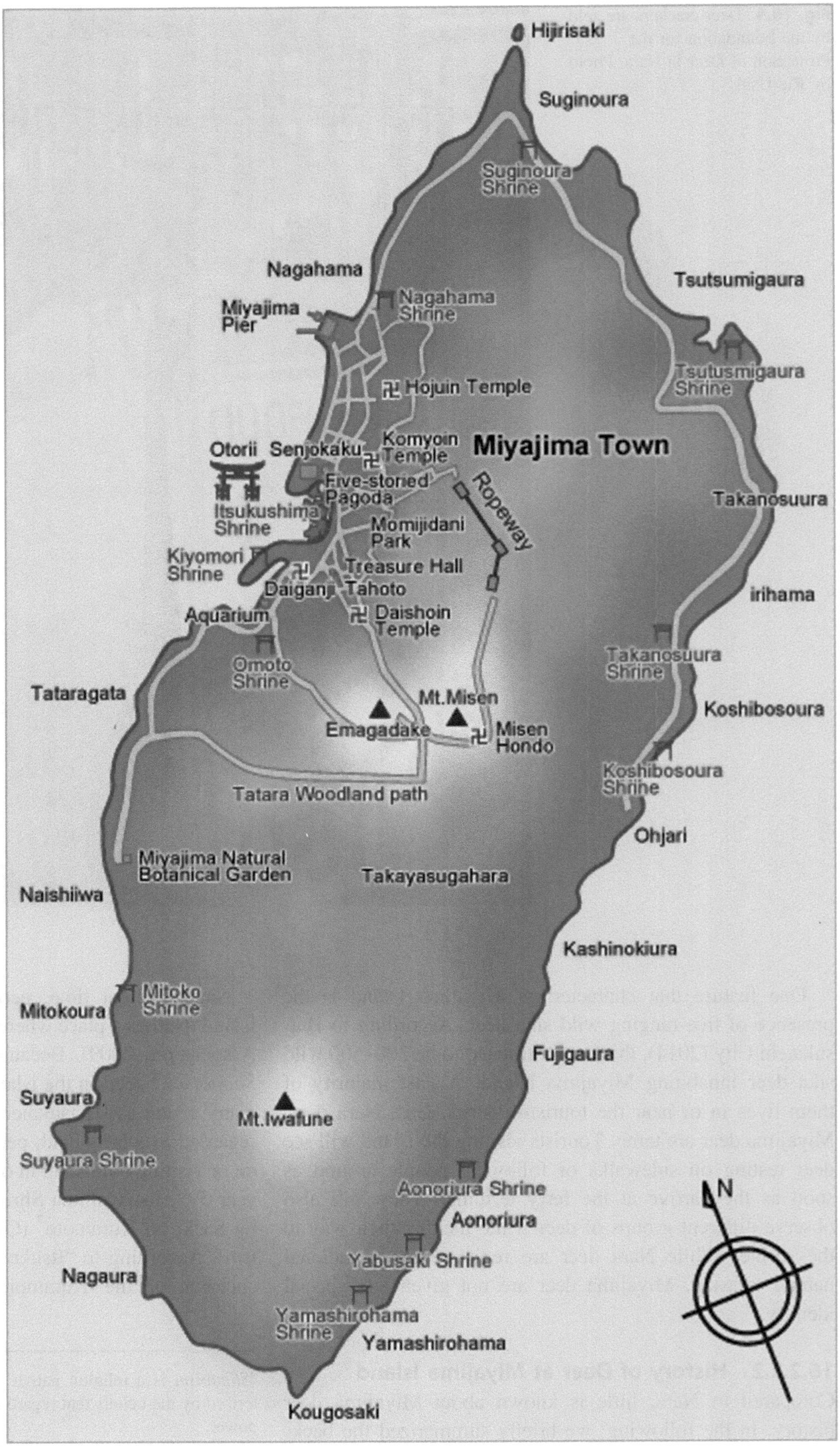

Fig. 16.5 A map of Miyajima Island from Miyajima Tourism Association (2009)

Fig. 16.6 Paintings of deer in Miyajima from Edo period. "Itsukushima Kuramazu" from interesting details of Itsukushima Zubyōbu (4) and (5) (Hiroshima Prefectural Art Museum 2012). Edo period refers to the time period between 1603 and 1868 in Japan

presented a hunted deer to Emperor Nintoku[9]–who was fond of the deer. Since then, an admonition to cherish the deer was passed on from one generation to the next in the Kuramoto clan. Hence, the deer on Miyajima Island were treated well for hundreds of years.[10]

The present form of Itsukushima Shrine was built by Taira no Kiyomori[11] in the 12th century. During the 14th century when Miyajima *Shintoism* belief permeated throughout society, shrine priests and Buddhist monks began living on the island. Later, a town was established and people started residing there. Eventually, the religious significance of the island was overshadowed by the new role of the island as a trade and commerce hub (Official Website of Miyajima Tourism, n.d.). As early as the 17th century, Miyajima Island developed into a popular travel destination and a large number of travellers gathered. The arts and travel records from this time period show that the deer on the island closely interacted with people (Fig. 16.6). These pictures depict deer receiving food from people and strolling the streets of Miyajima.

During *Meiji ishin*[12] (Meiji restoration) in the 19th century, the deer population decreased drastically. When the politics became stable, the deer were protected under a prefectural ordinance. A prohibition on deer hunting as well as dog keeping was issued in 1879 (Asano 2002). In this time period, human-deer relationships at Miyajima Island were recorded through the eyes of foreign visitors. For instance, Herbert George Ponting, a British photographer who visited Miyajima Island in the beginning of 1900s, wrote about Miyajima deer in his travel journal: Deer appear from the mountain and walk along the shore while licking salt on the rocks or eating the crackers that visitors gave (Ponting 2005).

According to Hatsukaichi City (2014), the population of deer dropped after WWII due to illegal hunting. The prefectural ordinance was abolished in 1949, but instead, the town of Miyajima enacted the ordinance for the protection of the deer (Asano 2002). In 1960s, the locals raised 48 remaining deer in captivity in order to increase their population ('Revival of Miyajia' 1966). Within ten years, the deer population became over abundant, which resulted in human-deer conflicts (Hatsukaichi City 2014).

16.2.2.3 Deer-Related Problems

Unlike Nara, Miyajima Island has no crop-raiding problem as there are no agricultural fields. Yet, other deer-related problems were generated as the population of deer recovered after WWII. Some serious issues that have been reported are: (1) physical injuries, (2) accidental intake of inappropriate food by deer, and (3) degradation of vegetation in the forest (Hatsukaichi City 2014). All these problems are partly attributable to the high concentration of deer in the tourism district. Through a long history of close relations with humans, deer are accustomed to inhabiting near human living space, which is located in the northern part of the island. Because at least a few thousand tourists visit Miyajima Island daily, tourists and deer interactions are likely to occur, which can result in physical injuries for both groups. One of the authors, Rie Usui (RU), observed several injured deer during the fieldwork. Some deer had broken antlers while others had broken legs. Although there is no large road on Miyajima Island as compared to Nara, taxis and large vans drive through narrow streets of the district. This may also be the cause of physical injuries for the deer.

As in Nara, accidental intake of inappropriate food is one of the issues on Miyajima Island. During the field observation on tourist-deer interactions conducted by the author (RU), deer were observed eating maps, brochures and/or paper cups in a large number of cases (Fig. 16.7).

[9]The 16th Japanese emperor who reigned in the forth century.

[10]No one lived on the island at this time period but those who are affiliated with the shrine was allowed to enter the island (Official Website of Miyajima Tourism).

[11]One of the influential leaders in the Japanese history.

[12]A reform of political and social systems.

Fig. 16.7 Deer eating a map. Photo by Rie Usui

Furthermore, many tourists were found feeding food that they purchased from the food vendors to the deer. The vendor foods typically consisted of chicken skewers, fried bread, French fries and ice cream, which are of very different composition to food normally eaten by deer.

Hatsukaichi City (2014) is concerned that the increased number of deer are causing damage to the vegetation of the Miyajima's primeval forests. According to research conducted by the Ministry of the Environment in 2006, the richness of plant species has been lost and tree bark has been heavily eaten by the deer. However, there is an opposing claim that the direct cause of the current forest condition was a result of human activity[13] (Dr. H Tsubota 2015, pers. comm., 14 July).

16.2.2.4 Management of the Deer for Tourism

Policy and protection laws

When an increased deer population became a problem after WWII, the first official meeting regarding deer management was held in 1978 (Hatsukaichi City 2014). Town officials of Miyajima agreed to introduce feeding stations in the forest in an attempt to entice the deer outside of the tourism district. This tactic continued until 1987, but ultimately failed to work (Hatsukaichi City 2014). In 1998, the officials proposed a ban on deer feeding. Unfortunately, they were unable to enforce the regulation and feeding remained a problem.

After the town of Miyajima merged with Hatsukaichi City in 2005, Hatsukaichi City officials gained jurisdiction over management of the deer. Despite the fact that the deer on Miyajima Island were habituated, the officials consider the deer on Miyajima Island wild animals. In 2007, tourists were advised not to feed the deer, but street stalls continued selling deer crackers to tourists until they were strictly banned in 2008 (Hatsukaichi City 2014). Currently, deer management is based on the Miyajima Region Deer Protection Management Plan (MRDPMP). In response to the problems caused by aggregation of deer in the tourism district, Miyajima Region Deer Protection Management Committee (MRDPMC) was organized in 2009. The members of the committee consist mainly of Hatsukaichi city officials and academics. Note there is no involvement of Itsukushima

[13]In 1984, there was a fire in Miyajima. It was put out with salt water and it damaged the soil.

Shrine. Although Hatsukaichi City (2014) claims that no single institution or person is responsible for the deer in Miyajima, city officials have jurisdiction over the management of them.

MRDPMP states that the deer on Miyajima Island are wild (Hatsukaichi City 2014). Therefore, human and deer living space should be separated (Hatsukaichi City 2014). The goal is to bring deer to the state where they will no longer rely on human-derived food, and to reduce the total number of deer on the island. Hatsukaichi City (2014) reported that feeding deer is the primary cause of all the problems they face. A prohibition on feeding is the ultimate solution to mitigate human-deer conflicts. Eradication of deer is the least likely management option at Miyajima Island as the island is a designated wildlife protection area (Asano 2002). Under this law, deer hunting is strictly prohibited (Hiroshima Prefecture 2011).

Onsite Management

Miyajima's onsite deer management can be described as a passive management style in contrast to the management model of Nara Park. Effort devoted to the management of deer at Miyajima Island is kept to a minimum. Regulatory management strategy (Orams 2002) is the main strategy that has been implemented: warning signs are posted in Japanese and English to inform tourists that the deer on Miyajima Island are wild and to advise them not to feed or touch deer. In addition, an audio announcement is played repeatedly to warn tourists to watch their belongings when deer approach. The warning signs and audio announcement state that any accidents resulted from interacting with the deer (e.g., deer eating tourists' tickets) are the tourists' sole responsibility.

No organizations that are similar to FPDN or NDSC exist at Miyajima Island, which makes it more difficult to keep the site free from human-deer conflicts. For instance, during their annual fireworks in August, a large number of tourists visit the island and fill the streets where deer usually hang around. Many of them leave garbage behind, which causes deer to eat the food or items that they cannot digest (Fig. 16.8).

A local resident of Miyajima reported to RU on 22 June, 2014 that if deer were found injured or dead, the local police station would handle such circumstances. Then, city officials or hired agency would take the deer to a place the locals call *ura* (on the other side of the island). *Tsunokiri* (antler-cutting) is performed but not as a public event like Nara Park. It is operated non-publicly by firefighters of Miyajima. Unlike Nara, pregnant female deer in Miyajima are not protected in a confinement. Another local mentioned that pregnant females would give birth in or near the tourism district. This is because staying closer to the human environment keeps fawns relatively safe from predators like crows. Young fawns in the tourism district are often seen being chased by tourists (Fig. 16.9), which raises concerns about stress they may experience.

Fig. 16.8 After fireworks. Photo by Rie Usui

Fig. 16.9 A fawn in the tourism district. Photo by Rie Usui

Although Hatsukaichi city officials have banned feeding of the deer since 2007, this rule does not have any lawful force and it only remains as advice, making it less effective. In fact, not only tourists, but also some locals have been observed feeding the deer. For example, workers at a photo store provision deer with food in order to offer tourists photo opportunities with them. Moreover, some restaurant owners have been observed giving deer discarded vegetables. The city officials' decision to ban deer feeding has received some pushback from people. They claim that the deer have been relying upon human food so much that it is cruel to terminate artificial feeding immediately. Challenges regarding the feeding of deer at Miyajima Island remain to this day (Hatsukaichi City 2014). Table 16.1 provides a comparison of the key aspects of Nara Park and of Miyajima Island in Japan.

16.3 Summary

The deer in both Nara Park and Miyajima have rich social and cultural histories. While Nara deer were used as a symbol of importance for the temples and shrines to maintain their social power, Miyajima deer were protected because of where they lived—the religious significance of their habitat protected them from being harmed. Both Nara and Miyajima deer have had close associations with humans throughout history and are well adapted to human modified environment. Nonetheless, the recent growth in tourism brought new challenges to each of them on how to coexist with the deer.

The review of deer management policy in Nara Park revealed that they attempted to coexist with the deer by acknowledging their cultural significance. Although feeding of wildlife is a controversial practice (Orams 2002), Nara deer are fed with deer crackers made exclusively for them by FPDN. This prevents the deer from dispersing to neighboring lands where farmers suffer from *Rokugai* and are in urgent need of solutions to crop-raiding problems. Nara Park made management of over a thousand free-ranging deer possible by working with onsite management organizations (i.e., FPDN, NDSC and NDCO).

In contrast to Nara's deer management policy, the management goal of Miyajima was to coexist by separating the living space of the deer and humans. MRDPMP committees have been attempting to ban deer feeding, but the regulations are hardly enforced and there is little commitment from

Table 16.1 A summary of comparisons between Nara Park and Miyajima Island

	Nara Park	Miyajima Island
Location	Nara city, Nara	Hatsukaichi city, Hiroshima
Area size	6.6 km^2	4.3 km^2 (tourism district)
Tourists number	14 million/year	4 million/year
Deer		
Population	1495 (July 16, 2015 census)	350–500 (August, 2014 report)
Density	226 deer/km^2	81–116 deer/km^2
Species and status	Sika deer (*Cervus nippon*), wild	Sika deer (*Cervus nippon*), wild
National natural treasure	Since 1957	No
Main issues	Crop raiding, traffic accident, accidental ingestion of food by deer, physical injuries	Traffic accident, accidental ingestion of food by deer, physical injuries
Policy	Nara Deer Conservation Management Plan since 2013	Miyajima Region Deer Protection Management Plan since 2009
Goal	Hundred years from now, the deer in Nara Park continue to inhabit healthy (Nara Prefecture 2016b)	To bring deer back to the mountain and separate human and deer living space
Management Committee	Nara prefecture, Nara city, Kasuga shrine, Foundation for Protection of Deer in Nara (onsite management), Nara Deer Consultation Office (onsite management), Academics	Hatsukaichi city, academics, local representatives
Onsite deer management	Warning signs, *Tsunokiri* (antler-cutting), *Rokuen* (deer garden), daily monitoring	Warning signs, audio announcement, *Tsunokiri* (antler-cutting)
Laws and land use regulation	Sacred deer killing prohibition area, law for the protection of cultural properties	Wildlife protection area

City of Nara (2014), Foundation for the Protection of Deer in Nara (2015), Hatsukaichi City (2008, 2014), Ikeda (2010), Nara Prefecture (2016a, b. Multiple sources, edited by Rie Usui

management. Their management practice is limited to regulatory management strategy (Orams 1996), which includes posting and announcing the rules about how not to interact with deer on the island. Like reported from other tourism sites (Fuentes et al. 2007; Ruesto et al. 2010), the warning signs are ineffective at Miyajima Island. As a consequence, "undomestication" of the semi-domesticated deer population seems to be a failure as many deer still remain in the tourism district and are observed begging tourists for food or eating garbage.

What is apparent from comparisons of Nara Park and Miyajima deer management models is that while both sites consider the deer to be wild, they attempt to coexist with deer with a different manner (Table 16.1). Hill (1997) states that management decisions and planning are made based on the policies that reflect the values of wildlife in communities. The deer in Nara Park and Miyajima Island utilize forest resources to certain extent, thus are considered to play an important ecological role (Maesako 2002; Okuda 1984). At the same time, their significance is rooted in cultural history at both sites. What can be interpreted from their management decisions is that Nara Park values the cultural significance of the deer and Miyajima Island values the ecological significance of the deer.

A full discussion on why they practice dissimilar management strategies is beyond the scope of this chapter, but one reasonable explanation stems from their geographical locations. Nara Park is surrounded by farmlands and an urban city center. Simply terminating artificial feeding will likely cause more problems in the surrounding areas. Miyajima Island, on the contrary, is able to take this course of management presumably because there is little to no concern about crop raiding from deer dispersion.

16.4 Concluding Remarks

The aim of this chapter was to introduce some case studies of wildlife tourism that are distinctive from the dominant wildlife tourism research that have been reported so far. Wildlife tourism, in a broad sense, includes all animals in a tourism context and therefore, we need to recognize there is diversification in the forms and places where wildlife tourism occurs. There should be a diversification of wildlife management models in tourism settings.

As we have seen in our case studies, both Nara Park and Miyajima Island have encountered problems with deer throughout their history and attempted to mitigate the problems in their own way. These case studies can further address ethical issues that cannot be ignored when animals are involved in tourism. The central challenge generated when managing animals is whether the goal should be protection of the ecosystem or individual wellbeing. Typically, the former goal is more emphasized when animals occur in natural settings whereas the latter goal is valued under captive settings. In the case of Nara Park and Miyajima Island, the deer do not fit exclusively into any one of these categories: wild or domesticated. In fact, these case studies illustrate two contradicting ideas. One is that even though these deer are tame and are fed by humans, they are still considered wild. The other aspect is that seemingly similar deer that are considered wild are managed differently. By no means are we trying to claim one management model is better than the other, but understanding the context of wildlife tourism and the history of the human and animal relationship there is one of the necessary components of more sustainable wildlife management for tourism.

Acknowledgements We sincerely thank all individuals who made our work possible. Special acknowledgement is made to Taoyaka Program Scholarship at Hiroshima University for the financial support throughout my academic career. We would also like to extend our acknowledgement to Dr. Ismar Borges de Lima and Dr. Ronda J. Green for giving us the opportunity to contribute to this book.

References

Apps K, Dimmock K, Lloyd D, Huveneers C (2016) In the water with white sharks (*Carcharodon carcharias*): participants' beliefs toward cage-diving in Australia. Anthrozoös 29(2):231–245

Argüelles MB, Coscarella M, Fazio A, Bertellotti M (2016) Impact of whale-watching on the short-term behavior of southern right whales (*Eubalaena australis*) in Patagonia, Argentina. Tour Manag Perspect 18:118–124

Asano T (2002) Potential for ecotourism development at Miyajima Island (Miyajima ni okeru ecotourism no kokoromi). Geogr Sci (*Chiri-Kagaku*) 57(3):194–207 (in Japanese)

Asano T, Funck C (2001) Development and alteration of Setouchi tourism region: case studies of Miyajia Island and Shimanami Kaidō (Setouchi kankō chiiki no keisei to henyō: Miyajima to Shimanami Kaidō o jirei toshite). Research enter for Regional geography special publication no. 36 (in Japanese)

Bach L, Burton M (2016) Proximity and animal welfare in the context of tourist interactions with habituated dolphins. J Sustain Tour, pp 1–17

Burgin S, Hardiman N (2016) Crocodiles and grey nomads: a deadly combination? Curr Iss Tour 19(1):60–63

City of Nara (2014) Report of incoming tourists of Nara City (Nara-shi kankō irikomikyakusū chōsa hōkoku). http://www.city.nara.lg.jp/www/contents/1347582712483/files/H26.pdf. Accessed 23 May 2016 (in Japanese)

Cong L, Newsome D, Wu B, Morrison AM (2014a) Wildlife tourism in China: a review of the Chinese research literature. Curr Iss Tour, pp 1–24. doi:10.1080/13683500.2014.948811

Cong L, Wu B, Morrison AM, Shu H, Wang M (2014b) Analysis of wildlife tourism experiences with endangered species: an exploratory study of encounters with giant pandas in Chengdu, China. Tour Manag 40:300–310

Foundation for the Protection of Deer in Nara (2010) Nara no shika: "shika no kuni" no hajimete no hon. Keihanna Jōhō Kyōiku Shuppan, Japan

Foundation for the Protection of Deer in Nara (2015) Nara deer population census of 2015 fiscal year. (Heisē 27 nendo "Nara no sika" seisoku tōsū chōsa). http://naradeer.com/common/img/aboutnaradeer/naratousu.pdf. Accessed 23 May 2016 (in Japanese)

Foundation for the Protection of Deer in Nara (2016) Problems. http://naradeer.com/problems/index.html. Accessed 23 May 2016 (in Japanese)

Fuentes A, Shaw E, Cortes J (2007) Qualitative assessment of macaque tourist sites in Padangtegal, Bali, Indonesia, and the Upper Rock Nature Reserve, Gibraltar. Int J Primatol 28:1143–1158

Fukuda N (1973) Itsukushima Zue. Miyajima Town, Japan (in Japanese)

Funck C (2013) International and Japanese tourists' behaviors at Miyajima Island. (Miyajima ni okeru gaikokujin to nihonjin kankōkyaku no kōdō). Studies of the Itsukushima Island (World Heritage Itsukushima, History and Culture of the Inland Sea Projects Research Center) 9:(1)–(12) (in Japanese)

Funck C, Maruyama N (2011) Nature-based tourism at Miyajima Island and Mt. Misen (Miyajima to Misen ni okeru shizen kankō). Studies of the Itsukushima Island (World heritage Itsukushima, history and culture of the inland sea projects research center) 7:(1)–(12) (in Japanese)

Granquist SM, Nilsson PÅ (2016) Who's watching whom?—An interdisciplinary approach to the study of seal-watching tourism in Iceland. J Clean Prod 111:471–478

Hatakama K (2010) From emergence of Nara deer to antler cutting. (Shinroku no tanjō kara tsunokiri e). In: Nagano H (ed) Nara deer: the first book of the deer kingdom (Nara no sika: "sika no kuni" hajimete no hon). Keihan Nara Jōhō Kyōiku Shuppan, Japan, pp 103–170 (in Japanese)

Hatsukaichi City (2008) Miyajima region deer protection management planning. https://www.city.hatsukaichi.hiroshima.jp/uploaded/attachment/8228.pdf. Accessed 31 May 2016 (in Japanese)

Hatsukaichi City (2014) Miyajima region deer protection management planning (second term). https://www.city.hatsukaichi.hiroshima.jp/uploaded/attachment/8229.pdf. Accessed 31 May 2016 (in Japanese)

Hatsukaichi City (2016a) National census (Kokuzei chōsa). https://www.city.hatsukaichi.hiroshima.jp/soshiki/10/10962.html. Accessed June 4 2016 (in Japanese)

Hatsukaichi City (2016b) Miyajima incoming tourists chart (Miyajima raitōshasū ichiranhyō). https://www.city.hatsukaichi.hiroshima.jp/uploaded/attachment/12566.pdf. Accessed June 4 2016 (in Japanese)

Hill M (1997) The policy process in the modern state. Prentice-Hall, London

Hiroshima Prefectural Art Museum (2012) Interesting details of Itsukushima Zubyōbu (4) and (5). http://www.hpam.jp/blog/?p=3885. Accessed June 23 2016

Hiroshima Prefecture (2011) Hunting license. https://www.pref.hiroshima.lg.jp/site/huntinglicense/hunter-map.html. Accessed 5 June 2016 (in Japanese)

Ikeda S (2010) Visiting Foundation For Protection of the Deer in Nara ("Nara no sika aigokai" o tazunete). In: Nagano H (ed) Nara deer: the first book of the deer kingdom (Nara no sika: "sika no kuni" hajimete no hon). Keihan Nara Jōhō Kyōiku Shuppan, Japan, pp 13–58 (in Japanese)

Ikeya K (2009) From the principle of exclusion to the principle of coexistence: a new relationship concerning wildlife protection area

(Haijo no ronri kara kyōzon no ronri e—doubutsu hogoku o meguru aratana kankei). In: Ikeya K, Hayashi Y (eds) Human-animal relationship studies the fourth edition: wild and environment (Hito to doubutsu no kankeigaku dai 4 kan: yasei to kankyō). Iwanami Shoten, Tokyo, pp 296–319 (in Japanese)

Knight J (2011) Herding monkeys to paradise: how macaque troops are managed for tourism in Japan. Brill, Leiden

Kurita H (2012) Provisioning and tourism in free-ranging Japanese macaques. In: Russon AE, Wallis J (eds) Primate tourism: a tool for conservation?. Cambridge University Press, UK, pp 44–55

Lemelin RH (2015) From the recreational fringe to mainstream leisure: the evolution and diversification of entomotourism. In: Markwell K (ed) Animals and tourism: understanding diverse relationships. Channel View Publications, Bristol, pp 229–239

Maesako Y (2002) Ecological issues relating to the conservation of Sika deer (*Cervus nippon* TEMMINCK) and the warm-temperature evergreen forest on Mt. Kasugayama, Nara, Japan. Vegetation Sci 19:61–67 (in Japanese)

Miyajima Tourism Association (2009) Home. http://www.miyajima.or.jp/english/index.html. Accesed 4 June 2016

Nara Deer Supporters Club (2016) Q&A. http://www.shikasapo.jp/qa.html. Accessed 4 June 2016 (in Japanese)

Nara Prefecture (2016a) About deer. http://www.pref.nara.jp/5819.htm. Accessed 23 May 2016 (in Japanese)

Nara Prefecture (2016b) Nara deer conservation management plan discussion committee. http://www.pref.nara.jp/33948.htm. Accessed 2 June 2016 (in Japanese)

Newsome D, Dowling RK, Moore SA (2005) Wildlife tourism. Channel View Publications, Clevedon

Official Website of Miyajima Tourism (2016) Summary of Miyajima history (Miyajima rekishi gaiyō). Accessed 12 June 2016. http://www.miyajima-wch.jp/index.html

Okuda T (1984) Food habits of sika deer (*Cervus nippon*) and their ecological influence on the vegetation of Miyajima Island. Hikobia Suppl 9:93–102

Orams MB (1996) A conceptual model of tourist-wildlife interaction: the case for education as a management strategy. Aust Geogr 27:39–51

Orams MB (2002) Feeding wildlife as a tourism attraction: a review of issues and impacts. Tour Manag 23(3):281–293

Paksuz S, Özkan B (2012) The protection of the bat community in the Dupnisa Cave System, Turkey, following opening for tourism. Oryx 46:130–136

Ponting HG (2005) Enoura bay and Miyajima Island (Enourawan to Miyajima). In: In Lotus-land Japan (Translated by Nagaoka S. Kodansha Academic Library, Tokyo), pp 302–323

Qingming C, Gonggang X, Wall G (2012) A cultural perspective on wildlife tourism in China. Tour Recreat Res 37(1):27–36

'Revival of Miyajima' (1966) Chūgoku Shinbun, 18 Sept, p 14 (in Japanese)

Ruesto LA, Sheeran LK, Matheson MD, Li JH, Wagner RS (2010) Tourist behavior and decibel levels correlate with threat frequency in Tibetan macaques (*Macaca thibetana*) at Mt. Huangshan, China. Primate Conserv 25:99–104

Saarinen J (2016) Wilderness use, conservation and tourism: what do we protect and for and from whom? Tour Geogr 18(1):1–8. doi:10.1080/14616688.2015.1116599

Sinha BC, Sinha SP (2008) Impact of religious tourism on Gir National Park, Gujarat, India. Indian Forester 134(5–8):667–673

Smith H, Samuels A, Bradley S (2008) Reducing risky interactions between tourists and free-ranging dolphins (Tursiops sp.) in an artificial feeding program at Monkey Mia, Western Australia. Tour Manag 29(5):994–1001. doi:10.1016/j.tourman.2008.01.001

Takahashi S (1999) People's attitudes and feelings towards deer in Nara Park. Sōgō Kenkyūsho Shohō, pp 91–100 (in Japanese)

Takeda Y (2010) The deer crackers that deer beg to get are additives-free and high quality (Shika mo atama o sagete hoshigaru shikasenbei wa mutenka yūryōhin nanoda). In: Nagano H (ed) Nara deer: the first book of the deer kingdom (Nara no sika: "sika no kuni" hajimete no hon). Keihan Nara Jōhō Kyōiku Shuppan, Japan, pp 59–68 (in Japanese)

Torii H, Tatsuzawa S (2009) Sika deer in Nara Park: unique human-wildlife relations. In: McCullough DR, Takatsuki S, Kaji K (eds) Sika deer: biology and management of native and introduced populations. Springer, Tokyo, pp p347–p363

UNESCO (2016) Historic monuments of ancient Nara. http://whc.unesco.org/en/list/870/. Accessed 23 May 2016

Usui R, Sheeran LK, Li JH, Sun L, Wang X, Pritchard AJ, DuVall-Lash AS, Wagner RS (2014) Park rangers' behaviors and their effects on tourists and Tibetan macaques (Macaca thibetana) at Mt. Huangshan, China. Animals 4(3):546–561. doi:10.3390/ani4030546

Watanabe S (2001) Dealing with the damage to the crops cause by a protected animal: a case with "Nara-no-shika" authorized as a national monument. Jpn Assoc Environ Sociol, pp 129–144 (in Japanese)

Watanabe S (2007) Idea and reality in measures of agricultural damage by NARA-NO-SHIKA -based on the questionnaire survey to the farmer around Nara Park- Bullten of Center for Natural Environment Education. Nara University of Education (Nara Kyōiku Daigaku Fuzoku Shizen Kankyō Kyōiku Sentā Kiyō) 8:23–41 (in Japanese)

Watanabe S (2010) Nara deer in modern days: seeking for "coexistence" and challenges (Kindai ni okeru Nara no sika: "kyōzon" e no mosaku to kon'nan). In: Nagano H (ed) Nara deer: the first book of the deer kingdom (Nara no sika: "sika no kuni" hajimete no hon). Keihan Nara Jōhō Kyōiku Shuppan, Japan, pp 171–214 (in Japanese)

Watanabe S (2012) The problems about provisions and capture in "free-randing" animals—why are there two provisions concerning the "Deer of Nara"? Bull Nara Univ Educ 61(1):109–119 (in Japanese)

'We often hear' (2010) Asahi Shinbun digital. http://www.asahi.com/kansai/travel/news/OSK201002190061.html. Accessed 21 June 2016 (in Japanese)

Wheatly BP (1999) The sacred monkeys of Bali. Waveland Press, Illinois

Will the Ark Sink? Captive Wildlife, Tourism and the Human Relationship to Nature: Demystifying Zoos

17

Dirk Reiser

Abstract

Apart from displaying many facets of the ambiguous relationship of humans and animals, zoos and aquariums are also a significant part of the tourism industry. Their supposed goals are visitor education, visitor entertainment, species conservation and behavioural animal research. It seems therefore reasonable to describe those institutions as an ark for endangered species. However, the role of zoos in modern societies is more and more called into question. Their main purpose is described as visitor entertainment (to create economic benefits) covered up by conservation, education and research claims. Arguably, some would like to see their demise, while others propose changes that include an alternative, but more sustainable version of zoos and aquariums: wildlife parks and wildlife sanctuaries that focus on the well-being of the animals.

17.1 Introduction

> '… zoo professionals have often labelled their institution as an 'ark', a vessel in which we could carry threatened species from around the world until the dangers to their habitat pass, at which time they could be released to replenish the earth.' (Mazur 2001, p. 1)

Animals are a significant tourism resource that is consumed by the tourist and offered by the tourism industry. As attractions, animals contribute in a variety of ways to tourism, e.g. as a form of transportation, as destination icons, as travel companions, as components of regional cuisine (Marxwell 2015, p. 1), as clothing, or as attractions for entertainment (Donaldson and Kymlicka 2013). Two of those attractions are zoos and aquariums.

Every year more than 700 million people visit zoos and aquariums world-wide (World Association of Zoos and Aquariums 2015). This high number of visitors is a clear indication for the popularity of those institutions (Frost 2011). Amongst their visitors are recreationists, domestic and international tourists for which the zoo industry caters.

D. Reiser (✉)
Rhine-Waal University of Applied Sciences, Kleve, Germany
e-mail: dirk.reiser@hochschule-rhein-waal.de

The industry itself describes its goals as educating visitors about animals, researching about animals, the conservation of animals, especially endangered species and entertaining visitors (Patrick and Tunnicliffe 2013). It can therefore not be surprising that zoo managers describe their institutions as an ark for endangered animals (Mazur 2001). At the same time, it is some of those endangered and rare species that visitors come to see, including travellers.

There are a variety of definitions of zoos around the world. In very general terms, a zoo is a 'tourism attraction displaying captive wildlife for visitors.' (Dobson 2015) In other definitions, the goals of those institutions are sometimes included in a definition. Jamieson (2006, p. 6) for example defines zoos as 'public parks which display animals, primarily for the purposes of recreation and education.' Additionally, definitions can also include a temporal dimension: The German Government defines zoos in the nature protection legislation (paragraph 42 nature protection law-'Naturschutzgesetz') as permanent institutions in which living animals of living species are displayed for at least seven days of a year (dejure.org 2016). Zoo tourism is centred around those permanent public or private institutions displaying captive animals in cages (spatial limitation of their free movement) at least for some time. The main aim is

I. Borges de Lima and R.J. Green (eds.), *Wildlife Tourism, Environmental Learning and Ethical Encounters*, Geoheritage, Geoparks and Geotourism, DOI 10.1007/978-3-319-55574-4_17

to show wild animals to humans, but it needs to be recognised that not all zoos are created or managed the same (Tribe 2004).

Despite the point that zoos are just one example of the human relationship with other animals, they nevertheless symbolise many aspects of the connection between humans and nature. The following chapter analyses some of those relationships before focusing on zoos in Germany. The article will show that animals are an important tourism resource around the world, including in Germany. They are part of the historically grown human-other animal relationship which is highly contradictory. In particular zoos sit at the nexus between animals as objects (e.g. livestock) and animals as subjects (e.g. pets). Their stated ideals of conservation, education and entertainment are hereby critically evaluated with regards to sustainability and ethics.

17.2 Humans Amongst Other Animals

The human relationship with other animals symbolises many aspects of the link between humans and nature. Humanity relates to animals in manifold ways, including as wild creatures for discovery, tame creatures for interactions, objects of exhibition, targets for shooting and fishing, a source of education, training or science, mythical or symbolic representation or as ancillary roles (Fennell 2012). Even though, just the thought about this connection moves humans out of their comfort zone as it calls a number of their ethical behaviours into question. Just writing "the human relationship with other animals" seems to be wrong, out of place, not correct as it questions the exceptionality and superiority of humans as well as their treatment of animals. However, the uniqueness of human beings within nature has continuously been eroded in the last few decades (Sommer 2012). The way humans treat other animals is consequently more and more questioned. This includes the caging of animals in zoos for a variety of reasons such as conservation, education, research and entertainment.

The relationship between humans and other animals has a long tradition. Historically, the first humans settled at least around 14,000 years ago and started to domesticate animals. Archaeological findings suggest that such domestications started between 12,000 and 14,000 years ago with wolves, followed by cats around 9000 years ago (Fennell 2012). Humans did this out of security considerations to protect their owners and out of socio-economic necessities as animals could be made available all year round (e.g. for food, to be exchanged) thereby making the survival of human kind much easier. This is the origin of the power relationship between humans and animals (Morgenthaler 2013)-humans see themselves as superior to animals. Throughout history those differences expressed itself in a variety of ways, for example to keep exotic or native animals in cages for human pleasure.

17.3 The Human Relationship with Animals in Captivity

For more than 150 years observing wild and domestic animals in captivity has been a prominent leisure activity, but keeping such animals for personal pleasure or as status symbols is much older. Philosophically, this was accepted. Aristotle argued that humans have the ability to speak and to build states while animals would not be able to do this. Later, Descartes believed that humans have a soul in contrast to animals, while Kant's main distinction was human reasoning and Hölderlin's differentiation was built on human's awareness of the own, inevitable death (Lieckfeld 2012). In general it can be summarized that for more than 2500 years human culture knew one central answer to the main difference between humans and animals: Humans are rational beings who can reason (Eilenberger 2012). Spirit, culture and language count as the sign of the uniqueness of humanity. Even so the molecular biologists Charles Sibley and John Alquist DNA analysis of human, chimpanzee and bonobo genetics in 1984 came to the surprising result that the difference between these species DNA is 1.6% (Morgenthaler 2013).

Being entertained by wild animals in captivity is therefore not a new concept. Historical texts (Jamieson 1985; Bostock 1993; Tribe 2004) indicate that as far back as the ancient Egyptian Pharaohs (e.g. Queen Hatshepsut), animals were kept in menageries for royal pleasure and as status symbols as were animals for the emperors of China (e.g. Wen, the first emperor of the Zhou dynasty) and the Aztec, Roman (Stevens and McAlister 2003; Fennell 2012) or German rulers (Emperor Charlemagne as a symbol of his power) (Miller 2005). In some cases the world's 'royal' houses collected these animals, while in others they were the gifts of "a grateful people", such as the Bengal tigers from Indian Maharajas as a gift for Queen Victoria. This 'tradition' continued well into the 20th century, when for example Yugoslavian leader Tito maintained a private zoo on Brioni Island (Fennell 2012).

However, as the perception that menageries were solely a royal prerogative changed, the idea that zoos should be for the 'people' emerged, leading to the creation of the concept of the public zoological garden (Tisdale 1993). Consequently, zoos as formal institutions developed away from exotic collections of Renaissance princes (Boehrer 2007) representing the absolute monarchy and its power over nature. This is closely linked to the conditions of European societies during the Age of Enlightenment, when ideas by Rousseau, the French Revolution and the works of Descartes and Newton lead to more educational components to the use of animals (Altick 1978). Fennell (2012, p. 77) therefore argues that 'the modern zoo has its foundation in the Versailles menagerie in France, although even before this a zoo had been established by Franz Stephan I of Lorraine at the Schönbrunn Palace of the Habsburg monarchy in Vienna'. It was opened to the public in 1765, containing a variety of exotic animals in 13 enclosures with an octagonal pavilion in the centre (Park 2015). London Zoo, the world's first zoo incorporating a scientific component into its mission opened in 1828 with a collection of exotic animals that were studied by the eminent scientists of the day. Even though, it did not open its doors to the public until 1847 (Tisdale 1993).

Following the success of the London Zoo, many major cities throughout Europe and the United States opened zoos based on the London Zoo model. Zoos continued in their general popularity but it was only after World War Two and the rise in the popularity of the use of the private car for excursions that one saw a dramatic increased interest in zoos (Reichenbach 2002). Around that time many zoos were on the outskirts of cities and so a zoo was seen as an excellent distance for the 1950s motorist to take the family 'for a ride'.

As zoos became more open to the public the scientific element was pushed aside in favour of the idea that zoos should become a place for recreation and amusement, as well as the showplace for what were once viewed as 'freaks of nature'. Thus the simple viewing of animals subsequently gave way to the use of animals for entertainment, such as the feeding of animals, animal rides, animal petting and animal performances. Zoos became pseudo entertainment venues, and because of the perceived exotic nature of zoological gardens, both public and elite social events were staged in the zoo's grounds, including sporting events and ballroom dancing (Tribe 2004).

Though gradually zoos changed their roles from providing privileged access to allowing the general public to gaze upon those animals. In conjunction, the expectation of the visitors and the offers of zoos changed as well. Today, the role of zoos and aquariums is based on a complicated concept that is riddled with contradictions closely linked with fast changing public attitudes and visitor perceptions that move between traditional, modern and post-modern values.

17.4 Current Issues and Debates

Zoos and aquariums have changed much in recent years. This is reflected in the great number of publications about the development of modern zoos of whom only some can be mentioned that cover the contradictory relationship of humans with caged zoo animals. Some examples include: Hancock's (2001), 'A different nature. The paradoxical world of zoos and their uncertain future', Mazur's (2001), 'After the ark', Donaldson and Kymlicka's 'Zoopolis' or Garrett's (2014), 'Why do we go to zoos'. While these books describe the general contradictions, there is also an increasing number of publications that relate to the specific contradictions present in the relationship of tourism and tourists with animals. Those publications include Mason's (2000) 'Zoo tourism: a need for more research', Fennell's (2012) 'Tourism and animal ethics', Frost's (2011) Zoos and tourism and Markwell's (2015) 'Animals and tourism'. In all of those publications, it becomes clear that it is difficult on generalize on institutions as diverse as zoos and aquariums.

With regards to modern zoos and aquariums it has to be clarified that standards of animal welfare and management, and conservation work in many zoo in the developed world has changed positively in the last few decades. Additionally, it is not possible to generally judge all of those institutions (Cunningham 2016) as there is a great variety of standards within the zoos of the world. Nevertheless, since the 1970s there was some change in the attitudes on the human relationships with other animals and in the common perception of the acceptance of the capture and confinement of animals for entertainment purposes that can be looked at. Many zoos and aquariums have responded to those changes by labelling themselves as agents for preservation, public education and research (Marino et al. 2010), including in Germany.

Today, the relationship between humans and non-human animals is a constant source of debate. Publications in German journals such as the Philosophy Journal 'How much animal am I?' (Wie viel Tier steckt in mir?) (2012), Geo Kompakt's 'How do animals think?' (Wie Tiere denken) (2012), Geo's 'What you really need to know about animals' (Was Sie über Tiere wissen sollten) (2013) or die Zeit's 'The ethic of the ham sandwich' (Die Ethik des Schinkenbrots) (2014a) confirm the relevance of this topic in today's societies.

Conceptually, the human-animal encounter is happening in different settings (e.g. natural or contrived) with different forms of engagement (e.g. non-interactive to relational), where animals are in diverse conditions (e.g. wild or trained to perform) and where the interaction could be mediated (e.g. by a tour guide) or not (Cohen 2009). In general, there are two main encounters in everyday life with pets as subjects on one side and with livestock as objects on the other, with zoos being somewhere in between.

These different encounters show much about the irrational relationship between humans and animals. While livestock is a thing that is objectified, pets are person-like subjects who are often perceived as a part of the family. On the one side, humans show atrocious disregard for the welfare of livestock, but see welfare as vitally important for pets. Moreover, livestock is part of an economic activity, where medical treatment is important to enhance productivity and to avoid economic loss to maximize profits. Those animals are solely judged on their economic value, an early death is the inevitable outcome of their lives. Pets, however, receive medical treatment to enhance and lengthen their lives, whereby cost is often not a consideration. They have friendship value and their death is a tragedy.

An additional encounter that is often occurring in tourism is with wild animals in their natural settings. Sometimes wildlife is treated without any regard for their welfare (e.g. hunting, fishing, the killing of Cecil the lion) at other times people relate to them very closely (e.g. Knut the polar bear, Cecil the lion), their welfare is vitally important and reported about in the media. They are used for economic activity (e.g. tourists paying for viewing wildlife), but at the same time their lives are lengthened through medical treatment and their death is a catastrophe (e.g. the death of Knut the polar bear). Accordingly, it could be argued that zoos as 'important and popular tourist attraction' (Frost 2011, p. 1) sit at the nexus between livestock and pets, also representing much of the irrational and ambivalent behavior of humans towards non-human animals. Consequently, they provide an important area of research in the realm of human-non-human relationships.

However, it needs to be recognized that there are a wide variety of opinions on zoos, captive animals and the ethics of those institutions. A good overview of those was published by van Tuyl (2008). In her edited book the authors of the chapters are of a great variety of backgrounds with different views on animals and zoos. Those voices include The Captive Animals' Protection Society, the Philadelphia Zoo, In Defense of Animals, The Association of Zoos and Aquariums, the Wildlife Conservation Union or the Elephant Sanctuary in Tennessee. This book makes the point that the opinion on animals in captivity does not exist and that there are a great array of institutions and organisations working in the field that may or may not support the modern goals of zoos: education, conservation, entertainment and scientific research.

17.5 Demystifying Zoos

There are a variety of definitions of zoos. Zoos could be described as institutions that house predominantly wild animals accessible for human observation for at least some time of the year on a limited area of 110 acres (around 0.45 km^2) or less with or without charging for admission (Webster Dictionary 2015; Hunter-Jones and Hayward 1998). Others define zoos as public parks that display animals for the purposes of recreation and education (Jamieson 2006, p. 6). The German Government defines zoos in the nature protection legislation (§42 Naturschutzgesetz) as permanent institutions in which living animals of living species are displayed for at least seven days of a year (de-jure.org 2016). The Association of Zoological Gardens in Germany (Verband der Zoologischen Gärten, VdZ) extends from those definitions to state that zoo (Old Greek for animal or being) is a short form for zoological garden that has different meanings. In everyday language it is mostly seen as a public park that is accessible and in which a higher number of animals of different species is held. The administrative definition of the EU is the same as in German nature protection legislation, while the self-understanding of zoos is that they are publicly accessible institutions in which animals of different species are held for the purposes of recreation, education, research and conservation (Verband der zoologischen Gärten 2016a). Zoo tourism is a form of tourism that is centred around those permanent institutions displaying captive animals in cages (spatial limitation of their free movement).

In the following, the main goals of zoos- conservation, education, research and entertainment- are described and evaluated. Consequently, the three myths that zoos provide conservation, education and research will be demystified to provide evidence about the cynical relationship that human animals, including tourists, have with caged animals.

17.5.1 Myth 1: Zoos Are There to Preserve Biodiversity (Conservation)

With regards to conservation, Gatland (1996) argues that there is 'overwhelming evidence that this concept [conservation through zoo] was flawed from the beginning and that zoos today are continuing to function not because they are conservation centres but on the strength of their own myths.' (Gatland 1996, online) The preservation of animals, especially the reintroduction of endangered species into the wild is often cited as a symbol for the successful conservation efforts of zoos. Culling (cited in Mount 2016) makes the point that attempts to reintroduce zoo breed animals into the wild are not possible. She exemplifies this with the argument that four hundred pandas were raised in captivity, but only five were released (of which three survived). An additional example is the fate of the last remaining northern white rhinos (they are extinct in the wild). For more than 40 years, so Culling, only four of them were born in captivity and the remaining females are unable to breed (Mount 2016).

Nevertheless, zoos continue to use conservation and the preservation of biodiversity as their main goal. The Association of Zoological Gardens (Verband der Zoologischen Gäarten 2016c) in Germany states in its vision that they help to preserve biodiversity. Yet, it is argued that zoos will have great difficulties to hold the potential flood of animal extinctions and fulfil their biodiversity pledge, because of the way humans relate to the environment.

Human beings have made a strong impact on the environment (e.g. habitat loss, habitat fragmentation, over-hunting). This is one of the main reasons that more and more species are becoming extinct. There are estimates that 70–200 animal and plant species become extinct every day. Many of those are not even studied yet. The United Nations Environmental Programme believes that the yearly extinction rate is currently 1000 times higher than it would be without the impact of humans. Other environmental organisations such as the International Conservation Union (IUCN) assess the situation as worse by arguing that the extinction rate could be up to 10,000 times higher than ever before in history (Bundeszentrale für politische Bildung 2011). Human stewardship of the animal world or conservation to protect the biodiversity on planet earth a one of the main arguments for the continued existence of zoos can hardly be supported.

Luckily, there is increasing debate about the human relationship with zoo animals, especially about its exploitative character purely for cultural and economic reasons. As more and more animal habitats are altered or destroyed by humanity, it started to dawn on humans that they need to change their behavior towards animals if they want to protect the biodiversity of planet earth.

However, the current picture is rather bleak. In Patterson (2002) described the animal-human relationship as 'Eternal Treblinka', where animals are exploited for clothing, food, leisure, research and industrial production. Today, one could easily agree with Gannon (2002, p. 589): 'What a strange relationship we humans have with our furry friends. We eat them and we cuddle them. We feed them, we play with them, we make money from them, we hunt them, we build tourism around them, we breed them, we teach them tricks and we pass legislations to protect them. In short, we use them for any purpose we wish …' But surprisingly there is not much more debate about this (Donaldson and Kymlicka 2013), especially in tourism (Fennell 2012), despite the enormous implications for every tourism business that is linked to animal viewing.

Even in more general terms, it remains questionable if humans should accept the loss of many species in the wild and recognize that their only chance of survival lies within the protective walls (or cages) of zoos and aquariums or if they will be able to recognize the need to change their relationship with the other animals to became the stewards of the natural world and the protectors of natural habitats instead of its destructor. Despite the fact that there were some successes such as the reintroduction of the Arabian Oryx of which the last wild on was killed in 1972 (1000 wild Arabian Oryx now live between Oman and Jordan) (Ganzert 2016) or the Przewalski horse in Mongolia (Palmer 2015), the overall picture remains rather bleak.

In a 2011 enquiry into zoos in Europe, the Born Free Foundation (2012) writes that only 13 of the species in European zoos were listed on the Red List of Threatened Species of the IUCN as globally threatened. Culling (cited in Mount 2016) confirms that 90% of zoo animals are not endangered. For the future, so Gatland (1996, online) criticizes the potential conservation efforts of zoos by contesting the possibility of the re-introduction of animals to the wild in years to come. He states that 'zoos need to provide conditions exactly mirroring natural habitats, enough space to maintain a continuing breeding pool without excessive inbreeding and an adequate flow of animals between zoos to maintain genetic diversity'. This is currently not the case. Even more, European zoos cull around 3000 to 5000 animals a year, because they fear a lack of genetic diversity or they do not have the space to house those animals (Mount 2016). The well-documented story of Marius, the giraffe that was killed at the Copenhagen Zoo, is just one example of such behavior.

Additionally, zoo animals are often kept in inappropriate social groups such as group animals like Gorillas held singularly as well as being incapable of showing normal behavior as the controlled zoo environments lack stimuli (Gatland 1996). The Great Ape Project (2012) for example argues that 33% of the 40 zoos in Germany keeping 450 great apes should not be allowed to keep any of them. But even more disturbing for the organization is that 10 zoos have just one or two apes equaling solitary confinement. This can hardly be seen as helping to protect biodiversity. Hancocks (2001) agrees by writing that only five species have been saved from extinction by zoos, but even worse the last animals of some endangered species like the Tasmanian Tiger or the Pinta Island tortoise Lonesome George suffered or died in zoos or zoo-like enclosures (Frost 2011).

If one looks into the future, it could be asked where the polar bears could go if their habitat is lost? Should they only remain as remnants of the past until the last will die in some zoo around the world? Shall we keep them in zoos as unique attractions for those visitors 'hungry' to see the last of their kind? Is it not a good opportunity to make a higher profit through those extinction tourists taking the last chance to see polar bears?

17.5.2 Myth 2: Zoos Are There to Educate Visitors (Education)

The mission statements of modern zoos normally include reference to conservation and to education. The National Zoo and Aquarium in Canberra, Australia for example states that its goal is 'to inspire and assist in the conservation of the natural world through innovative and exciting educational experiences.' (National Zoo and Aquarium 2015, online) Another institution, Hamilton zoo in New Zealand wants 'to inspire conservation action' (Hamilton Zoo 2015). However, it is highly questionable if zoos can achieve such ambitious goals.

In Sommer (1972) already wrote in an article in Natural History about 'What do we learn at the zoo?' that caged animals perpetuate visitor misconceptions and a unnatural notion of nature. Nearly 40 years later, this statement still holds true as there is 'no compelling evidence for the claim that zoos and aquariums promote attitude change, education, or interest in conservation in visitors' (Marino et al. 2010, p. 136). Packer and Ballantyne (2010) for instance researched the effects of zoo visits on the education of visitors. Their results confirming that a value or attitude change is rather unusual. Only 5% of 1000 visitors in their research believed that they had changed their values or attitudes following their visits and only 7% had taken any direct action to support conservation or pro-environment initiatives (Packer and Ballantyne 2010). Moreover, it seems to be logical to assume that one visit will not be enough to bring life-changing patterns of behavior nor that all zoo visitors with their great variety of backgrounds respond equally to the zoo experience (Grayal 2013; Graham 2015).

Consequently, Gatland (1996, online) argues that 'educational effort is a token gesture at best and is, in many cases, non-existent. Keeping animals in artificial conditions is not conservation education, in fact it is the opposite… zoo has themselves become part of, and in some respect intensified, the problem.' This argument is strengthened by Goldner (2015), who researched the average time a visitor spends at individual enclosures. Independent from species or number of animals inside a cage, visitors stay less than one minute. The only time they remain somewhat longer is during feeding times or if an infant animal is present. In an earlier research at the zoo in London the average time at each enclosure in the central mammal house was approximately 20 s per cage and in the ape house 46 s (World Society for the Protection of Animals and Born Free Foundation 1994). It is difficult to educate anyone in that time. Consequently, it could be said that that the primary purpose of zoos is entertainment (Tribe 2004; Frost 2011) and the public would be better educated through watching animal documentaries (Casamitjana 2004).

However, Marino et al. (2009) believe that sanctuaries and wildlife parks could provide a solution as they are very different to zoos and aquaria, because there the culture of viewing and entertainment is not contradictory to animal welfare. Lovelock and Lovelock (2013) reason in a similar direction that from an animal rights position, zoos need to be rejected, while sanctuaries that provide reconstructed habitats accessible for visitors to educate them about the value of species and habitat protection may be ethical even for endangered species.

17.5.3 Myth 3: Zoos Are There to Do Research into Animal Species

Research is sometimes used as another argument for the continued existence of zoos. In general, so the World Association of Zoos and Aquariums (WAZA) (2016, online), 'zoos and aquariums are uniquely placed to contribute to conservation-based research', especially in the areas of pure and applied biological sciences (e.g. behaviour, diseases), in situ conservation (e.g. field surveys) and research aimed at developing other roles (e.g. visitor learning). In that respect, there are often conflicts between animal welfare concerns and 'the ethical imperative to understand and conserve a population or ecosystem through research and management intervention', but also unpredictable environmental and societal changes outside of zoos (Minteer and Collins 2013, p. 41, 48). From the earliest days of modern zoos, research was not just impacted upon by outside forces, but also an important point for the justification of these institutions as indicated by specific publications.

Some of the earliest zoo research journals are the Journal of Zoological Research (1916) and Der Zoologische Garten (the zoological garden) (1859–1922; 1928-today) (Elsevier 2016), the official organ of the World Association of Zoos and Aquariums. On their website, they name five scientific journals in zoo and aquarium research: Zoo Biology, Journal of Zoo and Aquarium Research, Der Zoologische Garten, Journal of Zoo and Wildlife Medicine and International Zoo Yearbook (WAZA 2016). But not only in Europe did zoo research start very early after the formation of zoos. The first US zoo, Philadelphia Zoo (opened in 1874), saw research as one of its objective, reporting as early as 1876 about it (Snyder 1974). Recently, zoo research about animal health or animal behaviour has become prominent again. The contemporary introductions of three new tourism research journals are a clear indication: Journal of Zoo and Aquarium Research (JZAR) in 2013 and the American Journal of Zoological Research (AJZR) in 2013.

However, a number of publications in journals and newspapers question the role of research in zoos as yielding any useful results for the wild counterparts of captive animals. Reneau (2016), for example argues that many zoo animals show signs to a phenomenon called 'zoochosis' (obsessive and repetitive behaviour such as circling, pacing, self-mutilation), a behaviour that is not shown by those species in the wild. In some cases the conditions are so mentally exhausting that animals turn to cannibalism, such was the case in Heilongjiang Zoo where a penguin was filmed eating its own young (ITN 2012). In other research on the endangered brow antlered deer Sing came to the conclusion that those animals change their behaviour when they are taken out of the natural habitat as they become more vigilant and alert when visitors are around (Press Trust of India 2016). This clearly shows that animals in captivity are behaving unnaturally (Reneau 2016). Any research on captive zoo animals is accordingly only of limited scope.

Apart from behavioural issues in regards to the transferability of research of captive animals to wild animals, there are also biological processes that constrain this. Research by Clayton et al. (2016, online) on the faeces of primates provides evidence that captive primates 'lose their native microbes and become colonized with Prevotella and Bacteroides, the dominant genera in the modern human gut microbiome.' In other words, the contact with humans is changing the biological set-up of primates, thereby giving some further indication that zoo research is of limited scope of wild animal populations. In short, there is little hope that captive animals will contribute directly to the conservation of their wild population as they are too far detached from their wild origins (Graham 2015).

Certainly, there are some areas that produce useable results, but they are limited to animals in captivity. Such research includes animal behaviours and welfare of captive animals, the impact of humans (visitors and keepers) on the zoo animals (Ward 2016) or the impact of viewing captive animals on the visitor behaviour and attitudes.

There is therefore some agreement that research can only play a very limited role as a justifiable goal for zoos (Frost 2011) as zoo animals live in contrived, unnatural circumstances. A comparison with animal behaviour in their natural environment is consequently difficult, if not impossible (tierschutz.org 2016). As the reality remains the fourth goal: zoos are for entertainment. Visitors come to see big, charismatic animals like elephants, polar bears, pandas or big cats, but those do not do very well in zoos (Mount 2016). Visitors are often passionate about them, but there are not many visitors who are passionate about protecting spiders, snakes, frogs or other amphibians. Yet the potential for preventing the loss of many those species has much more chance for success, and zoos could indeed an instrumental role on their protection (Ward 2016), including in Germany.

17.6 Zoos in Germany

Zoos in Germany are major recreational and tourist attractions. In 2014, there were as many visitors as never before (Hucklenbroich 2015) to the 865 zoos in Germany (Zoo-Infos.de 2016). The Association of Zoos in Germany puts the annual growth rate in the last 10 years to 2% (Verband der zoologischen Gärten 2016d).Overall, there are between 36 million (Goldner 2014) and 65 million (Verband der zoologischen Gärten 2016d) visitors to German zoos per year. This signifies that even if there are no reliable statistics it can be assumed that a significant part of those are tourists.

Major developments of zoos for the general public in Germany started in the 19th century. The oldest continuously run zoo in Germany is the Zoo in Berlin that was opened in 1844 (Haufe 2013), followed by the Wilhelma in Stuttgart in 1846, the zoo in Frankfurt in 1858, the zoo in Cologne in 1860 and the Dresden zoo in 1861(Verein der zoologischen Gärten 2016b). Today, Berlin Zoo is not just the oldest, but also the largest zoo in Germany regarding animal (20,365) and species (1500) numbers. It also had the highest number of visitors with 3.25 million people in 2014 (Statista 2016a, b). Why it is difficult to judge the number of zoo visitors from abroad, the German Bureau of Statistics measures the visitor frequency of the German population. In 2015, the majority of German adults above 14 years visit a zoo at least sometimes (23.9 million. A further 19 million visit a zoo either once a year (10.6 million), twice a year (6.1 million) or more than trice (2.4 million), while 25.1 million never visited a zoo (Statista 2016d). Many of those German visitors, especially families with children visit zoos for nice weather events to buy some peace and quiet time (Hucklenbroich 2015). Unfortunately, the number of international tourists to German zoos or the zoo image is not recorded (Hachmeister 2014).

The ambivalent behaviour of these German visitors to zoos becomes obvious by looking at the results of a representative survey by the German Bureau of Statistics in 2014 (Statista 2016c). While 78% thought that zoos protect animals from extinction, 62% believed that wild animals can only protected in their natural environment. Furthermore, 60% agree with the statement that animals should not be kept in enclosures. Only 20% stated that animals have a better live in zoos than in the wild (Statista 2016c). Such inconsistencies also show in the behaviour towards different species ranging from complete ignorance to the well-being of objectified agricultural animals to the pampering of the subjectified pets. In 2012, around 754 million animals in Germany were slaughtered for human consumption, especially chicken (~628 million), pigs (~58 million), turkeys (~38 million), ducks (~25 million) and cattle (~3 million). At the same time, around 3.1 million animals were used in animal testing (in 2000 just 1.83 million), mainly mice (~2 million) and rats (~400,000) and 121,000 were kept in zoos and aquaria (56,100 fish, 27,540 birds, 22,950 mammals, 7650 reptiles and 4590 amphibians) (Die Zeit 2014b). Those animals and their well-being is either completely ignored or selectively regretted, but the fate of the majority of the 33.3 million pets (12 million cats, 8 million small mammals, 7 million dogs) (Die Zeit 2014b) is of outmost importance. A representative survey of 520 pet owners in Germany (yougov 2016) revealed that 50% spend less than 50 €/month, 33% between 50 €/month and 100 €/month and 10% more than 100 €/month. However, more important is the way that pets are treated: 95% give their pets a name and 91% describe pets as part of the family. Other results of this study include agreements to the statement that the animal replaces friends (27%), kids (25%) or partners (14%) (yougov 2016). It appears as Germans have a hierarchy of animals in their head (and their hearts) that some are worth being treated as having rights and some as having no rights. This argument is strengthened by a study by Chrismom (2015) about which animals Germans see as killable. The results confirm the contradictions of human behaviour: 84% agreed that flies can be killed, 72% that pigs can be killed, 54% that mice can be killed, but only 11% that cats can be killed. Only 7% stated that no animals should be killed (Chrismom 2015).

However, there seems to be a change happening with regards to zoo. The result of a survey amongst readers of the weekly newspaper Die Zeit (2015) revealed that 94% of them believe that the great apes have a consciousness (dolphins 91%, pigs 86%, mice 77%, earthworms 37%), and a great majority also believes that some animals do not belong in zoos (dolphins 82%, polar bears 76%, big cats 73%, great apes 72%). Accordingly, 63% of them also believe that there is an urgent need for action with regards to zoos. Even more, 66% state that wild animals suffer in enclosures and that someone who needs to learn something about them could watch documentaries. Only 26% follow the statement that zoo bring humans into contact with wild animals which they would not have gotten to know (Die Zeit 2015). Maybe humanity is on its way to recognise 'If we place the interests of animals over the interests of humans… and this is not the case in zoos or any other activity that compromises the wellbeing of animals for pleasure—we will have made a decisive steps towards the realisation of a more ethical ecotourism [tourism] industry.' (Fennell 2012, p. 11)

17.7 The End of Zoos and the Beginning of Wildlife Sanctuaries?

Animals are significant tourism resources around the world, including in Germany. The relationship expressed in zoos and aquariums is thereby one element of the extremely contradictory behaviour of humans towards other animals. This relationship has grown out of a long history that started more than 10,000 years ago with the domestication of wolves. Since then, especially animals in captive settings, first as a privilege to pharaos, emperors, kings and other leaders and for around 200 years as a source for public enjoyment, display much of the anthropocentric connection between human and non-humans animals. However, the nowadays claimed stewardship of zoos through conservation, education, research and entertainment is increasingly called into question. Many researchers believe that zoos as attractions are mainly in the business of entertainment covered by the 'curtains' of conservation, research and education. Even more, zoos are showing clear tendencies towards Disneyfication (Beardsworth and Bryman 2001, 83). It could therefore be said that if humanity is serious about issues like sustainability and ethical behaviour there is no future for traditional and modern zoos, but wildlife parks/sanctuaries could be a solution, where people could see native animals in their natural environment and where the breeding of endangered species is more than just a mere attraction that 'extinction' tourists will come to see.

But even for sanctuaries a word of caution is important. There are a number of publications that criticise the current treatment of captive animals within institutions that call themselves or part of their institutions as sanctuaries. People for the Ethical Treatment of animals (PETA) for example write: 'Many roadside and travelling zoos operate under the guise of nonprofit sanctuaries, preying on people's sympathy while exploiting animals in their care. Animals "rescued" from one tragic situation are sentenced to another when they end up in pseudo-sanctuaries.' (PETA 2016, online) There were a number of events that support their argument. In 2009, the Captive Animals' Protection Society (CAPS) made an incident public where Noah's Ark Zoo in Bristol, UK, renamed its Tiger enclosure the Big Cat Sanctuary despite breeding and loaning tigers to Great British Circus, a practice that was still continuing in 2013 (CAPS 2013). Similarly, other examples of the mistreatment of animals in zoo tourism are reported, e.g. the mistreatment of elephants in Thai tourism (Cohen 2015) or the abuse of tigers in a Buddhist centre, where investigators found 40 frozen tiger cubs in a freezer (Holmes and Vidal 2016). Such events have led to the formation of organisations that inform people about 'real' sanctuaries, which could be the future avenue for zoos. At least, we should 'be honest with ourselves about the way we choose to treat our fellow earth-dwellers and acknowledge how little respect we give to animals' (Reneau 2016).

References

Altick RD (1978) The shows of London. Belknap Press, London

Beardsworth A, Bryman A (2001) The wild animal in late modernity: the case of the Disneyization of zoos. Tourist Studies 1(1):83–104

Boehrer B (2007) Introduction: The animal renaissance. In: Boehrer B (ed) A cultural history of animals in the renaissance. Berg, Oxford, pp 1–26

Bostock SC (1993) Zoos and animal rights: the ethics of keeping animals. Routledge, London

Bundezentrale für politische Bildung (2011) Globalisierung. Bundeszentrale für politische Bildung, Bonn

Captive Animals Protection Society (CAPS) (2013) Noah's Ark Zoo Farm does business with (another) notorious circus trainer. http://www.captiveanimals.org/news/2013/06/noahs-ark-zoo-farm-does-business-with-another-notorious-circus-trainer. Accessed 10 Sept 2016

Casamitjana J (2004) Aquatic zoos: a critical study of UK public aquaria in the year 2004. The Captive Animals Protection Society, London

Chrismom (2015) Was meinen sie welche Tiere darf der Mensch töten? www.chrismon.de/umfragen. Accessed 08 July 2015

Clayton JB, Vangay P, Huang H et al (2016) Captivity humanizes the primate microbine. PNAS 113(37). http://www.pnas.org/content/113/37/10376.full.pdf. Accessed 28 Sept 2016

Cohen E (2009) The wild and the humanized: animals in Thai tourism. Anatolia 20(1):100–118
Cohen E (2015) Young elephants in Thai tourism. In: Markwell K (ed) Animals and tourism: understanding diverse relationships. Channel View, Bristol, pp 163–179
Cunningham A (2016) The case for zoos. A scientist perspective. Zoological Society of London (ZSL). https://www.zsl.org/blogs/wild-science/the-case-for-zoos-a-scientists-perspective. Accessed 28 Sept 2016
Dejure.org (2016) Bundesnaturschutzgesetz §42 Zoos. https://dejure.org/gesetze/BNatSchG/42.html. Accessed 12 July 2016
Die Zeit (2014a) Die Ethik des Schinkenbrots. http://www.zeit.de/2014/21/tierrechte-nutztiere-fleisch-essen/komplettansicht. Accessed 13 July 2016
Die Zeit (2014b) Ich bin wie du. Die Tierzählung. 15.04.2014. Vol. 21: 38
Die Zeit (2015) Wir und das Tier. Was Zeit Leser an Landwirtschaft, Tierversuchen und Zoos kritisieren, vol 51, p 44. 17 Dec 2015
Dobson J (2015) Zoo. In: Cater C, Garrod B, Low T (eds) The encyclopedia of sustainable tourism. CABI, Wallingford, p 554
Donaldson S, Kymlicka W (2013) ZooPolis. Eine politische Theorie der Tierrechte. Suhrkamp Verlag, Berlin
Eilenberger W (2012) Wie viel Tier steckt in mir? In: Philosophie Magazin, vol 6, Oct./Nov. 2012
Elsevier (2016) Der Zoologische Garten. http://www.journals.elsevier.com:80/der-zoologische-garten. Accessed 23 Sept 2016
Fennell DA (2012) Tourism and animal ethics. Routledge, New York
Frost W (2011) Rethinking zoos and tourism. In: Frost W (ed) Zoos and tourism: conservation, education, entertainment?. Channel View, Bristol, pp 1–12
Gannon F (2002) Animals on the menu. European Molecular Biology Organization (EMBO) reports 3(7):589
Ganzert R (2016) Zoos are not prisons. They improve the lives of animals. http://time.com/4364671/zoos-improve-lives-of-animals/. Accessed 28 Sept 2016
Garrett EA (2014) Why do we go to the zoo?. Communication, animals, and the cultural-historical experience of zoos. Fairleigh Dickinson University Press, Madison
Gatland A (1996) Zoos: the conservation myth. Animals Today 3(4). http://www.wildlife.org.nz/zoocheck/myths.htm. Accessed on 07 Nov 2015
Geo (2013) Was sie wirklich über Tiere wissen sollten. Geo Magazin 04/13
Goldner C (2014) Zoos manipulieren Besucherzahlen. Humanistischer Pressedienst. http://hpd.de/artikel/10124. Accessed 11 July 2016
Goldner C (2015) Lebenslänglich hinter Gittern: Die Wahrheit über Gorilla. Orang Utan & Co in deutschen Zoos, Alibri, Aschaffenburg
Graham N (2015) Conservation education in zoos and aquariums. The benefits of these institutions and their impact on conservation ethic according to staff. Research paper submitted for the Master of Teaching at the University of Toronto. https://tspace.library.utoronto.ca/bitstream/1807/68682/1/Graham_Nathalie_LM_201506_MT_MTRP.pdf. Accessed 27 Sept 2016
Grayal A (2013) Zoos as ecotourism experiences. In: Ballantyne R, Packer J (eds) International handbook of ecotourism. Edward Elgar, Cheltenham, pp 464–469
Great Ape Project (2012) Menschenaffenkarte. http://www.nationalgeographic.de/static/images/evt-07-2012/menschenaffen-karte-1200.jpg. Accessed 14 July 2016
Hachmeister P (2014) Gefängnis oder Arche? Zur medialen Darstellung der Institution Zoo in Deutschland. Lit Verlag, Münster
Hamilton Zoo (2015) Conservation. http://hamiltonzoo.co.nz/conservation/. Accessed 09 Nov 2015
Hancocks D (2001) A different nature: The paradoxical world of zoos and their uncertain future. University of California Press, Berkley
Haufe A (2013) Die Geschichte des Zoos- was bisher geschah. http://www.ndr.de/ratgeber/reise/tierparks/Die-Geschichte-des-Zoos-was-bisher-geschah,zoofeier100.html. Accessed 13 July 2016
Holmes O, Vidal J (2016) Fourty dead tiger cubs found in freezer at Thai temple. The Guardian. 01 June 2016. https://www.theguardian.com/world/2016/jun/01/40-dead-tiger-cubs-found-in-freezer-thai-temple-wildlife-trafficking. Accessed 10 Sept 2016
Hucklenbroich C (2015) Tiergärten im Wandel der Zeit. Und jeden Nachmittag ins Paradies. Frankfurter Allgemeine Feuilleton. 27 Apr 2015. http://www.faz.net/gqz82ixx. Accessed 11 July 2016
Hunter-Jones P, Hayward C (1998) Leisure consumption and the United Kingdom zoo. In: Ravenscroft N, Philip D, Bennett N (eds) Leisure. Culture and commerce, LSA, Brighton, pp 97–107
ITN (2012) Penguin eats young in Chinese zoo [video clip]. In: The guardian. 20 Dec 2012,https://www.theguardian.com/science/video/2012/dec/20/penguin-eats-young-china-zoo-video. Accessed 03 April 2017
Jamieson D (1985) Against Zoos. In: Singer P (ed) In defence of animals. Blackwell, Oxford, pp 108–117
Jamieson D (2006) Against Zoos. In: Singer P (ed) In defense of animals: the second wave. Blackwell Publishing, Malden, MA, pp 132–143
Lieckfeld A (2012) Wie Tiere denken. In: Geo Kompakt 33:138–144
Lovelock B, Lovelock KM (2013) The ethics of tourism: critical and applied perspectives. Routledge, Milton Park
Marino L, Bradshaw G, Malamud R (2009) The captivity industry: the reality of zoos and aquariums. Best Friends Magazine March/April:25–27
Marino L, Lilienfeld SO, Malamud R et al (2010) Do zoos and aquariums promote attitude change in visitors? A critical evaluation of the American zoo and aquarium study. Soc Anim 18(2):126–138
Marxwell K (2015) Birds, beasts and tourists: human-animal relationships in tourism. In: Markwell K (ed) Animals and tourism: understanding diverse relationships. Channel View, Bristol, pp 1–23
Mason P (2000) Zoo tourism: the need for more research. J Sustain Tourism 8(4):333–339
Mazur N (2001) After the ark? Environmental policy making and the zoo. Melbourne University Press, Melbourne
Miller I (2005) Didactic nature: exhibiting nation and empire at the Ueno Zoological Gardens. In: Pflugfelder G, Walker B (eds) JAPANimals: history and culture in Japan's animal life. University of Michigan, Ann Arbor, pp 273–313
Minteer BA, Collins JP (2013) Ecological ethics in captivity: balancing values and responsibilities in zoo and aquarium research under rapid global change. Inst Lab Anim Res (ILAR) 54(1):41–51
Morgenthaler K (2013) Das forsche Fräulein und der König. Greenpeace Magazin, Tier&Wir: Wie die Menschen mit Haustieren und Wildtieren umgehen 4(13)
Mount H (2016) Is it time to close our zoos? The Telegraph. http://www.telegraph.co.uk/tv/2016/04/17/is-it-time-to-close-our-zoos/. Accessed 28 Sept 2016
National Zoo and Aquarium (2015) Education. http://www.nationalzoo.com.au/education.htm. Accessed on 09 Nov 2015
Packer J, Ballantyne R (2010) The role of zoos and aquariums in education for a sustainable future. New Dir Adult Continuing Educ 127:25–34
Parks C (2015) The World's oldest zoo is a modern attraction with a storied past. The Austrian destination has a lot more to offer than exotic animals. http://www.smithsonianmag.com/travel/zoo-endured-inside-worlds-oldest-zoo-180957191/?no-ist. Accessed 28 Sept 2016
Palmer J (2015) The last truly wild horse is making a comeback, BBC Erath. 11 Nov 2015,http://www.bbc.com/earth/story/20151111-the-worlds-last-truly-wild-horse-is-making-acomeback. Accessed 12 Aug 2016

Patrick PG, Tunnicliffe SD (2013) Zoo talk. Springer Verlag, Dordrecht
Patterson C (2002) Eternal Treblinka: our treatment of animals and the Holocaust. Lantern Books, New York
People for the Ethical Treatment of Animals (PETA) (2016) Pseudo-sanctuaries. http://www.peta.org/issues/animals-in-entertainment/zoos-pseudo-sanctuaries/pseudo-sanctuaries/. Accessed 10 Sept 2016
Press Trust of India (2016) Study on brow antlered deer at Delhi zoo. Business Standard. http://www.business-standard.com/article/pti-stories/study-on-brow-antlered-deer-at-delhi-zoo-116090400131_1.html. Accessed 29 Sept 2016
Reichenbach H (2002) Lost menageries—why and how zoos disappear. Part I. Int Zoo News 49(3). http://www.izn.org.uk/Archive/316/Izn-316.htm. Accessed 27 Sept 2016
Reneau R (2016) Zoo animals do not belong in captivity. The University Star. https://star.txstate.edu/2016/09/20/zoo-animals-do-not-belong-in-captivity/. Accessed 23 Sept 2016
Snyder RL (1974) Introduction. In: Centennial symposium on science and research, Philadelphia Zoo, pp 1–3
Sommer R (1972) What do we learn at the zoo? Nat Hist 81:26–29
Sommer V (2012) Tierethik. In: Geo Kompakt, vol 33, pp 138–144
Statista (2016a) Ranking der größten Zoos in Deutschland nach Zahl der Tiere im Jahr 2014. http://de.statista.com/statistik/daten/studie/314716/umfrage/groesste-zoos-in-deutschalnd-nach-anzahl-der-tiere/. Accessed 13 July 2016
Statista (2016b) Ranking der größten Zoos in Deutschland nach Anzahl der Besucher im Jahr 2014. http://de.statista.com/statistik/daten/studie/248643/umfrage/ranking-deutscher-zoos-nach-besucherzahl/. Accessed 13 July 2016
Statista (2016c) Zustimmung zu Aussagen zur Tierhaltung in Zoos in Deutschland. http://de.statista.com/statistik/daten/studie/316077/umfrage/zustimmung-zu-aussagen-zur-tierhaltung-in-zoos-in-deutschland/. Accessed 13 July 2016
Statista (2016d) Umfrage in Deutschland zur Häufigkeit des Besuchs von Zoos und Tierparks bis 2015. http://de.statista.com/statistik/daten/studie/290972/umfrage/umfrage-in-deutschland-zur-haeufigkeit-des-besuchs-von-zoos-und-tierparks/. Accessed 13 July 2016
Stevens PMC, McAlister E (2003) Ethics in Zoos. Int Zoo Yearb 38 (1):94–101
The Born Free Foundation (2012) The EU zoo enquiry 2011. Born Free Foundation, Horsham
Tierschutz.org (2016) Zoo. http://www.tierschutz.org/tierschutz/problembereiche/wildtiere/zoo.php. Accessed 01 May 2016
Tisdale S (1993) London zoo review. American Way 1:74–76
Tribe A (2004) Zoo tourism. In: Higginbottom K (ed) Wildlife tourism: impacts, management and planning. Common Ground Publishing, Victoria
Van Tuye C (2008) Zoos and animal welfare. Gale Cengage Learning, Farmington Hills
Verband der Zoologischen Gärten (VdZ) (2016a) Zoo. http://www.zoodirektoren.de/index.php?option=com_k2&view=item&id=904:zoo&Itemid=216. Accessed 13 July 2016
Verband der Zoologischen Gärten (VdZ) (2016b) Zoogründungen im 19.Jahrhundert. http://www.zoodirektoren.de/index.php?option=com_k2&view=item&id=4717:zoogruendungen-19-jahrhundert&Itemid=339. Accessed 13 July 2016
Verband der Zoologischen Gärten (VdZ) (2016c) VdZ Faktenblatt. http://www.zoodirektoren.de/images/Wir_ueber_uns/VdZ_Faktenblatt.pdf. Accessed 13 July 2016
Verband der Zoologischen Gärten (VdZ) (2016d) Fragen und Antworten zu Zoos im deutschsprachigen Raum. http://www.zoodirektoren.de. Accessed 14 July 2016
Ward S (2016) In defence of zoos: how captivity helps conservation. The conversation. https://theconversation.com/in-defence-of-zoos-how-captivity-helps-conservation-56719. Accessed 29 Sept 2016
Webster Dictionary (2015) Zoo. http://www.merriam-webster.com/dictionary/zoo. Accessed 28 Oct 2015
World Association of Zoos and Aquariums (WAZA) (2015) Zoos and aquariums around the world. http://www.waza.org/en/site/zoos-aquariums. Accessed 23 Sept 2016
World Association of Zoos and Aquariums (WAZA) (2016) Science and research. http://www.waza.org/en/site/conservation/science-and-research. Accessed 23 Sept 2016
World Society for the Protection of Animals and Born Free Foundation (1994) The Zoo enquiry. World Society for the Protection of Animals, Oxford
Yougov (2016) Case study: Pet Owners. Aktuelle Umfrageergebnisse auf der Basis des Omnibus PetOwner. https://yougov.de/loesungen/infocenter/white-paper/case-study-petowner/. Accessed 14 July 2016
Zoo-infos.de (2016) Die Zoo-Datenbank für Deutschland. http://www.zoo-infos.de/. Accessed 18 Aug 2016

Hunting Tourism: The Case of Canadian Prairie Waterfowl Hunters

18

Farhad Moghimehfar, Howard W. Harshaw, and Lee Foote

Abstract

Hunting tourism plays important roles in the conservation of wildlife; hunters provide financial support for conservation programs and habitat protection, assist in the monitoring of wildlife populations, and play important roles in population management (Heffelfinger et al. in Int J Environ Stud 70(3):399–413, 2013). These conservation roles indirectly benefit host communities through the stewardship of wildlife populations and the protection of habitats. The hunting of waterfowl is a popular form of hunting tourism; however, the number of North American waterfowl hunters has been declining since the mid-1970s (NAWMP Revised objectives: an addendum to the 2012 North American Waterfowl Management Plan. Fish and Wildlife Service, Washington, 2014). Although a number of conservation and waterfowl hunting organizations and government programs offer educational and mentorship programs to promote waterfowl hunting and retain hunters, declines in the number of waterfowl hunters continues. This chapter examines constraints to hunting tourism and explores waterfowl hunters' motivations in an effort to understand what influences decisions to participate in waterfowl hunting as a nature-based tourism activity. Using a deductive approach, this chapter employs self-determination theory (Deci and Ryan in J Mind Behav 1(1):33–43, 1980), the theory of planned behavior (Ajzen in Organ Behav Hum Decis Process 50(2):179–211, 1991), and leisure constraints theory (Crawford et al. in Leisure Sci 13(4):309–320, 1991) to guide a thematic analysis. Thirty-four waterfowl hunters, representing a range of skill levels and commitment to the activity, were interviewed in three Canadian provinces (Alberta, Saskatchewan, and Manitoba) about the factors that influenced their decisions to hunt waterfowl. These factors included structural, interpersonal, and intrapersonal constraints, and attitudes, subjective norms, and perceived behavioral control. We contrast the theoretical and managerial implications of these findings. Findings of this qualitative research inform tourism, outdoor recreation, and wildlife managers and planners to develop strategies for the recruitment and retention of hunting tourists and help retain the local ecological and economic benefits of waterfowl hunting.

F. Moghimehfar (✉)
Outdoor Recreation and Tourism Management, Ecosystem Science and Management, University of Northern British Columbia, 8-244, 3333 University Way, Prince George, BC V2N 4Z9, Canada
e-mail: farhad.moghimehfar@unbc.ca

H.W. Harshaw
University of Alberta, 2-130J Van Vliet Complex (University Hall), Edmonton, AB T6G 2H9, Canada
e-mail: harshaw@ualberta.ca

L. Foote
Devonian Botanic Garden, University of Alberta, GSB 7-51, Edmonton, AB T6G 2H1, Canada
e-mail: lfoote@ualberta.ca

18.1 Introduction

The controversial yet compelling claims of Wilson (1984, 2007) hold that people have an innate, emotional affiliation to other species, and that people's interaction with other species increases their understanding, appreciation and valuation of nature. Wildlife tourism provides opportunities (and infrastructure) for human interactions with wildlife that can facilitate this understanding and appreciation. In its

I. Borges de Lima and R.J. Green (eds.), *Wildlife Tourism, Environmental Learning and Ethical Encounters*, Geoheritage, Geoparks and Geotourism, DOI 10.1007/978-3-319-55574-4_18

simplest conception, wildlife tourism involves visiting destinations to observe wildlife (Higginbottom 2004; Reynolds and Braithwaite 2001). More broadly, wildlife tourism can be defined as "an area of overlap between nature-based tourism, eco-tourism, consumptive use of wildlife, rural tourism, and human relations with animals" (Reynolds and Braithwaite 2001, p. 32). Wildlife tourism can involve non-extractive appreciation of wild animals, through activities like viewing, and extractive activities such as hunting.

Hunting tourism, specifically the case of waterfowl hunting, is the focus of this chapter. Hunting plays important roles in connecting people to nature and fostering a conservation ethic among participants as well as supporting individuals' desires and needs for leisure. In return, hunters provide financial support for conservation programs and habitat acquisition, protection, restoration and enhancement; hunters also play important roles in waterfowl population monitoring and management (Heffelfinger et al. 2013). Waterfowl hunters make direct financial contributions through licensing to the acquisition and management of waterfowl habitat. Annual revenue of $24 million (USD) from North American waterfowl licences (i.e., Duck Stamps) have been used to develop and support a network of wetland habitat through the North American Waterfowl Management Plan (NAWMP). Waterfowl hunters make other financial contributions outside of the licensing framework through an environmental non-governmental organization, Ducks Unlimited: since 1938, Ducks Unlimited Canada has conserved 6.4 million acres; in 2014, Ducks Unlimited Inc. conserved 4.9 million acres of wetlands and waterfowl habitat in the US; and Ducks Unlimited Mexico, has protected 1.9 million acres of waterfowl habitat since 1974.

Waterfowl are a unique group of game animals composed of web-footed ducks, geese and swans of the taxonomic Family designated Anatidae. They possess life history characteristics evolved to keep their populations robust in the face of withering natural mortality, often exceeding 80% annually (Klett et al. 1982) and adult mortality rates exceeding 30% annually (Lauckhart 1956). The ability to accommodate large mortality rates is essential for a resource that is managed for sustainable human off-take. Waterfowl compensate for this in several ways; many ducks lay clutches in excess of 8 eggs per nesting attempt. Should a nest be lost to predation for example, Mallards (*Anas platyrynchos*) will typically attempt a second nest and examples exist of up to 4 nesting attempts made. Geese are long lived, sometimes exceeding 18 years in the case of banded, known-aged Snow Geese (*Chen caerulescens*). With 6-egg clutches laid, that would suggest one pair of Snow Geese could theoretically produce over 70 offspring to replace themselves as well as produce a large harvestable surplus.

Scientific estimates indicate that the numbers of North American ducks in 2015 was 49.5 (±0.8) million and 48.4 (±0.8) million in 2016. These numbers are up 38% from the long term average (1955–2016). Snow Goose numbers currently sit at over 5 million birds and increasing; White-fronted Goose (*Anser albifrons*) numbers are over a million and up 45% in recent years, and Canada Goose numbers are over 7 million and increasing (US Fish and Wildlife Service 2016).

While wetland habitats have dwindled across much of North America, the agricultural basis of food availability has tremendously increased availability of corn, wheat, peas and rice exists along their migratory routes. Consequently, goose health, clutch size and numbers today sit at levels higher than any recorded before in the 61-year continuous and systematic North American waterfowl surveys conducted by the US Fish and Wildlife and the Canadian Wildlife Services.

Indeed, Snow Geese have reached epidemic proportions creating crop damage, gosling mass starvation on their arctic breeding grounds and hazards to aircraft. Individual flocks in excess of 150,000 are not uncommon on the wintering grounds of Louisiana and Texas, USA. Adaptive management attempts have encouraged hunters to act as agents of Snow Goose population control through springtime hunting, limits of 30 birds per day, electronic calling devices and other "control" efforts but to little avail. Snow Goose numbers continue to climb. Canada Geese (*Branta canadensis*) numbers too have swelled as they have recolonized most of their former range and many places never found before. Canada Geese have spread into the nuisance range and urban geese now inhabit many city parks, golf courses, and city waterfronts. Altercations are common as geese actively defend nests and offspring. They represent a concern for fouling water supplies and disease transmission.

As with other outdoor recreation activities that occur in natural settings, waterfowl hunting can have negative impacts on wildlife (to waterfowl and other un-hunted species). Negative wildlife impacts that result from human interaction include: the habituation of wildlife to people; behaviour modification, such as the flushing of animals (and other avoidance strategies) from nests, dens, and shelter (and associated impacts to the animals' energy budgets); and impacts to breeding success as a result of the mere presence of people in an area (Green and Giese 2004; Knight and Cole 1995; Hammitt and Cole 1998). Wildlife and landscape managers have successfully implemented strategies to reduce these impacts, including implementing temporal zoning to separate people from wildlife during critical periods such as breeding seasons, limiting the spatial extent of some activities to avoid known critical habitat, and the limiting of some activities (e.g., motorized activities) in areas where wildlife populations are vulnerable (e.g., wintering ungulates and snowmobiles).

Waterfowl hunting present a different set of impacts to wildlife, including the killing of individual animals. As

Green and Giese (2004) note, uncontrolled hunting which is synonymous with poaching (i.e., hunting without a license, and/or hunting species outside of a specified season, area, time of day etc.) can have considerable impacts to wildlife by contributing to population declines. The consequences of the killing of animals as a result of poaching are that population numbers are not limited to the removal of scientifically designated animals from the ecosystem, leading to anomalous and undesirable demographics. The presence of humans during breeding season can deter birds (Beale and Monagahan 2004) yet waterfowl seasons are set to avoid breeding seasons.

The majority of landscapes are managed landscapes. Even remote parks and protected areas are managed landscapes, though the management emphasis often focuses on ecological integrity. As a result, elements of landscapes like flora and fauna are managed also. Often the management of landscapes and constituent parts are deliberately managed in response to habitat fragmentation (due to roads and development) and loss. These degraded landscapes often cannot support large populations of wildlife due to carrying capacity limitations. In North America, the controlled hunting of wildlife (i.e., managed hunts) plays a critical role in maintaining an ecological and human welfare balance whether it is Lyme disease from overly abundant white-tailed deer, predation of elk, waterfowl or large predators on agricultural products, or ancillary threats to endangered species such as boreal caribou. Hunting is one of the management levers that can steer wild populations toward a more sustainable, diverse, and socially acceptable level.

Waterfowl are a special case in point. No endangered waterfowl are hunted in North America and as a result of abundant agricultural forage, goose populations are at the highest levels since record keeping started 6 decades ago.

In some places, such as the Canadian prairies, agricultural crops and wetland habitats would be degraded without the managed hunting of waterfowl.

As Knight and Cole (1995) note, the type and magnitude of recreationists' and tourists' impacts on wildlife is dependent on the type of activity and individual behaviour. Although hunting is not the only recreation activity that can disturb waterfowl (birdwatching can disturb waterfowl as well), it can have the highest impact of recreation activities: in one study of waterfowl hunting in Québec, Canada, hunting-related activities were responsible for 30% of all flight disturbances (Bélanger and Bédard 1995). There is evidence that hunting can alter the behaviour, population structure, and distribution patterns of wildlife, such as waterfowl; unhunted wildlife populations behave differently from hunted ones. Yet, fishing, nature viewing, backcountry hiking, cross-country skiing, horseback riding, boating, motorized activities, and recreation development (i.e., infrastructure) can disturb wildlife too, and impacts include increased aggressive behaviour, stress, habituation to people, and displacement from breeding/nesting areas (Knight and Cole 1995).

Managed hunting is distinct form poaching as population and habitat characteristics are inventoried, measured, and modeled; the results of these inventories and models are used to determine appropriate levels of hunting for defined wildlife management units (i.e., a recognition that management prescriptions need to be locally relevant). It is also possible that without managed hunting, some wildlife populations, such as waterfowl, would increase their numbers past what could be supported by their habitats; starvation and stress could impact individual animals, and could also lead to population declines. Finally, Dobson et al. (1997) showed that habitat loss and fragmentation were the greatest threats to endangered species. Waterfowl hunting provides abundant financial and political incentive for wetland habitat creation, thus driving the support of wetland-dependent endangered species in the desired direction. Without wetlands there can be no wetland dependent species to be disturbed.

As noted by Bauer and Herr (2004), hunting has many characteristics of tourism: it is a leisure pursuit that is culturally significant; it entails travel to and from particular destinations; it may involve overnight stays at these destinations; and an active of service industry (e.g., outfitters, guides, skinners, trackers, cooks, drivers) both provides and receives support for the activity. Yet, hunting participation has declined in North America. For example, the number of waterfowl hunters in Canada has declined since the late-1970s from 500,000 to an average of 178,000 between 1999 and 2013 (NAWMP 2014). This decline in the numbers of waterfowl hunters has negatively affected the realization of conservation benefits and has also affected rural and wildlife tourism: waterfowl hunters' total daily expenditures were $83 (CAD) (or $609 annually) for a total of $327 million (CAD) in 2011 (Federal, Provincial, and Territorial Governments of Canada 2014); the loss of these expenditures affects accommodation-providers and other supporting businesses. Explanations for the decline in hunting participation include increased urbanization, agricultural development, road network expansion, and natural resource development that have caused people to travel further to realize hunting opportunities (Bauer and Herr 2004). It is also possible that people have substituted other (perhaps more accessible) recreation activities for waterfowl hunting. The difficulty has been that hunting tourism lacks information about visitor characteristics, such as their needs, desires, attitudes, and preferences (Reynolds and Braithwaite 2001).

Methodologically, it is worth noting that this chapter does not have data and interviews from non-hunters and anti-hunters. This project is focused on understanding *waterfowl hunters*' behaviour and constraints from a qualitative perspective. As this is the goal of the paper, data and interviews were not collected from non-hunters and anti-hunters. Yet, the chapter does not discuss hunting within a societal perspective of right or wrong. This should be a post hoc use of data collected for another purpose, internal inspection of an existing sub-discipline of hunting, and to debate anti-hunting issues with data collected in the research would disqualify the paper. In fact, no outdoor activity is entirely positive in its practice. There are always unintended or accidental consequences whether it is hoof, boot and tire erosion from trail users, nesting bird disruption from bird watchers or pesticide runoff from golf courses. Hunting is no different. Some very small number of firearm accidents occur, however there is a lower injury risk/participant than in tennis or bicycling and occasionally a non-game species will be injured or killed, but this appears to be very rare. Fortunately, waterfowl hunting seasons do not overlap with any breeding season or area. The support for wildlife enforcement is drawn largely from hunter support, thus, poaching and improper shooting is somewhat self-regulating. The largest conservation effects appear to emerge from public awareness, and increasing the valuation and outspoken advocacy for habitat preservation, species recovery, appropriateness of harvest level, and occasionally, population reduction where species numbers or sex ratios are skewed unfavorably (such as burgeoning Snow Goose populations, Mallards hybridizing with Black Ducks, or Mallard sex ratios causing hen stress). To note, no waterfowl in the last six decades has been threatened by over hunting. This study uses a *qualitative approach* to investigate waterfowl hunters' behaviours and constraints. Creswell (1998), Mayan (2009), Pattons (2002), and several other qualitative researchers discussed that biased samples can be used in qualitative inquiries and actually encouraged biased sampling in different qualitative methods.

Our objective was to understand why the number of waterfowl hunters has decreased; thus, we examined factors that influenced the (lack of) participation including constraints (i.e., structural, interpersonal, and intrapersonal), attitudes, subjective norms, and perceived behavioral control. Knowing more about the factors that influence waterfowl hunters' participation can assist wildlife tourism providers and wildlife management agencies in retaining existing participants, recruiting new participants, and reengaging with lapsed participants. This chapter employs a qualitative approach using the theory of planned behavior as the guiding framework to explore waterfowl hunting behaviors in the Canadian Prairies. In-depth interviews with waterfowl hunters in the Canadian provinces of Manitoba, Saskatchewan, and Alberta were conducted to identify factors influencing peoples' decision to participate in this tourism and recreation activity.

18.2 Theoretical Perspectives of Participation in Waterfowl Hunting

Examinations of people's participation in nature-based tourism activities have been of interest over the last few decades. The influence of several factors such as attitudes (e.g., Ardoin et al. 2015; Marques et al. 2015), values and beliefs (Apps et al. 2016), knowledge (e.g., Ardoin et al. 2015), social surroundings (e.g., Luo and Deng 2007), motivations (e.g., Luo and Deng 2007; Meng et al. 2008), self-efficacy (e.g., Hung and Petrick 2012) as well as things that constrain people's leisure and outdoor recreation behavior have been examined (Kleiber et al. 2011). Investigation of these factors in different contexts has elicited several theories to explain and predict people's behavior. The theory of planned behavior (Ajzen 1991) has been employed in several tourism and outdoor recreation studies (Kouthouris and Spontis 2005; Huang et al. 2014; Peng et al. 2014; Lee et al. 2015) including hunting research (Hrubes et al. 2001; Shrestha 2013; Shrestha et al. 2012; Shrestha and Burns 2016). In this chapter, this theory was extended by recognizing and incorporating motivations and constraints to engagement in outdoor recreation activities, self-determination theory (Deci and Ryan 1980), and leisure constraints theory (Crawford et al. 1991). This study also explored the constraints and motivations to waterfowl hunting based on these theories.

18.2.1 The Theory of Planned Behavior

The theory of planned behavior (TPB) seeks to explain and predict people's behavior based on their response(s) to a particular situation (Ajzen 1991). The readiness to participate in an activity is held to be the most accurate predictor of volitional behaviors. Theoretically, the ability to predict people's intentions allows prediction of their behaviors. The TPB employs attitudes, subjective norms, and perceived behavioral control as the three predictors of intention (Fig. 18.1). Attitudes, driven by values and beliefs, reflect the degree to which people positively or negatively value an action. The TPB considers both cognitive (i.e., attitudinal components that reflect people's thoughts and ideas regarding an activity) and affective (i.e., component that reflect personal person's feelings or emotions) aspects of individuals' attitudes. Subjective norms refer to the influence of social surroundings on people's behavior, and includes injunctive (i.e., social approval of the action) and descriptive (i.e., popularity of the activity among those whose opinions are important for the person) aspects of social norms in their measurement model. The third predictor is perception of control over the action: the TPB posits that people who are confident that they can perform a behavior are more likely to

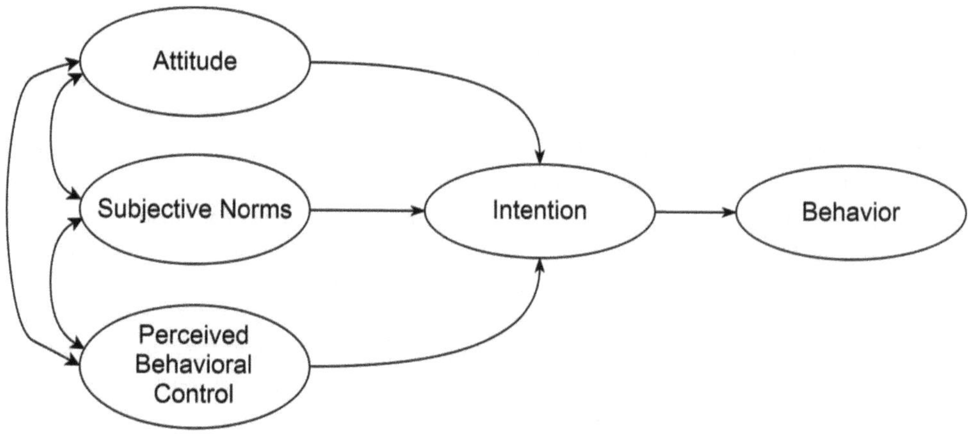

Fig. 18.1 The theory of planned behavior (Ajzen 1991)

participate in the activity. For instance, it is more likely that a person who believes that they are capable of hitting a target would participate in hunting than an individual who is not confident of their shooting abilities. The TPB considers both people's self-efficacy and their control over an activity.

The TPB has been used to understand people's participation in outdoor recreation, tourism, and hunting. For instance, Hrubes et al. (2001) employed the TPB to predict hunter intentions and behavior, and found that intentions to hunt significantly predicted self-reported hunting frequency. Attitudes, subjective norms, and perceived behavioral control were reported as strong predictors of hunting intentions. Sherstha et al.'s (2012) study of deer hunting behavior explored the mediating effect of attitudes, subjective norms, and perceived behavioral control on the association between hunting constraints and hunting intention. Their findings confirmed the direct influence of attitudes, subjective norms, and perceived behavioral control on intention and also the role of these variables as mediators. Although significant, attitudes and subjective norms showed little influence on intention; perceived behavioral control was strongly associated with intentions. Sherstha et al. (2012) concluded that enhancing the influence of these three variables reduces the negative effect of constraints on people's hunting intentions. As the perceived behavioral control showed the strongest mediating effect on the association among constraints and intention, they suggested that increasing people's self-efficacy (skills and competence) mitigates the negative influence of constraints on hunting behaviors. Using a similar approach, Shrestha and Burns (2016) investigated hunter behaviors and constraints to hunting participation, and found that perceived behavioral control was a predictor of hunting behavioral intention. Rossi and Armstrong (1999) compared the theory of reasoned action (Ajzen and Fishbein 1975) and the TPB in a hunting context. Their results indicated that the TPB was a better framework to predict hunting behavior; however, they argued that the theory's original items were not able to predict a considerable amount of variance in intention. Therefore, they suggested the inclusion of other variables to improve the predictive power of the theory.

18.2.2 Motivations

Wildlife researchers have examined motivations to explain the declining number of hunters in past few decades; motivation "concerns energy, direction, persistence and equifinality-all aspects of activation and intention" (Ryan and Deci 2000, p. 69). Enck et al.'s (1993) investigation of waterfowl hunters' motivations demonstrated that appreciative-oriented motivations (i.e., getting away from daily routines and problems) were the most important motivations for continued participation, followed by affiliative motivation (e.g., being afield with important others) and achievement motivations (e.g., being a successful waterfowl hunter). These results were congruent with findings of Decker and Connelly's (1989) research on deer hunting motivations. Hayslette et al. (2001) tested a multiple-satisfaction model of hunting indicating hunters' satisfaction was a combination of harvest- and non-harvest-based elements. Their results suggest that wildlife viewing, companionship, nature/aesthetics, exercise, challenge, tradition, escape, and knowledge aspects of hunting motivations made contributions to satisfaction; however, success-based motivations, such as harvesting the daily hunting limit, did not. Everett and Nelson's (2016) study of young waterfowl hunters' satisfaction examined their interest to pursue hunting. Their findings highlight the importance of interest in the hunt, happiness, and intrinsic motivation in youth participation in waterfowl hunting. Among these factors, intrinsic motivation earned the highest score from the viewpoint of waterfowl hunters in their research. Self-determination theory (Deci and Ryan 1980) is known as one of the most successful motivation theories in social psychology. In this study we employed self-determination theory's approach as the framework to explore waterfowl hunters' motivations. This theory is elaborated below.

18.2.2.1 Self-determination Theory

Deci and Ryan (1980) explained human motivations to engage in activities in terms of the basic psychological needs of autonomy, competence, and relatedness. Self-determination theory (SDT; Deci and Ryan 1980) suggests that there are two types of motivations: intrinsic motivations, which refer to an internally driven feeling of satisfaction that result from the performance of a behavior, such as children's play; and extrinsic motivations, which are triggered by external factors such as prize, punishment or social norms. According to SDT, people internalize externally informed factors, such as norms and knowledge that they gain from their interactions with the environment. Through this process individuals self-regulate behavior to bring it under their autonomous control; this type of behavior is called self-determined behavior. According to Danner (2009), during this process (called internalization) people synchronize their externally regulated rules with their cognitive structure to make them their own personal values and norms. Within SDT there are four different types of regulation that derive extrinsically motivated behaviors (i.e., external, introjected, integrated, and identified) and one intrinsic regulation that regulates intrinsically motivated behaviors (Fig. 18.2).

External regulators are those behaviors performed because of the outcomes of the behavior; people perform behaviors to avoid an unpleasant outcome as the result of the behavior (i.e., punishment) or to obtain a reward from the performance of the behavior; for example, people adhere to hunting regulations to avoid being punished. Introjected regulation is actively avoiding feelings of guilt or to boost ones' sense of pride; for example, a hunter may feel guilty over an inability to retrieve an injured animal; similarly, a hunter may feel proud after a successful hunt. Identified regulation refers to behaviors that are congruent with individuals' goals and values, but have not been fully internalized; for example, children in hunting families may participate in hunting activities as it is congruent with their family values—even though they may have more interest in other recreational activities, they participate in hunting because it is a family norm. Integrated regulation refers to behaviors that are fully integrated with individuals' personal goals and values; people who have strong bond with nature and see hunting as an activity that connects them to the nature represent examples of regulated behavior. Finally, intrinsic regulation refers to behaviors performed for fun.

Although self-determination theory has been known as one of the most important motivation theories in social psychology, it has not been widely employed in hunting and hunting tourism research. We employ SDT to identify peoples' motivations to participate in waterfowl hunting.

18.2.3 Leisure Constraints Theory

Leisure constraints theory (Crawford et al. 1991) is a tool for understanding constraints that influence people's participation in different leisure activities. Jackson (2000) defined constraints as those factors that inhibit or prohibit peoples' participation in/enjoyment of a leisure activity. Leisure constraints theory identifies three types of constraints: intrapersonal, interpersonal, and structural. Intrapersonal constraints reflect personal and psychological factors that influence people's participation in an activity, such as introverted and extroverted personalities, lack of interest in an activity, lack of knowledge, or laziness. Interpersonal constraints refer to barriers that are perceived by people as a result of their interactions with the social surroundings, and include lack of family support and lack of time due to family responsibilities. Structural constraints refer to the physical environments and contexts that inhibit people's participation in an activity; examples include difficult weather conditions or lack of nearby waterfowl hunting opportunities. The leisure constraints theory posits that these types of constraints influence people's decision-making in a hierarchical order (Fig. 18.3); motivation is an important component of the theory as it influences every stage of people's decision-making process.

Metcalf et al. (2015) utilized leisure constraints theory to explore the influence of hunting constraints on female

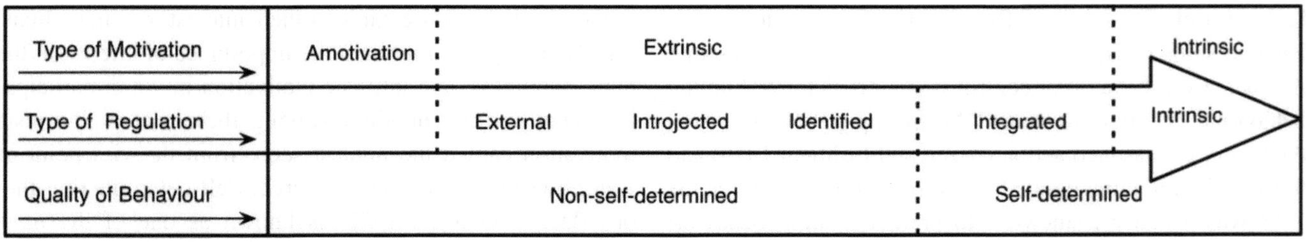

Fig. 18.2 Self-determination theory (Deci and Ryan 1980)

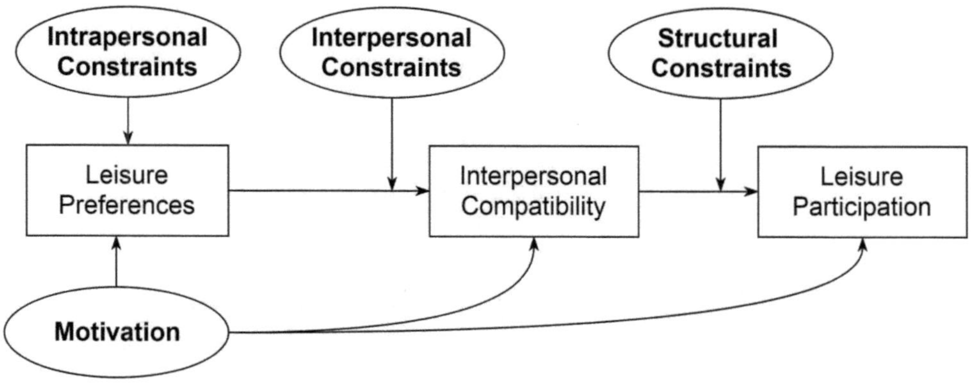

Fig. 18.3 Leisure constraints theory (Crawford et al. 1991)

hunters' experiences using a hunter typology based on Manfredo et al.'s (1996) recreation experience preferences. Different types of constraints across these groups of female hunters were compared; overall, participants of their study did not perceive themselves as being highly constrained.

Constraints inform the theory of planned behavior in hunting research. For instance, Shrestha and Burns (2016) hypothesized that constraints are antecedent to predictors of intentions (i.e., attitude, subjective norms, and perceived behavioural control) Thus, constraints influence people's attitudes toward hunting, their perception of control over hunting (perceived behavioural control), and the influence of social surrounding on their hunting behavior (subjective norms). These three factors, in turn, influence people's intention to participate in hunting activities. Therefore, adding constraints to the theory of planned behavior improved the predictive power of this theory in their study of hunters.

18.3 Case Study: Waterfowl Hunting in the Canadian Prairies

18.3.1 Methods

Thirty-six in-depth interviews (semi-structured, qualitative) were conducted with waterfowl hunters. Two different groups of waterfowl hunters were approached: experienced waterfowl hunters that had mentored other people to learn waterfowling; and beginner hunters who were introduced to waterfowl hunting through a mentorship program. The initial list of participants was obtained from membership and participation records of the Delta Waterfowl Foundation and the Alberta Conservation Association. The study was conducted in the Canadian provinces of Alberta, Saskatchewan, and Manitoba (Fig. 18.4).

Participants were contacted by email; hunters who agreed to share their waterfowl hunting experiences were interviewed in person between January and March, 2016. A semi-structured interview protocol was used and participants were asked about their experiences with waterfowl hunting, methods of engagement in the activity, constraints encountered, motivations to hunt waterfowl, and their intentions to participate in future waterfowl hunting. Each interview lasted between 30 and 90 min before being transcribed and coded to identify themes and categories.

18.3.2 Results and Discussion

Of 36 participants of this study 13 waterfowl hunters were from Manitoba, 5 were from Saskatchewan, and 17 were from Alberta; the majority were male (75%), and their average age was 42 years old. Regarding highest education levels held, 37% held a bachelor's degree, 31% had a graduate degree, 25% had college diplomas, and 8% had high school diplomas. The average duration of waterfowl hunting experience was 20 years.

We incorporated quotes and qualitative descriptions from interviews to demonstrate respondent types. These narratives help make clear the results of the analyses. Results are organized in three sections: motivations to participate in waterfowl hunting; perceived constraints to engage in waterfowl hunting; and predictors of intentions (i.e., attitude, subjective norms, and perceived behavioral control) to hunt waterfowl.

18.3.2.1 Waterfowl Hunters' Motivations

Intrinsic regulation was identified as an important reason to hunt waterfowl for the participants of this study. A majority of participants expressed excitement about the moment of shooting waterfowl and other hunting activates. Gordon stated, "[Waterfowl hunting] is purely fun. It is just so much fun; tons and tons of fun." When asked "what aspects of waterfowl hunting are the most important for you" he replied,

> The most important aspect is the family relation and the camaraderie. We have so much fun as a group. That's the most

Fig. 18.4 Location of study area in Canada; *shaded area* indicate the provinces of (from west to east) Alberta, Saskatchewan, and Manitoba

> important part. Seeing the animals, seeing the waterfowl, watching nature is the secondary part and the third part is harvesting the wild waterfowl - shooting the ducks and the geese.

Being with friends and family was of great importance for many participants and was considered to be the most important motivation for engagement in waterfowl hunting. Participants who grew up in hunting families highly valued hunting with their parents and grandparents and enjoyed spending time with their family during holidays, such as Thanksgiving during the hunting season. Their childhood memories, or their memories of hunting with their own children, were mentioned as major reasons to go on a family hunt.

Participants also indicated that 'connecting to nature' was an important motivation. For instance, one of the participants, John, revealed that the most important aspect of waterfowl hunting was

> [...] connecting to nature. It is being outside and all the great thing that you see. That's probably my biggest one. It is just being outside and finding kind of a peace with the world and the connection to Mother Nature and the earth I guess. [...] I am not a religious person but that's kind of my religion I guess it is Mother Nature.

Connecting to nature was not only important during the hunting season but throughout the year. Many participants indicated that hunting introduced them to the beauty and peacefulness of nature, and that a considerable amount of free time was spent viewing wildlife. Bob, an experienced hunter indicated that:

> It is about getting out in nature and enjoying the outdoors and enjoying the friendship and fellowship of being out in nature and that sort of thing. So it means quite a bit to me. I spend quite a bit of time out in the outdoors – not only hunting, but also paddling around the marsh and enjoying nature on the off-season. I do quite a bit of photography – wildlife

photography – in the springtime and summertime for waterfowl and that relates back to my first experience with hunting. That is why I got into photography was that first experience and the appreciation of nature and the outdoors.

When asked about integrated regulations, most participants identified waterfowl hunting as being a large part of their identities, and described how it had influenced their decisions and their relationships. John said:

Yeah that's a big part of who I am. If I am meeting someone for the first time I will talk about hunting. It will come out very quickly. That's kind of who I am and is a big part of me and doesn't take long to learn I am a hunter.

Participants explained that hunting was important to them, and how hunting activities and culture were intertwined with their different aspects of their lives; this is an indicator of identified regulation. For example, some participants documented each hunting trip they took, and kept hunting journals for over forty years. Tom described the role of hunting in his life:

Primary, I have these philosophical conversations with friends a lot and so I think of it from an evolutionary perspective [...]. I spend a lot of my leisure time hunting. I spend a lot of my discretionary spending on hunting. I spend every day cooking food that I got from hunting. So it is very important to me.

Most participants were proud to be hunters; however, this feeling of pride was a minor reason given for hunting. Although some mentioned that they feel proud of themselves and would like to speak of it out loud, other participants were hesitant to express their feeling about hunting freely. For instance, John said:

I am very proud to be a hunter and I am very outspoken about it and I always try to introduce new people and provide opportunities [...]. When you're outspoken about it and able to talk to people you just met about hunting, it is amazing how many people are interested in going. So if I had nothing but time, that's what I'd be doing.

In contrast, Martin, another interviewee indicated:

Well, I don't know. I have thought about this one a few times. I really enjoy it. It is something I enjoy. [...] About 95% of people being nice and 5% [are not so nice]. I think that number is probably a little bit higher in our area of people that don't seem to have the same kind of respect for hunting or even the whole system. [...] It's the people that abuse the system, but also I have been noticing as hunting becomes more and more commercialized; like you see TV shows and stuff. That really makes me less proud to do it. Because that is not what it is about. It is not meant to be a commercial thing where you are trying to get attention off of it. It is something that you do because you enjoy it and you like to provide food and be with your friends. So yeah, that question about: am I proud? It is hard to answer right now. I used to be and as I get older I like doing it but I don't like kind of the stigma around what everybody or the rest of society thinks of it. Because I find that there is a lot of negativity towards hunting now that makes you think twice about being a part of it.

A major reason for feelings of guilt during a hunt was not being able to retrieve an injured bird. Most participants highlighted this as one of the worst moments during hunting. Having a retriever dog was mentioned as the best way to deal with this issue. Generally, introjected regulation (feeling of guilt as a result of injuring a bird) did not seem to be an important motivation for waterfowl hunters (see Sect. 18.2.2.1).

Many participants mentioned the food provided through hunting as the most important reward of this activity; this is consistent with external regulation. However, it was not considered to be the strongest motivation to hunt. In conclusion, among all the motives mentioned by the hunters, camaraderie (being with friends and family) was the most important motivation for waterfowl hunting participation. Most participants discussed time spent together preparing food after the hunt as an important part of the experience; these times were generally shared with companions; connecting to nature was also identified as an important motivation. Although hunting success, the experience of shooting waterfowl, and the food provided through hunting were frequently mentioned, these factors were not highly valued among participants. These findings are similar to other research (e.g., Enck et al. 1993; Hayslette et al. 2001). Wildlife, outdoor recreation, and tourism managers should consider the importance of social aspects of waterfowl hunting in hunter recruitment and retention plans. Promoting waterfowl hunting as a family activity could support youth recruitment to waterfowling.

18.3.2.2 Perceived Constraints

Several different constraints to waterfowl hunting were identified by participants and most of these constraints were structural. The most commonly identified constraint was related to land access permission. Many participants indicated that gaining permission from farmers and land owners for access to their property could be problematic; this issue was more salient for Albertan participants than for participants from Saskatchewan and Manitoba. Richard from Saskatchewan mentioned:

We have a huge amount of land that is accessible, especially since I grew up on the land and I know the local farmers; but it is probably the biggest limitation for other people and it occasionally has been for me when there has been some competition, but it is a small factor for me.

In contrast, an Albertan hunter, Steve, indicated:

I think the thing I enjoy the least are the hard times when it's difficult to find the waterfowl or when it's difficult to find access. We are pretty fortunate in Alberta to have a lot of land and a lot of waterfowl; but if you are told by a landowner that they don't permit hunting, I can respect that, but it puts a bit of a negative tone on it. Because you feel like it's a lost opportunity almost. Sometimes the cost is a little annoying, because I do have to travel a long distance to a hunt; but if that's the biggest negative in my hunt, I am still quite happy about it. I think if access became a greater issue to the point where it all of a sudden seemed to be unattainable – you know if private land was just inaccessible either for reasons of no hunting permitted or hunting was exclusive to somebody else, that would make it difficult. But I think more on a fundamental level, even if it became incredibly difficult, I would still want to try, if nothing else, to purchase the license because with my role and knowing all of the reasons people stop hunting, and the implications that has on just being represented as a hunter, it would take some pretty excessive legislative changes or changes to my ability to stop hunting altogether. I suppose if I had a life-limiting illness, that would probably do it too; but if it was up to me, as long as I was able-bodied and safe, I will continue to hunt.

As noticed, some land-owners have denied access to hunting on their properties, and it was not possible to identify their reasons for it; it was not also possible to find any published material, including government and hunting organization reports, addressing this issue. Hunting related land use regulations and culture in Canada are quite different from US and from other countries.

As for Steve's experiences, they were congruent with the opinions of many other participants. This issue is important for several reasons: first, he clearly indicated the importance of access to land as an invaluable opportunity; Steve also mentioned the influence of limited access to farms as a cost associated with hunting. Second, he compared access against other constraints such as illness and legislative changes. Many other participants were as determined as Steve to hunt and mentioned serious physical illness or age as major constraints to hunting.

Distance from home to hunting opportunities was also identified as an important issue. Some hunters in this study lived or were planning to live close to a place where they could hunt easily. Most participants preferred places that do not require driving long distances. Jack from Alberta indicated,

You know we are looking for our next house and I don't care about the house so much as the property and what's it near. And my girlfriend wants to hunt too so we might buy a house that we wouldn't really want to afford; but oh wow – look at the location; maybe its got land and bush lots that we can hunt or a wetland or something.

A salient constraint for younger hunters was lack of waterfowl hunting equipment (e.g., waterfowl decoys, and firearms). Most participants agreed that it was important to have a large number of decoys; however, most experienced hunters did not consider this to be a relevant constraint. Although they mentioned a need for more decoys—or higher quality decoys—experienced hunters believed they possessed enough equipment for hunting. Bill said:

I think I have everything I need right now. Like if I wanted to keep hunting for ten years, I have enough right now that I can keep going, other than ammunition. Ammunition is an expected and acceptable cost to continue hunting. But if I found myself in a financial situation where I couldn't buy any more stuff, I could probably afford to sell some duck hunting equipment actually.

Catharine, a college student who recently started waterfowl hunting stated,

I find that people who waterfowl hunt are usually the older gentlemen that have money to spend on it and they are looking – instead of going to the bar and having beers – they are going out duck hunting and probably not having beers but drinking coffee. It's the same idea – it's just social hanging out with the guys kind of thing. But they have the money to put into decoys and things like that. As a student, it is very difficult to get your own set of decoys; to get all of the product and things like that.

She pointed out two important issues. First waterfowl hunting is seen as a masculine activity for older adults. Second, she perceived costs associate with hunting as a great barrier to her hunting participation; this was also mentioned by other younger participants and beginners.

Interpersonal constraints were generally not relevant for experienced hunters. Less experienced hunters, however, indicated a need to find a hunting group or their dependency on other experienced hunters. Some participants mentioned their partners—who taught them to hunt—or their children—with whom they hunted—as their main reason to take up waterfowl hunting or to reengage in this activity. Tammy, a mother of two children, indicated that her daughter's interest in waterfowl hunting influenced her decision to learn about waterfowl hunting:

My daughter actually asked one of the instructors about waterfowl hunting because she was interested and then the instructor approached me and said there were mentored hunts and then we applied for a mentored hunt and we were selected to go.

Although family responsibilities are identified as interpersonal constraints to waterfowl hunting, families could help people engage in waterfowling. In general, interpersonal constraints were not identified as major barriers to waterfowl hunting among participants.

Overall, several constraints to waterfowl hunting were identified in this research. Wildlife managers should consider the negative influences on the decision to hunt; constraints such as access to hunting fields, decoy expenses and firearms. Introducing novice hunters to hunting clubs and organizations can be a good strategy that may help them overcome these barriers of lack of hunting gear and companions.

18.3.2.3 Predictors of Hunting Behavioral Intentions

As mentioned earlier, the theory of planned behavior employs intention as an immediate predictor of behavior. Almost all the participants expressed strong intentions to hunt waterfowl during the upcoming hunting season. An investigation of affective and cognitive attitudes of participants toward hunting revealed that all participants agreed that, although hunting can be difficult due to uncomfortable conditions, hunting waterfowl was a pleasant activity (i.e., affective attitude). Regarding cognitive attitudes toward hunting, most participants identified responsible hunting as an activity that contributes to the conservation of the species. They believed that "hunting and conservation go hand-in-hand", indicating that they highly valued waterfowl hunting; this was supported by mention of the financial contributions that hunters provide through the purchase of hunting licenses and donations to conservation organizations. Jay, an experienced waterfowl hunter, said:

> The way I look at it is as a hunter you can't keep taking from the resource. You have to give something back. So how you give back is you become a member of conservation group […]. You get into mentorship programs, teach kids or new hunters how to hunt safely/ethically; so conservation is intertwined or highly-linked to hunting obviously. Hunting is a management tool for wildlife management. We use it to harvest animals. I guess the analogy is we harvest the surplus, keep the populations health. So there is that perspective as well. But I think the key thing in the conservation of hunting relationship is it is kind of a symbiotic relationship […]. The hunter just can't keep taking from conservation or the wildlife resource without giving back in some fashion. And I think just through the mere action of buying a license, you are actually contributing to conservation because a lot of that money is going to habitat programs and that kind of thing. So hunters, I think, a lot of them unknowingly as soon as they buy a license they are contributing to conservation. And then what you do after that point also reflects on that relationship – becoming a mentor/joining a group.

Consistent with camaraderie and social activities as important motivations to hunting waterfowl, subjective norms were important factors for participants. Although the societal support for hunting was important for some participants, the opinions and support of the friends and family seemed to be of greater importance. Jay believed:

> I hunt because legally I am allowed to hunt. It is not a right that I can hunt. I don't wait for somebody's permission or approval to hunt. Like what do you call it? I don't feel like I have to make a decision based on what a non-hunter thinks. I shouldn't say I don't care what they think but it is important to me that people like an anti-hunter or a non-hunter, I would want them to know that hunters aren't bad people. So in my actions and activities as well, I try to portray a good image – being ethical and safe and all that. But I don't sit there and wait for their approval to partake in the activity. I would do that anyways.

Many other participants did not find non-hunters or anti-hunters opinions to be important. Our findings revealed that for most participants, injunctive subjective norms are less important than descriptive subjective norms: family and friends' support for hunting was more important. These results are congruent with finding of Kramer et al. (2016) that emphasized the influence of peers on hunters' harvest decision.

Although many participants did not consider waterfowl hunting to be an easy activity due to weather conditions and difficulty of setting up for a hunt, they mostly expressed great control over planning a hunting trip activity, an indication of perceived behavioral control. Thus, waterfowl hunting trips are usually organized by more experienced people. However, less experienced waterfowl hunters seemed confident about organizing a hunting trip. For instance, Arizona, one of the female novice hunters revealed, "I think that anybody who thinks hunting is easy has never tried it. It takes a lot of skill and a lot of knowledge. I mean I guess if you are in the right place at the right time maybe it's really easy." However, when asked earlier whether she sees herself capable of organizing a waterfowl hunting trip she responded: "I think so. I have planned my own trips say when I go on travel or something—and I have done lots of project management so I should be able to." Many experienced waterfowl hunters expressed similar feelings about this issue. Cameron, an experienced hunter, said: "You wear hip waders and you sweat and it's too hot and you are too noisy; and you weren't careful enough in sneaking up; and you didn't shoot well; and they are missed. You know, it's not an easy activity." However, he was pretty confident about organizing a hunting trip: "I plan [hunting trips]. And usually if I go with friends, usually I either plan it or participate seriously into decision. They come with me essentially."

Our findings about subjective norms and perceived behavioral control indicated that waterfowl hunters are confident about their abilities to participate and organize this type of tourism and recreation activity. Moreover, it can be concluded that subjective norms are not of considerable importance for waterfowl hunter participants. Although with a qualitative approach, these finding are congruent with Sherstha et al. (2012) and Shrestha and Burns (2016) findings regarding the association among these factors. Their body of research reported that perception of control considerably influenced hunters' intention. However, subjective norms were not as important for hunters. Table 18.1 briefly synthesizes findings of this research.

18.4 Conclusion

These findings will likely resonate with readers for different reasons, predicated largely on their degree of familiarity or history with waterfowl hunting. For example, two of this chapter's authors (HH and FM) are recreation and tourism scholars who found explanatory power in existing theories and congruence with concepts of leisure activity choices. The other author (LF), is a natural scientist with 44 years of waterfowling experience and he found explanatory power through contrasting the patterns of waterfowling groups up to the current qualitative analysis.

Table 18.1 Analytical table of the results

Theory	Hypothesis	Abstracted results	Qualitative fit
TPB[a]	The following factors explain waterfowl hunting participation: – Attitude (affective and cognitive) – Subjective norms – Perceived behavioral control	– All the participants expressed strong intention to continue hunting – Affective attitude: all participants agreed that hunting is a difficult but pleasant activity – Cognitive attitude: participants believed hunting was beneficial due to the financial supports hunters provide as well as its contributions to the conservation of species – Opinions of hunters' close friends and family (injunctive subjective norms) were important for most hunters. Descriptive subjective norms such as opinions of non-hunters or anti-hunters, however, were not considered important for the participants of this research – Most participants did not consider waterfowl hunting to be an easy activity. However, they expressed great control over planning a hunting trip, an indication of perceived behavioral control	Overall model: high congruence with original hypotheses – Attitude: strong predictor of hunting intention – Subjective norms: moderate predictor of hunting intention – Perceived behavioral control: strong predictor of hunting intention
SDT[b]	The following motivation regulations influence waterfowl hunting participation: – Intrinsic motivations – Integrated motivations – Identified motivations – Introjected motivations – Extrinsic motivations	– Intrinsic motivation was identified as an important motive for hunting (e.g., enjoying shooting time, spending time with family and friends, and connection with nature) – Most participants reported hunting as an important part of their identity. They valued the contributions of hunting to the society and ecosystem. Therefore, integrated and identified motivations were important drives for hunters – Although, feeling of pride for being a hunter and feeling of guilt as a result of harvesting waterfowl were mentioned by the hunters, introjected motivation was not identified as an important motive for waterfowl hunters – External rewards such as food provided through hunting was not considered as a great motive for hunting waterfowl among interviewees	Overall model: moderate congruence with original hypotheses; some motivations were found more important for hunters. – Intrinsic motivation: strong motive for waterfowl hunting – Integrated: strong motive for waterfowl hunting – Identified: moderate motive for waterfowl hunting – Introjected: weak motive for waterfowl hunting – Extrinsic: weak motive for waterfowl hunting
LCT[c]	The following constraints influence people's participation in waterfowl hunting: – Structural constraints – Interpersonal constraints – Intrapersonal constraints	– Structural constraints were the most common barrier to hunting from the view point of participants (e.g., land access permission, distance, costs, etc.) – Interpersonal constraints were identified in this research but did not considerably influenced hunters' decisions to engage in hunting activities. Some hunters engaged/reengaged in hunting because of their family (e.g., their children) – Intrapersonal constraints were not identified as important barriers to hunting	Overall model: moderate congruence with original hypotheses; structural constraints were found as stronger restraining factors – Structural constraints: strong barriers to waterfowl hunting – Interpersonal constraints: weak barriers to waterfowl hunting – Intrapersonal constraints: not an important barrier to waterfowl hunting

Notes [a]*TPB* theory of planned behavior (Ajzen 1991); [b]*SDT* self-determination theory (Deci and Ryan 1980); [c]*LCT* leisure constraints theory (Crawford et al. 1991)

The incentives for participation (intrinsic regulation) we found are powerful and central to hunters' personal identities involving sharing experiences with family and friends, connection to nature, provision of special and meaningful food, and the positioning of their relationship to the wild world. Their participation was tempered by barriers identified (perceived constraints) such as land access, debilitation through infirmities, costs, aloneness, or family responsibilities. Still, the aspiration to hunt (behavioral intention) was strong and helped overcome the barriers to yield rewards such as pleasant activity, a moral satisfaction in contributing to conservation efforts, and a signature activity that is both tangible and visible in portraying their world view of man and nature.

Importantly, most interviewees were relatively immune to negative social pressures about hunting (subjective norms) even though some recognized the existence of anti-hunting sentiments. Conversely, some did admit to sensitivity to opposition offered by family and close friends. As was mentioned earlier, this study explored active waterfowl hunters; therefore, non-hunters and lapsed hunters were excluded from our investigation. Non-hunters and lapsed hunters might perceive different levels of social pressure about hunting compared to active hunters. This issue needs to be considered while interpreting the findings.

What are we to make of these findings? Results that confirm or reinforce commonly held beliefs are far less interesting than the unexpected results. There was a surprisingly strong and recurrent theme on the importance of being connected with other hunters, the desire for a hunting group or the admiration of the relationships of established hunting companions, all of which reinforce the importance of the social fabric that undergirds hunting. Viewed differently, what waterfowlers appear to seek first and foremost is the belonging to a group with a common goal (of pursuing and killing some ducks and geese) and once a participant, the actual harvest of animals is less important.

This attitude informs those who would recruit new waterfowlers. Hunting is not a highly predictable activity like driving, reading, or video gaming (activities that are typically learned then easily practiced solitarily). At its core, hunting is an exercise of managed uncertainty; therefore, new hunters find reassurance through shared experiences even as experienced hunters find satisfaction in sharing. In this way, hunting may be more akin to mountain climbing, sky-diving, or even team sports, all of which face outcome uncertainties and the perception of establishing and managing a physical relationship with the "other" whether it is a mountain, gravity or another team. The process of actively managing a climb, dive, game or hunt benefits from the decision-validation and reality checks of companions, thus, hunting becomes a group endeavor. One cannot easily train a hunter in the classroom and then realistically expect them to solitarily adopt the activity. Mentorship, whether formal or informal, is vital to create a sense of confidence and belonging.

If belonging is so crucial however, why are hunters largely immune to mild social disapproval? The sense of belonging to a committed and responsible community is possibly even enhanced by the "other" of those opposing hunting. A sense of in-group versus out-group quite possibly strengthens the sense of belonging and group cohesiveness. The responses we heard toward non- and anti-hunters varied; some sought engagement to propound their positions while others were just put off by opposition and chose to quietly continue to hunt "below the radar'. Still, they pursued their hunting avocation with commitment.

Clearly, the rewards of hunting, once experienced, are sufficiently profound to propel hunters from intention onward to action even in the face of large start-up inconvenience, perceived structural barriers, and some amount of social opposition. Defining and describing the intricate reward and reinforcement system points strongly toward an opportunity to cultivate (a) a sense of belonging, (b) mentor relationships, (c) mild reduction in costs and barriers and (d) preparation to understand those who would oppose the choice to hunt. Each of these may be addressed in a complete system of hunter education that offers mentorship programs as well as equipment loaning services for new hunters.

Acknowledgements Funding for this research was provided by the Delta Waterfowl Foundation and the Alberta Conservation Association.

References

Ajzen I (1991) The theory of planned behavior. Organ Behav Hum Decis Process 50(2):179–211

Ajzen I, Fishbein M (1975) Belief, attitude, intention and behavior: an introduction to theory and research. Philos Rhetoric 10(2):130–132

Apps K, Dimmock K, Lloyd D, Huveneers C (2016) In the water with white sharks (*Carcharodon carcharias*): participants' beliefs toward cage-diving in Australia. Anthrozoös 29(2):231–245

Ardoin NM, Wheaton M, Bowers AW, Hunt CA, Durham WH (2015) Nature-based tourism's impact on environmental knowledge, attitudes, and behavior: a review and analysis of the literature and potential future research. J Sustain Tour 23(6):838–858

Baur J, Herr A (2004) Hunting and fishing tourism. In: Higginbottom K (ed) Wildlife tourism: impacts, management and planning. Common Ground, Altona, Victoria, pp 57–77

Beale CM, Monaghan P (2004) Human disturbance; people as predation-free predators? J Appl Ecol 41(2):335–343

Bélanger L, Bédard J (1995) Hunting and waterfowl. In: Knight RL, Gutzwiller KJ (eds) Wildlife and recreationists: coexistence through management and research. Island Press, Washington, pp 230–241

Crawford DW, Jackson EL, Godbey G (1991) A hierarchical model of leisure constraints. Leisure Sci 13(4):309–320

Creswell JW (1998) Qualitative inquiry and research design. Sage Publications, Thousand Oaks, CA

Darner R (2009) Self-determination theory as a guide to fostering environmental motivation. J Environ Educ 40(2):39–49

Deci EL, Ryan RM (1980) Self-determination theory: when mind mediates behavior. J Mind Behav 1(1):33–43

Decker DJ, Connelly NA (1989) Motivations for deer hunting: implications for antlerless deer harvest as a management tool. Wildlife Soc Bull (1973–2006) 17(4):455–463

Dobson AP, Rodriquez JP, Roberts WM, Wilcove DS (1997) Geographic distribution of endangered species in the United States. Science 275(5299):550–553

Enck JW, Swift BL, Decker DJ (1993) Reasons for decline in duck hunting: insights from New York. Wildl Soc Bull 21(1):10–21

Everett MW, Nelson CM (2016) 'Flow'and satisfaction of Michigan youth waterfowl hunters: implications for hunter retention. Loisir et Société/Society and Leisure 39(2):1–13

Federal, Provincial, and Territorial Governments of Canada (2014) 2012 Canadian nature survey: awareness, participation, and expenditures in nature-related recreation, conservation, and subsistence activities. Canadian Councils of Resource Ministers, Ottawa, ON, p. 181

Green R, Giese M (2004) Negative effects of wildlife tourism on wildlife. In: Higginbottom K (ed) Wildlife tourism: impacts, management and planning. Common Ground, Altona, Victoria, pp 81–97

Hammitt WE, Cole DN (1998) Wildland recreation: ecology and management. Wiley, New York

Hayslette SE, Armstrong JB, Mirarchi RE (2001) Mourning dove hunting in Alabama: motivations, satisfactions, and sociocultural influences. Hum Dimens Wildl 6(2):81–95

Heffelfinger JR, Geist V, Wishart W (2013) The role of hunting in North American wildlife conservation. Int J Environ Stud 70 (3):399–413

Higginbottom K (ed) (2004) Wildlife tourism: impacts, management and planning. Common Ground Publishing, Australia

Hrubes D, Ajzen I, Daigle J (2001) Predicting hunting intentions and behavior: an application of the theory of planned behavior. Leisure Sci 23(3):165–178

Huang LF, Chiang CC, Chen HC (2014) Willingness to pay of visitors for the nature-based public park: an extension of theory of planning behavior (TPB). J Inf Optim Sci 35(5–6):405–429

Hung K, Petrick JF (2012) Testing the effects of congruity, travel constraints, and self-efficacy on travel intentions: an alternative decision-making model. Tour Manag 33(4):855–867

Jackson EL (2000) Will research on leisure constraints still be relevant in the twenty-first century? J Leisure Res 32(1):62

Kleiber DA, Walker GJ, Mannell RC (2011) A social psychology of leisure. Venture Pub, NY

Klett AT, Johnson DH (1982) Variability in nest survival rates and implications to nesting studies. USGS Northern Prairie Wildlife Research Center, Paper 191

Knight RL, Cole DN (1995) Wildlife responses to recreationists. In: Knight RL, Gutzwiller KJ (eds) Wildlife and recreationists: coexistence through management and research. Island Press, Washington, DC, pp 55–72

Kouthouris C, Spontis A (2005) Outdoor recreation participation: an application of the theory of planned behavior. Sport J 8(3)

Kramer DB, Mitterling A, Frank KA (2016) Understanding peer influence in hunter harvest decisions using social network theory and analysis. Hum Dimens Wildl 21(5):1–13

Lauckhart JB (1956) Calculating mortality rates for waterfowl. The Murrelet 37(3):31–34

Lee TH, Tseng CH, Jan FH (2015) Risk-taking attitude and behavior of adventure recreationists: a review. J Tour Hosp 4(149):1–3

Luo Y, Deng J (2007) The new environmental paradigm and nature-based tourism motivation. J Travel Res 46:392–402

Manfredo MJ, Driver BL, Tarrant MA (1996) Measuring leisure motivation: a meta-analysis of the recreation experience preference scales. J Leisure Res 28(3):188

Marques C, Reis E, Menezes J, Salgueiro MD (2015) Modelling preferences for nature-based recreation activities. Leisure Stud 1–19

Mayan MJ (2009) Essentials of qualitative inquiry. Routledge, NY

Meng F, Tepanon Y, Uysal M (2008) Measuring tourist satisfaction by attribute and motivation: the case of a nature-based resort. J Vacat Market 14(1):41–56

Metcalf EC, Graefe AR, Trauntvein NE, Burns RC (2015) Understanding hunting constraints and negotiation strategies: a typology of female hunters. Hum Dimens Wildl 20(1):30–46

NAWMP (2014) North American Waterfowl Management Plan Committee. Revised objectives: an addendum to the 2012 North American Waterfowl Management Plan. Fish and Wildlife Service, Washington, DC

Pattons (2002) Qualitative research and evaluation methods. Sage Publications, Thousand Oaks, CA

Peng N, Chen A, Hung KP (2014) Including pets when undertaking tourism activities: incorporating pet attachment into the TPB Model. Tour Anal 19(1):69–84

Reynolds PC, Braithwaite D (2001) Towards a conceptual framework for wildlife tourism. Tour Manag 22(1):31–42

Rossi AN, Armstrong JB (1999) Theory of reasoned action vs. theory of planned behavior: testing the suitability and sufficiency of a popular behavior model using hunting intentions. Hum Dimens Wildl 4(3):40–56

Ryan RM, Deci EL (2000) Self-determination theory and the facilitation of intrinsic motivation, social development, and well-being. Am Psychol 55(1):68

Shrestha SK (2013) Predicting deer hunting participation using theory of planned behavior and constraint integrated theory of planned behavior models: a study of Oregon big game hunters. West Virginia University

Shrestha SK, Burns RC (2016) Integrating constraints to the theory of planned behavior in predicting deer hunting participation. Hum Dimens Wildl 21(5):1–15

Shrestha SK, Burns RC, Pierskalla CD, Selin S (2012) Predicting deer hunting intentions using the theory of planned behavior: a survey of Oregon big game hunters. Hum Dimens Wildl 17(2):129–140

US Fish and Wildlife Services (2016) Waterfowl population status report 16. https://www.fws.gov/migratorybirds/pdf/surveys-and-data/Population-status/Waterfowl/WaterfowlPopulationStatusReport16.pdf. Accessed 5 Oct 2016

Wilson E (1984) Biophilia. Harvard University Press, Cambridge

Wilson E (2007) Biophilia and the conservation ethic. In: Penn DJ, Mysterud I (eds) Evolutionary perspectives on environmental problems. Transaction Publishers, New Brunswick, pp 249–257

Conclusion

19 Wildlife Tourism, a Multidisciplinary Field of Inquiries and Insights: Final Considerations

Ronda J. Green and Ismar Borges de Lima

Abstract

This conclusion chapter offers a summary of key contributions of each chapter, gathering the main conceptual approaches and major issues of the case studies presented in the volume. A recurring theme is the importance of quality interpretation for visitors and tourism businesses in the wildlife tourism sector. As readers will have noted, the chapters encompass a range of wildlife topics to varied audiences to convey information, a sense of wonder and concern for conservation. The authors have sought to present their topics bias-free by balancing ecological and anthropocentric views, but the welfare and conservation of wildlife used as tourism attractions, whether captive, semi-captive or in the wild, has been of major concern. Ethical and moral issues pervade the discussions on human-animal encounters, particularly in situations where wildlife has apparently been over-exploited for entertaining visitors. While extreme cases such as obvious abuses of animal welfare and practices that threatened endangered species are easy to condemn, the debates are often not so easy to resolve with the simple answers that many hope for. Emotions are often high when pros and cons of various tourism operations are discussed, but there remains much that we do not know about pressures on the ecological needs of wildlife and the seriousness of stress imposed on individual animals. Much further research related to some of the impasses and deadlocks related to wildlife tourism planning and management would be valuable for future publications. For example, some argue that the 'end of zoos' should become a future reality, or that zoos should be replaced by sanctuaries, while others point to valuable work by zoos supporting conservation, caring for rescued wildlife, educating the public and conducting research. There will also be many readers who will be uncomfortable with the chapter on hunting, while others will maintain that although we must remain vigilant on the welfare and conservation implications, without hunters many areas that retain most of their natural attributes would never have been protected and may not remain protected into the future if all hunting is banned.

R.J. Green (✉)
Environmental Futures Research Institute, Griffith University, Nathan, Australia
e-mail: chair@wildlifetourism.org.au

R.J. Green
Wildlife Tourism Australia Inc., Rathdowney, Australia

I. Borges de Lima
School of Business and Tourism, Southern Cross University, Gold Coast, Australia
e-mail: ismarborgeslima@gmail.com; ismarlima@yahoo.com.br

I. Borges de Lima
MultiAmazon Lab, Recinatur Intl Foundation, UNIFESSPA, Roraima State University (UERR), Boa Vista, Brazil

I. Borges de Lima and R.J. Green (eds.), *Wildlife Tourism, Environmental Learning and Ethical Encounters*, Geoheritage, Geoparks and Geotourism, DOI 10.1007/978-3-319-55574-4_19

Management of both animals and people is crucial to truly sustainable wildlife tourism, to mediate the interaction with wildlife in tourism contexts and incorporate compassionate and conservation-based ideology. Integration of different disciplines is essential, which may be a novel approach for many and call for innovative goals, research and management plans.

The study of wildlife tourism is ideally a multi-disciplinary pursuit involving many fields of inquiry, including ecology of animals important to tourists and other species that share their habitat, conservation biology and management of both animals and people, stress in wild and captive animals, attitudes and needs of tourists, sociology of local human residents, regional economic influences, the politics of decision making and other issues. It is hoped that this volume will both assist readers working within the field covered by particular authors and also provide insights by stepping outside their own specialties.

Examples are presented from many parts of the world, and these can be of great use both to readers interested in the particular regions and to those who would like to consider how to apply the methods, results and insights to investigations or plans for other geographic areas across the planet.

The Sheldon et al. chapter recognises the need for integrating different disciplines, and will probably demonstrate a novel approach for many readers. They attempt to locate wildlife tourism, applied ecology, environmental education and interpretation within a philosophical framework dealing with notions of nature and the environment and to then link this with the analytical approach of political ecology. Taking a particular project, the reintroduction of the yellow-eyed penguin (*Megadyptes antipodes*) at Long Point, New Zealand, where habitat is managed with particular species in mind and with tourism infrastructure to be developed before the penguins and other birds are re-introduced, they explore the 'political ecology' of the process, demonstrating the unavoidable entanglement of political economy with ecological concerns. They briefly discuss the increasing integration of tourism and conservation, the rearrangement and renegotiation of the power relationships between various groups, and other aspects of the project, and conclude that unravelling the political dimensions of site and species management is an endless task, and that the ability to engage politically in a multi-faceted way makes political ecology a useful analytic approach to the study of wildlife tourism, applied ecology and environmental education and interpretation.

Macoll and Tribe also consider an inter-disciplinary approach to the potential for the tourism industry, wildlife management, research, visitor educational training of tertiary students to be intertwined. They present a particularly interesting case on a large, wooded mountain property in subtropical Australia, where research and education facilities plus six very large aviaries at currently (i.e. at time of writing) being constructed with a view to rehabilitating endangered wildlife to the property and involving the owners and staff, a university, local residents and of course the tourists. A survey showed a considerable interest by current visitors, even while staying at the accommodation for different reasons, in seeing and learning about the animals and the rehabilitation process (while acknowledging that this has yet really been tested). They caution that the commercial success of any such venture is linked to entertainment value rather than education or conservation, but that conservation messages can be delivered in an entertaining way, and that on-going research will explore actual behaviour and attitudes of guests as well as outcomes of rehabilitation projects. Other establishments wanting to combine conservation, tourism, research and education could do well to watch the progress of this quite elaborate program.

Many regions have the potential to diversify their tourism offerings, and an understanding of both the attractions and the prospective tourists is important for effective planning. Werdler's chapter explores the potential for Rwanda, famous for its primate trekking, to attract birdwatching tourists, especially from the Netherlands. After surveying the literature, including African bird guides, national park reports, research articles and a Dutch journal aimed at birdwatchers, and actually participating in birding tours within Rwanda, the author concludes the country has great potential to become a prime African birdwatching destination, and recommends the wider dissemination of information on Rwanda's birds through various channels and the specialised training of local guides for birdwatching tourism.

A country already famous for attracting tourists with its rich biodiversity as well as tropical climate is Brazil. Lanzer et al. explore the tourism potential as well as human impacts on one of the country's less famous regions, the coastal southern lakes. They conclude that past tourism activities as well as other human pressures have tended to exert negative impacts on the ecosystem, but that the rich diversity of animals and plants, including endangered species, makes the region eminently suitable for the development of more ecologically sustainable tourism pursuits such as educational and science tourism. They recommended careful management to avoid further environmental damage, and efforts to educate the local community about the region's biodiversity, not just of the larger and more conspicuous animals but including the rich invertebrate and plant life.

Moswete et al. examine a common problem of impact often intensifying in the more popular areas and efforts to spread the tourist load to other sites in the overall tourist destination. Specifically they consider the Chobe National Park in Botswana, where tour operators and self-drive tourists tend naturally to congregate around the best wildlife-viewing areas along the river, causing much congestion, and neglect other parts of the Park. They conclude that the simple provision of other roads has not helped alleviate the problem, and that provision of facilities and a shift in marketing approach are needed.

Quality interpretation is a recurring theme, and it is a primary focus of several chapters. Mayes notes that wildlife watching has an emotional context as well as an intellectual one, and analyzes the relationship between intensity of wildlife encounter and interpretation by the guides when watching, feeding or swimming with dolphins. She concludes that interpretation during an emotionally intense event is not effective, advising it is best to let the ocean speak' and provide a very different kind of education, but that as the intensity decreases the interpretation becomes increasingly important for the learning experience and visitor satisfaction.

Lima's chapter notes the history and complexities of Australia's system of protected areas, and points to a gap in the literature regarding interpretation in the parks, which tends to be carried out more by tour guides, educators and volunteers than by rangers. He presents a framework for introducing wildlife topics for interpretation to different age groups, discusses learning theories and presents information on some iconic Australian animals. He recommends that Kolb's experiential learning theory could be more widely adapted for rangers to present interpretation on a range of wildlife topics to varied audiences to convey information, a sense of wonder and concern for conservation, and that more research is needed in this field.

Hassan and Sharma consider the potential for better educational signage and guiding focussed on the Bengal tiger in India and Bangladesh, and present the results of a survey of attitudes of visitors, local residents and tourism staff. They conclude that much potential for education is currently missed, and recommend adding more kinds of information on signs, in English and other languages in addition to Hindi and provision of better access roads and amenities in the sites, and propose a multi-day educational tour that could add to the educational experience regarding the tigers and also other species that share their ecosystems. They also point to the need for further research in this area.

Moreira et al. outline a project (TAMAR) focussed on marine turtle research and conservation in a leading eco-tourism destination in Brazil. They discuss exciting opportunities for visitors to participate in various aspects of this research, both by day and night, on land and in the water, and thus gain deeper understanding of turtle biology, conservation and research techniques. They conclude that there is economic benefit to the local community but that the most important benefits are public education and turtle conservation.

Harman and Dilek review the history of whale and dolphin watching, and analyse recent visitor comments on cruises in the context of sensory impressions, emotional affinity, reflective response and behavioural response, concluding that behavioural response (i.e. intended behaviour change as a result of the experience) was low. They recommend that tour companies should concentrate more on the quality of education, and use sensory and emotive messages. They also recommend research into the reasons behind their findings, and in continents other than North America.

Ayazlar describes Turkey's current and potential wildlife tourist attractions, including wildlife viewing and photography, hunting and visits to captive wildlife attractions, and notes that wildlife tourism as a sector has not been given sufficient attention in this country. He lists a number of species of interest, including various mammals, birds, turtles and fish, conservation management in respect of minimising tourism impacts, and visitor impressions of a zoo and its learning opportunities. He concludes that there needs to be further research on motivation of tourists and the tourism potential of different regions within Turkey.

Ethical issues are an important aspect of wildlife tourism, including responsibilities towards tourists, tour operations, local communities and the wildlife themselves. Burns highlights some of the ethical complexities worthy of deeper consideration in the context of wildlife tourism and proposes a way forward for responsible human engagement with wildlife. She integrates various theories from the field of ethics, emphasises the importance of recognises management of both animals and people as crucial and combined components in sustainable wildlife tourism, and proposes an ethically responsible approach to interacting with wildlife in tourism contexts that incorporates compassionate conservation ideology.

Green considers the impacts of disturbance by tourists of wildlife in both wild and captive settings and explores some of what we already know and what we need further research on in regards to how much stress could be a serious factor in individual animal welfare and to what extent this could lead to local population declines. She also considers ethical considerations of tourist disturbance of wildlife that may affect local human residents, other tourists and tour operators. She stresses the need for far more research on the ecological impacts of both the animals the tourists are seeking and other animal they may inadvertently disturb in the process, on the kinds of disturbance that are trivial or overly stressful to free or captive wildlife, and the effects of close encounters of tourists' subsequent understanding of and feelings towards animals (so that optimal outcomes may be achieved where compromises are needed).

Reiser examines the claims of zoos to support conservation, research and education and finds them wanting, recommending there should instead be a shift to large sanctuaries.

Moghimehfar et al. consider the controversial topic of recreational hunting, lamenting that it is declining in Canada partly due to barriers to access to hunting grounds. They analyse interviews with hunters and consider that hunting plays an important role in conservation, and that results of their analysis can inform tourism, outdoor recreation, and wildlife managers and planners to retain the local ecological and economic benefits of waterfowl hunting.

This volume thus provides readers with much to think about, with new insights that we hope will lead to practical improvements within the industry, as well as inspiring some much-needed further research.

MIX
Papier aus verantwortungsvollen Quellen
Paper from responsible sources
FSC® C105338

If you have any concerns about our products,
you can contact us on
ProductSafety@springernature.com

In case Publisher is established outside the EU,
the EU authorized representative is:
Springer Nature Customer Service Center GmbH
Europaplatz 3, 69115 Heidelberg, Germany

Printed by Libri Plureos GmbH
in Hamburg, Germany